Current Scenario in Biotechnology

Current Scenario in Biotechnology

EDITORS

P. Ponmurugan
Puniethaa Prabhu
G. Ayyappadasan
R. Ganesh Babu

BLOOMSBURY
LONDON • NEW DELHI • NEW YORK • SYDNEY

First Published, 2013
First Reprint, 2013

BLOOMSBURY PUBLISHING INDIA PVT. LTD.
London New Delhi New York Sydney

ISBN: 978-93-82563-27-3

Published by Bloomsbury Publishing India Pvt. Ltd.
VISHRUT Building, DDA Complex, Building No. 3, Pocket C-6 & 7
Vasant Kunj, New Delhi 110 070

Laser Typeset by SHINU
A-5/56, Moti Nagar, New Delhi 110 015

Printed at Anvi Composers
1 DDA Market, Block A-1B, Pachim Vihar, New Delhi 110 063

Contents

Section–II: Shri K.N. Narasimhaiah Award

Section–III: Kunnath Pharmaceuticals Award

Section–V: Prof. Edathil Vijayan Award

Section–VI: KAPL Award

Section–VII: Abstracts

Foreword

The K.S. Rangasamy College of Technology, started its educational service from 1994 with the aim of providing world class education under one roof. The approval of AICTE, New Delhi, and ISO 9001–2000 certification and the National Board Accreditation (NBA), New Delhi, are some of the notable milestones in our quality service.

I am pleased to know that the Department of Biotechnology of the K.S. Rangasamy College of Technology is organizing an Annual Meeting of the Society for Biotechnologists (India) and National Conference on "*Current Scenario in Biotechnology*" during 27–29 September 2012 at our campus in a big way.

I am proud of the Department head, faculty members and the students, who have made strenuous efforts to make this upcoming conference cum annual meet a success. I greet all the participating students, attending this wonderful function.

The conference is organized by the Department of Biotechnology, K.S.R.C.T. in collaboration with the Society for Biotechnologists (India).

I am certain that this conference will definitely yield new ideas, which will be worthwhile for the younger generation, academicians and researchers.

I appreciate Dr. K. Thyagarajha for hosting successful events continuously with his dedicated service.

I wish everyone success in "*Current Scenario in Biotechnology*".

Lion Dr. K.S. Rangasamy
MJF, *Chairman*
K.S.R. Educational Institutions

Preface

Bill Gates, the President of Microsoft, wrote that if he wasn't in the computer field, he'd be working in biotechnology. "I expect to see breathtaking advances in medicine over the next two decades, and biotechnology researchers and companies will be at the center of that progress. I'm a big believer in information technology.... But it's hard to argue that the emerging medical revolution, spearheaded by the biotechnology industry, is any less important to the future of human kind. It, too, will empower people and raise the standard of living". Gates' chief rival in the software industry, Oracle's Larry Ellison, agrees. "If I were 21 year old, I probably wouldn't go into computing. The computing industry is about to become boring", he said. Rather, "I'd go into genetic engineering".

The obvious growth of biotechnology was recognized by such innovators and this is the time for the students, researchers and the academicians to take this field into directions that give rise to well being.

The annual meeting of the Society for Biotechnologists (INDIA) and the National Conference on "*Current Scenario in Biotechnology*" is going to be hosted by the Department of Biotechnology, K.S. Rangasamy College of Technology, Tiruchengode.

I felicitate the Organizing Chair, Dr. P. Ponmurugan, Professor and Head, Organizing Secretary Mrs. Puniethaa Prabhu, Associate Professor and also the Committee Members for their indefatigable work. Finally I wish for the grand success of the conference.

Dr. K. Thyagarajah
Principal
K.S.R.C.T.

Prologue

Biotechnology is a field of life sciences that has recently emerged as a booming area of research. Myriad technologies and methods are being developed and are being successfully harnessed for the benefit of mankind. Biotechnology has been there since the evolution of humans in various forms. It has been an integral part of all aspects encompassing agriculture, textile, manufacturing processes, etc. The efforts of various researchers, academicians and industrialists have contributed to the growth of this field. Drops together form an ocean. Improvisation and development in biotechnology have been brought about by the strenuous efforts of individuals from all backgrounds.

The annual meeting of the Society for Biotechnologists (India) and the National Conference on "*Current Scenario in Biotechnology*" (SBTI – CSB 2012)" is being hosted during 27–29 September 2012 in the Department of Biotechnology, K.S. Rangasamy College of Technology. The event is organized by the Department of Biotechnology of the K.S. Rangasamy College of Technology (Autonomous) in collaboration with the Society for Biotechnologists (India).

The event is an effort to provide a platform for budding scientists to showcase their talents and an opportunity for the research, academic and industrial community to share their knowledge and innovations in the field of biotechnology. We sincerely hope that all the delegates and the participants would find the sessions productive and we wish this event would stir the imagination of the researchers to come up with effective technologies and solutions for current issues.

The souvenir for the event bears enormous information comprising of various topics ranging from cancer biotechnology to agricultural biotechnology. It is comprises research articles that will be discussed during the technical sessions. The book would thus be of great use to everyone involved in research and serve not only as a storehouse of information, but also as a remembrance of the conference.

We, at the SBTI would like to laud the onerous efforts taken by the Organizing Chair Dr. P. Ponmurugan, the Organizing Secretary Ms. Puniethaa Prabhu and all the organizing committee members. The event would not have been accomplished without their industrious efforts. We wish this conference would be an inspiring and delightful event!

Dr. Niranjali Devarajan, *President, SBTI*
Dr. C.S. Paulose, *Secretary, SBTI*
Dr. V. Oommen, *Vice President, SBTI*
Dr. K.E. George, *Treasurer, SBTI*
Dr. E. Vijayan, *Emeritus Professor, SBTI*

Acknowledgements

The Annual Meeting of the Society for Biotechnologists (INDIA) and the National Conference on "*Current Scenario in Biotechnology*" (SBTI – CSB 2012)" is a joint venture between the Department of Biotechnology of the K.S. Rangasamy College of Technology and the Society for Biotechnologists (INDIA). This meet provides a platform for researchers, academicians and industrialists to share their perspectives about recent trends in the vast field of biotechnology.

We are thankful to our honourable Chairman Lion Dr. K.S. Rangasamy, MJF, Founder and President, Principal Dr. K. Thyagarajah, Secretary Shri R. Srinivashaan and Assistant Secretary Smt. Kavithaa Srinivashaan, K.S.R. Educational Institutions for their constant encouragement and support.

We would also like to convey our gratitude to the President Dr. Niranjali Devarajan, the Secretary Dr. C.S. Paulose, the Vice President Dr. Oommen V. Oommen, the Treasurer Dr. K.E. George, Emeritus Professor E. Vijayan, Society for Biotechnologists (SBTI).

We would like to convey our gratitude to the faculty members for their efforts in the various committees. We are thankful to Mr. I. Muthiah for his wonderful support in managing the finances for the entire event. We also appreciate Ms. M. Nithya for distributing the brochures and invitations throughout the nation. We are thankful to Mr. G. Ayyappadasan for effectively taking care of the tidiness activities. Thanks to Ms. K. Sindhu for her assiduous efforts in arranging the reception and collecting feedbacks. We are grateful to Dr. K. Poongothai for the brilliant supervision of the registration and help desk. We appreciate Ms. R. Arutselvi and Ms. S. Anita for conducting the inaugural and valedictory functions in a magnificent manner. Thanks to Ms. B. Kalpana and Ms. M. Ramya for making oral technical sessions informative and interesting. We would like to thank Ms. B. Bagyalakshmi for designing impressive certificates and awards.

We are thankful to Dr. V. Balakrishnan for efficiently taking care of the VIP hospitality. Thanks to Mr. G. Mohan and Mr. R. Ganesh Babu for arranging the wonderful lunch and to Dr. J. Philip Robinson and Mr. H. Pandey for arranging the refreshments. We appreciate Mr. S. Venkatesa Prabhu for looking after the public relations and photos. We are thankful to Dr. M. Balasubramanian and Ms. S. Poornima for arranging the accommodations of the delegates in the gents and ladies hostel respectively. We are thankful to Mr. R. Subbaiya and Mr. M. Abishek for managing the catering arrangements for dinner. Thanks to Mr. M. Karthikeyan for maintaining discipline and taking care of mobilization of delegates. We are thankful to Mr. R. Arulvel for arranging an entertaining and enjoyable cultural programme for the conference.

We would also like to convey our thanks to the following students of the final year B.Tech, biotechnology for their laborious efforts in making of the souvenir – C. Monica, S. Niranjana Kamalapriya, R. Nanthini, S. Ramya, S.M. Brinda Shree, R.P. Janani, D. Vaishnu Devi, R. Kaviya Divya.

P. Ponmurugan
Puniethaa Prabhu
G. Ayyappadasan
R. Ganesh Babu

SECTION–I

Shri C.V. Jacob Award

Muscarinic Receptor Subtypes Functional Regulation in the Pancreas of Multiple Low Dose Streptozotocin Induced Diabetic Rats

Naijil George, Ajayan M.S. and Paulose C.S.*

Molecular Neurobiology and Cell Biology Unit, Centre for Neuroscience,
Department of Biotechnology, Cochin University of Science and Technology, Kerala
E-mail: cspaulose@gmail.com; cspaulose@cusat.ac.in

INTRODUCTION

Diabetes is a severe health problem and its incidence is increasing at alarming proportions. Recent estimates indicate that around 200 million people suffer from diabetes mellitus, making it the most common serious metabolic disorder worldwide. Diabetes mellitus is characterized by hyperglycemia due to an absolute or relative insulin deficiency. Pancreatic islets receive nerve innervations from autonomic nervous system and pancreatic endocrine secretion is partly controlled by the neurotransmitters released by autonomic neurons (Liu *et al.*, 2001). Acetylcholine, a major neurotransmitter of the peripheral parasympathetic nervous system helps to facilitate the release of insulin in a glucose-dependent mode. This activity has been shown to be mediated by the activation of muscarinic acetylcholine receptors located on the pancreatic beta cells (Iismaa T.P. *et al.*, 2004). Our present study on muscarinic receptor subtypes functional regulation in the pancreas of diabetic rats will definitely enlighten the novel therapeutic possibilities for diabetes.

MATERIALS AND METHODS

[^{3}H] QNB and [^{3}H] 4-DAMP were purchased from NEN Life Sciences Products Inc., Boston, USA. Streptozotocin, pirenzepine, 4-DAMP mustard, and curcumin from

*Corresponding author.

Sigma Chemical Co., USA. Tri-reagent kit was purchased from MRC, USA. Real-time PCR Taqman probe assays on demand were from Applied Biosystems, Foster City, CA, USA. All other reagents of analytical grade were purchased locally.

Animals and Experimental Design

Male Wistar rats of 90–110 g body weight were used for all experiments All animal care and procedures were done in accordance with the Institutional and CPCSEA guidelines Animals were divided into two groups: i) control ii) diabetic. Diabetes was induced according to the procedure of Like and Rossini, 1976 by intra-peritoneal injection of multiple low dose STZ (MLD-STZ, 40 mg/Kg BW) for five consecutive days. STZ freshly dissolved in 0.1 M citrate buffer, pH 4.5. Control rats were injected with citrate buffer alone. Diabetes was confirmed on the 14th day by checking the fasting blood sugar and plasma insulin. Insulin assay was done according to the procedure of BARC radioimmunoassay kit. Rats were sacrificed on 15th day by decapitation. The pancreas was dissected out quickly over ice and the tissues collected were stored at −80°C until assayed.

Total Muscarinic, Muscarinic M1 and M3 Receptor Binding Studies in the Pancreas

Radio receptor assay in pancreas was done according to the modified procedure of Yamamura and Synder (1981). Total muscarinic receptor binding parameter assays were done using different concentrations, i.e. 0.1–2.5 nM of [^{3}H] QNB. Nonspecific binding was determined using atropine. Muscarinic M1 receptor binding assays were done using [^{3}H] QNB and pirenzepine. Muscarinic M3 binding assay was done using [^{3}H] 4-DAMP and 4-DAMP mustard. Bound radioactivity was counted with cocktail-T in a Wallac 1409 liquid scintillation counter. The receptor binding parameters were determined using Scatchard analysis (Scatchard, 1949).

Analysis of Gene Expression by Real-Time PCR

RNA was isolated from the pancreas of experimental rats using the Tri-reagent according to the procedure of Chomczynski and Sacchi, 1987. Total cDNA synthesis was performed using ABI PRISM cDNA archive kit in 0.2 ml microfuge tubes. Real-time-PCR assays were performed in 96-well plates in an ABI 7300 Real-time-PCR instrument (Applied Biosystems). The specific primers and probes were purchased from Applied Biosystems.

Statistics

Data are presented as means with their standard errors. Statistical evaluations were done by ANOVA using In Stat (version 2.04a; GraphPad Software Inc.). Relative Quantification Software (Relative Quantification Software) was used for analysing real-time PCR results.

RESULTS

Fasting Glucose and Insulin Concentrations

Fasting blood glucose and insulin levels of all rats before MLD-STZ administration were within the normal range. STZ administration leads to a significant increase ($p < 0.05$) in blood glucose level and significant a decrease ($p < 0.05$) in plasma insulin levels of diabetic group compared with control (Table 1).

Table 1: Fasting Blood Glucose and Insulin Concentrations Receptor in Experimental Rats

Experimental Groups	*Blood Glucose Level (mg/dL)*	*Insulin Concentration (μU/ml)*
Control	96.5 ± 3.4	69.2 ± 4.3
Diabetic	396.3 ± 9.4[a]	21.4 ± 1.9[a]

Total Muscarinic Receptor Analysis

The Scatchard analysis showed that the Bmax of the [^{3}H] QNB receptor binding decreased significantly ($p < 0.001$) in the pancreas of diabetic rats compared to control group. The ligand binding affinity of total muscarinic receptor didn't show any significant changes (Table 2).

Table 2: Scatchard Analysis of Total Muscarinic, M1 and M3 Receptor in the Pancreas of Experimental Rats

Experimental Groups	*Total Muscarinic Receptor*		*Muscarinic M1 Receptor*		*Muscarinic M3 Receptor*	
	B_{max} (fmoles/ mg protein)	*K_d (nM)*	*B_{max} (fmoles/ mg protein)*	*K_d (nM)*	*B_{max} (fmoles/mg protein)*	*K_d (nM)*
Control	428.48 ± 25.01	1.75 ± 0.16	56.57 ± 4.64	0.97 ± 0.08	121.64 ± 10.56	0.87 ± 0.07
Diabetic	273.98 ± 22.59[a]	1.66 ± 0.15	38.69 ± 3.58[a]	0.93 ± 0.07	83.32 ± 7.61[a]	1.01 ± 0.9

Values are mean ± S.E.M of 4–6 separate experiments. [a]$p < 0.001$ when compared to Control.

Muscarinic M1 and M3 Receptor Analysis

The Scatchard analysis showed that the Bmax of muscarinic M1 and M3 receptors of pancreas were decreased significantly ($p < 0.001$) in diabetic condition compared to control group. The binding affinity of M1 and M3 receptors doesn't show any significant change (Table 2).

Real-Time PCR Analysis of Muscarinic M1 and M3 Receptor

Real-time PCR analysis showed that the muscarinic M1 and M3 receptor gene expression in the pancreas was decreased significantly ($p < 0.001$) in diabetic condition when compared with control (Figure 1).

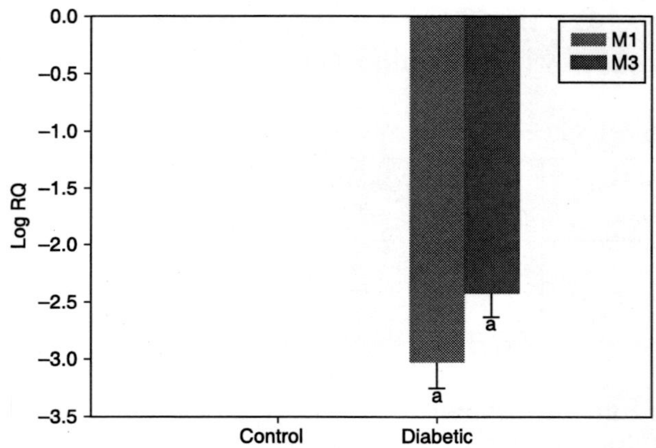

Values are mean ± S.E.M of 4–6 separate experiments. [a]$p < 0.001$ when compared to Control.

Fig. 1: Real-Time PCR Amplification of Muscarinic M1 and M3 Receptor Mrna in the Pancreas of Experimental Rats

DISCUSSION

One of the key pathological features of diabetes is that pancreatic beta-cells fail to secrete sufficient amounts of insulin to overcome elevated blood glucose levels (Kahn, 2003). Several studies have shown that acetylcholine, the major neurotransmitter of peripheral parasympathetic nerves, exerts a pronounced stimulatory effect on pancreatic insulin release (Gilon and Henquin, 2001). In the present study MLD-STZ induced diabetic rat pancreas showed a significantly decreased muscarinic receptor number and mRNA levels when compared to control without any significant changes in ligand binding affinities. This will leads to the overall reduction in the muscarinic receptor mediated activation of Gq-type G proteins (Zawalich *et al.*, 1989). Muscarinic receptors induced pancreatic insulin release is mediated by two major mechanisms, one depending on elevated intracellular calcium levels and another involving a protein kinase C-mediated increase in the efficiency of calcium on insulin exocytosis (Gilon and Henquin, 2001). Decreased muscarinic receptor number leads to the impairment of these two pathways and reduce insulin releases from beta cells. Pancreatic muscarinic receptors also play a major role in maintaining normal glucose homeostasis by fine-tune the glucose-induced insulin secretion (Zawalich *et al.,* 1989). In pancreatic islets acetylcholine serves as feed-forward signal to keep the beta cell responsive to any changes in blood glucose levels and hence limiting plasma glucose fluctuations. The intracellular signalling pathways activated by acetylcholine via muscarinic receptors promote long-term survival of pancreatic beta cells (Kiba *et al.,* 1996) reported that the activation of peripheral vagal nerves has a stimulatory effect on beta-cell differentiation and proliferation. Acetylcholine is the major neurotransmitter stored in vagal nerve endings and the muscarinic receptors are the predominant cholinergic receptor expressed by pancreatic beta-cells. Diabetes associated reduction in muscarinic receptor will directly influence the functional beta-cell mass and contribute to the decreased insulin output from the pancreas.

The role of muscarinic receptors in marinating glucose homeostasis, insulin release, beta cell death and proliferation suggested that the stimulation of beta-cell muscarinic receptors will be beneficial in the treatment of diabetes. Our data suggest that the use of selective M1 and M3 receptor agonists will offer therapeutic benefits in the long-term treatment of diabetes.

ACKNOWLEDGEMENTS

This work was supported by research grants from DBT, DST, ICMR, Govt. of India and KSCSTE, Govt. of Kerala to Dr. C.S. Paulose. Naijil George thanks CSIR for JRF.

REFERENCES

Chomczynski, P. and Sacchi, N. (1987). Single-step method of RNA isolation by acid guanidinium thiocyanate–phenol–chloroform extraction. *Analytical Biochemistry* 162: 156–159.

Gilon, P. and Henquin, J.C. (2001). Mechanisms and physiological significance of the cholinergic control of pancreatic beta-cell function. *Endocr Rev.* 22: 565–74.

Iismaa, T.P. *et al.* (2004). Quantitative and functional characterization of muscarinic receptor subtypes in insulin secreting cell lines and rat pancreatic islets. *Diabetes* 9: 392–398.

Kahn, S.E. (2003). The relative contributions of insulin resistance and beta-cell dysfunction to the pathophysiology of Type 2 diabetes. *Diabetologia* 46: 3–19.

Kiba, T., Tanaka, K., Numata, K., Hoshino, M., Misugi, K. and Inoue, S. (1996). Ventromedial hypothalamic lesion-induced vagal hyperactivity stimulates rat pancreatic cell proliferation. *Gastroenterology* 110: 885–893.

Like, A.A. and Rossini, A.A. (1976). Streptozotocin-induced pancreatic insulitis: new model of diabetes mellitus. *Science* 193 (4251): 415–417.

Liu, Y.Q., Montanya, E. and Leahy, J.L. (2001). Increased islet DNA synthesis and glucose derived lipid and amino acid production in association with beta cell hyperproliferation in normoglycaemic 60% pancreatectomy rats. *Diabetologia* 44: 1026–1033.

Scatchard, G. (1949). The attractions of proteins for small molecules and ions. *Ann NY Acad Sci* 51: 660–672.

Yamamura, H.I. and Synder, G. (1981). Binding of [H] QNB in rat brain. *Proc Natl Acad Sci USA,* 71: 1725–1729.

Zawalich, W., Zawalich, K. and Rasmussen, H. (1989). Cholinergic agonists prime the beta-cell to glucose stimulation. *Endocrinology,* 125: 2400–2406.

Effect of Humic Acid along with Growth Regulators in *in vitro* Propagated Banana Plants

S. Dhanapal* **and D. Sathish Sekar**
Department of Biotechnology, Arignar Anna College (Arts and Science), Krishnagiri
E-mail: dhanapalu@gmail.com

INTRODUCTION

Humic acid is a principal component of humic substances, which are the major organic constituents of the soil (humus), peat, coal, many upland stream, dystrophic lakes, and ocean water. Humic acid is a long chain molecule which is high in molecular weight, dark brown and soluble in alkali solution. In soil, humic acids are formed through the chemical and biological humification of plant and animal matter and through biological activities of microorganism. Humic acid is also found in coal, and the low grade coal leonardite is particularly the good source of humic acid. Leonardite is an organic matter and it is the end product of the humification process. Humic acid has been recognized to play a major role in producing morphological and physiological effect on plants. The application of humic substances to nutrient solution, to soil, or sand showed that they enhanced significant growth responses. Foliar spray of humic acid promoted the growth in many plants such as tomato, cotton and grapes (K. Obuswan, S. Namchote *et al.*, 2011).

Humic acids are not a significant source of plant nutrient but are a soil stimulant and a transporation vehicle for carrying nutrient into plants (Dr. Boris Levinsky, 2001). Once the humic acid, enter plant cells several biochemical changes occur in membrane and various cytoplasmic component of plant cells. Excessive concentration of humic acids can kill young seedlings. Humic substance has a pronounced influence on the growth of plant (Robert E. Pettit, 2007).

---

*Corresponding author.

Humic compounds can be absorbed by the plant roots and transported to the shoots, thus enhancing the growth of the whole plant (Rautham and Schnitzer, 1981). The organic matter in soil is divided into that which is partially decomposed matter and that which is fully decomposed, and termed 'humus' (Lynette Morgan, 2012).

Tissue culture is important to establish and/or maintain a "virus-free" stock somatic hybridization induction and selection of mutants and germplasm conservation (Conger, 1980; Vuylsteke, 1989). Micro-propagated banana plants have two major advantages: (1) their size and septic quality make easier to introduce new varieties into production areas and (2) their nematode-free status makes possible to improve soil parasite control and reduce the use of nematicides (Ganry, 1986; Navarro *et al.,* 1994).

Banana (Musa spp.) is the fourth most important food crop in the world as well as in India (Ganapathi *et al.,* 1991). Musa paradisiaca, which is very popular and with high commercial value, has a high demand on markets due to its sweet aroma, taste and higher post harvest life. However, the shortage of planting material and synchronization of fruit ripening are two major bottlenecks that cause unavoidable trouble to local banana growers. There is thus a need to establish a micropropagation protocol for this banana cultivar (Olivia Saha Roy *et al.,* 2010). There have been scarce reports on the effect of HA on plant growth and development in vitro propagation, especially at low nutrient level in which the sources of plant food were limited (Kullanart Obsuwan *et al.,* 2011). Presently, banana is grown in around 150 countries across the world on an area of 4.84 million ha producing 95.6 million tones. Asia, Africa and Latin America are the major banana producing continents. Among the major producers, India alone accounts for 27.43% (26.2 million tonnes). The present study was aimed to investigate the effect of various concentrations of Humic acids obtained (Alkali extraction with 4N KOH) from its native ore on the growth of of *Musa* sp. Cv Grand Naine Banana Seedlings in tissue cultures.

MATERIALS AND METHODS

Extraction of Humic Acid from Leonardite

Sample Collection

The coal sample (leonardite) was collected from Mines II of Neyveli lignite corporation, Neyveli.

Humic Acid Extraction

10 gm of coal (leonardite) sample was weighed and ground and humic acid was extracted by alkali extraction method (www.humicsubstances.org) the extracted humic acid

was estimated colorimetrically by DTPA (Diethylene triamine pentaacetic acid) method using Titan Biotech Humic acid as standard.

In vitro Propagation of Grand Naine

Sample Collection

The multistage culture of Grandnaine (*Musa accuminata*) was collected from Genewin Biotech, Hosur.

Micropropagation

Trial 1: Multi Stage—Surface sterilized suckers were inoculated in MS Medium (Murashige, T. and F. Skoog, (1962) supplied with growth regulators sucrose (3%), NAA (0.1%), 6-BAP (0.1%), and humic acid in the range from 0.1% to 0.4%. Trial bottles are labeled as GW04-NA, 0.1, 0.2, 0.3, 0.4 respectively and taken to incubation room or growth room.

Trial 2: Shooting Stage—Samples were inoculated in MS Medium supplied with growth regulators sucrose (3%), NAA (0.1%), 6-BAP (0.1%), kinetin (0.1%) and humic acid in the range from 0.1% to 0.4%. Pure culture bottles of multistage belonging to batch accession No. GW04 of week 45/2 were got from production lab of Genewin biotech. Open the culture bottle cap; slowly pick a culture (group of 3 small shoots) carefully. Place it on the utility sheet and trim them to suit the shooting condition.

Trial 3: Rooting Stage—Pure culture bottles of shooting stage belonging to batch accession No. GW04 of week 45/2 were got from production lab of Genewin biotech. Samples were inoculated in MS Medium supplied with growth regulators sucrose (3%), NAA (0.l%), adenine sulphate (0.1%), kinetin (0.1%) and humic acid in the range from 0.1% to 0.4%. Open the culture bottle cap; slowly pick a culture (group of 3 small shoots) carefully. Place it on the utility sheet and trim them to suit the rooting condition. Trial bottles are labeled as GW04-C, 0.1, 0.2, 0.3, 0.4 respectively and taken to incubation room or growth room.

RESULTS AND DISCUSSIONS

Humic acids are extracted by using alkali KOH. The extracted Humic acids are used to propagate Grandnaine *(Musa accuminata)* and compared with growth regulator supplemented plants as control.

Humic acid has direct effect on plant cell membrane. It increases the permeability, ease by which mineral element move back and forth through the all membrane, resulting in a increased transport of various mineral nutrient to site of metabolic need. When Humic acid is applied to plant leaves the chlorophyll content of those leaves increases (Robert E. Pettit, 2007). In the present study humic acid was extracted by using alkali extraction method from Leonardite and its effect on plant growth was checked using *Musa* sp. Cv Grand Naine Bananas along with growth regulators. In trial I humic acid with the concentration of 0.4% gave 5X highest multiplication compare to other concentration (Figure 1). The positive effects of humic substances on the growth of plants in Gramineae family has been well documented (Chen and Aviad, 1990). Dormaar in 1975 reported increased N uptake by rough fescue (Festuca scabrella Torr.) in response to application of humic substances extracted from 3 soils, while P, K, Ca, Mg and Na uptake was unaffected. Humic acid have long recognized that play a major role in producing morphological and physiological effects in plants (Nardi S. *et al.*, 2002 and Eyheraguibel *et al.*, 2008).

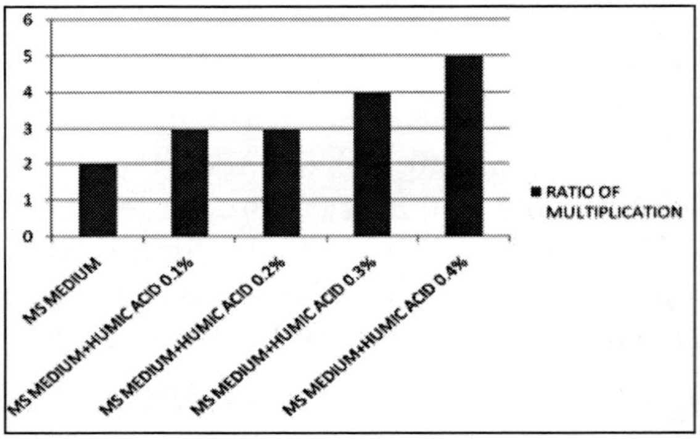

Fig. 1: Effect of Humic Acid on Plant Multiplication

In trial 2, predominant results were observed in the humic acid supplemented plants 0.4% humic acid supplied plants had good stem thickness (3 mm) and increased biomass density (88%), when compared with all the other concentrations (Figures 2 and 3). It has been reported that humic acids are able to stimulate or inhibit plant growth depending on their differences in origin, nature and concentration. The application of humic substances to nutrient solution, to soil or sand has been documented and the results showed that they enhanced significant growth responses (Lulakis M.D and Petsas S.I., 1995).

In trail 3 the rooting occurs densely in both GW04 NA and GW04 HU, and there is no prominent difference between them. This suggested that HAs at the concentration of 0.4% could improve the growth of *Musa* sp. Cv Grand Naine Banana seedlings in tissue culture.

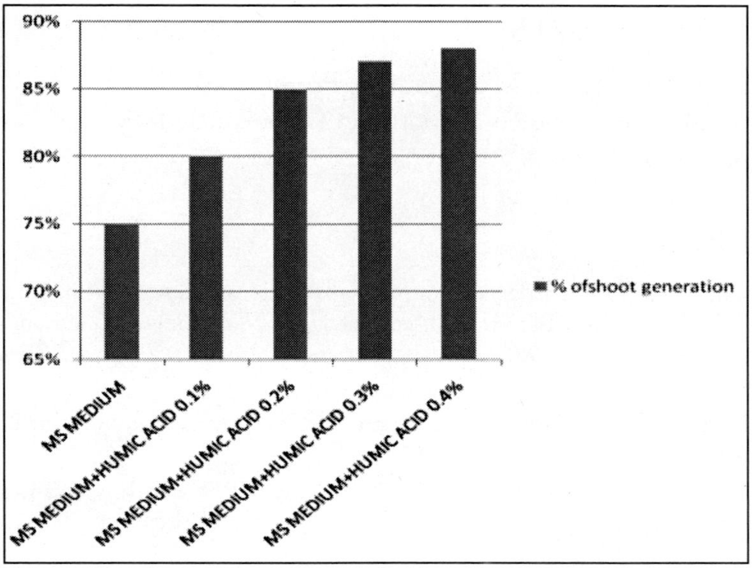

Fig. 2: Effect of Humic Acid on Shoot Generation

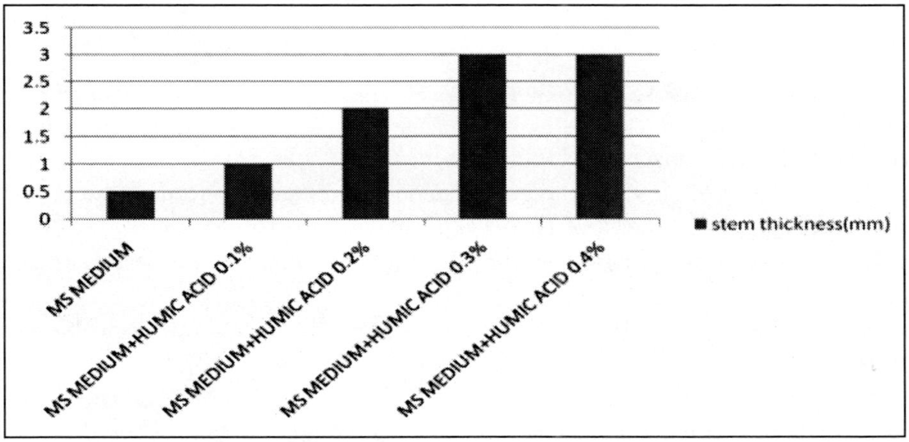

Fig. 3: Effect of Humic Acid on Stem Thickness

Musa sp. Cv Grand Naine Banana seedlings were grown on MS Medium supplied with growth regulators and MS Medium supplied with growth regulators along with humic acid in the concentration range from 0.1% to 0.4% was taken to ensure the effect of humic acid on the plant growth (multistage, shooting and rooting). Humic acid with the concentration of 0.4% gave good results compare to others. To conclude that 0.4% of humic acid is best for the commercial production of *Musa* sp. Cv Grand Naine Banana seedlings in plant tissue culture.

ACKNOWLEDGEMENTS

The authors are grateful to the Management, Principal of Arignar Anna College (Arts and Science), Krishnagiri, for their support and Genewin Biotech, Hosur, for providing necessary laboratory facilities.

REFERENCES

Chen, Y. and Aviad, T. (1990). Effects of humic substances on plant growth. In: MacCarthy P., Clapp C.E., Malcolm R.L., Bloom P.R., editors. Humic Substances in Soil and Crop Science: Selected Readings. *Madison, WI: American Society of Agronomy Inc., Soil Science of America, Inc.* pp. 161–186.

Conger, B.V. (1980). Cloning Agricultural Plants Via In Vitro Techniques, p. 273. *CRC press, Inc., Boca Raton, Florida.*

Eyheraguibel, B., Silvestre, J. and Morard, P. (2008). Effects of humic substances derived from organic waste enhancement on the growth and mineral nutrition of maize. *Bioresource Technology* 99, 4206–4212.

Ganapathi, T.R., Suprasanna, P., Bapat., Kulkarni, V.M. and Rao, P.S. (1999). somatic embryogenesis and plant regeneration from male flower buds in banana. *Current Science*, Vol. 76, No. 9, pp. 1228–1331.

Ganry, J. (1986). Improving banana phytosanitation and banana production excised stem tips from aseptically cultured terminal floral apex of banana. *American Journal of Ann. Bot.*, 53: 321–328.

Levinsky, Boris (2001). Humates and Humic Acids.

Lulakis, M.D. and Petsas, S.I. (1995). Effect of humic substances dfrom vinecanes mature compost on tomato seedling growth, *Biosource Technology*, Vol. 54, 179–182.

Morgan, Lynette (2012). Growing Green for Australia Maximum Yield-Indoor Gardening—the Pythium Predator.

Murashige, T. and Skoog, F. (1962). A revised medium for rapid growth and bioassays with tobacoo tissue culture. *Physical Pl.*, 15: 473–97.

Nardi, S.; Pizzeghello, Diego; Muscolo, Adele and Vianello, Angelo (2002). Physiological effects of humic substances on higher plants. *Soil Biology and Biochemistry* 34, 1527–1536.

Navarro, C., Teissonc., Cote, F. and Ganry, J. (1994). Effect of light intensity and CO_2 concentration on growth of banana plants (Musa AAA, cultivar "petite Naine") in vitro and subsequent growth following acclimatization. *Sci. Hort.*, 60: 41–54.

Obsuwan, K. (2011). Effect of various concentration of humic acid on growth and development of eggplant seedling in tissue culture at low nutrient level, World academy of science, *Engineering and Technology*, 80, 276–278.

Obsuwan, Kullanart, Namchote, Suluck, Sanmanee, Natdhera, Panishkan, Kamolchanok and Dharmvanij, Sirichai (2011). Effect of various concentrations of Humic acid on growth and development of Eggplant seedlings in tissue cultures at low nutrient level. *World academy of science, engineering and technology*, pp. 276–278.

Pettit, Robert E. (2007). Organic matter, humus, humate, humic acid, Emeritus Associate Professor, Texas A&M University.

Rauthan, B.S. and Schnitzer, M. (1981). Effects of a soil fulvic acid on the growth and nutrient Content of *Cucumis Sativus* Plants. *Int. Plant and Soil.*, Vol. 63, p. 491.

Roy, Olivia Saha, Bantawa, Pranay, Ghosh, Swapan Kumar, Jaime, A., Silva, Teixeira da, DebGhosh, Partha and Mondal, Tapan Kumar (2011). Micropropagation and Field Performance of Malbhog: Popular Banana Cultivar with High keeping quality of North East India. In: Tree and Forestry Science and Biotechnology. Global science books.

Vuylsteke, D.R. (1989). Shoot-tip Culture for the Propagation, Conservation through shoot-tip culture. In: Technical Centre for Agricultural and Rural Cooperation and Caribbean Agricultural Research and Development Institute (Editors), Improving Citrus and Banana Production in the Caribbean through Phytosanitation, pp. 234–6. Seminar Proc. 2–5 December 1986, St. Lucia, W.I tissue of triploid plantains (Musa ABB). Plant cell Reproduction, 2: 289–291.

Isolation and Screening of *Bacillus* Spp. and Antagonist Activity against *Fusarium* Spp. Causing Rhizome Rot Disease in Turmeric Plants

A. Mohamed Halith*, G. Velayutham, R. Ganesh Babu, P. Ponmurugan, C. Citra[1] and N. Mathivanan[1]

Department of Biotechnology, K.S. Rangasamy College of Technology, Tamil Nadu

[1]Biocontrol and Microbial Metabolites Lab, Center for CAS in Botany, University of Madras, Tamil Nadu

E-mail: velayuthamg2@gmail.com

INTRODUCTION

Turmeric (*Curcuma longa*) is a major spice crop and having more medicinal properties. Turmeric is grouped under the Zingiberaceae family; there are more than 100 species of *Curcuma* around the world. Turmeric is native to India. In India, highest cultivation of turmeric is Andhra Pradesh, Tamil Nadu, Orissa, West Bengal and some of the important states cultivated in India. The turmeric plants affected by number of disease like Leaf Blotch, Leaf spot, Rhizome rot, Shoot borer and Nematode pest.

The natural products are found to be more effective with least side effects as compared to commercial fungicides so that reason they are used an alternated remedy for the treatment of various infections (Tepe *et al.,* 2004).

Turmeric is a mild digestive, being aromatic, a stimulant and carminative turmeric is one of nature's most powerful healers. Turmeric was first used as a dye. Long known

---

*Corresponding author.

for its anti-inflammatory properties, recent research has revealed that turmeric is a natural wonder, providing beneficial in the treatment of many different health conditions from cancer to Alzheimer's disease. An ointment base on the spice is used as an antiseptic in India. Turmeric helps to regulate the female reproductive system and purifies the uterus and breast milk, and in men it purifies and builds semen, which is counterintuitive for a pungent bitter (Debjit Bhowmik, 2009).

Biological control by antagonistic microorganisms is widely recognised as a promising method for control of post-harvest plant diseases (Janisiewicz and korsten, *2002*). Several studies report the effectiveness of microbial antagonists against a wide range of post-harvested pathogens growing on a variety of harvested fruits and vegetables (Droby and chalutz, 1994; Droby *et al.,* 1993; Fravel, 2005).

The genus *Pythium* belongs to Oomycetes which was discovered by Pringsheim (1858); Martin (1991). It is often known as a pathogen causing root rot and damping-off disease. The taxonomy of the genus *Pythium* is mainly based on the morphological descriptions and the keys provided by Middleton (1943), Waterhouse (1968), and van der Plaats-Niterink (1981).

One of the major diseases is rhizome rot disease caused by fungi called *Pythium aphanidermatum, Pythium graminicolum, Fusarium* spp. (Ganesh Babu, 2012). The fungal pathogens completely decay the rhizome of turmeric. The pseudostem of turmeric was affected then complete plant was dried and the plat growth was stopped, so the yield of turmeric rate will decrease. The disease will control by using of PGPRs like *Pseudomonas* and *Bacillus* spp. The farmers applied the fungicides (Mannose) for controlling this disease. The fungicides affect the nature of soil and human health.

Plant Growth-Promoting Rhizobacteria (PGPR) are beneficial bacteria that colonize plant roots and enhance plant by a wide variety of mechanisms. Soil application of PGPR was evaluated for various plant growth parameters such as root length and plant height, survival rate, intermodal length, seedling strength, flowering, callus branches, etc.

To avoid the use of fungicides PGPRs formulation will apply to the field for controlling the disease of Rhizome rot. PGPRs act as a plant protection, Biofertilizer and Biostimulants. *Pseudomonas* and *Bacillus* spp. having the biocontrol activity for the control of rhizome rot disease. The presence study is PGPRs improve the yield of rhizome of turmeric and control the rhizome rot disease caused by *P. aphanidermatum* and *P. graminicolum* by in vitro and *in vivo* methods. Rhizome root rot disease causing pathogen family.

Crop losses to the tune of 50 percent and above have been reported in some parts of Telungana and formers resort to distress harvest to salvage the remaining crop once the disease starts appearing.

MATERIALS AND METHODS

Isolation of *Fusarium* Spp.

Sample Collection

The infected turmeric rhizomes were collected from the production site of Erode, Kodumudi, Karur, Namakkal, Salem areas of Tamil Nadu. Infected rhizome of turmeric plant variety PTS8 (*Paradeep Turmeric Station*) were used in the study. The pathogen was isolated by the tissue segment method (Rangeswamy, 1958) on Potato Dextrose Agar (PDA) medium. The infected rhizome of turmeric plants were washed with running tap water for 15 minutes, to remove the soil present in the surface of infected rhizome. Tissues from the rotted rhizome were surface sterilised by using 70% alcohol and 0.1% mercuric chloride for 60 seconds. Sterilized root pieces were washed twice with sterile water for 60 seconds, plated on PDA plates, and incubated at room temperature for 3 days (Reddy *et al.,* 2005). White cottony mycelial growth from the infected root pieces was transferred to plates with 1.5% water agar. The growing hyphal tips were transferred to PDA plates. The pathogen was identified as *Fusarium* spp. by using lacto phenol cotton blue staining method.

Media Composition of PDA

Potato extract – 100 ml
Dextrose – 2 g
Agar – 2 g

Isolation of PGPRs from Turmeric Soil

Collection of Soil Samples

Turmeric soil samples were collected from the turmeric fields of Erode, Kodumudi, Salem, Karur, Namakkal districts. Isolation of Bacillus from rhizosphere soil uptained from different turmeric fields by using serial dilution techniques. Isolation of plant growth promoting rhizobacteria from different soil, *Bacillus cereus* (1000 ml): peptone 1.0 g, mannitol 10.0 g, sodium chloride 2.0 g, magnesium sulphate 0.1 g, potassium dihydrogen phosphate 0.25 g, disodium dihydrogen phosphate 2.5 g, bromothymol blue 0.12 g, sodium pyruvate 10.0 g, agar15.0 g.

Bacillus polymyxa (1000 ml): yeast extract 5.0 g, peptone 5.0 g, starch soluble 10.0 g, Agar 20.0 g. *Bacillus coagulans* (1000 ml): glucose 5.0 g, protease peptone 5.0 g, yeast extract 5.0 g, potassium hydrogen phosphate 4.0 g, magnesium sulphate 10.0 g, calcium chloride 10.0 g. *Bacillus racemilaticus* (1000 ml): yeast extract 5.0 g, calcium carbonate 5.0 g, glucose 5.0 g, peptone 5.0 g, agar 15.0 g. *Bacillus megatarium* (1000 ml): sucrose 10.0 g, potassium dihydrogen phosphate 2.5 g, potassium hydrogen phosphate 2.5 g, ammonium hydrogen phosphate 1.0 g, magnesium sulphate 0.01 g, ferric sulphate 0.01 g, manganese sulphate 0.007 g, agar 20.0 g. *Bacillus thermoglucocidasials* (1000 ml): soluble starch 10.0 g, peptone 5.0 g, beef extract 3.0 g, potassium dihydrogen phosphate 3.0 g, yeast extract 3.0 g, agar 20.0 g.

Antagonist Activity

Dual Culture Method

The dual culture test was carried out by using dual culture medium. The composition of dual culture medium for 100 ml: potato broth-100 ml, dextrose-2 g, peptone-0.5 g, sodium chloride-0.5 g, yeast extract-0.3 g, beef extract-0.3 g, agar-2 g. The bacterial isolates were streaked on the dual culture medium 10 mm from the edge of the plate by using sterile inoculating loop. 5 mm mycelial disc of 5 day old *Fusarium* spp. culture was placed at the opposite side of the edge the petri dish to the bacterial streak plates were incubated at 28 ± 2°C, and growth of the fungal mycelia towards the bacterial colony and inhibition zone was measured after 72 hours of incubation. Among the 30 different bacterial strains four isolates were efficient. The percentage of inhibition was measured by the following formula:

$$\text{Mi} = \frac{Mc - Mt}{Mc} \times 100$$

Where,

Mi = Inhibition of Mycelial growth
Mc = Mycelial growth in Control
Mt = Mycelial growth in Treated plates

RESULTS AND DISCUSSION

Isolation of PGPRs

Soil samples were collected from the turmeric various fields. *Bacillus* spp. was isolated from the soil sample collected from the different turmeric fields, by using serial dilution techniques. Finally 30 isolates of *Bacillus* spp. were isolated by using different types of selective medium. Isolated *Bacillus* spp. was stored at 4°C for further studies.

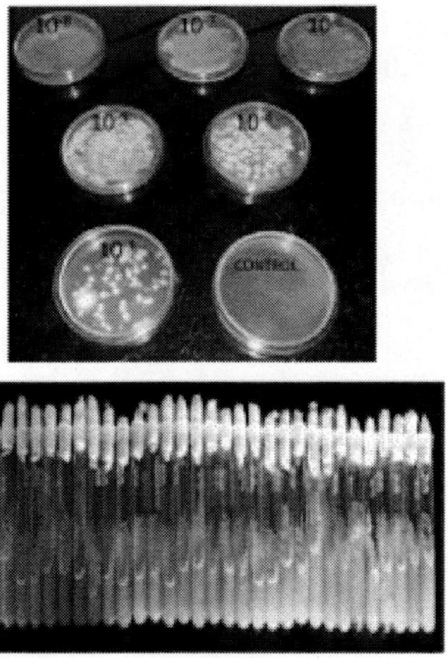

Fig. 1: Isolation of *Bacillus* Spp. from Different Fields of Turmeric Plants

Isolation of Pathogen

The *Fusarium* spp. was isolated from the infected rhizome of turmeric plants which is collected from different turmeric fields of Tamil Nadu.

Fig. 2: Symptoms of Rhizome Rot Disease in Turmeric Plants

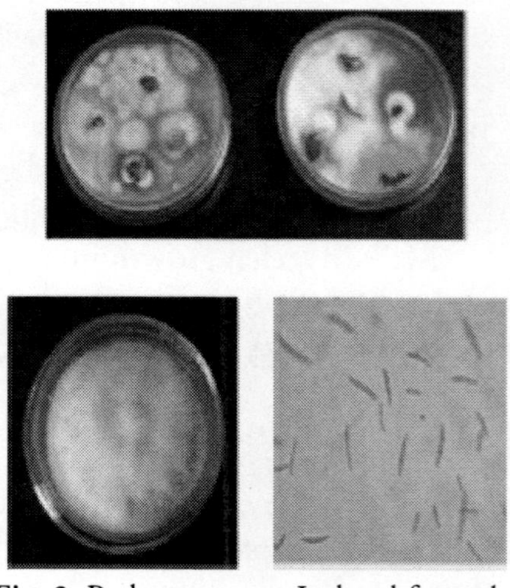

Fig. 3: Pathogens were Isolated from the Infected Part of Turmeric Plants and Microscopic Image of *Fusarium* Spp.

The isolated pathogens were identified as *Fusarium* spp. by lactophenol cotton blue staining method. The spores of the fungus having 4 septates. Which was conformed that the isolated fungal spp. is *Fusarium* spp. Isolated *Fusarium* spp. was observed cottony white in colour on the PDA plates.

Screening of *Bacillus* Isolates

Four isolates of *Bacillus* spp. were screened by using dual culture method. The screened bacterial isolates were showed maximum inhibition rate against pathogen. Four bacterial strains were screened for antifungal activity against *Fusarium* spp. The bacterial strains were B1, B6, B11, and B14.

Table 1: Screening of *Bacillus* spp. by Using Dual Culture Test against *Fusarium* spp.

S. No.	*Bacillus Strains*	*Growth Rate (cm) on 5th day*			*Percentage (%) of Inhibition*
		Control	*Bacillus Spp.*	*Fusarium Spp.*	
1.	B1	6.8	3.5	1.3	80.88
2.	B2	6.8	3.3	2.5	63.23
3.	B3	6.8	2.8	3.0	55.88
4.	B4	6.8	3.0	2.9	57.35
5.	B5	6.8	3.2	2.3	66.17
6.	B6	6.8	2.7	1.9	72.05
7.	B7	6.8	3.0	3.4	50.02
8.	B8	6.8	3.7	2.2	67.64
9.	B9	6.8	3.0	2.6	61.76
10.	B10	6.8	2.9	3.0	55.88
11.	B11	6.8	3.5	1.7	75.00
12.	B12	6.8	3.4	3.0	55.88
13.	B13	6.8	3.2	2.5	63.23
14.	B14	6.8	3.6	1.2	82.35
15.	B15	6.8	4.0	3.0	55.88

Bacillus spp. was effective against *Fusarium* spp. and may have the efficiency to control the rhizome rot disease in turmeric plants. Our further studies will focus on the study of defence related enzymes (proteins) present in the infected rhizome rot of turmeric plants. The present investigation has shown that *Bacillus* spp. progressively inhibited the radial mycelial growth of *Fusarium* spp. on potato dextrose agar. The *Bacillus* strains such as B1, B6, B11, B14 are inhibited mycelial growth by 80.88%, 72.05%, 75.00%, and 82.35% respectively. Inhibition rate of mycelium was compared with the mycelial growth of control plates. The effectiveness of ridomil at relatively low concentration suggests that it could be used in combination with other very effective fungicides against very resistance fungi (Fernando and Linderman, 1994). *Fusarium* spp. infection of turmeric roots resulted in marked increase in all the nitrogen fractions including total nitrogen, protein nitrogen, soluble nitrogen, and amino nitrogen. Higher amount of total proteins was observed in infected tissues as compared to the healthy tissues. It is possible that this increase of proteins in infected tissues may be due to the accumulation of nitrogenous components in the tissues as a result of changes in nitrogen metabolism of infected plants (Reddy *et al.*, 2005). The pathogen was isolated by the tissue segment method (Rangeswamy, 1958) on potato dextrose agar (PDA) medium.

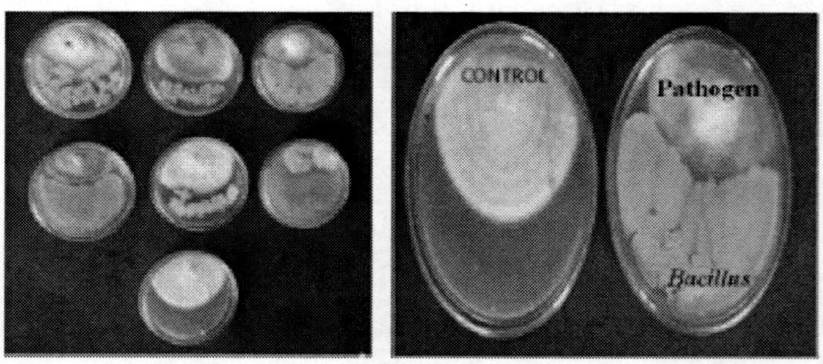

Fig. 4: Dual Culturte Assay of *Bacillus* Spp. against *Fusarium* Spp.

The bacterial strains were screened for antifungal activity against *Fusarium* spp. These were B1, B6, B11, and B14. The Bacterial strains were isolated by using serial dilution method. And the antagonist activity of isolated *Bacillus* spp. was confirmed by the dual culture methods.

REFERENCES

Debjit, Bhowmik, Chiranjib, K.P., Sampath, Kumar, Margret, Chandira and Jayakar, B. (2009). Turmeric: A Herbal and Traditional Medecine 1(2), 86–108 0975–508.

Droby, S. and Chalutz, E. (1994). Mode of action of biocontrol agents for post harvest diseases. In: Wilson C.L., Wisniewski M., (Eds), Biological control of post harvest diseases of fruits and vegetables—Theory and practice. CRC Press, Boca Raton, FL, pp. 63–75.

Droby, S., Chalutz, E., Wilson, C.L. and Wisniewski, M, (1993). Biological control of post harvest diseases: A promising alternative to the use of synthetic fungicides, *Phytoparasitica,* 20: 149–153.

Fernando, W.G. and Linderman, R.G. (1994). Chemical control of stem and root rot of cowpea caused by *Phytopthora vignae. Plant Disease*, 78: 967–971.

Ganesh Babu, R. (2012). Isolation and Characterization of PGPRs for Management of Rhizome rot disease in *Curcuma longa. Biozoom*, Vol. 04, 2250–0693.

Janisewicz, J.W. and Korsten, L. (2002). Biological control of post harvest diseases of fruits. *Annual Review of Phytopathology*, 40, 411–441.

Martin, F.N. (2000). Phylogenetic relationships among some *Pythium* species inferred from sequence analysis of the mitochondrially encoded cytochrome oxidase II gene. Mycologia 92: 711–727.

Middleton, J.T. (1943). The taxonomy, host range, and geographic distribution of genus *Pythium. Memoirs Torrey Bot Club* 20: 1–171.

Rangeswamy, G. (1958). An agar block technique for isolating soil microorganisms with special reference to Pythiaceous fungi. *Science and Culture*, 24: 85.

Reddy, M.N., Sridevi, N.V. and Charitha Devi, M. (2005). Changes in the Nitrogen fractions and aminoacid metabolism of turmeric (*curcuma longa* L.) Roots infected with *fusarium solani. Plant Pathology Bulletin* 14: 221–226, 2005.

Tepe, B., Daferera, D., Sokmen, M., Polissiou, M. and Sokmen, A. (2004). *In vitro* antimicrobial and antioxidant activities of the essential oils and various extracts of thymus. *J. Agri and Food Chemistry*, Vol. 52, 1132–1137.

Van der Plaats-Niterink, A.J. (1981). Monograph of the genus *Pythium. Stud Mycol*, 21:1–242.

Waterhouse, G.E. (1967). Key to *Pythium* Pringsheim; commonwealth Mycological Institute. *Mycol paper* 109: 1–15.

Formulation of Green Nano-Fertilizer to Enhance the Plant Growth through Slow and Sustained Release of Nitrogen

R. Subbaiya, M. Priyanka and M. Masilamani Selvam[1*]
Department of Biotechnology, K.S. Rangasamy College of Technology, Tiruchengode
[1]Department of Biotechnology, Sathyabama University, Jeppiaar Nagar, Chennai
E-mail: ramsubbubio@gmail.com

INTRODUCTION

Nitrogen is naturally present in the soil but it cannot be completely utilized by the plants. So, it is supplemented to the plants artificially with the help of fertilizers. Basically the nitrogen applied using conventional fertilizers, with particle size dimensions greater than 100nm, is lost to the soil due to leaching, so nitrogen utilization efficiency (NUE) by plants is low (Monreal *et al.,* 1986). This leaching occurs due to water–soluble nitrates, emission as ammonia and nitrogen oxides and soil microorganisms–mediated incorporation into soil organic matter. To reduce the leaching and to improve the NUE the surface area of the fertilizer to nanosize. And this strategy involves the slow–release of fertilizer also (Sastry *et al.,* 2010).

Controlled-Release Fertilizers

Controlled-release fertilizers provide an attractive alternative to granular fertilizers. These are fertilizer 'cocktails' that slowly release nutrients to the substrate. The release depends on water availability or soil temperature. Controlled-release fertilizers are more expensive than the more common water soluble fertilizers, but they have several advantages:

- The danger of over-fertilizing is reduced as the release of fertilizers occurs gradually.
- Fertilizing is necessary only occasionally, sometimes only once in a season.

*Corresponding author.

- A balanced fertilizer mixture is provided at all times as the plants get what they need at different growth stages.
- Nutrients do not leach from the substrate so the plants receive all the nutrients applied (Cabrera, 1997).

Nanourea

Nanourea is prepared by reducing its size using some of the reducing agents like Tri Sodium Citrate (TSC). This could reduce the size of the urea and effect of urea to the soil, due to this the pH of the soil will be maintained as such. So the soil does not loss its fertility.

Hydroxyapatite

Hydroxyapatite ($Ca_{10}(PO_4)_6(OH)_2$) is an important component which provides phosphorus nutrient to the plants. Mostly the application of HAp is focused on its biomedical applications due to its excellent biocompatibility and bioactivity (Teng *et al.,* 2008). While it will not be adequately applied in agriculture. The HAp can be synthesized by several methods like sol-gel method (Balakrishnan *et al.,* 2009). HAp is mainly used for bone-replacement because it consist of calcium and phosphorus. Hydroxyapatite (HAP) microspheres with peculiar spheres-in-sphere morphology were prepared by using oil-in-water emulsions and solvent evaporation techniques (Ramesh *et al.,* 2005).

Composite Preparation

Here the surface of the HA is modified with chemically synthesised nanourea. To perform slow release this nanocomposite is prepared. Here two types of composites were prepared. They were as follows: 1. Hydroxyapatite nanoparticle + nanourea (chemical synthesis), 2. Biologically synthesis of nanourea. In biological synthesis two types of concentrations were followed: they are, 0.4 mM concentration, 200 mg.

MATERIALS AND METHODS

Collection of *Rhodococcus* Culture

Rhodococcus spp-2891 was obtained from National Collection of Industrial Microorganism, Pune, India. The obtained species was inoculated in MGYP media which acts as growth media for its growth and proliferation.

Preparation of Hydroxyapatite Nanoparticles

HA nanoparticles are to be synthesized by using aqueous solution of $Ca(OH)_2$ and H_3PO_4. First take 19.29 gm of $Ca(OH)_2$ are to be dissolved in 250 ml of distilled water then prepare 0.6M H_3PO_4(250 ml). Add H_3PO_4 to the $Ca(OH)_2$ suspension in dropwise, while stirring it vigorously under mechanical agitation (1000 rpm).

$$6H_3PO_4 + 10Ca\,(OH)_2 \rightarrow Ca(PO_4)_6(OH)_2 + 18H_2O$$

The synthesized HA nanoparticles are to be refrigerated overnight to form precipitate.

Biological Synthesis

Urea solution were prepared at different concentrations like 0.2 mM, 0.4 mM, 0.6 mM, 0.8 mM and 1 mM. The *Rhodococcus* species were inoculated and kept at incubation in an incubatory shaker at 37°C for 2–3 days. Then the UV-spec reading was taken for that solution to analyze the reduction of the urea particle.

Synthesis of Nanourea

0.005 mol of urea molecule are to be mixed with 5% of 10 ml Trisodium citrate.

- Mixture is heated to 70°C for 30 minutes.
- Color gets changed into ash color solution.
- It indicates the presence of Nanourea molecule.
- It is further confirmed by using UV spectrophotometer (580 nm).

Composite Preparation

About 0.05 gm of nanourea is to be added to the 100 ml of HA nanoparticles and this mixture is subjected to ultrasonic mixing (30 KHz for 1hr). The resulting mixture is then allowed to settle and the excess liquid is drained off. Now the nanourea modified HA particles are subjected to lyophilization to remove the moisture content present in it. The nanourea modified HA nanoparticles are characterized using UV-Spectrophotometer, FTIR and SEM.

Characterization Studies

The characterization studies were performed through UV-Spectrophotometer, FTIR and SEM analysis.

Encapsulation

The nanocomposite mixture was then encapsulated into the hydroxyapatite using ultrasonicator to enhance the slow release of nitrogen to the plants.

Field Studies

The soil is collected from K.S. Rangasamy College of Technology and the green gram are to be treated with chemically synthesized and biologically synthesized nanourea in different concentrations and they are to be potted in a small pot.

Estimation of Nitrogen (AOAC, 1990)

Measure one gram of sample in 100 ml conical flask and add 3 ml of salicylic acid (3.2% in concentrated sulphuric acid) and a pinch of sodium thiosulphate. Digestion is done on a hot plate after adding 5 ml of hydrogen peroxide till the digest become colourless. Simultaneously, a blank is also run. The contents are neutralized with sodium hydroxide and transferred to a 100 ml volumetric flask, then add one ml of 10% sodium potassium tartarate and one ml of Nessler's reagent. The solution is made up to the mark with distilled water and the absorbance is measured at 420 nm using a spectrophotometer. Ammoniacal nitrogen (ammonium sulphate) is used as the standard.

Estimation of Urea by DAM Method

0.2 to 1ml of standard was taken in the test tubes marked as S1-S5. The given unknown solution was made upto 100 ml with distilled water and take 0.5 and 1ml in test tubes marked as T1 and T2, Simultaneously the blank is also prepared which is marked as B. All the test tubes were made upto 1 ml with distilled water. 3 ml of orthophosphoric acid solution. The content of the test tubes followed by 3 ml of orthophosphoric acid solution. The content of the test tubes was mixed well and kept in boiling water bath for 30°C for colour development and cooled. Then the colour developed was read calorimetrically at 540 nm. Draw a graph by taking the concentration in X-axis and the optical density on Y-axis. From which calculate the amount of urea present in the whole of the given unknown solution.

Peroxidase Assay

Pipette out 3 ml of already prepared Phosphate buffer (pH 7.0) solution in the series of test tubes marked as T_1, T_2, T_3 and T_4. 0.05 ml of Guiacol solution was added to each test tube. Add 0.03 ml of H_2O_2 solution into each test tube. As soon as the solution is transferred into the cuvette, 0.1 ml of respective enzyme extract was added, mixed well and the absorbance was measured at 436 nm. The presence of enzyme activity was detected by reading the absorbance for every 10 min.

The enzyme activity was calculated using the formula,

$$\text{Enzyme Activity (U/ml)} = \frac{\Delta 436 \times 4 \times Vt \times \text{Dilution Factor}}{\text{Min.} \times \text{Molecular Extension Coefficient} \times Vs}$$

Where,

Vt : Volume of total sample

Vs : Volume of test sample

RESULTS AND DISCUSSION

The study on chemical and biological synthesis of urea nanoparticles by the *Rhodococcus* spp., and its field studies were carried out in this work using *Vigna radiata.* For chemical synthesis the appearance of ash colour formation during heating suggested the formation of urea nanoparticles. The synthesized nanourea particles were encapsulated inside the hydroxyapatite particles by performing ultra sonication. UV-Spectrophotometer reading was taken for the conformation of the nanourea particles. The biologically synthesized nanourea have shown the turbidity and for further conformation UV-Spectrophotometer reading was taken, then SEM&FTIR reading was taken and then finally the nanourea was observed in reduced size.The FTIR spectrum of the encapsulated material clearly indicated that the structural integrity of nanourea-modified HA nanoparticles was maintained.Then prepared composite was lyophilized, then the particle size and morphology of the synthesized samples were studied using a Scanning Electron Microscopy (SEM). The chemical nature and molecular bonding of the synthesized samples were studied using Fourier Transform Infra Red Spectroscopy (FTIR) in a range from 600 to 4000 cm^{-1} using attenuated total reflectance (ATR) technique. The field studies were performed using green gram (*Vigna radiata*) as a plant model.

Inoculation

The *Rhodococcus* spp., was inoculated into the SCN media with different concentration of urea solution and kept for incubation (3–4 days). The 0.4 mM shows the maximum growth and high absorbance at 324 nm. So, it was then mass cultured, filtered using membrane filtration and the sample was lyophilized. The lyophilized sample was characterized using FTIR, SEM. Then the field studies was performed using the lyophilized sample.

The pellet at the concentration of 200 mg shown the maximum growth and high absorbance at 206 nm. Then the culture was lyophilized and the characterization studies were performed using FTIR, SEM and the field studies were carried out using this composite.

0.4 mM Supernatant Solution

The supernatant solution does not shows any colour difference in the tubes this indicates the absence of nanourea synthesis by the *Rhodococcus* spp, extracellurally.

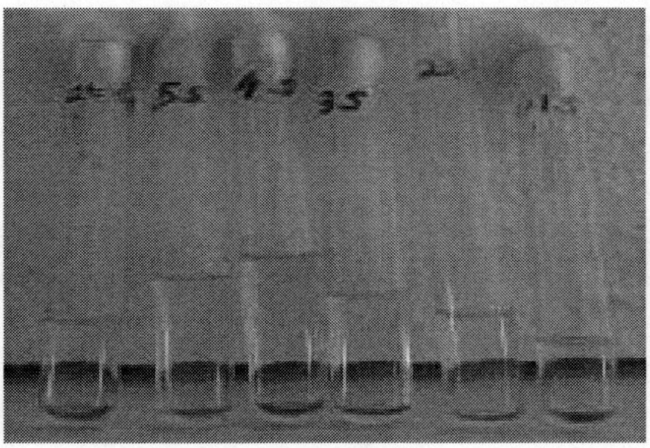

Fig. 1: Supernatant Solution

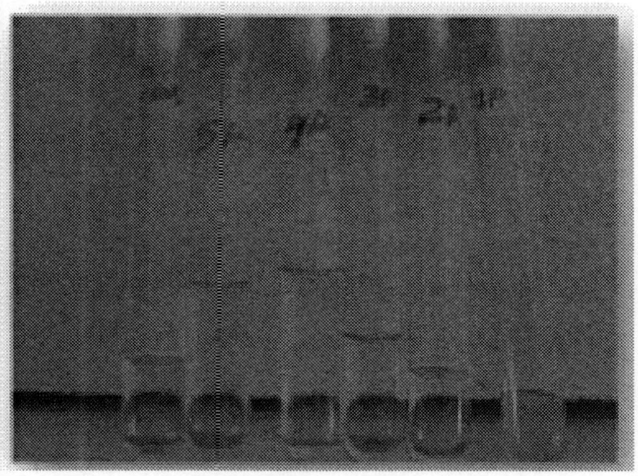

Fig. 2: Pellet Solution

0.4 mM Pellet Solution

The 0.4 mM pellet solution shows turbidity in its test tubes this indicates the reduction of urea size (i.e.) nanourea synthesis by the *Rhodococcus* spp., intracellurally.

Table 1: Peroxidase Assay

Test Sample	*OD at 436 nm*		*Enzyme Activity (U/ml)*
	Initial	*After 10 min*	
Sample1	0.084	0.101	0.00848

Calculation for Peroxidase Assay

$$\text{Enzyme Activity (U/ml)} = \frac{\Delta 436 \times 4 \times \text{Vt} \times \text{Dilution Factor}}{\text{Min.} \times \text{Molecular Extension Coefficient} \times \text{Vs}}$$

Where,

Vt : Volume of total sample

Vs : Volume of test sample

Characterization Studies

UV-Spec Reading for the 0.4 mM Rhodococcus Solution

The 0.4 mM solution where the *Rhodococcus* was inoculated has shown the maximum growth and production of nanourea at a 324 nm.

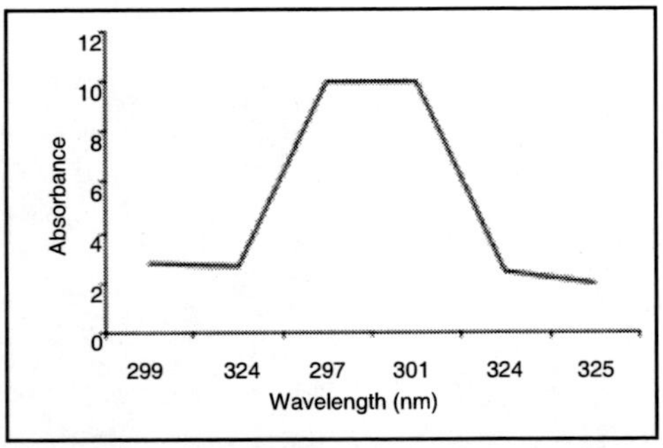

Fig. 3: UV-Spectrophotometer Reading for 0.4 mM

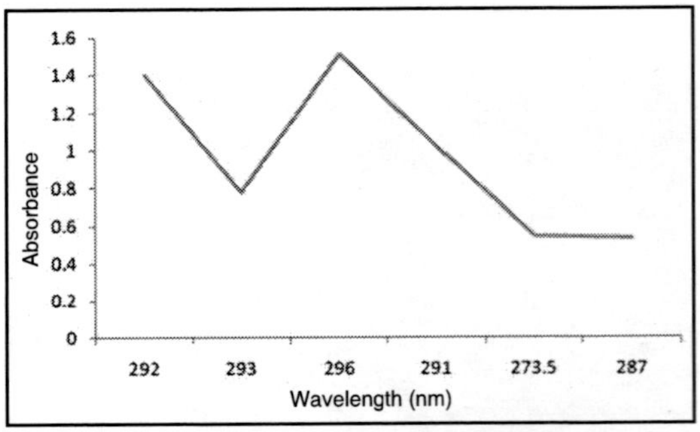

Fig. 4: UV-Spec Reading for Pellet Solution

UV-Spectrophotometer Reading for Pellet Solution

The 0.4 mM pellet solution has shown the maximum growth and production of nanourea at a 295 nm.

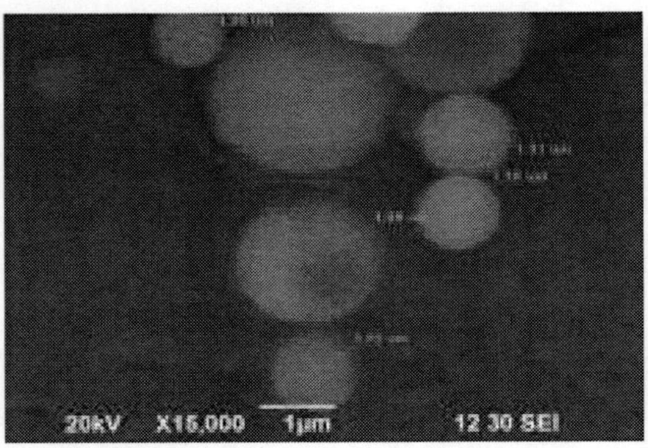

Fig. 5: SEM Image of Nanourea Particle

Fig. 6: SEM Image for Hydroxyapatite Particles

Fig. 7: SEM Image of Biologically Synthesised Nanourea Particle for 200 mg

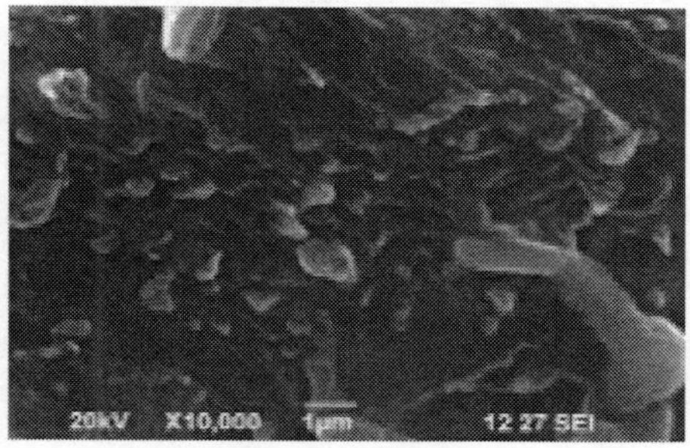

Fig. 8: SEM Image of Biologically Synthesised Nanourea Particle for 0.4 mM

FTIR Studies

The chemical nature and molecular bonding of the synthesized samples were studied using Fourier Transform Infra Red Spectroscopy (FTIR).

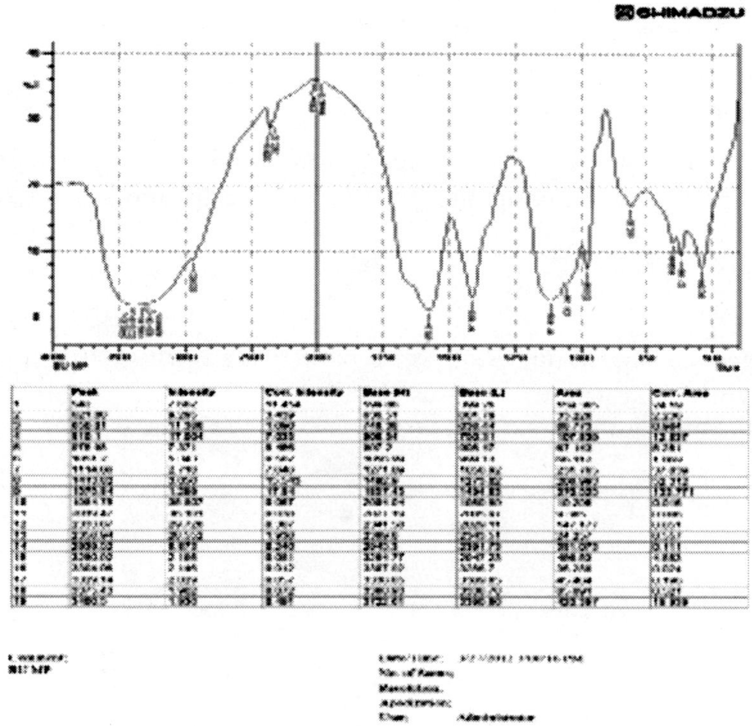

Fig. 9: SEM Image of Biologically Synthesised Nanourea Particle for 200 mg Fourier Transform Infrared Spectra (FTIR) of Lyophilized 0.4 mM Nanourea Sample

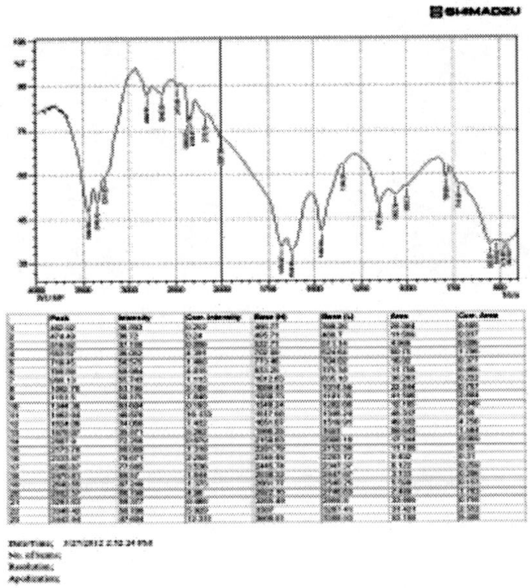

Fig. 10: Fourier Transform Infrared Spectra (FTIR) of Lyophilized 200 mg Nanourea Sample

Different Forms of Composites

The composites was prepared in different forms (i.e.) chemically synthesised and encapsulated into the hydroxyapatite particle,biologically synthesised nanourea are given below:

POT Studies

The pot studies was carried out using *Vigna radiata* as a plant model. Chemically and biologically synthesised nanourea was used and the results obtained are given below:

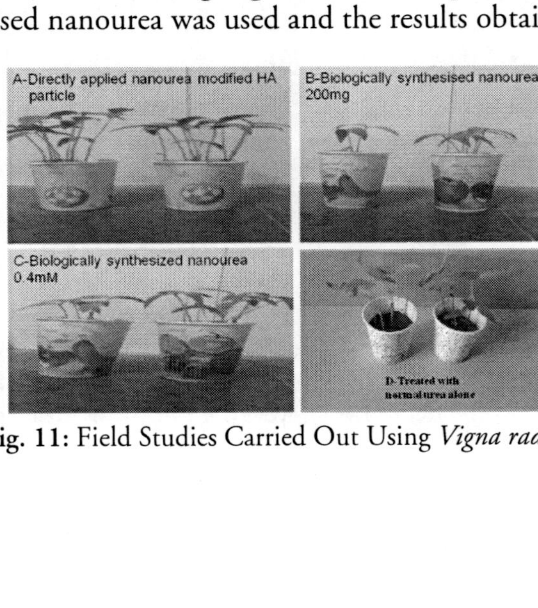

Fig. 11: Field Studies Carried Out Using *Vigna radiata*

Table 2: Measurement of Growth of *Vigna radiata* by Applying Different Forms of Prepared Nanourea

S. No.	*Days*	*Biologically Synthesized Nanourea Using 0.4 mM Filterate (Cm)*	*Chemically Synthesized NA-modified HAp (Cm)*	*Biologically Synthesized Nanourea Using 200 mg of Pellet Solution (Cm)*	*Normal Urea (Cm)*	*Control (Cm)*
1.	From 4th day	3.6±0.212	3.9±0.188	3.06±0.144	2.9±0.124	2.8±0.169
2	5th day	3.80±190	4.1±0.190	3.3±0.169	3.13±0.151	2.9±0.178
3.	6th day	4.1±0.222	4.5±0.216	3.56±0.190	3.33±0.151	3.1±0.178
4.	7th day	4.4±0.216	4.7±0.241	3.83±0.196	3.6±0.169	3.3±0.205
5.	8th day	4.6±0.241	5.1±0.249	4.06±0.222	3.8±0.169	3.4±0.196
6.	9th day	4.9±0.242	5.3±0.268	4.33±0.276	4.03±0.196	3.6±0.196
7.	10th day	5.2±0.222	5.6±0.294	4.56±0.268	4.36±0.222	3.7±0.190
8.	11th day	5.5±0.223	5.8±0.284	4.8±0.294	4.56±0.237	3.9±0.216
9.	12th day	5.8±0.241	6.2 ±262	5.0±0.329	4.73±0.259	4.06±0.237
10.	13th day	6.1±0.241	6.5±0.241	5.26±0.347	5.03±0.259	4.23±0.259
11.	14th day	6.4±0.243	6.8±0.212	5.46±0.349	5.26±0.284	4.36±0.237
12.	15th day	6.8±0.243	7.2±0.188	5.66±0.351	5.46±0.284	4.53±0.259

Table 3: Estimation of Urea by DAM Method

Contents	*Concentration (mg/g)*
Biologically synthesized nanourea (0.4 mM)	150
Biologically synthesized nanourea (200 mg)	100
Chemically synthesized Nanourea modified –HA particles	258
Treated with normal urea alone	090
Control	050

Table 4: Estimation of Nitrogen AOAC, 1990

Contents	*Concentration (mg/g)*
Biologically synthesized nanourea (0.4 mM)	29
Biologically synthesized nanourea (200 mg)	14
Nanourea -modified HA nanoparticles	35
Treated with urea alone	11
Treated with distilled water	8

Different types of nanourea was prepared by chemical and biological methods. *Rhodococcus* spp., was used for biological method. The chemically synthesized nanourea was encapsulated into the Hap (Hydroxyapatite particle) to enhance the property of slow and sustained release of nitrogen to the plants. The prepared nanocomposites were then characterized using UV-Spectrophotometer, SEM and FTIR studies. Thus of treating the seeds with the nanocomposites, the germination percentage of green gram seeds were all over 99%, which were similar to distilled water. The results showed that using this nanocomposites were safe to the germination of green gram seeds. Applying the 3 kinds of treated slow/controlled release fertilizer were used. The effects of new type of fertilizer on emergence of green gram seedlings and growth of seedlings was researched under the conditions of pot experiment. Results showed that the emergence rates of green gram seedlings for all treatments were over 95%, which were similar to chemical fertilizer of equal amount of urea fertilizer and no fertilizer:

- The major advantage of the above slow release fertilizer is to improve the crop efficiency and higher crop yield by providing the nutrients through slow release to the plants.
- Due to its slow release property it results in reduced environmental damage from leaching of nitrogen, compared to conventional water soluble fertilizers.
- This fertilizer composition may maximize the NUE while minimizing the adverse effects to the environments due to use of large quantities of fertilizer in agriculture.

ACKNOWLEDGEMENT

Authors are thankful to the management of KSR College of Technology and Head of the Department Dr. P. Ponmurugan for providing the facilities to carry out the research work.

REFERENCES

Abdel-Megeed, A., Al-Rahma, A.N., Mostafa, A.A. and Husnu Can Baser, K. (2011). Biochemical characterization of anti-microbial activity of glycolipids produced by *Rhodococcus erythropolis. Pak. J. Bot,* Vol. 43, No. 2, pp. 1323–1334.

Abdul, K., Hiroshi, F and Tetsushi, H. (2003). Photosynthetic performance of *Vigna radiata* L. leaves developed at different temperature and irradiance levels", *Science Direct,* Vol. 64, No. 5, pp. 451–458.

Balakrishnan, A., Sanosh, K.P., Cheolchu, M., Kim, S. and Seong-jai che, T.N. (2009). Preparation and characterization of nano-hydroxyapatite powder using sol-gel technique. *Bull. Mater. Sci,* Vol. 32, No. 5, pp. 465–470.

Bell, K.S., Philp, J.C., Awea, D.W.J. and Christofi, N. (1998). The genus *Rhodococcus. Journal of Applied Microbiology,* Vol. 85, No. 4, pp. 195–210.

Cabrera, R.I. (1997). Comparative evaluation of nitrogen release patterns controlled-release fertilizers by nitrogen leaching analysis. *Horti Sci,* Vol. 10, No. 2, pp. 232–234.

Corradini, E., Moura, M.R and Mattoso, L.H.C. (2010). A preliminary study of the incorparation of NPK fertilizer into chitosan nanoparticles. *EXPRESS Polymer Letters,* Vol. 4, No. 8, pp. 509–515.

Miller, J.H and Jones, N. (1995). Organic and compost-based growing media for tree seedling nursery. *Forestry Series,* Vol. 6, No. 4, pp. 77–82.

Monreal, C.M., McGill, W.B. and Nyborg, M. (1986). Spatial heterogeneity of substrates: Effects of hydrolysis, immobilization and nitrification of urea-N. *Journal of Soil Science,* Vol. 66, No. 4, pp. 499–511.

Nosheen Elahi, N., Saima Mustafa, P. and Javed Mirza, I. (2004). Growth and nodulation of mungbean (*Vigna radiata* [L.] Wilczek) as affected by sodium chloride. *Journal of Research (Science),* Vol. 15, No. 2, pp. 139–143.

Paulo da Silva, K., Marcelo Miyata, S., Daisy Nakamura Sato, S., Adolfo Carlos, B., Natália, H. and Clarice Queico, F. (2010). *Rhodococcus equi* isolation from sputum of patients with suspected tuberculosis. *Journal of Microbiology,* Vol. 55, No. 6, pp. 234– 267.

Phanrawee Sriprapha, J., Sukum Eitssayeam, A., Uraiwan, I., Tawee, T., Gobwute, R. and Kamonpan, P. (2011). Preparation of Hydroxyapatite Nanoparticles withVarious Shapes. *Journal of the Microscopy Society of Thailand,* Vol. 4, No. 2, pp. 120–122.

Pradeesh, T.S., Sunny, M.C., Varma, H.K. and Ramesh, P. (2005). Preparation of microstructured hydroxyapatite microspheres using oil in water emulsions. *Bull. Mater. Sci,* Vol. 28, No. 5, pp. 383–390.

Prashant Mohanpuria, H., Nisha Rana, K. and Sudesh Kumar, Y. (2008). Biosynthesis of nanoparticles: Technological concepts and future applications. *Journal of Nanobiotechnology,* Vol. 7, No. 3, pp. 145–176.

Ramesh Chand, P. and Pandey, L.M. (2008). Fertiliser Growth, Imbalances and Subsidies; Trends and Implications. *NPP Discussion Paper,* Vol. 5, No. 2, pp. 22–32.

Sastry, R.K., Rashmi, H.B., Rao, N.H. and Ilyas, S.M. (2010). Integrating nanotechnology into agri-food systems research in India: A conceptual framework. *Technology Forecast Science,* Vol. 77, No. 7, pp. 639–648.

Shelly Sinha, S., Ieshita, P., Pompee, C. and Sukanta, K. (2009). Nanoparticles fabrication using ambient biological resources. *Journal of Applied Biosciences,* Vol. 19, No. 4, pp. 1113–1130.

Shiying, S., Zhirui, G., Zhang, A., Song, Z., Wang, T. and NingGu, H. (2007). Biosynthesis of gold nanoparticles using the bacteria Rhodopseudomonas capsulate. *Science Direct,* Vol. 6, No. 6, pp. 3984–3987.

Teng, S.H., Lee, E.J., Wang, P., Jun, S.H., Han, C.M and Kim, H.E. (2008). Functionally gradient chitosan/hydroxyapatite composite scaffolds for controlled drug release. *J. Biomed,* Vol. 90, No. 5, pp. 275–282.

Antioxidant Studies on Healthy and Rhizome Rot Infected Turmeric Plants Caused by *Pythium* Spp.

N. Sivaram*, R. Ganesh Babu, C. Citra[1], P. Ponmurugan[1] and N. Mathivanan[1]

Department of Biotechnology, K.S. Rangasamy College of Technology, Tiruchengode, Tamil Nadu

[1]Biocontrol and Microbial Metabolites Lab, Center for CAS in Botany, University of Madras, Chennai, Tamil Nadu

E-mail: sivaramksrbt@gmail.com

INTRODUCTION

Turmeric (*Curcuma longa*) belongs to Zingiberaceae family is a rhizomatous herbaceous perennial plant. Turmeric (*Curcuma longa* L. Syn. *C. domestica* Valet.) is one of the important spice crops. It is cultivated in almost all the states in India. The plant is a native of Southern Asia (probably India) and is grown on a large scale in India, China and East Indies. It is called "Indian saffron"because of its orange yellow colour. It is a versatile natural product combining the propertiesof a spice, colourant, cosmetic and drug. As a spice, turmeric is used for its colouring properties and aroma. Important constituents of turmeric rhizomes arecurcumin (2–8%) and volatile oil (5–6%). Turmeric is used for dying wool and cotton fabrics. It is also employed as a colouringmaterial in pharmacy, confectionary and food industry, aswell as in paints and varnishes. Traditionally turmeric is being used in Indian system of medicine as stomachic, carminative, blood purifier, vermicide and antiseptic. Wound healing antiseptic property ofturmeric is well known to Indians since long. Curcumin is the main biologically activephotochemical compound of turmeric with wide range

*Corresponding author.

of therapeutic effects. In recent yearspharmacological properties and actions of curcumin have been widely researched and itsbeneficial effects have been well established. Curcumin is known to have a marked anti inflammatory effect. It accomplishes thisby reducing histamine levels and possibly by increasing production of natural cortisone by theadrenal glands. Curcumin also protects liver from a number of toxic compounds. It has alsobeen shown to reduce platelets from clumping together, which improves circulation and helpsto protect against atherosclerosis. Studies also have shown that curcuminsubdues NF-Kappa B, a powerful protein that promotes abnormal inflammatory response inthe body and excess of it can lead to cancer (Mohammed *et al.,* 1995) cells (Kawamori *et al.,* 1999). Curcumin is also one of the natural inhibitors of enzyme COX-2that helps in making carcinogens more active in the body and cancerous cells (Anonymous, 2001). Turmeric suffers from foliar and rhizome diseases. Rhizome diseases mainly include rhizome rot and storage rot of turmeric. Rhizome rot caused by *Pythium graminicolum, P.aphanidermatum* and *Fusarium* spp., have been recorded from Andhra Pradesh and Tamil Nadu. The infected plants first show progressive drying up of leaves, then spots on aerial shoots and then the roots and rhizomes become discolored and soft and begin to rot. The turmeric samples were collected from various districts of Tamil Nadu (Karur, Erode (yellow city), Salem, Namakkal and Dharmapuri).

An antioxidant is a molecule capable of slowing or preventing the oxidation. They are the substances that may protect cells from the damage caused by unstable molecules known as free radicals. Free radical damage may lead to cancer. Unnikrishnan and Rao (Sharma, 1976; Ruby, 1995; Sugiyama *et al.*, 1996) studied the antioxidative properties of curcumin and its three derivatives (demethoxy curcumin, bisdemethoxy curcumin and diacetyl curcumin). The antioxidant activity of curcumin could be mediated through antioxidant enzymes such as superoxide dismutase, catalase and glutathione peroxidase. Another fact proposed in the literature is that the antioxidant activity increases when the phenolic group with a methoxy group is at the ortho position (Motterlini *et al.*, 2000). The role of antioxidant is to remove free radical. One important mechanism through which this is achieved is by donating hydrogen to free radicals in its reduction to an unreactive species. Addition of hydrogen would remove the odd electron feature which is responsible for radical reactivity.

MATERIALS AND METHODS

Preparation of Crude Extract

The rhizome of turmeric both healthy and infected were shade dried and milled by using a mixer. 10 g of the powdered sample was taken in a filter sheet and inserted in thimble.

250 ml of ethanol was used as solvent and taken in round bottom flask. The temperature was maintained at 50°C for 5–8 hours. Cold water was given to the condenser for the condensation of solvent. The pooled extract was collected in round bottom flask and filtered. It is concentrated under vaccum using a rotary evaporator at 50°C.

Antioxidant Assays

DPPH Radical Scavenging Activity

Crude extract (1 ml) was aliquoted in test tubes to obtain a concentration of 20, 40, 60, 80 and 100 μg/ml. Methanol was added to these test tubes to make a final volume of 2.5 ml. 500 μl of 0.1 mM DPPH was added. The mixture was kept for 20 min. incubation at dark and the absorbance was recorded at 517 nm. (Khalaf *et al.*, 2008; Ravikumar *et al.,* 2008). The experiment was performed in triplicate and then compared with standard antioxidant L-Ascorbic acid. The scavenging activity of the extract against the stable DPPH* was calculated using the equation,

$$\text{Radical scavenging activity (\% inhibition)} = \left[\frac{(\text{Absorbance}_{\text{control}} - \text{Absorbance}_{\text{sample}})}{\text{Absorbance}_{\text{control}}}\right] \times 100$$

Ferric Ion Reducing/Antioxidant Power (FRAP) Assay

Different concentrations of methanolic extracts and tannic acid were mixed in separate tubes with 2.5 ml of phosphate buffer (200 mM, pH 6.6) and 2.5 ml of 1% potassium ferricyanide. The mixture was placed in a water bath for 20 min at 50°C, cooled rapidly and mixed with 2.5 ml of 10% trichloroacetic acid and 0.5 ml of 0.1% Ferric chloride. The amount of iron (II)—ferricyanide complex formed was determined by measuring the formation of Perl's Prussian blue at 700 nm after 10 min. The higher absorbance of the reaction mixture indicates increased reducing power (Oyaizu, 1986).

Assay for Estimation of Total Phenolic Compounds

Aliquoted test samples were added in test tubes to obtain a concentration of 20, 40, 60, 80 and 100 mg/ml. 0.5 ml of 2 N Folin's reagent and 3 ml of sodium carbonate (200 mg/ml) was added subsequently. The reaction mixtures were diluted with 10 ml of deionized water. The white precipitate that formed was removed by centrifuging for 5 min. Absorbance of the supernatant was measured at 725 nm. Catechol was used as a standard.

Estimation of Total Antioxidant Capacity

BHT (1 ml) was aliquoted in test tubes to obtain a concentration of 20, 40, 60, 80 and 100 mg/ml. 1ml of the reagent (0.6 Msulphuric acid, 28 mM sodium phosphate,

4 mM ammonium molybdate) was added. The test tubes were incubated at 95°C for 90 minutes. After cooling to room temperature the absorbance of the aqueous solution was measured at 695nm against a blank. BHT was used as a standard.

Curcumin Analysis

0.2 to 0.5 g of turmeric extract was taken with 250 ml of absolute alcohol. 1 to 2 ml of this mixture was added along with 10 ml of absolute alcohol. The absorbance was read at 425 nm.

$$\text{Curcumin content} = \frac{(0.0025 \times \text{Absorbance at 425 nm} \times \text{Dilution Factor} \times 100)}{[0.42 \times \text{Weight of Sample} \times 1000]}$$

GC-MS Analysis

The sample extract was given to SITRA (South Indian Textile Research Association), Coimbatore and the various compound analysis obtained was shown in the table.

Table 1: Compounds Differentiation between Healthy and Infected Turmeric Plants by Using GC-MS Analysis

S. No.	*Compound Name*	*Healthy Rhizome*			*Infected Rhizome*		
		RSI	*Area (%)*	*Relative Abundance*	*RSI*	*Area (%)*	*Relative Abundance*
1.	1-Eicosanol	–	–	–	933	1.88	28.57
2.	1H-3a,7-Methanoazulene	846	1.04	18.72	–	–	–
3.	1-Hexadecane	–	–	–	943	1.88	28.57
4.	1-Octadecanol	–	–	–	928	4.13	24.82
5.	3-Chloromethylfuran	780	0.53	10.03	–	–	–
6.	Alpha-Tumerone	**927**	**10.83**	**21.45**	**909**	**11.54**	**21.41**
7.	Alpha-Bisabolene	863	0.85	18.30	–	–	–
8.	Alpha-Citronellol	859	0.85	18.30	–	–	–
9.	Alpha-Thujene	837	0.93	19.63	–	–	–
10.	Anthroquinone	901	0.94	32.00	–	–	–
11.	Ar-Tumerone	**887**	**10.83**	**21.07**	**906**	27.79	**21.01**
12.	Atlantone	920	23.70	2.29			
13.	Chromium	–	–	–	739	2.28	31.78
14.	Cinnamic Acid	–	–	–	810	0.84	43.29
15.	Cyanamide	877	0.35	8.04	–	–	–
16.	Decanone	923	0.73	12.26	–	–	–
17.	Heptanol	–	–	–	840	2.96	12.23

S. No.	*Compound Name*	*Healthy Rhizome*			*Infected Rhizome*		
		RSI	*Area (%)*	*Relative Abundance*	*RSI*	*Area (%)*	*Relative Abundance*
18.	Myrcene	821	0.35	8.04			
19.	Nerolidol Isomer	**862**	**0.85**	**18.30**	**933**	**0.98**	**19.60**
20.	Pentanol	–	–	–	927	2.28	10.01
21.	Sabinene	848	0.93	19.63			
22.	Styrene	–	–	–	741	0.94	03.80
22.	Tetradecene	–	–	–	846	5.32	16.41
23.	Vitamin K oxide	–	–	–	833	1.57	34.29
24.	Zingiberene	**908**	2.77	15.77	–	–	–

RESULTS AND DISCUSSION

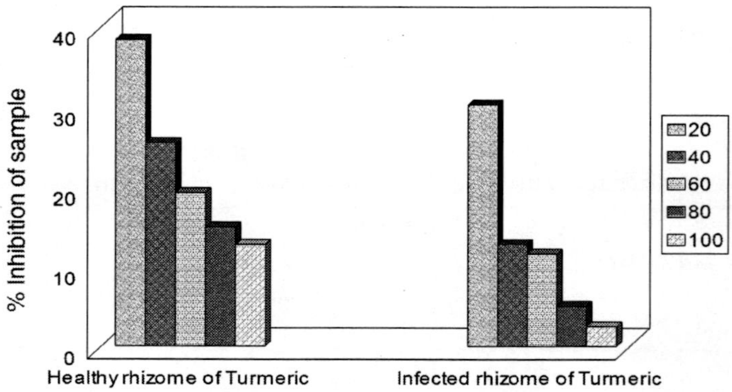

Fig. 1: DPPH Assay for Free Radical Scavenging

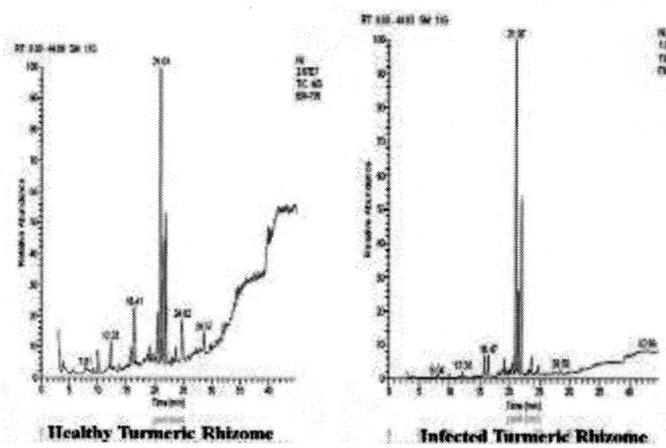

Fig. 2: GC-MS Analysis—Rhizome of Turmeric (Ethanolic Extract)

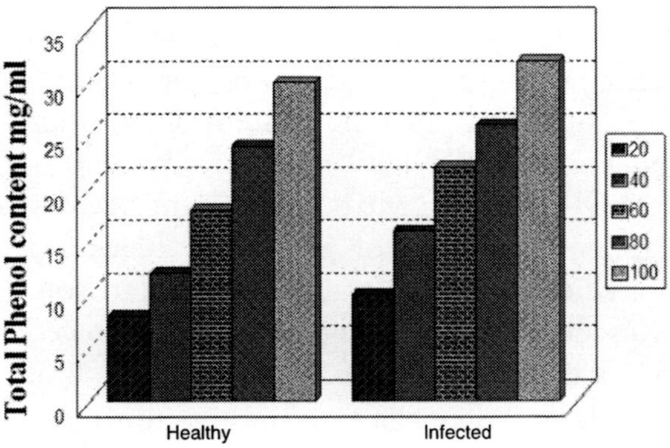

Estimation of Total Phenol Compounds

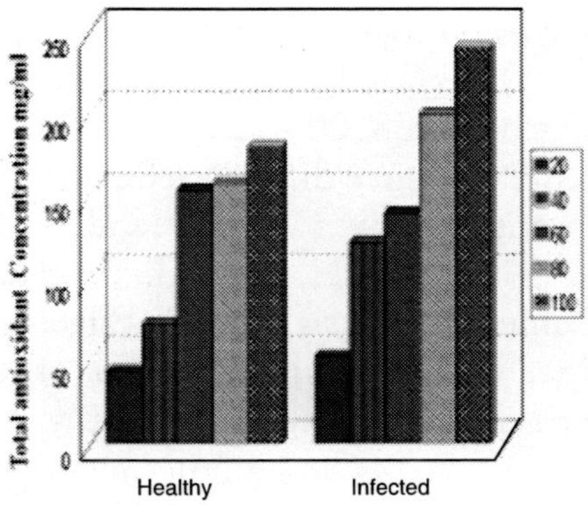

FRAP Assay

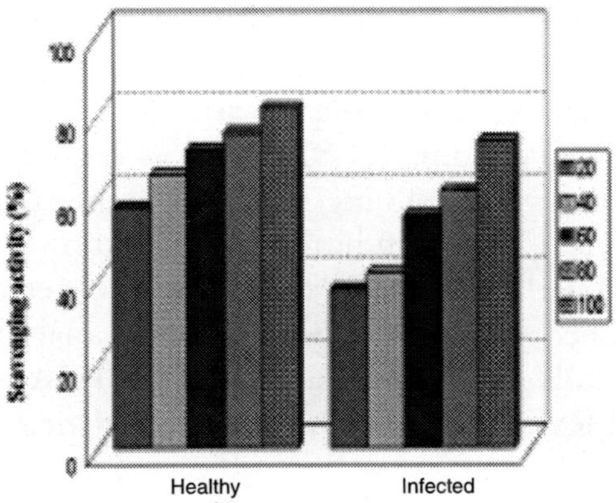

Total Antioxidant Assay

Fig. 3

The data show that all the samples increased their reducing ability when the concentration of extracts was increased. This result is similar to that reported by Gülçin *et al.* (2003) and Noriham *et al.* (2004), who demonstrated antioxidative activity on Pimpinella anisum seeds extracts and four types of Malaysian plants. At 200 ppm, BHA had the highest ability to reduce Fe (III) and had no significant difference with P. minus. The yield of plant methanolic extracts and the concentration of total phenolic content (mg/100 g fresh weight) M. koenigii shows the highest extraction yields (1.65%) and total phenolic content (38.6%) among the samples. The percentage of extraction yields of *C. caudatus, P. minus, C. asiatica* and *O. javanica* were 1.13, 0.80, 0.71, and 0.66 mg TAE, respectively. Based on Herode *et al.* (2003), the percentage of extraction yields will increase with the particle size of sample, temperature extraction and the ratio of solvent and sample extraction. From the analysis, it also showed that O. javanica had the lowest total phenolic content.

Studies on total phenolic content had been published in several papers. Yen and Hsieh (1998) reported that the total phenolic content in Du-Zong (Eucomnia ulmoides) ranged from 8700 to 21000 mg GAE/dry weight. Total phenolic content of Moringa oleifera in three different climates (India, Nicaragua and Niger) ranged from 2940— Among the eight extracts and standard tested for the in vitro antioxidant activity using the DPPH method, the crude methanolic extracts of green tea, black tea (Camellia sinensis Linn.), Eugenia caryophyllus (Spreng.) Bullock and Harrison, Piper cubeba Linn., Zingiber officinale Roscoe and Piper nigrum Linn. Showed antioxidant activity, with IC50 values of 6.7 ± 0.1, 9.7 ± 0.1, 9.9 ± 0.2, 11.3 ± 0.3, 65.1 ± 1.7, 144.1 ± 2.2 μg/ml, respectively (Table). Trigonella foenum graecum Linn. and Elettaria cardamomum (Linn.) It is being observed that the percentage inhibition of healthy is 40% as high and 4% as low. While for infected the highest is 32% and the lowest is 4% total phenolic content for healthy is 30 mg/ml and for infected is 8mg/ml. The antioxidant capacity obtained 250 mg/ml as high and 52 mg/ml as low in healthy, while 32mg/ml as high and 9mg/ml as low in infected sample. Ferric ion reducing/antioxidant power obtained is 82mg/ml as high and 58mg/ml as low in healthy, while 74 mg/ml as high and 38 mg/ml as low in infected sample. The curcumin content is obtained as 7, 10, 17.1 mg. in healthy sample from 1, 1.5, 2 ml concentration of the extract in 10 ml of absolute alcohol. Also it is being observed that GC-MS analysis shows the comparision between the various compound analysis between healthy and infected that the compounds like Alpha-Tumerone, Ar-Tumerone, Nerolidol Isomer, Zingiberene has reduced RSI, % Area and relative abundance as shown in Table 2.

Table 2: Curcumin Analysis

S. No.	*Volume of Sample in ml*	*Healthy (mg)*	*Infected (mg)*
1.	1	7	5.2
2.	1.5	10	8.1
3.	2	17.1	15

REFERENCES

Benzie Iris, F.F. and Strain, J.J. (1999). The Ferric Reducing Ability of Plasma (FRAP) as a Measure of Antioxidant Power the FRAP assay analytical moiety in the antioxidative mechanism of tetrahydrocurcumin. *Biochem. Pharmacol,* 52(4): 519–525.

Gulcin, I., Oktay, M., Kırecci, E. and Kufrevio lu, I. (2003). Screening of antioxidant and antimicrobial activities of anise (*Pimpella anisum* L.) seed extracts. *Food. Chem.* 83: 371–382.

Manjunatha, H. and Srinivasan, K. (2006). Protective effect of dietary curcumin and capsaicin on induced oxidation of low-density lipoprotein, iron-induced hepatotoxicity and carrageen an-induced in• ammation in experimental rats. *FEBS,* 273(19): 4528–4537.

Nooman, A. Khalaf, Ashok K. Shakya, Atif Al-Othman, Zaha el-Agbar, Husni FARAH (2007). Antioxidant activity of some common plants. *Turk J. Biology* 32: 51–55.

Noriham, A., Babji, A.S. and Aminah, A. (2004). Determination of antioxidative activities of selected Malaysian plant extracts. *Asean. Food. J.,* 13: 193–199.

Osawa, T. and Namiki, M. (1981). A novel type of antioxidant isolated from leaf wax of Eucalyptus Leaves. *J. Agric Food. Chem.* 45: 735–739.

Ottolenghi, A. (1959). Interaction of ascorbic acid and mitochondrial lipids. *Arch. Biochem. Biophysics,* 79: 355–358.

Oyaizu, M. (1986). Studies on products of browning reactions: antioxidative activities of products of browning reaction prepared from glucoamine. *Jpn. J. Nutr,* Vol. 4, 307–315.

Prieto, P.M. and Pineda M.M., Aguilar (1999). "Spectrophotometric quantification of antioxidant capacity through the formation of a phosho molybdenum complex; specific application to the determination of vitamin E", *Analytical Biochemistry,* 269: 337–341.

Ravikumar, Y.S., Mahadevan, K.M., Kumarasamy, M.N., Vaidya, V.P., Manjunatha, H., Kumar, V. and Satyanarayana, N.D. (2008). *Environ Toxicol Pharmacol,* 26(2), 142–146.

Sharma, O.P. (1976). Antioxidant activity of curcumin and related compounds. *Biochem. Pharmacol,* 25(15): 1811–1812.

In vitro Seed Germination and Protocorm Development of *Cymbidium aloifolium* (L.) Sw. from Eastern Ghats of Tamil Nadu

Philip Robinson J.* **and Jyothi Prasad Kakati**
Department of Biotechnology, K.S. Rangasamy College of Technology, Tiruchengode, Tamil Nadu
E-mail: philiprobin81@gmail.com

INTRODUCTION

Orchids are outstanding in many ways, like diverse shapes, forms and colours. Orchids are marketed both as plants and as cut flowers and their production has been increased in recent years due to demand in the floriculture field (Tokuhara *et al.,* 1993; Chang *et al.,* 2000). *In vitro* culture is an important tool to save the species which faces the extinction through anthropological activity (Hey *et al.,* 1966). Increasing population pressures on land and other natural resources can aggravate the intensity of natural disasters such as flood, drought and extinction of species. Depletion of the Earth's biological diversity has much more profound consequences than some other environmental threats. The human species has evolved biologically and culturally in a highly diverse world (Kumar, 1999). Therefore the need for conservation is exceptionally high and of paramount importance to preserve heritage for posterity of a plant. *In vitro* propagation is one of the important tool for plant conservation and utilization for various uses. The Current progress in plant genetics and biotechnology is highly adopted upon the use of *in vitro* cultures, hence the establishment of effective *in vitro* plant regeneration systems enabling a rapid production of fertile, genetically 'solid' plants, is of great interest to plant biotechnologists. As the plant is horticulturally important, it is collected

*Corresponding author.

indiscriminately from the nature and as such it has been vulnerable. The present study was undertaken with a view to developing and establish an efficient protocol for mass propagation and consequently commercial exploitation and conservation of this important indigenous orchid Cymbisiym aloifolium (L.) Sw.

Hard and intelligent decisions are needed to save the valuable taxa from the vastly changing landscape. As part of an *ex situ* conservation strategy, artificial propagation including micropropagation is successfully employed for recovering certain other species having specific problems in conventional horticulture (Fay, 1992; Fay 1994). The advantage of plant tissue culture for the conservation of endemic and endangered species lies in the fact that it makes use of small units without sacrificing the mother plant, takes pressure off the wild populations and makes available large number of plants for restoration as well as for horticultural applications. Symbiotic germination methods were only used for a short time before being replaced by non-symbiotic germination procedures. With non-symbiotic germination technology, thousands of seedlings could be raised to maturity from a single seed capsule. Besides the obvious potential in breeding, this technology was also extremely important for vegetative propagation (Griesbach, 2002). Propagation *in vitro* has been successfully employed for the conservation of crop genetic resources, particularly with those crops which are vegetative propagated or which have recalcitrant seeds, which cannot be stored under conventional seed bank conditions (Dodds, 1997; Villalolos *et al.,* 1991; Wilkins, 1882) while tissue culture is essentially used as a crisis management tool for the multiplication of the existing genotypes.

Dendrobium aqueum is an exquisite epiphytic orchid of the Kolli Hills (Eastern ghats) of Tamil Nadu. It is fast disappearing from its natural habitats due to extensive collections by the orchid enthusiast. The plants produce very minute seeds, which lack endosperm. Seeds are wind dispersed. Naturally they require mycorrhizal association for their germination. Hence, a fast method of growing and conserving them in the green houses is an urgent need.

In vitro cultures of orchid seeds have shown that different species require different and often specific medium composition for optimum germination and growth. However, the sequential steps of development and histomorphological changes from embryo to seedling have been traced only in a few species, including those of Vanda (Rao *et al.,* 1964), Cattleya (Shushan, 1959), Cymbidium (Vij, 1986) and Geodorum (Jon *et al.,* 2001).

Purpose of this article is to report our success in aseptic germination of seeds of Dendrobium aquem and to trace the sequence of developmental changes that take place from the onset of seed germination and development of protocorm-like bodies (PLBs).

MATERIALS AND METHODS

The mature undehised capsules were collected from Kolli Hills (Eastern ghats) of Tamil Nadu. For surface sterilization the freshly collected undehised capsules were first rinsed in 90% (v/v) ethanol for 90 sec, followed by soaking them in 1.0% (w/v) mercuric chloride solution for 20 min. and finally washed thrice with sterile distilled water. After opening the capsules with a scalpel, about 150 seeds inoculated in each of the 150 ml conical flask containing 50 ml culture medium. There were six replicate flasks per treatment, three for each capsule. The cultures were maintained at 25°C±2°C under a 12 hr photoperiod provided by Philips white fluorescent lights of 3000 lux intensity.

The germination media used are Knudson C and half MS medium. In KC media 2% sucrose and half MS medium 3% sucrose served as carbon source. In addition, two different plant growth regulators (PGRs) were supplemented to each type of medium in different combinations (Table 1) the pH adjusted to 5.3 in Kn C medium and 5.7 in half MS medium and the media were solidified with 0.9% agar to Knudson C medium and 0.8% agar to half MS medium and autoclaved at 121°C for 15 min. duration.

Table 1: Effect of Different Media on the Germination and Protocorm Development in Vanda sp.

Media Employed for Seed Germination	*% of Germination*	*Time Taken for Initiation of Germination (days)*	*Shoot Growth (%)*
MS-B	–	–	–
MS-CW	–	–	–
MS-P	–	–	–
HMS-B	50	85	10
HMS-CW	30	80	–
HMS-P	20	90	–
KnC-B	**90**	**65**	**50**
KnC-CW	80	70	50
KnC-P	70	75	40
VW-B	–	–	–
VW-CW	–	–	–
VW-P	–	–	–

The onset of germination and the percentage of seed germination on two basal media were recorded periodically at every 10 day interval after the day of initial inoculation. Seed germination was first observed after 40 days, 50 days, and 60 days of inoculation on KnC, VW and HMS media respectively. The highest seed germination of 90% was observed by the 30th day on KnC basal VW media, whereas on HMS and VW media the germination percentages were 40% in HMS and 30% in VW respectively. In KnC basal medium complete seed germination was achieved by the 60th day in both the cases.

For the induction of root, the regenerated multiple shoots of *D. primulinum* were excised and a single shoot was cultured on HMS, KnC medium individually fortified with IBA (0.1 to 0.5 mgl^{-1}). The observations were recorded with regular intervals of one week up to 16 weeks of culture and the obtained root number and their length was recorded.

RESULTS AND DISCUSSION

The first sign of germination on the 30th day on KnC basal medium was that the embryo turned into spherical form and was enclosed in dark yellow color. After a few days some adventive tissues appeared on their tips and the embryo ruptured at one of the poles and after a few days it showed 4–5 fold increase in size with abundant chloroplasts and starch cells. The embryos by the 50th day exhibited a prominent zone of promeristematic cells and developed a pair of leaf primordials. One of the primordials developed more rapidly than the other to produce an unequal pair of first embryogenic photosynthetic leaves. Simultaneously with the development of embryonic leaves at the proximal end of the embryo the marginal cells at the distal end of the embryo started giving rise to tubular and unicellular rhizoids.

The seeds are thin and transparent. The cells of seed coats are varied in size and shape. Such variations that were observed in the present study are in agreement with the observations noted by the previous workers (Clifford and Smith 1969; Ekanthappa, 1981). It has been earlier established that the nature of seed coat is of great taxonomic value within the orchidaceae (Stoutamire, 1963).

Among the different concentrations tested in the present study, BAP (1.0 mg/l) and NAA (1.0 mg/l) were found to be more suitable for shoot induction and multiplication. The previous investigations of several researchers also reported that the combination of BAP and NAA was favorable for the induction of multiple shoots.

Thus the early concept that the orchid seeds are sterile and or atleast incapable of germination is no more valid (Arditti, 1967) That the germination of seeds of epiphytic orchids posed no problem has been pointed out by various workers (Stoutamire, 1964a; Stoutamire, 1964b; Warcap, 1971; Mc Intyre *et al.,* 1972; Mc

Intyre *et al.*, 1974). However they have indicated that difficulty was encountered in the germination of terrestrial orchid seeds in contrast to those of epiphytic taxa.

Table 2: Effect on BAP and NAA on the Development of Protocorm into Microshoots and Regeneration on Kn-C Medium

Hormone Concentration		*Mean Number of Shoots (M±SD)*	*Mean Length (cm) of Shoots (M±SD)*	*Mean Number of Roots (M±SD)*
BAP	*NAA*			
0.5	–	1.3±1.34 lm	0.72±0.21 kl	0
1.0	–	2.0±1.34 i	0.54±0.37 mn	0
1.5	–	1.4±1.01 k	0.57±0.39 m	1.3±0.64 de
2.0	–	1.9±1.3 ij	0.80±0.30 h	1.4±0.48 cd
0.5	0.5	2.7±1.41 gh	0.73±0.34 ij	0
1.0	0.5	2.8±1.46 fg	1.32±0.14 de	0
1.5	0.5	4.8±1.46 b	1.46±0.28 c	1.6±0.19 c
2.0	0.5	3.2±2.00 ef	1.06±0.55 g	1.4±0.80 cd
0.5	1.0	4.5±1.24 bc	1.70±0.15 ab	2.5±1.02 b
1.0	1.0	6.1±1.13 a	1.79±0.19 a	3.1±1.64 a
1.5	1.0	3.8±0.97 d	1.21±0.31 f	1.6±0.66 c
2.0	1.0	3.4±1.42 de	1.40±0.26 cd	0

Table 3: Effect on IAA and IBA on Root Induction in HMS Medium after 25th Day of Inoculation

Medium + Hormone	*Concentration (mg/l)*	*Mean Root Number (M±SD)*	*Mean Root (cm) Length (M±SD)*
IAA	0.5	1.7±0.90 cd	0.3±0.35 hi
	1.0	2.5±0.80 c	0.58±0.45 ef
	1.5	3.8±0.97 a	0.06±0.38 cd
	2.0	1.4±0.91 de	1.74±0.20 a
	2.5	1.2±0.60 ef	0.41±0.33 gh
IBA	0.5	2.0±0.77 c	0.51±0.31 fg
	1.0	3.8±0.97 a	1.3±0.13 b
	1.5	1.4±0.8 de	0.68±0.21 e
	2.0	0.6±0.66 g	0.26±0.16 jk
	2.5	0.3±0.64 gh	0.16±0.16 bc

The well-developed shoots were removed from the culture tubes and transferred to the rooting media. About 90% of the shoots rooted well after transfer to the HMS and KnC medium supplemented with 0.2 mg/l IBA. Rooting also took place in the two basic media, but duration was longer when compared with the hormone-supplemented media. Well developed rooted shoots were taken out from the culture vessels and washed under running tap water then the plantlets were washed with fungicide solution. Individual regenerates were placed onto net pots containing equal proportion of charcoal (0.5–1.0 cm each), brick gravels and coconut husk as supporting materials. The plantlets were sprayed with balanced fertilizers (1 g/l 17N-17P-17k) twice weekly. After six months the plants were transferred to orchid pots. The plantlets thus obtained grew vigorously and flowered after two years of transfer in the pot. The hardened and gowned plantlets were relocated in natural habitats.

In the present work 90% of the shoots of the three species were cultured in to basal and IBA supplemented medium, which was essential for root regeneration. Similar results were obtained in *Cymbidium kanran* (Shimasaki and Uemoto 1990) and *Cymbidium sinense* (Chang and Chang 2000). In *Vanda coerulea* the rooting efficiency varied with different concentrations of auxin such as IAA, IBA and NAA. The highest percentage of rooting was significant in IAA supplemented half strength Vacin and Went medium (Malabadi, 2004). In 1922, Lewis Knudson at Cornell University reported that the fungus was not required for germination if the seed was sown on agar containing appropriate salts and sugars. His Knudson C medium is still being used to germinate the seed of some species.

The present investigation was mainly aimed at understanding the mode of seed germination and organogenesis of the developing seedling. Although different workers have suggested several nutrient media for orchids, three well-known media were tried in the present study to probe into the germination response of seeds selected species. While Knudson medium promoted the germination of *Dendrobium aqueum.*

ACKNOLEDGEMENT

Authors are grateful to Head, Department of Biotechnology, K.S.R. College of Technology, Tiruchengode for his encouragement and Dr. K.S. Rangasamy, Chairman, K.S.R. Group of Institutions for his support.

REFERENCES

Arditti, J. (1967). Factors affecting the germination of orchid Seeds. *Bot. Rev.*, 33: 1–97. Cell Dev. *Biol.,* 28:1–4.

Chang, C. and Chang, W.C. (2000). Effects of thidiazuron on bud development of *Cymbidium sinense* Wild in vitro. Plant Growth Regulator, 30: 171–175.

Chang, C. and Chang, W.C. (2000). Micropropagation of *Cymbidium ensifolium* var. *misericors.*

Clifford, H.T. and Smith, W.K. (1969). Seed morphology and classification of orchidaceae. *Phytomorphology,* 19: 133–139

Dodds, J.H. (1997). Introduction; conservation of plant genetic resources—the need for tissue culture; in Dodds JH (Ed) in vitro methods for conservation of plant genetic resources, Chapman and Hall, London, pp. 1–9.

Ekanthappa, K.G. (1981). Studies in the orchidaceae. Ph.D Thesis. University of Mysore, Mysore, India.

Fay, M.F. (1994). In what situation is in vitro culture appropriate to plant conservation; *Biodiversity and Conservation,* 3: 176–183.

Griesbach, R.J. (2002). Development of Phalaenopsis Orchids for the Mass-market, Trends in new crops and new uses, J. Janick and A. Whipley Ed., ASHS press, Alexandria, VA.

Hey, G.L. and Hey, M.G. (1966). Raising rare orchids from seeds. *In:* DE Garmo LR (ed) *Proc. 5th World Orchid Conf.*, pp. 35–38. Long Beach, USA.

Jon, R.J. and Nirmalya, B. (2001). Cultural requirements for in vitro seed germination, protocorm growth and seedling development of Geodorum denciflorum (Lam.) Schltr. *Indian Journal of Experimental Biology* 39: 1041–1047.

Kumar, H.D. 1999. Biodiversity and Sustainable Conservation. Oxford and IBH Publishing Co. Pvt. Ltd. New Delhi and Culcutta, 29.

Malabadi, R.B., Mulgund, G.S. and Nataraja, K. (2004). Efficient regeneration of *Vanda coerulea,* an endangered orchid using thidiazuron. *Plant Cell, Tissue and Organ Culture,* 76: 289–293.

Mc Intyre, D.K., Veitch, G.J. and Wringlem, J.W. (1972). Australian terrestrial orchid from seed. Amer. *Orchid. Soc. Bull.* 41: 1093–1097.

Mc Intyre, D.K., Veitch, G.J. and Wringley, J.W. (1974). Australian terrestrial orchid from seed-II. Improvement in technique and further success. Amer. *Orchid. Soc. Bull.* 43: 52–53.

Rao, A.N. and Avadhani, P.N. (1964). Some aspects of invitro Culture of Vanda seeds. *Proc. 4th World Orchid Cornference*, Singapore.

Shimasaki, K. and Uemoto, S. (1990). Micropropagation of a terrestrial *Cymbidium* species using rhizomes developed from seeds and pseudobulbs. *Plant Cell Tiss. Organ Cult.* 23: 237–244.

Shushan, S. (1959). Developmental anatomy of an orchid, Catteya Trimos in the Orchids: A scientific survey (Ed. C.L. Withener), Ronald Press, New York.

Stoutamire, W.P. (1963). Terrestrial orchid seedlings. *Australian Plants,* 2: 119–122.

Stoutamire, W.P. (1964a). Terrestrial orchid seedlings-II. *Australian Plants,* 2: 264–26.

Stoutamire, W.P. (1964b). Seeds and seedlings of native orchids. *MICH. BOT.* 3: 107–119. through callus-derived rhizome, *In vitro* Cell. Dev. *Biol. Plant.* 36: 517–520.

Culture from shoot tips excised from flower stalk buds of *Phalaenopsis* (orchidaceae). *In vitro* Cell Dev. *Biol.-Plant,* 37: 457–461.

Vij, S.P. (1986). Studies on growth and development of Cymbidium aloifolium Sw. seedlings in vitro. Biology, Conservation and culture of Orchids, pp. 423–427.

Villalolos, V.M., Ferreira, P. and Mora, A. (1991). The use of biotechnology in the conservation of tropical germplasam, *Biotech. Adv.,* 9: 197–215.

Warcap, J.H. (1971). Symbiotic germination of some Australian terrestrial orchids. *New Physiol,* 72: 387–392.

Influence of PGPR on Pigment Concentration on *Glycine max* (L). Merr.

R. Anisha, A. Mohamed Halith and V. Balakrishnan*

Department of Biotechnology, K.S. Rangasamy College of Technology, Tiruchengode, Tamil Nadu

E-mail: palanivbalu@gmail.com

INTRODUCTION

Modern agriculture is facing new challenges in which ecological and molecular approaches are being integrated to achieve higher crop yields while minimizing negative impacts on the environment. Diverse groups of soil-borne microbes (e.g., root endophytic fungi, mycorrhizal fungi, plant growth promoting fungi and rhizobacteria) exert positive effects on plant growth and survival.

Bacteria especially *Pseudomonads* and *Bacilli* found in the rhizosphere of various leguminous crops have been found to assist in root colonization by *Rhizobia* and in suppressing soil-borne plant pathogens (Senthilkumar *et al.,* 2009). The interactions between these PGPR and rhizobia may be synergistic or antagonistic and beneficial effects of such interaction may be exploited for enhancing the biological nitrogen fixation and crop yield. There is also report of the presence of plant growth promoting *Bacillus* strains in the root nodules of soybean plants (Yu Ming *et al.,* 2002). The effect of nine PGPR isolates on plant growth, yield and nutrient uptake of peanut in pots as well as under field conditions for 3 years have been evaluated by Dey *et al.* (2004). The observed increases in synthesis of essential oils can be considered as a defensive response to colonization by microorganisms, since several essential oils have antimicrobial properties (Erika *et al.,* 2008). The present objective is to evaluate PGPR stains are how will increase the pigment content in soybean plant.

---

*Corresponding author.

MATERIALS AND METHODS

Bacterial Strains

Plant growth promoting rhizobacteria such as *Pseudomonas fluorescens* and *Bacillus subtilis* were used for the present study. The cultures were collected from Tamil Nadu Agricultural University, Coimbatore.

Soybean Plant

The soybean (*Glycine max* L. Merr) seeds were used for the present study. It was obtained from agro based farm near Erode, Tamil Nadu.

Bacterial Growth and Seed Inoculation

The rhizobacterial cultures *Pseudomonas fluorescens* and *Bacillus subtilis* were streaked on nutrient agar medium and preserved as slant cultures for future reference. The soybean seeds were inoculated based on the method followed by Sher *et al.* (2010). The nutrient broth containing the culture was centrifuged at 8000 rpm at 4°C for about 10 minutes under chilled conditions. Supernatant was discarded and collect the pellets from all the centrifuge tubes and dissolve those pellets in 5 ml of sterile distilled water. Serial dilution was prepared upto 10^{-8} dilution. The soybean seeds were surface sterilized by dipping in 95% ethanol solution for 30 seconds, following a dip in 0.2% mercuric chloride solution for 2 minutes. The treated seeds were washed thoroughly with sterilized water to remove disinfectant completely from the seeds. The seeds were inoculated by immersing them for 15 minutes in the broth culture of individual strains containing 10^{-8} cfu ml^{-1}.

Sowing of Seed

The soil was also treated with *Pseudomonas fluorescens* and *Bacillus subtilis* by mixing the nutrient broth containing the cultures with it and the treated and untreated seeds were sown at a depth of about 2 cm. The soil mixed with nutrient broth without any culture to which untreated seeds were sown is denoted as $T_{1.}$ The soil mixed with nutrient broth inoculated with *Pseudomonas fluorescens* to which treated seeds were sown is denoted as $T_{2.}$ The soil mixed with nutrient broth containing *Bacillus subtilis* to which treated seeds were sown is denoted as T_3. The soil mixed with nutrient broth co-inoculated with *Pseudomonas fluorescens* and *Bacillus subtilis* to which treated seeds were sown is denoted as T_4.

Soybean Germination Test

The germination rates of soybean seeds were calculated for T_1, T_2, T_3 and T_4 respectively. From this the high percentage of germination rate can be reported.

Shoot and Root Measurements

The soybean shoot and root length, wet and dry weight of shoot and root, number of leaves, number of nodes, leaf area were measured. For measuring the dry weight, the samples were dried in a oven.

Estimation of Photosynthetic Pigments

Chlorophyll

Chlorophyll contents was estimated by following the method of Moran and Porath (1980) using the formulae suggested by Inskeep and Bloom (1985). Fresh leaf discs of 100 mg were cut and placed in a test tube containing 10 ml of N',N-dimethylformamide (DMF) and stored for 24 hours at 4°C.

Carotenoid (Ikan, 1969)

Absorbance values of pigment extracts at 480, 647 and 666 nm were used to find the corrected absorbance (A) for carotenoids.

Estimation of Absorbing Pigments

Anthocyanin (Mancinelli et al., 1975)

Anthocyanin content was extracted in acidified methanol (Methanol: HCl – 99:1) in 100 mg of leaf material. Extract was kept at 0°C for 24 hours. After 24 hours, the content was made upto 10 ml and the absorbance was read at 530 nm. The content of anthocyanin is expressed in absorbance unit.

Flavanoids (Mirecki and Teramura, 1984)

Hundred milligrams of leaves were placed in 80 percent acidified methanol (Methanol: Water: HCl – 80: 20: 1) for 12 hours in dark at 4°C to extract flavonoids. Absorbance was read at 315 nm and the content of flavanoid is expressed as absorbance units.

Estimation of Soluble Protein (Bradford, 1976)

Extraction: Five hundred milligram of leaves were ground in a mortar and pestle with 10 ml of 20 percent Trichloro acetic acid (TCA). The homogenate was centrifuged for 15 minutes at 800 g. The supernatant was discarded and to the pellet, 5 ml of 0.1 N NaOH (400 mg of NaOH was dissolved in distilled water and made upto 100 ml) was added and centrifuged at 800 g for 15 minutes. The supernatant was made up to 10 ml with 0.1 N NaOH and used for estimation of protein content.

Assay: After complete solubilization, 0.1 ml of the extract was added to 5 ml of protein reagent and the contents were mixed by overtaxing. The absorbance at 595 nm was measured after 2 minutes in 3 ml cuvettes against a reagent blank prepared from 0.1 ml of 0.1 N NaOH and 5 ml of protein reagent. Bovine serum albumin fraction V was used as standard.

Protein Reagent: Dissolve 100 mg of Commassie brilliant blue G 250 in 50 ml of 95 percent ethanol. To this solution, 100 ml of 85 percent phosphoric acid was added. The resulting solution was diluted to a final volume of one litre. Final concentration in the reagent was 0.01 percent (w/v) Commassie brilliant blue G 250, 4.7 percent (w/v) ethanol and 8.5 percent (w/v) phosphoric acid.

RESULTS AND DISCUSSION

The inoculated and the uninoculated soybean seeds were sown in the soil and its growth characteristics in terms of shoot length, root length and number of leaves was observed (Table 1).

Table 1: Growth Characteristics of Soybean Plant

Plant	*Shoot Length (cm)*	*Shoot Weight (g)*	*Root Length (cm)*	*Root Weight (g)*
Control	21	2.04	10	1.02
T_1	23	5.49	15	3.85
T_2	25	10.54	14	5.02
T_3	56	15.38	30	5.25

Control: Soybean plant without any bacterial culture inoculated.
T1: Soybean plant with *Bacillus subtilis* inoculated.
T2: Soybean plant with *Pseudomonas fluorescens* inoculated.
T3: Soybean plant co-inoculated with *Bacillus subtilis* and *Pseudomonas fluorescens.*

This study (Table 1) indicates that dual inoculation with *Pseudomonas fluorescens* and *Bacillus subtilis* can increase the plant weight, leaf number as well as the length of the soybean plants more than the uninoculated one.In this connection, Gholami *et al.* (2009) proved that *Pseudomonas fluorescens* and *Bacillus subtilis* can play an important role in promoting the growth of the plants. The plant pigments such as chlorophyll a, chlorophyll b, total chlorophyll (Moran and Porath, 1980; carotenoid (Ikan, 1969) present in the soybean plant were determined and the results were as tabulated below (Table 2).

Table 2: Pigments in the Soybean Plant

Plant	*Chlorophyll a (mg/g leaf fr.wt)*	*Chlorophyll b (mg/g leaf fr.wt)*	*Total Chlorophyll (mg/g leaf fr.wt)*	*Carotenoid (mg/g leaf fr.wt)*
Control	3.33	1.84	5.01	0.13
T_1	4.49	1.88	5.81	0.16
T_2	6.87	2.15	10.26	0.18
T_3	7.24	3.39	16.07	0.27

The data in Table 2 showed that the chlorophyll content of the soybean plant co-inoculated with *Pseudomonas fluorescens* and *Bacillus subtilis* had been increased when compared with the control. This finding supported results from previous studies reporting that nitrogen-fixation by the rhizobacteria can help in improving the chlorophyll content of the soybean plant (Wu *et al.*, 2006). The protein content of the soybean was determined and the amounts were tabulated below (Table 3).

Table 3: Protein Content of the Soybean

Plant	*Protein Content (mg/g)*
Control	2.05
T_1	3.84
T_2	4.05
T_3	4.26

Plant growth promoting rhizobacteria such as *Pseudomonas fluorescens* and *Bacillus subtilis* are common beneficial microbes of leguminous plants. It is frequently suggested that these rhizobacteria may improve phosphorous nutrition, enhance nitrogen uptake, and improve disease resistance in their host plants or adaptation to various environmental stresses. The principal objective of this work was to promote the growth of the soybean plant using PGPR *Pseudomonas fluorescens* and *Bacillus subtilis* and also to find whether there is any change in the protein content of soybean.

The agriculturally important microorganisms play a remarkable role in nutrient (nitrogen and phosphorous) acquisition for plants. In pursuit of that goal, various workers (Dey *et al.*, 2004; Saleh Al-Garni, 2006) have used PGPR as single inoculants and in combination with each others in various plants. These symbiotic organisms have high ability to increase nitrogen, phosphorous and potassium as well as other nutrients in inoculated plants.

Plant growth promoting rhizobacteria such as *Pseudomonas fluorescens* and *Bacillus subtilis* are common beneficial microbes of leguminous plants. In this study, the effect of PGPR strains *Pseudomonas fluorescens* and *Bacillus subtilis* when inoculated onto the soybean plant had been demonstrated. The soybean plant had been singly and co-inoculated with the above mentioned strains. The co-inoculation with *Pseudomonas fluorescens* and *Bacillus subtilis* resulted in increased shoot and root length, plant dry weight, soil nutritive content and also increased protein content of the soybean.

ACKNOWLEDGEMENT

The authors are thankful to the Management, Principal and Professor and Head of the Department of Biotechnology for providing an opportunity to do this project successfully.

REFERENCES

Bradford, M.M. (1976). A rapid and sensitive method for the quantitation of microgram quantities of protein utilizing the principle of protein dye binding. *Anal. Biochem,* 72: 248–253.

Dey, R., Pal, K.K., Bhatt, D.M. and Chauhan, S.M. (2004). Growth promotion and yield enhancement of peanut (*Arachis hypogaea* L.) by application of plant growth-promoting rhizobacteria. *Microbiological Research,* 159: 371–394.

Erika, Banchio; Pablo, C.; Bogino, Julio Zygadlo and Giordano, Walter (2008). Plant growth promoting rhizobactetria improve growth and essential oil yield in *Origanum majorana L. Biochemical Systematics and Ecology,* 36: 776–771.

Gholami, A., Shahsavani, S. and Nezarat, S. (2009). The effect of plant growth promoting rhizobacteria on germination, seedling growth and yield of Maize. *World Academy of Science, Engineering and Technology,* 49: 156–162.

Ikan, R. (1969). In: Natural products. A laboratory guide. Academic Press, New York.

Inskeep, W.P. and Bloom, P.R. (1985). Extinction co-efficient of chlorophyll 'a' and 'b' in N, N-dimethyl formamide and 80% acetone. *Plant Physiol,* 77: 483–485.

Mancinelli, A.L., Yang, C.P.H., Lindquist, P., Anderson, O.R. and Rabino, I. 1975, Photoregulation. *Plant Soil.* 102: 10–18.

Mirecki, R.M. and Teramura, A.H. (1984). Effects of ultraviolet-B irradiance on soybean. V. The dependence of plant sensitivity on the photosynthetic photon flux density during and after leaf expansion. *Plant Physiol.* 74: 475–480.

Moran, R. and Porath, D. (1980). Chlorophyll determination in intact tissues using N', N-dimethylformamide. *Plant Physiol*, Vol. 478–479.

Saleh, M. Saleh Al-Garni (2006). Increased heavy metal tolerance of cowpea plants by dual inoculation of an arbuscular mycorrhizal fungi and nitrogen-fixer rhizobium bacterium. *African Journal of Biotechnology*, 5: 133–142.

Senthilkumar, M., Madhaiyan, M., Sundaram, S.P. and Kannaiyan, S. (2009). Intercellular colonization and growth promoting effects of *Methylobacterium* sp. with plant growth regulators on rice (*Oryza sativa* L. Cv CO–43). *Microbial Research*, 164: 92–104.

Sher, M., Shahzad., Azeem Khalid., Muhammad Arshad, Jibran Tahir and Tariq, Mahmood (2010). Improving nodulation, growth and yield of *Cicer arietinum* L. through bacterial ACC-deaminase induced changes in root architecture. *European Journal of Soil Biology*. 46: 342–347.

Wu, S.C., Cheung, K.C., Luo, Y.M. and Wong, M.H. (2006). Effects of inoculation of plant growth promoting rhizobacteria on metal uptake by *Brassica juncea*. *Environmental Pollution*, 140: 124–135.

Yuming Bai, Bo Pan, Trevor, C., Charles and Donald, L. Smith (2002). Co-inoculation dose and root zone temperature for plant growth promoting rhizobacteria on soybean [*Glycine max* (L.) Merr] grown in soil-less media. *Soil Biology and Biochemistry*, 34: 1953–1957.

Occurrences of *Clostridium perfringens* Type A in Fresh Water Fishes

Arunava Das* and S. Hemanth

Department of Biotechnology, Bannari Amman Institute of Technology, Sathyamangalam, Tamil Nadu

E-mail: drarunavadas@rediffmail.com, arunava20@gmail.com

INTRODUCTION

The pisci-culture sector in India has been witnessed a phenomenal growth in terms of production from 1947 and has recorded double digit growth in fish production between 1999 and 2010 (*The UNFAO,* 2011). As food items, fresh water fish contribute to country's economy and people's well being by meeting the dietary protein needs of the common man. However, due to high perishable characteristic, fish and fish based products often get spoiled and contaminated by the food borne pathogens particularly bacteria if not properly stored. When such contaminated foods are consumed either raw or inadequately cooked by human population, there is a detrimental impact on public health.

Among the food borne pathogens that thrives in high protein and canned food items, *Clostridium perfringens* always comes foremost. *C. perfringens* is ubiquitous, anaerobic, gram positive rod shaped bacteria which causes numerous enteric diseases caused by food infections to human, wildlife and domestic animals (Songer, 1996). *C. perfringens* thrives in high-protein foods of animal origin such as meat and meat products, meat dishes, stews, soups, gravies, and milk. On the basis of production of four major toxins (alpha, beta, iota and epsilon), isolates of *C. perfringens* are classified into five genotypes A to E (Songer and Meer, 1996). Along with four major lethal toxins, enterotoxin and beta2 toxin produced by types of C. perfringens also considered as

*Corresponding author.

important toxins for enteric diseases in domestic animals. However, it is not so easy to differentiate between different strains of *C. perfringens* from clinical cases. Therefore, PCR has been used to detect the presence of toxin genes and to identify the specific strains of *C. perfringens* (Yamagishi *et al.*, 1997).

In last five to six years, the detection of *C. perfringens* type A from animal and human have been reported from India. But none of the clostridial cases from India have been reported from fresh water fishes. In literature review, it has been noticed that the research on the involvement of *C. perfringens* in fresh water fish is limited (Wen and McClane, 2004 and Cai *et al.,* 2008) and therefore, it becomes necessary to determine the types of *C. perfringens* prevalent in fish species in India. The present study is therefore designed to identify the prevalence of *C. perfringens* types in freshwater fishes for human consumption.

MATERIALS AND METHODS

Collection of Fish Samples and Isolation of *C. Perfringens*

Two hundred and sixty seven fresh water fishes obtained from common carp (90), cat fish (91) and silver carp (86) were collected from the randomly selected dams, freshwater lakes, ponds, retail shops and local markets across Tamil Nadu, India. The intestine samples from the same number of fishes were collected scientifically processed for the microbial investigation. Intestinal contents were inoculated aseptically in Robertson's cooked meat (RCM) medium (HiMedia, Mumbai) with neutral oil overlay and incubated at 37°C for 48 hr. The inoculums from the RCM media was seeded onto 10% goat blood agar and incubated anaerobically for 24 hr at 37°C.

Biochemical and Phenotypical Identification of *C. Perfringens*

Bacterial colonies from blood agar were subjected to various morphological, cultural and biochemical characters as per the Bergey's manual of determinative bacteriology (Holt *et al.*, 1994) with little modification. These tests included, motility of the bacteria, Gram's and endospore staining, gelatin liquefaction, litmus milk fermentation, nitrate reduction, urease production, cytochrome oxidase, catalase, indole production, hydrogen sulphide (H_2S) production, Simmon's citrate utilization, triple sugar iron agar for the fermentation of sugar like glucose, lactose and sucrose. All the isolated bacteria were evaluated for the phenotypical tests like deoxyribonuclease (DNase) enzyme, lecithinase enzyme and hemolysin production as per the standered protocol. Scanning electron micrography of the isolated bacteria was performed according to standered protocol. The photography was done under the SEM (JEOL JSM-6360, Tokiyo, Japan) and was analyzed with the help of the specific software.

Detection of Virulence Genes by Polymerase Chain Reaction

A single colony of *C. perfringens* was pilled from blood agar plate taken in 100 µl of Milli-Q water, gently vortexed and boiled at 100°C for 10 min in water bath. The cell debris was removed by centrifugation at 10,000 rpm for 5 min at 4°C and the top clear supernatant was used as source of template DNA. The PCR amplification was carried out in iCycler (BioRad, USA) in 25 µl reaction volume 12.5 µl of 2 × PCR master mix [4 mM $MgCl_2$; 0.4 mM of each dNTPs (dATP, dCTP, dGTP, dTTP); 0.5 units/µl of *Taq* DNA polymerase; 150 mM Tris-HCl PCR buffer (pH 8.5)]; desired µM of each forward and reverse primers and 2.5 µl of template DNA. The details of primer pairs for *cpa, cpb, etx, iA, cpe* and *cpb2* genes (Das *et al.*, 2008) were commercially synthesized (Eurofins, Bangalore). After initial denaturation at 94°C for 4 min, the amplification cycle had denaturation, annealing and extension at 94°C, 55°C and 72°C for 1 min each respectively. Final extension was done at 72°C for 10 min. *C. perfringens* (ATCC 13124) and *Escherichia coli* (MTCC 723) were used as positive and negative controls respectively. The PCR amplicons (5µl) were separated by electrophoresis in 1.5% Agarose (Promega, USA) gel with TAE 1× (Tris-Acetate-EDTA; pH 8.0) running buffer at 60 V. The gel was stained with 0.4µg/ml ethidium bromide (Pharmacia Biotech), visualized and photographed in gel documentation system (Universal Hood, BIORAD, Italy).

RESULTS AND DISCUSSION

Isolation and Identification of *C. perfringens*

C. perfringens is an important food poisoning organism and is ubiquitous in nature, however, there is limited understanding for such obligate and facultative anaerobes in fresh water fishes. On 10% sheep blood agar, after 24 hr of anaerobic incubation at 37°C, the young colonies of *C. perfringens* were observed as hemolytic (both alpha and beta hemolytic), round, 1–3 mm in diameter. The bacterial colonies were observed to be typical umbonate, centrally dense and round with rhizoid periphery and were picked up and purified by repeated streaking and restreaking on fresh blood agar media plates until the pure cultures were obtained. All the isolates of *C. perfringens* were found to be non-motile, gram-positive, thick rods shaped on Gram's staining and sub-terminal oval shaped spores on malachite green spore staining.

All the isolates were found to be liquefied gelatin; stormy fermentation, acidity, reduction and coagulation of skim milk; negative in the production of oxidase, catalase, indole, hydrogen sulphide (H_2S) and Simmon's citrate. However, both the nitrate reduction and urease production were found to be variable in reactions. All the isolates characteristically produced acid (characterized by yellow/yellow color in butt/slant) and gas after fermentation of glucose, lactose and sucrose in triple sugar iron agar slants. After detailed microbiological investigation, 49 (18.35%) isolates (17 from common carp, 12 from silver carp and 20 from cat fish) were tentatively identified as *C. perfringens*. The ultrastructure study of *C. perfringens* in SEM was observed to be in clusters of thick rods (Figure 1). The rods were observed to be variable in length; sometimes occur either single or in pairs and occasionally in short chains. Similar isolation and identification methods of *C. perfringens* were also reported by other researchers (Das *et al.*, 2008).

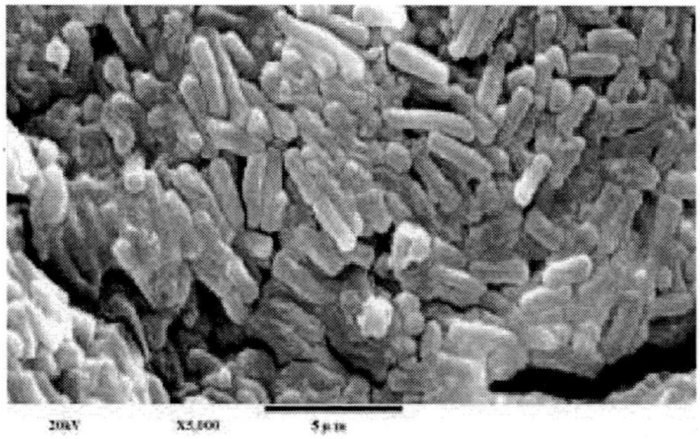

Fig. 1: Ultra-Structure of *C. Perfringens* Type A under SEM

Detection of Virulence Genes by PCR

In PCR, all the 49 (100%) suspected isolates of *C. perfringens* irrespective of source and place of isolation amplified the primers specific for alpha toxin gene (*cpa*) of products 324 bp (Figure 2). However, 29 (59.18%) isolates were also positive for additional beta2 toxin gene (*cpb*2) of 567 bp (Figure 2). The other virulence gene specific primers of *cpb*, *etx*, *i*A and *cpe* genes did not amplify any of the isolates. The virulence gene detection by PCR assay used in this study revealed that all the isolates are belonged to *C. perfringens* type A.

In a similar study, Cai *et al.* (2008) reported that out of 75 isolates of *C. perfringens* from fresh water fish, 13 (17.3%) isolates were found positive for toxin type A, 58 (77.3%) isolates for type C and 4 (5.3%) isolates were type B, and 47 (62.7%) isolates with additional beta 2 toxin gene (seven from type A, two from type B, and 38 from type C).

In this study, none of the isolates of *C. perfringens* were found positive for *cpe* gene or any other major virulence genes, since, the presence of *cpe* gene in *C. perfringens* is very

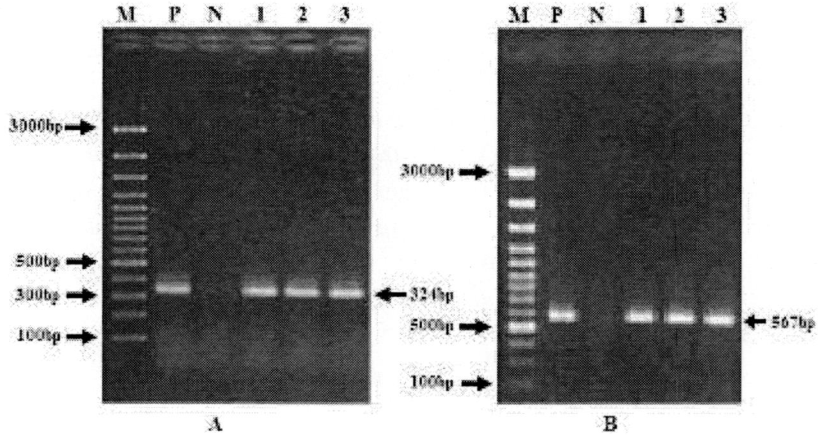

A: Detection of *cpa* gene of 324 bp fragment
B: Detection of *cpb2* gene of 567 bp fragment
Lane M: High range DNA ruler (Genei, India)
Lane P: Positive control
Lane N: Negative control
Lane 1–3: Field isolates positive for virulence genes

Fig. 2: Detection of Virulence Genes from *C. perfringens* Type A from Fish by PCR

uncommon, and <5% of global *C. perfringens* type A isolates are *cpe* positive (Czeczulin *et al.,* 1993). In congruence to present result, the *cpe*-negative *C. perfringens* type A from the clinical cases of gastroenteritis was reported previously (Das *et al.,* 2008b; Das *et al.,* 2012). Wen and Mcclane (2004) reported that out of 34 isolates of *C. perfringens* from fish samples 31 were positive for type A (only alpha toxin producer) and 3 isolates were type A with additional enterotoxin gene, but none of the type A isolates were positive for *cpb2* gene. In contradiction, present study reported the occurences of *cpb2* positive isolates in 59.18% *C. perfringens* type A also suggested the frequent distribution of this gene along with the predominantly distributed *cpa* gene in fresh water fishes in Tamil Nadu India.

The present study suggested that PCR is a reliable DNA based technique and useful tool for the detection of virulence genes typing of *C. perfringens* isolates from the fresh water fish samples. The presence of alpha toxin gene in all the isolates of *C. perfringens* type A irrespective of source of isolation and geographical location also suggested that the primer specific to alpha toxin gene in PCR assay could be useful for the rapid identification of *C. perfringens.* The bacterium is spore forming and both alpha and beta2 toxin can cause these diseases in animal and human by entering in to the food chain. Thus, the consumption of uncooked fish for human has a crucial impact on public health.

REFERENCES

Cai, Y., Gao, J., Wang, X., Chai, T., Zhang, X., Duan, H., Jiang, S., Zucker, B.A. and Schlenker, G. (2008). *Clostridium perfringens* toxin types from freshwater fishes in one water reservoir of Shandong province of China, determined by PCR. *Deutsche Tierarztliche Wochenschrift*, 115: 292–294.

Czeczulin, J.R., Collie, R.E. and Mcclane, B.A. (1996). Regulated expression of *Clostridium perfringens* enterotoxin in naturally *cpe*-negative Type A, B and C isolates of *C. perfringens. Infection and Immunity*, 64: 3301–3309.

Das, A., Mazumder, Y., Dutta, B.K., Sharma, G.D., Shome, B.R. and Bujarbaruah, K.M. (2008a). *Clostridium perfringens* type A from broiler chicken with necrotic enteritis. *International Journal of Poultry Science*, 7: 601–609.

Das, A., Mazumder, Y., Dutta, B.K., Shome, B.R., Bujarbaruah, K.M. and Sharma, G.D. (2008b). *Clostridium perfringens* type A beta2 toxin in elephant (*Elephas maximus indicus*) and pygmy hog (*Sus salvanius*) with haemorrhagic enteritis in Assam, India. *African Journal of Microbiology Research*, 2:196–201.

Das, A., Mazumder, Y., Dutta, B.K., Shome, B.R., Bujarbaruah, K.M. and Kumar, R. (2012). Molecular typing of *Clostridium perfringens* isolated from diarrhoeic cattle. *Journal of Animal Science Advances*, 2: 226–229.

Faoun (2011). Fishery and Aquaculture Country Profiles: India. Food and Agriculture Organization of the United Nations. http://www.fao.org/fishery/countrysector/FI-CP_IN/3/en

Holt, J.G., Krige, N.R., Sneeth, P.H.A., Staley, J.T. and Williams, S.T. (1994). Bergey's Manual of Determinative Bacteriology. Ninth edition. Williams and Wilkins Baltimore, USA, 749–755.

Songer, J.G. (1996). Clostridial enteric diseases of domestic animals. *Clinical Microbiology Review*, 9:216–234.

Songer, J.G. and Meer, R.R. (1996). Genotyping of *Clostridium perfringens* by polymerase chain reaction is a useful adjunct to diagnosis of clostridial enteric disease in animals. *Anaerobe*, 2: 197–203.

Wen, Q. and McClane, B.A. (2004). Detection of enterotoxigenic *Clostridium perfringens* Type A isolates in American retail foods. *Applied and Environmental Microbiology*, 70: 2685–2691.

Yamagishi, T., Sugitani, K., Tanishima, K. Nakamura, S. (1997). Polymerase chain reaction test for differentiation of five toxin types of *Clostridium perfringens. Microbiology and Immunology*, 41: 295–299.

Probiotic Effect of Marine Bacteria on Shrimp Growth

M. Masilamani Selvam*, S. Kiran Nivedh and A.N. Nishanth
Department of Biotechnology, Sathyabama University, Rajiv Gandhi Salai, Chennai
E-mail: masilamaniselvam@gmail.com

INTRODUCTION

Probiotics are live microbial feed supplements that improve the health of the host animals (Parker, 1974; Fuller, 1989; Tannock, 1997). The research of probiotics for aquatic animals is increasing with a demand for environment-friendly aquaculture. The gastro-intestinal microbiota of fish and shellfish are dependent on the external environment, due to the water flow passing through the digestive tract. Most bacterial cells are transient in the gut, with continuous intrusion of microbes coming from water and food. Some commercial products are referred to as probiotics, though they were designed to treat the rearing medium, not to supplement the diet. Most attempts to propose probiotics have been undertaken by isolating and selecting strains from aquatic environment. These microbes were *Vibrios*, *Pseudomonads*, lactic acid bacteria, *Bacillus* spp. and yeasts. Three main characteristics have been searched in microbes as candidates to improve the health of their host: (1) antagonistic potential to pathogens; (2) colonization potential; and (3) induction of disease resistance to hosts. Many other beneficial effects may be expected from probiotics and they are competition with pathogens for nutrients or for adhesion sites, and stimulation of the immune system (Gate soupe, 1999).

Traditional shrimp farming has been practiced in India from time immemorial. The production through farming increased substantially to one lakh ton in the year 1995–1996. After this period an outbreak of the white spot disease has done

*Corresponding author.

considerable damage to the development of shrimp industry. The disease outbreak and production losses have been linked to deteriorating water quality conditions. Hence, to control the water quality problems, application of probiotic is necessary. Probiotics in the way of applying through water is beneficially acted upon improving the water quality increasing growth of the shrimps and making the animals healthy and active. Very limited works are available on the use of probiotics on the shrimp *Penaues indicus*. Hence, the present investigation was made.

MATERIALS AND METHODS

Isolation of Marine Bacteria

Soil samples adhering to the root system were collected carefully from *Rhizophora mucronata* from a 5 year old mangrove plantation site, raised along the Vellar estuary. The soil samples were brought to the laboratory immediately. A portion of the samples was analyzed immediately for microbial counts. For isolation of phosphate solubilizing bacteria, the Pikovskaya's medium (Modified by Sundara Rao and Sinha, 1963) was prepared by using 50% seawater.

Actinomycetes

The sediment samples collected were air-dried for one week and kept at 45°C for 1 h to minimize the bacterial contaminants. About 1g of sample was transferred to an Erlenmeyer's flask containing 99 ml of sterile 50% seawater. The sediment suspension was further diluted up to 10^{-5} level. One ml of the diluted suspension was spread over the surface of starch-casein agar medium (Nadaraj, 1996) prepared in 50% seawater to enhance the isolation of marine actinomycetes. The *p*H of the selected media used was adjusted to 7.2. To prevent the fungal and bacterial contaminants, cycloheximide (100 mg/l) and nalidixic acid (20 mg/l) were added to the medium. The Petri plates were then incubated at 28 ± 2°C, and the colonies were observed from fifth day onwards for one month. Strains of marine actinomycetes were picked out and purified by repeated streaking on ISP 2 medium. The pure cultures of the actinomycetes were transferred to yeast extract-malt extract agar (ISP 2) slants and preserved at 4 ± 2°C.

Probiotic Effect of Marine Bacteria on Shrimps

Collection and Acclimatization of Test Animal

The juvenile tiger shrimp *Penaeus indicus* weighing of 0.70±0.05 g animal^{-1} were collected from a Shrimp Hatchery. The animals were visually sorted out for their healthy and active conditions and were stocked in plastic trough (40 l) for three days for acclimatization to laboratory conditions.

Ten animals were used for each treatment in filtered seawater with aeration. For each treatment, duplicates were maintained prior to stocking. Juveniles were measured for total length and weight. The experiment was conducted for 30 days in 40 liter plastic tank containing 35 litre of filtered estuarine water collected from the estuary. Aeration was provided using air-stones. Salinity was maintained at 30 ppt. The culture of PB-14 and Rhi-1 with the density of 10^8 cells per ml were inoculated into the tank with the 20 litre of seawater and aerated for 5 h. Then the culture inoculum with the cell density of 10^6 per ml was splashed over the experimental tanks. The inoculation was made once in five days for upto one month. Water exchange was made for 20% of the culture water every morning. Feeding was given at a rate of 10% of the body weight, twice daily.

Prawn Growth Analysis

Total length was measured both initial and final with the help of a steel graduated scale. Fresh weight was taken by weighing the animal in live condition in an electronic balance initially, after one month final weight was also recorded.

RESULTS AND DISCUSSION

In this study, the mangrove soil has yielded many bacterial strains. The strains include 24 phosphate solubilizing bacteria and 40 actinomycetes. Of these, the most beneficial strains were selected for their possible utility as probiotics based on the following tests: (a) Evaluation based on shoot growth in *Rhizophora mucronata,* (b) Evaluation based on phosphate solubilization, (c) Evaluation based on ammonia production, (d) Antmicrobial activity.

Analysis of Shrimp Growth

The results obtained from the experiment are presented in Figures 1 and 2. The growth in terms of length and weight was more in the experimental animals than in the control. The experimental observation revealed that the animals maintained in the bacteria inoculated tanks were more active and healthier than those in the control. The growth in length as well as in weight increased substantially. The growth in terms of length and weight was observed for the animals inoculated with *Bacillus megaterium* and *Streptomyces fradiae* in the form of water and feed probiotics, whereas the control tank showed low growth of the animal.

Linear Growth of *Penaeus indicus*

The linear growth was higher in shrimp culture inoculated with bacillus (7.02 cm) than control tank (5.28 cm). The animals inoculated with streptomyces showed growth rate of 6.75 cm, which is comparable with the shrimps grown in control tanks (Figure 1).

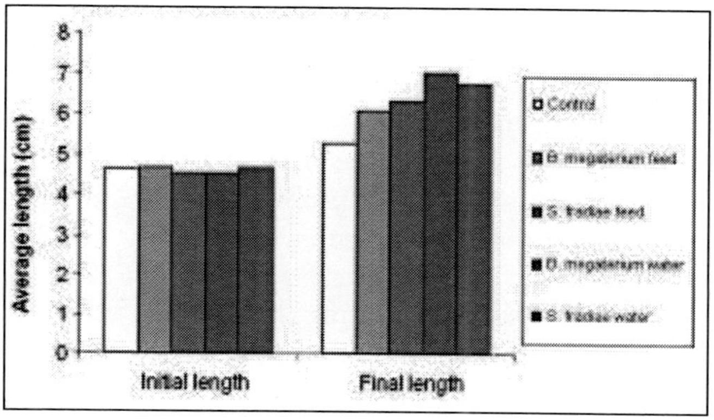

Fig. 1: Linear Growth of Penaous Indicus Fed for 30 Days

Wet Weight Growth of *P. indicus*

The mean growth (in wet weight) obtained for the animals inoculated with bacillus mixed with water, streptomyces (mixed with water and feed), bacillus mixed with feed and control were found to be 2.35, 2.30, 1.70, 1.53 and 1.17 g respectively (Figure 2).

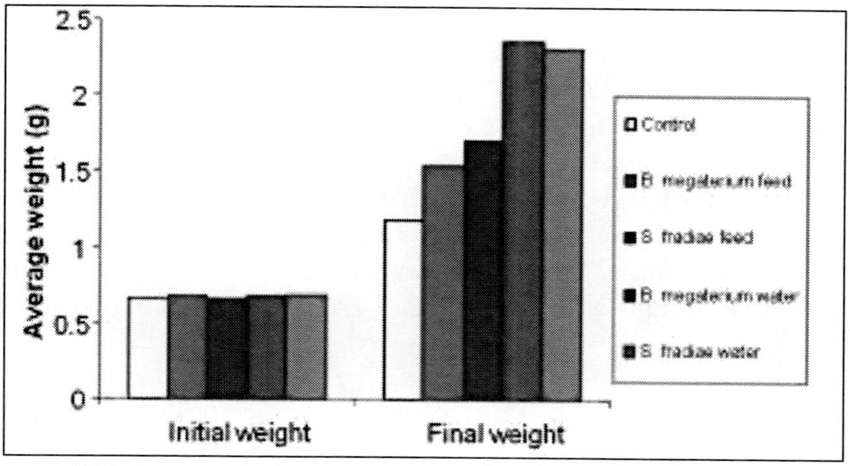

Fig. 2: Wet Weight Growth of *Penaeus indicus* Fed for 30 Days

In general, water probiotic treated shrimps showed better growth response than the feed probiotic application. Among these two species bacillus is proved to be a good probiotic than streptomyces.

In aquaculture practices, the water quality deteriorates mainly due to accumulation of metabolic wastes. When the water quality is impaired, the cultured stocks become vulnerable to diseases due to stress. In such a situation, it has been suggested that some commercial preparations can be added like probiotics, which are a group of beneficial microorganisms (Nitrosomonas, Cellulomonas, Bacillus subtilis and Nitrobacter). These microorganisms are helpful in maintaining the water quality parameters (Fuller and Turvey, 1971; Bruce *et al.,* 1979). By maintaining the water quality parameters at the optimum level, probiotics are reported to be helpful in enhancing the growth rate among the cultured organisms. The first application of probiotics in aquaculture seems relatively recent (Kozasa, 1986) but the interest in such environment-friendly treatment is increasing rapidly. In this context, beneficial microbes like Phosphate solubilizing bacteria and N_2 fixing azotobacters were studied in the present study for their effect on shrimp and its environment.

The first successful report was of Maeda and Liao (1992). They found a microbial strain, isolated from the rearing water of larvae *Penaeus monodon,* was effective in producing shrimps with good survival and moulting rate. The effect was attributed to the efficiency of microbes in (1) antagonism to pathogens (2) gut colonization, with possible adhesion to intestinal mucus (*Evelyn, 1996;* and to (3) increased resistance of the host to pathogens.

Moriarty (1998) noted an increase of prawn survival in pond water that was inoculated with *Bacillus* spp. This was attributed to the water quality improved by the bacterial inoculation, as a result of degradation of organic matter. This treatment was also found to decrease the pathogenic luminous *Vibrio* spp. and other bacteria (Maeda *et al.,* 1997). The strains of *Bacillus* further were known to compete for nutrients and for space for pathogenic forms. These mechanisms of *Bacillus* spp., may even prevent the emergence of resistant strains, a well-known risk of antibiotic treatments (Moriarty, 1998). Antagonistic activity of bacteria like *Bacillus* sp may be mediated not only by antibiotics, but also by many other inhibitory substances, for example: organic acids, hydrogen peroxide (Ringø and Gatesoupe, 1998) and Siderophores (Gram and Melchiorsen, 1996).

REFERENCES

Bergey's Manual of Determinative Bacteriology (1974). (R.E. Buchanan and N.E. Gibbons, eds) 8th ed., The Williams and Wilkins company, Baltimore, 253–255.

Bruce, B.B., Gilliland, S.E., Bush, L.J. and Staley, T.E. (1979). Influence of feeding cells of *Lactobacillus acidophilus* on the faecal flora of young dairy calves. Oklahoma. *Anim. Sci. Res.*, p. 207.

Cappuccino, James G. and Sherman, Natalie (2001). Microbiology: A laboratory manual, 6th ed., Benjamin Cummings, San Francisco, CA. 133–198.

Evelyn, T.P.T. (1996). Infection and disease. In: G. Iwama and T. Nakanishi (Eds.,). The Fish Immune System. Organisms, Pathogen and Environment. *Fish Physiol Series*, p. 15.

Fuller, R. (1989). Probiotics in man and animals. *J. Appl. Bacteriol.*, 66: 365–378.

Fuller, R. and Turvey, A. (1971). Bacteria associated with the intestinal wall of the fowl (*Gallus domesticus*) *J. Appl. Bacteriol.*, 34: 617.

Gatesoupe, F.J. (1999). The use of probiotics in aquaculture: A review. *Aquaculture*, 180: 147–165.

Gram, L. and Melchoirsen, J. (1996). Interaction between fish spoilage bacteria *Psedomonas* sp. and *Shewanella putrefaciens* I fish extracts and on fish t issue. *J. Appl. Bacteriol*, 80, 589–595.

Kozasa, M. (1986). Toyocerin *(Bacillus toyoi*) as growth promotor for animal feeding, *Microbiol. Aliment, Nutrition*, 4: 121–135.

Maeda, M. and Liao, I.C. (1992). Effect of bacterial population on the growth of a prawn larva, *Penaeus monodon. Aquaculture* 21: 25–29.

Maeda, M., Nagami, K., Kanematsu, M. and Hirayama, K. (1997). The concept of biological methods in aquaculture. *Hydrobiologia* 358: 285–290.

Moriarty, D.J.W. (1998). Control of luminous *Vibrio* sp. In penaeid aquaculture ponds. *Aquaculture* 164: 351–358.

Nadaraj, P. (1996). Isolation and characterization of an antifungal agent from an indigenous Streptomyces sp. M.Sc., Thesis. University of Malaya, Kuala Lumpur, Malaysia.

Nonamura, H. (1974). Key for the classification and identification of 458 species of streptomycetes included in the ISP. *J. Ferment Technol*, 52, 78–92.

Okazaki, T. and Okami, Y. (1972). Studies on marine organisms. II. Actinomyctes in Sagami Bay and their antibiotic substances. *J. Antibiotics*, 25, 461–466.

Parker, R.B. (1974). Probiotics, the other half of the antibiotic story. *Animal nutrition and Health*, 29: 4–8.

Porubcan, R.S. (1991b). Reduction in chemical oxygen demand and improvement in *Penaeus monodon* yield in ponds inoculated with aerobic Bacillus bacteria, Program and Abstracts of the 22nd Annual Conference and Exposition, 16–20 June 1991, Sari Juan, Puerto Rico, World Aquaculture Society.

R.S. (1991a). Reduction of ammonia nitrogen and nitrite in tanks of *Penaeus monodon* using floating biofilter containing processed diatomaceous each media pre-inoculated with nitrifying bacteria. Program and Abstracts of the 22nd Annual Conference and Exposition, 16–20 June, 1991, San Juan Puerto Rico. World Aquaculture Society.

Ringø, E. and Gatesoupe, F.J. (1998). Lactic and bacteria in fish: A review. *Aquaculture* 160, 177–203.

Shende, S.T., Apte, R.G. and Singh, T. (1977). Influence of *Azotobacter* on germination of rice and cottonseeds. *Curr. Sci.*, 46(19): 675–676.

Shirling, E.B. and Gottileb, D. (1966). Methods for characterization of *Streptomyces* species. *Int J Syst Bacteriol,* 16: 313–340.

Strickland, J.D.H. and Parsons, T.R. (1972). A practical handbook of seawater analysis. *Fish. Res. Bd. Canada,* 167 p.

Sundara Rao, W.V.B. and Sinha, M.K. (1963). Phosphate dissolving organisms in the soil and rhizosphere. *Indian J. Agri. Sci.,* 33: 272–278.

Tannock, G.W. (1997). Modification of the normal macrobiotic by diet, stress, antimicrobial agents and probiotics. In: R.I. Mackie, B.A. With and R.E. Isaacson (Eds.), Gatro intestinal Microbiology, Vol. 2. Gastrointestinal Series, International Thomson Publishing, New York, 434–465.

Vikineswary, S., Nadaraj, P., Wong, W.H. and Balabaskaran, S. (1997). Actinomycetes from a tropical mangrove ecosystem—Antifungal activity of selected strains. *Asia Pacific J. Mol. Biol. Biotech* 5: 81–86.

Micropropagation, Plant Regeneration with Addition of Charcoal for *in vitro* Root Enhancement of *Bacopa monnieri* (Linn) Pennell

P. Arjun*, M. Krishnamoorthy, R. Kalpana, J. Saibaba and K. Balasubramanian

Plant Tissue Culture, Food safety and Microbiology lab, National Agro Foundation, Anna University Taramani Campus, Taramani, Chennai

E-mail: naf.ptc@gmail.com

INTRODUCTION

Bacopa monnieri (L) Pennell belongs to the family Scrophulariaceae is an amphibious plant of tropics, normally found growing on the banks of the rivers and lakes. It is commonly called as brahmi in India. The herb mainly used for nerve and brain cells and therefore has played a very important role in Ayurvedic therapies for the treatment of cognitive disorders of aging, anti-inflammatory, analgesic, antipyretic, epilepsy, insanity, anticancer and antioxidant activities (Tripathi *et al.,* 1991, Russo and Borrelli, 2005, Bafna and Balaraman, 2005, Ernst, 2006, Sinha and Saxena, 2006). It is also used to treat asthma, hoarseness, enlargement of spleen, snake bite, rheumatism, leprosy, eczema and ring worm, it is also used as a diuretic, appetitive and cardio tonic (Basu and Walia, 1994). The term plant tissue culture is an important tool, is the aseptic culture of cells, tissues, organs and their components under defined physical and chemical conditions *in vitro*. It is an important tool in both basic and applied studies, as well as in commercial applications (Thrope, 2007, Nagarajan *et al.,* 2009).

---

*Corresponding author.

MATERIALS AND METHODS

Leaf and nodes were collected from well grown plants, washed thoroughly under running tap water without damage the tissues. The leaf and nodes were surface-sterilized with (1%) sodium hypochlorite for 15 min followed by (0.05%) mercuric chloride for 3 min and finally 3 to 5 times in sterilized distilled water under aseptic conditions. The leaf and nodes were cut into small pieces (0.5–1 cm) and inoculated into Murashige Skoog (MS) medium (Murashige and Skoog, 1962, Nagarajan *et al.*, 2009) with different concentrations of auxin with the combination of cytokinin, addition of 0.8% (w/v) agar and 3% (w/v) sucrose. The culture bottle and tubes were capped with cotton plugs and autoclaved at 121°C for 15 min. The cultures were incubated under 16 h light/8 h dark cycle and were inspected regularly for the induction of callus. The callus, once formed on the MS medium (with agar), was sub-cultured on MS medium containing with addition of activated charcoal, every 3 to 4 weeks subculture the cultured callus and *in vitro* plants. All experiments were carried out in triplicate.

RESULTS AND DISCUSSION

Plant regeneration in *in vitro* is depends the manipulation of inorganic and organic constituents in the medium, plant environment, type of explants and species. Biotechnological tools are important for multiplication, genetic enhancement of the medicinal plants by adopt of the techniques such as *in vitro* plant regeneration (Tripathi, Tripathi, 2003). The establishment of a reliable *in vitro* regeneration protocol is a prerequisite for application of genetic engineering techniques to enhance production of medicinally important secondary metabolites (Nikam *et al.*, 2009).

Callus Formation from Leaf and Nodal Explants

Undifferentiated mass of parenchymatous cells are known as callus, for callus induction, surface sterilized explants (0.5–1.0 cm) segments were placed on the semi-solid MS basal medium supplemented with 2, 4-D (0.5–3.0 mg/L), IAA (0.5–3.0 mg/L), NAA (0.5–3.0 mg/L) and BAP (0.5–3.0 mg/L) incubated under dark for 5 d, then transferred into light condition for callus formation. 11th day onwards callus formation was observed, maximum callus (90.23%) was obtained on MS+NAA 3 mg/L, followed by MS+NAA 2.5 mg/L medium (72.11%) from the leaf explants (Figure 1). Maximum callus (78.65%) was obtained on MS+2,4-D 2 mg/L, followed by MS+NAA 2.0 mg/L medium (52.60%) from nodal explants of *B. monnieri*, (Table 1). Subculture was performed on the 4th weeks after *in vitro* callus maintaining the culture, with the medium having the same and different composition as MS medium.

The combination of 5.0 μM BA and 2.5 μM 2,4-D produced slightly more callus than the combinations of 5.0 μM BA/Kn with 2.5 μM NAA. Good callus growth in 5.0 μM BA/Kn with 2.5 μM IAA has been reported by *Nikam et al.* (2009). According to Asghari and Lockwood (2002), in *Peganum harmala*, another member of Zygophyllaceae, the combination with cytokinin was very effective for callus proliferation. Callus formation at the basal cut ends of the node and shoot-tip explants of *Leptadenia reticulata* on medium enriched with BA or Kn is in concomitant with the report on *Gymnema sylvestre* and *Holostemma ada-kodien* (Komalavalli and Rao, 2000; Martin, 2002).

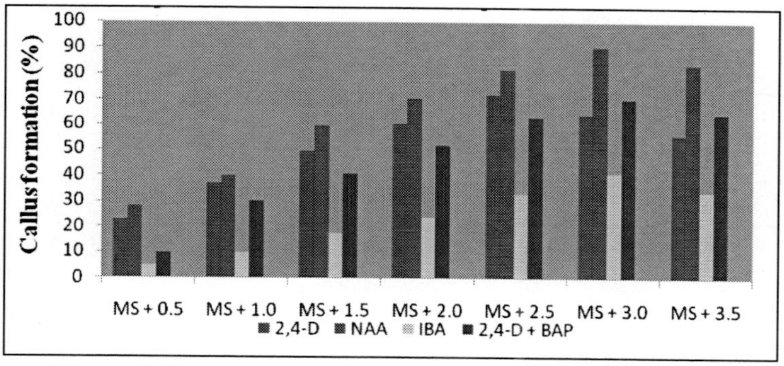

Fig. 1: Callus Formation from Leaf Explants of *B. monnieri*

Table 1: Influence of Different Hormones on Callus Formation from Node Explants of *Bacopa monnieri*

Con of Hormone (mg/L)	*Callus Formation (%)*			
	2,4-D	*NAA*	*IBA*	*2,4-D + BAP*
MS + 0.5	41.65 (± 2.0)	18.33 (± 1.5)	9.67 (± 1.4)	8.11 (± 1.4)
MS + 1.0	52.52 (± 1.5)	30.42 (± 1.6)	17.55 (± 1.7)	16.20 (± 2.0)
MS + 1.5	64.24 (± 1.0)	41.59 (±1.1)	23.42 (± 0.9)	23.48 (± 1.6)
MS + 2.0	**78.65** (± 1.7)	**52.60** (± 2.0)	**34.34** (± 1.3)	**34.65** (± 1.5)
MS + 2.5	65.70 (± 2.5)	40.76 (± 1.0)	23.60 (± 1.5)	22.57 (± 1.1)
MS + 3.0	52.58 (± 1.4)	29.68 (± 1.1)	15.54 (± 2.0)	11.70 (± 2.0)

Note: Column means (n = 3) sharing common letter are not significant at $P < 0.05$ Number in parentheses shows standard deviation (SD).

Shoot Formation from the Callus

Well developed callus transferred to MS basal medium supplemented with auxin and cytokinins, maximum shoot formation (74.43%) was observed from callus in combination of MS+IBA+BAP 3 mg/L, followed by (61.43%) MS+BAP 2 mg/L and lowest concentration (51.32%) was observed MS+IAA+BAP 3 mg/L from the callus. The elongated shoots were transferred to fresh MS basal medium without hormones and incubated in the light for another 4 weeks.

Table 2: Effect of Different Hormones on Shoot and Multiple Shoot Formation from Callus of *Bacopa monnieri*

S. No.	*Hormone con (mg/L)*	*Shoot Formation (%)*	*Multiple Shoot Formation (%)*
1.	MS + BAP (0.5)	34.33 (± 2.2)	16.32 (± 1.0)
2.	MS + BAP (1.0)	47.44 (± 2.4)	21.36 (± 2.1)
3.	MS + BAP (2.0)	**61.56** (± 2.2)	**38.34** (± 2.5)
4.	MS + IBA (1.0)	60.1 (± 2.0)	35.83 (± 1.7)
5.	MS + IBA (2.0)	**70.3** (± 2.4)	**45.24** (± 2.1)
6.	MS + IBA (1.0) + BAP (1.0)	49.5 (± 2.0)	23.54 (± 2.0)
7.	MS + IBA (2.0) + BAP (2.0)	62.21 (± 2.0)	38.7 (± 2.0)
8.	MS + IBA (3.0) + BAP (3.0)	**74.43** (± 2.0)	**51.1** (± 2.0)
9.	MS + IAA (1.0) + BAP (1.0)	34.50 (± 2.6)	17.12 (± 2.4)
10.	MS + IAA (2.0) BAP (2.0)	42.62 (± 1.5)	24.57 (± 2.7)
11.	MS + IAA (3.0) BAP (3.0)	**51.32** (± 1.8)	**37.14** (± 1.6)

Note: Column means (n = 3) sharing common letter are not significant at P<0.05 Number in parentheses shows standard deviation (SD).

Table 3: Root Formation from *in vitro* Shoots of *Bacopa monnieri*

S. No.	*Hormone and Concentration (mg/L)*	*Root Formation MS without Charcoal (%)*	*No. of Roots (%)*	*Root Formation MS without Charcoal (%)*	*No. of Roots (%)*
1.	MS+IAA (0.5)	46.54 (± 1.4)	35.22 (± 1.6)	29.51 (± 1.2)	9.12 (± 1.1)
2.	MS+IAA (1.0)	63.38 (± 2.2)	43.34 (± 2.2)	42.23 (± 2.3)	16.33 (± 1.8)
3.	MS+IAA (2.0)	**71.47** (± 2.4)	**36.65** (± 2.4)	**56.54** (± 2.8)	**23.55** (± 2.2)
4.	MS+NAA (0.5)	29.35 (± 2.5)	16.65 (± 1.0)	19.21 (± 2.6)	7.66 (± 0.5)
5.	MS+NAA (1.0)	43.23 (± 2.6)	28.32 (± 2.3)	30.32 (± 2.4)	12.66 (± 1.1)
6.	MS+NAA (2.0)	**51.42** (± 2.1)	**29.44** (± 2.7)	**41.54** (± 2.2)	**22.33** (± 2.1)
7.	MS+IBA (0.5)	64.65 (± 2.3)	31.75 (± 2.4)	32.35 (± 1.5)	11.42 (± 1.4)
8.	MS+IBA (1.0)	70.46 (± 2.6)	43.52 (± 2.2)	45.44 (± 1.8)	21.75 (± 2.2)
9.	MS+IBA (2.0)	**81.52** (± 2.8)	**45.55** (± 2.5)	**52.22** (± 2.1)	**26.44** (± 2.1)

Note: Column means (n = 3) sharing common letter are not significant at P < 0.05 Number in parentheses shows standard deviation (SD).

Fig. 2: Callus Formation and Plant Regeneration: (a) Callus Formation from Leaf Explant, (b) Callus Formationfrom Nodal Explant, (c) Multiple Shoot Formation without Charcoal, (d) Multiple Shoot Formation Addition of Activated Charcoal

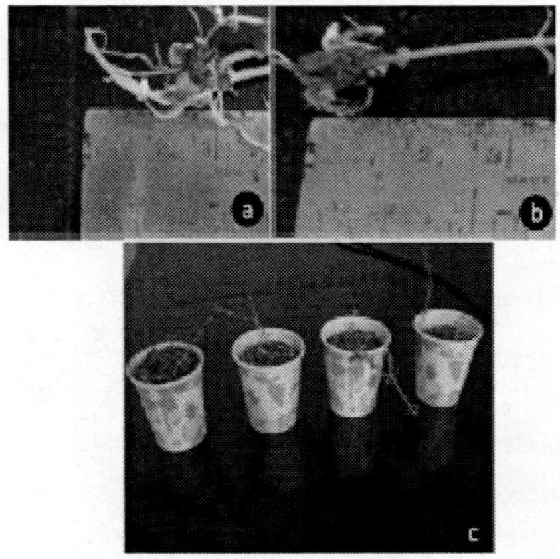

Fig. 3: Rooted Shoots and Regenerated Plants: (a) Roots Measured Additon of Activated Charcoal, (b) Roots Measured without Charcoal, (c) Plants Kept in Cups Under Lab Condition

MS medium supplemented with 2 mg/L BAP showed the highest shoot regeneration frequency and number of regenerated shoots (6.0 ± 0.14). The frequency of shoot proliferation was relatively low and there were fewer shoots per explants when medium was supplemented with KIN (60%, 4.06 ± 0.44) or TDZ (45%, 3.30 ± 0.48) instead of BAP (Singh, Tiwari, 2011). The superiority of BAP over KIN for shoot regeneration has been reported in several systems including *Andrographis paniculata* and *Pterocarpus marsupium* (Chand, Singh, 2004, Purkayastha *et al.,* 2008). TDZ is very commonly used for high frequency shoot regeneration from different explants in several species (Tiwari *et al.,* 2001, Lata *et al.,* 2009), but it is not effective in case of *Clitoria ternatea.* Medium supplemented with 1 mg/L or more TDZ failed to induce shoot regeneration from DEAs. Similarly, shoot buds developed on medium containing TDZ were stunted and often fascinated and subsequently failed to elongate in several plant species such as *Balanites aegyptiaca* (Anis *et al.,* 2010), *Rhododendron* spp. and *Dalbergia sisso* (Preece and Imel, 1991, Pradhan *et al.,* 1998). Thus, BAP when added alone in the medium was the most effective plant growth regulator indicating the cytokinin specificity of DEAs of *Clitoria ternatea* for multiple shoot regeneration (Singh, Tiwari, 2011).

Root Formation from Shoot Plants

The micropropagated shoots were transferred to the MS basal medium supplemented with different hormone concentrations. With addition of activated charcoal, maximum root formation (81.52%) was observed from shoots in combination of MS+IBA 2 mg/L, followed by (71.47%) MS+IAA 2 mg/L and lowest concentration (51.42%) was observed MS+IAA 2 mg/L from the shoots. Without charcoal maximum root formation (56.54%) was observed MS+IAA 2 mg/L and lowest (41.54%) MS+NAA. At the end of the experiment, the percentage of shoot elongation and root lengths were measured.

Plant regeneration has been achieved from leaf callus of *Cephaelis ipecacuanha* on MS medium supplemented with 4.5 mg/L KIN and 0.1 mg/L NAA (Rout *et al.,* 1992). Ghosh, Sen (1994) established plant regeneration via callus cultures from different explants of *Asparagus cooper*. Joshi and Dhawan (2007) reported 0.5 g/L AC had been employed along with 1/2 MS medium supplemented with NAA for rooting in *Swertia chirayita* micropropagation. In a Rutaceae shrub *Fortunella crassifolia*, the shoots formed from epicotyl explants exhibited maximum rooting (75%) on ½ MS medium supplemented with NAA, Kn, and 0.5 g/l AC (Yang *et al.*, 2006). Roots were induced on shoots of *Arachis hypogea* derived from regenerated virus free ELISA tested calli. According to Radhakrishnan *et al.* (1999) About 60–80% of the shoots rooted when cultured on MS medium containing 0.04 g/L AC and 0.2 g/L CH, another extra additive commonly used in plant tissue culture. In *Thapsia garganica* (Apiaceae) rooting of the stock cultures kept on MS medium supplemented with BA and NAA were rooted on ½ MS medium fortified with IBA prior to 0.5 g/L AC treatment. 50% of the plantlets rooted and an average number of 6 roots per shoot were observed in each shoot reported by Makunga *et al.* (2006). Nirmal Babu *et al.* (2003) reported Successful rooting of camphor tree (*Cinnamomum camphora*) shoots derived from shoot tips and nodal segments were reported on WPM medium supplemented with AC, IBA and NAA. 100% of the shoots rooted on IBA and 2 g/L AC containing medium. In this study activated charcoal played major role for root induction and formation from *in vitro* shoots.

Hardening

Regenerated plantlets carefully removed without disturbance, roots and shoots washed in sterilized dist. Washed plantlets were transferred to plastic cups containing sterilized vermiculite and soil (2:1). Initially the hardened plantlets were closed with polythene bags to prevent transpiration, 100% survivability was observed, kept under 16/8 h photoperiod and half strength MS salt solution as poured regularly.

Statistical Analysis

The data were subjected to One-way Analysis of Variance (ANOVA) to evaluate the significant of difference of means of various treatment groups, using SPSS statistical software package (Version: 10). The values are presented as mean ± S.D and $P < 0.05$. Shoot and multiple shoot formation.

CONCLUSION

The present investigations demonstrate with the addition of activated charcoal showed very thin shoots and leaf size was reduced the charcoal, it increase the root growths, improves the *in vitro* morphogenic response of tissues in several ways. It is necessary to focus more studies on the exact mechanism of action of AC in plant tissue culture.

REFERENCES

Anis, M., Varshney, A. and Siddique, I. (2010). *In vitro* clonal propagation of *Balanites aegyptiaca* (L.) Del *Agroforest. Syst.*, 78: 151–158.

Asghari, G. and Lockwood, B.G. (2002). Stereospecific biotransformation of phenylethyl propionate by cell cultures of *Peganum harmala* L. *Iran Biomed. J.*, 6: 43–46.

Bafna, P.A. and Balaraman, R. (2005). Antioxidant activity of DHC-1, an herbal formulation, in experimentally induced cardiac and renal damage. *Phytother. Res.*, 19: 216–221.

Basu, N.K. and Walia, J.S. (1994). The chemical investigations of the leaves of *Herpestis monniera. Indian J. Pharm.*, 4: 84–85.

Chand, S. and Singh, A.K. (2004). *In vitro* shoot regeneration from cotyledonary node explants of a multipurpose leguminous tree *Pterocarpus marsupium* Roxb. *In vitro Cell. Dev. Biol. Plant*, 40: 167–170.

Ernst, E. (2006). Herbal remedies for anxiety—A systematic review of controlled clinical trials. *Phytomedicine*, 13: 205–208.

Ghosh, B.E. and Sen, S. (1994). Micropropagation of *Asparagus cooperi* as affected by growth regulators. *Biologia Plantarum*, 36: 527–34.

Joshi, P. and Dhawan, V. (2007). Assessment of genetic fidelity of micropropagated *Swertia chirayita* plantlets by ISSR marker assay. *Biol. Plant*, 51: 22–26.

Komalavalli, N. and Rao, M.V. (2000). *In vitro* micropropagation of *Gymnema sylvestre*—A multipurpose medicinal plant. *Plant Cell Tiss. Organ Cult.*, 61: 97–105.

Lata, H., Chandra, S., Khan, I. and ElSohly, M.A. (2010). Thidiazuron-induced high-frequency direct shoot organogenesis of *Cannabis sativa* L. *In vitro Cell. Dev. Biol., Plant*, 45: 12–19.

Makunga, N.P., Jager, A.K. and Van Staden, J. (2006). Improved *in vitro* rooting and hyperhydricity in regenerating tissues of *Thapsia garganica* L. *Plant Cell Tissue Organ Cult.*, 86: 77–86.

Martin, K. (2002). Rapid propagation of *Holostemma ada-kodien* Schult., a rare medicinal plant, through axillary bud multiplication and indirect organogenesis. *Plant Cell Rep.*, 21: 112–117.

Murashige, T. and Skoog, F. (1962). A revised medium for rapid growth and bioassays with to bacco-tissue cultures. *Physiol. Plant.*, 15: 473–497.

Nagarajan, S., Mohan Das, T., Arjun, P. and Raaman, N. (2009). Design, synthesis and gelatin studies of 4, 6-*O*-butylidene-α, β-unsaturated-β-*C*-glycosidic ketones: application to plant tissue culture. *J. Mater. Chem.,* 19: 4587–4596.

Nikam, T.D., Mohammad Ali Ebrahimi and Patil, V.A. (2009). Embryogenic callus culture of *Tribulus terrestris* L. a potential source of harmaline, harmine and diosgenin. *Plant Biotechnol. Rep.,* 3: 243–250.

Nirmal Babu, K., Sajina, K.A., Minoo, D., John, C.Z. Mini, P.M. and Tushar, K.V. (2003). Micropropagation of camphor tree (*Cinnamomum camphora*). *Plant Cell Tissue Organ Cult.*, 74: 179–183.

Pradhan, S., Kar, S. Pattnaik, P.K., Chand. (1998). Propagation of *Dalbergia sisoo* Roxb. through *in vitro* shoot proliferation from cotyledonary nodes. *Plant Cell Rep.*, 18: 122–126.

Preece, J.E. and Imel, M.R. (1991). Plant regeneration from leaf explants of Rhododendron 'P.J.M. hybrids' *Sci. Hortic.*, 48: 159–170.

Purkayastha, J., Sugla, T., Paul, A., Solleti, S. and Sahoo, L. (2008). Rapid *in vitro* multiplication and plant regeneration from nodal explants of *Andrographis paniculata*: a valuable medicinal plant *In Vitro Cell. Dev. Biol. Plant,* 44: 442–447.

Radhakrishnan, T., Murthy, T.G.K., Desai, T. and Bandyopadhyay, A. (1999). Meristem culture of interspecific hybrids of groundnut. *Biol. Plant,* 42: 309–312.

Rout, G.R., Saxena, C. Samantaray, S. and Das, P. (1992b). Rapid clonal propagation of *Plumbago zeylanica* Linn. *Plant Growth Reg.,* 28: 1–4.

Russo, A. and Borrelli, F. (2005). *Bacopa monnieri,* a rapid nootropic plant; an overview. *Phytomedicine,* 12: 305–317.

Singh, J. and Tiwari, K.N. (2011). *In vitro* plant regeneration from decapitated embryonic axes of *Clitoria ternatea* L. An important medicinal plant. *Industrial Crops and Products,* 35: 224–229.

Sinha, S. and Saxena, R. (2006). Effect of iron on lipid peroxidation, and enzymatic and non-enzymatic antioxidants and bacoside—A content in medicinal plant *Bacopa monnieri* L. *Chemosphere,* 62: 1340–1350.

Thrope, A. (2007). History of plant tissue culture. *Mol. Biotechnol.*, 37: 169–180.

Tiwari, V., Tiwari, K.N. and Singh, B.D. (2001). Comparative studies of cytokinins on in vitro propagation of *Bacopa monniera Plant Cell Tiss. Org. Cult.*, 66: 9–16.

Tripathi, Y.B., Chaurasia, S., Tripathi, E. Upadhyay, A. and Dubey, G.P. (1996). *Bacopa monnieri* Linn. as an antioxidant mechanism of action. *Indian J. Exp. Bio.*, 4: 523–526.

Yang, L., Xu, C.J., Hu, Chen, K.S. (2006). Direct shoot organogenesis and plant regeneration in *Fortunella crassifolia. Biol. Plant,* 50: 729–732.

Evaluation of Toxic Effects of Effective Microorganism (EM) Fermentation Extracts on Suitable Biofertilizers, Biocontrol Agents and Biopesticides

R.S. Arvind Bharani*, S. Karthick Raja Namasivayam[1] and Sreesha B.[1]

Microbial Technology Division, Centre for Bioresource and Development (C-BIRD), Sathyabama University, Chennai

[1]Department of Biotechnology, Sathyabama University, Chennai

E-mail: arvindbharani@gmail.com

INTRODUCTION

The concept of "effective microorganisms" was developed by Japanese agronomist Teruo Higa, from the University of the Ryukyus in Okinawa who reported in the 1970s that a combination of approximately 80 different microorganisms is capable to positively influence decomposing organic matter such that it reverts into a "life promoting" process. EM technology is used to maintain sustainable practices such as farming and sustainable living, and to support human health and hygiene, animal husbandry, compost and waste management, disaster clean-up, and generally used to promote functions in natural communities. Effective Microorganisms is also called "EM technology" by the proponents of this concept. "EM technology" is a scientifically unconfirmed method to improve soil quality and plant growth using a mixture of microorganisms consisting mainly of lactic acid bacteria, purple bacteria, and yeasts which co-exist for the benefit of whichever environment they are introduced. Effective Microorganisms (EM) a mixture of lactic acid bacteria, photosynthetic bacteria, yeasts and fungi are used mainly in agriculture and organic

*Corresponding author.

waste treatment which can be applied as inoculants to increase the microbial diversity of soils and plants by producing various metabolites (Formowitz *et al.*, 2007, Namasivayam and Kirithiga, 2010). The use of EM for increasing the crop productivity is now being extensively used as major component in agriculture. In Peanut crops, the biomass increased by 34%. In soybean crops, the biomass increased by 40%. Rice yield increases in the first year of application. Maize and sugar cane grew taller by 33%. Papayas income increased by 150%. Cashew yield increased by 47.5%. Fruit orchards were much healthier. Bigger sizes were recorded in sweet potatoes and tropica. Recently they have also been added to water and feed for animals, as well as to processing their excrements into compost and to eliminate the stench. EM is used to treat the persistent substances contained in waste water (Buerkert, 2007) and a variety of phenol degrading bacteria in EM degraded trichloroethylene (Tang *et al.*, 2009; Chui *et al.*, 2004) also reported the anti tumour activity of EM fermentation extract on selected cancer cell line EM is also introduced as an alternative to antibiotics in broiler diets (Javaid *et al.*, 2008). The production of organic food requests for methods that use non-chemical inputs. In this production system, soil fertility, pests and disease causing microorganisms are essentially managed by biological agents, where the biofertilizers *Rhizobium* sp, *Azospirillum* sp *Phosphobacteria* sp, biocontrol agents *Pseudomonas fluorescens,Gliocladium* sp and *Trichoderma harzianum* entomopathogenic biopesticides *Beauveria bassiana* and *Metarhizium anisopliae.* Samson are considered important factors of soil fertility and reduction of insect pests and disease causing pathogens.So there is the necessity of biocontrol and biofertilizers conservations,if it occurs naturally, are applied or introduced with the objective of controlling disease and pests.However,for their conservation,it is important to know the biocontrol agents and biofertilizers compatibility with other agricultural practices to avoid losses of control efficiency (Duarte *et al.*, 1992; Hirose, 2001).

Though EM are found to benefit to the crops it is necessary to study the impact of EM and their metabolites on non target beneficial micro biota in the environment, In this point of view, the present work was carried out to determine the non target effects of ethyl acetate, acetone, hexane and petroleum ether cell free extracts of effective microorganisms against *Alternaria alternata* adapting solid plate spot assay, dual liquid assay and spore germination inhibition assay. The results from the present studies shown no distinct inhibition on all tested bio fertilizers, bio control agents, bio pesticides and plant pathogen.

MATERIALS AND METHODS

Effective Microorganism (EM)

Effective microorganism (EM) used in this study was supplied by Manidharma Biotech as EM stock liquid culture containing a mixture of lactic acid bacteria *Lactobacillus planetarium* (1.0×10^3) yeast with 1.0×10^4 CFU/ml *Candida utilis*, actinomycetes *Streptomyces albus* (3.0×10^2 CFU/ml), fermenting fungi *Aspergillus oryzae* (1.1×10^5 CFU/ml). EM solution is a yellowish liquid with a pleasant odour and sweet sour taste with a pH of 3 and stored in cool place without refrigeration.

Preparation of EM Extracts

EM stock was diluted in sterile distilled water as per the instruction of manufactures and the diluted EM-MA (500 ml) was extracted with the double the volume of ethyl acetate, acetone, hexane and petroleum ether separately at 40°C, the concentrated crude extracts were collected and dissolved in 1ml of DMSO, kept in sterile screw cap vials at room temperature.

Evaluation of Toxic Effects of EM Extracts on Bio Fertilizers, Bio Control Agents and Bio Pesticides

Bacterial Biocontrol Agents and Biofertilizers

Rhizobium leguminosarum, Azospirillum lipoferum, Phosphobacterium sp and Pseudomonas *fluorescens* and fungal biocontrol agents were selected in the present study. Respective biofertilizers, bio control agents *Gliocladium* sp and *Trichoderma harzianum* entomopathogenic fungal biopesticides *Beauveria bassiana* and *Metarhizium anisopliae* (F.) Samson were obtained from Manidharma Biotech, Porur, Chennai. All the cultures were maintained on Nutrient agar slants (bacteria) and PDA (fungi).

Solid Plate Spot Assay

Nutrient Agar plates and PDA plates were prepared and sterilized by autoclaving. After sterilization, 100 µl of respective extract prepared from respective solvent was added into the media. The media was mixed thoroughly for complete mixing of metabolites. The well mixed media was poured into sterile petriplates.

The respective microorganisms was inoculated on the centre of the solid media as spot inoculation with sterile inoculation loop (about 8–10 mm).Control and replications were maintained. The seeded plates were incubated at 37°C for 48 hrs (bacteria) and

28°C for 7 days (fungi). After incubation the diameter of the respective organism was measured and percent inhibition was calculated as per Kulkanni and Lingappa (2001) method.

$$\text{Percent Inhibition} = \frac{\text{Growth in Control} - \text{Growth in Treatment}}{\text{Growth in Control}} \times 100$$

Dual Liquid Assay

100 ml of Nutrient broth and Potato Dextrose broth were prepared in 250ml conical flask sterilized by autoclaving. After sterilization 0.1 ml of 24 hrs respective bacterial cultures and fungal spore suspensions were added. All the inoculated flasks were incubated under shaking condition at 37°C (bacteria – 24 hrs), 28°C (fungi – 4 days). After the incubation the dry weight of the biomass was determined by filter paper method. Control and replications were maintained. The percentage of inhibition was also carried out by the formula as described earlier.

Spore Germination Inhibition Assay

The spore germination inhibition assay was carried out with the fungal organisms with the respective extracts. In this test, 0.3 ml of sterile molten PDA media was added on the centre of the clean grease free glass slide. 0.1ml of respective extract was mixed with the media gently with tooth pick; 0.1ml of the respective fungal spore suspension was added. The preparation was transferred to the sterile petriplate containing sterile moist filter paper with a layer of cotton moistened with sterile distilled water. The plates were closed with lid, kept at 28°C for 12–16 hrs. After the incubation period, the slides were stained with lacto phenol cotton blue stain and observed under microscope for spore germination.

RESULTS AND DISCUSSION

Significant reduction was not observed in the growth of all the tested biofertilizers and biocontrol agents in all the treatments except ethyl acetate and petroleum ether extract in which 20% and 10% growth inhibition was observed in *Rhizobium leguminosarum*, *Azospirillum lipoferum*,and *Pseudomonas.fluorescens* (Table 1). Both the extracts reveals the colony diameter of *Rhizobium leguminosarum*,*Azospirillum lipoferum* and *Pseudomonas flurescens* in ethyl acetate treatment as 1.6 cm, 1.6 cm, 1.6 cm and 2.0 cm and the percentage of growth inhibition for the respective organisms was found to be 20%, 20%, and 10% (Table 1). Interestingly hexane extract stimulated the vegetative growth of *Rhizobium leguminosarum* (20.0%) and *Phosphobacterium* (10.0%)

Acetone extract did not cause any toxic effect on all the tested bacterial biofertilizer and biocontrol agent. The colony diameter of 1.8, 1.8, 1.5 and 1.8 cm was recorded in *Rhizobium leguminosarum, Azospirillum lipoferum, Phosphobacterium* sp and *Pseudomonas flurescens* in acetone treatment whereas 1.8cm of colony diameter was recorded in control But, growth inhibition was not recorded in all the tested fungal biocontrol agents. There was no significant difference in colony diameter was recorded in respective extracts (Table 2). In the case of fungal biocontrol agents and biopesticides, the vegetative growth was not inhibited in all the tested extracts (Table 2). The vegetative growth of all the tested fungi in respective extracts treatment revealed the colony diameter of untreated control of respective fungi. Dual liquid assay reveals similar toxic and growth stimulation effect that was noticed in solid plate assay. Dry weight of bacterial biomass of *Rhizobium leguminosarum, Azospirillum lipoferum* and *P. fluorescens* was reduced in ethyl acetate treatment which reveals 0.123 g, 0.119 and 0.127 g. of dry weight of biomass. 0.141 g, 0.147 g. and 0.139 g. of dry weight of bacterial biomass was recorded in untreated control of respective bacterial biofertilizer and biocontrol agent (Tables 3 and 4). The dry

Table 1: Colony Diameter and Growth Inhibition (%) of Bacterial Biofertilizers Treated with Different Cell Free Extracts of EM

Sl. No.	*Tested Organism*	*Colony Diameter (cm)*					*Growth Inhibition (%)*				
		C	EA	AC	HX	PE	C	EA	AC	HX	PE
1.	Rhizobium	1.8	1.6	1.8[a]	2.0	1.8	0.0	20.0	0.0	0.0	0.0
2.	Azospirillum	1.8	1.6	1.8[a]	1.8[a]	2.0	0.0	20	0	0	0
3.	*Phosphobacterium*	1.5	1.6	1.5[a]	1.6	1.5[a]	0.0	0	0	0	0
4.	*Pseudomonas flurescens*	1.8	2.0	1.8[a]	1.8[a]	1.7	0.0	0	0	0	0

'a' in the column, the mean carries the alphabet is not statistically significant ($p > 0.05$ level by DMRT).

Table 2: Colony Diameter and Growth Inhibition (%) of Bacterial Biofertilizers Treated with Different Cell Free Extracts of EM

Sl. No.	*Tested Organism*	*Colony Diameter (cm)*					*Growth Inhibition (%)*				
		C	EA	AC	HX	PE	C	EA	AC	HX	PE
1.	*Trichoderma harzianum*	4.0	4.0[a]	4.0[a]	4.0[a]	4.0[a]	0	0	0	0	0
2.	*Alternaria alternate*	4.0	4.0[a]	4.0[a]	4.0[a]	4.0[a]	0	0	0	0	0
3.	*Gliocladium*	2.5	2.5[a]	2.5[a]	2.5[a]	2.5[a]	0	0	0	0	0
4.	*Metarhizium anisopliae*	4.0	4.0[a]	4.0[a]	4.0[a]	4.0[a]	0	0	0	0	0
5.	*Beauveria bassiana*	3.0	3.0[a]	3.0[a]	3.0[a]	3.0[a]	0	0	0	0	0

'a' in the column, the mean carries the alphabet is not statistically significant ($p > 0.05$ level by DMRT).

Table 3: Biomass (dry weight) of Biofertilizers Treated with Different Cell Free Extracts of EM

Sl. No.	*Tested Organism*	*Biomass (dry weight) in Grams*				
		C	*EA*	*AC*	*HX*	*PE*
1.	Rhizobium	0.210	0.200	0.227	0.183	0.240
2.	Azospirillum	0.300	0.228	0.290	0.310	0.320
3.	*Phosphobacterium*	0.125	0.128	0.187	0.188	0.217
4.	*Pseudomonas flurescens*	0.150	0.570	0.165	0.227	0.112

Table 4: Biomass (dry weight) of Fungal Biocontrol Agents, Plant Pathogen and Biopesticide Treated with Different Cell Free Extracts of EM

Sl. No.	*Tested Organism*	*Biomass (dry weight) in Grams*				
		C	*EA*	*AC*	*HX*	*PE*
1.	*Trichoderma harzianum*	0.354	0.330	0.300	0.391	0.482
2.	*Alternaria alternata*	0.143	0.150	0.160	0.120	0.180
3.	*Gliocladium*	0.067	0.079	0.320	0.149	0.070
4.	*Metarhizium anisopliae*	0.129	0.320	0.320	0.320	0.430
5.	*Beauveria bassiana*	0.158	0.160	0.157	0.158	0.190

Table 5: Percentage of Spore Germination Inhibition of Fungal Biocontrol Agents, Biopesticides and Plant Pathogen Treated with all Cell Free Extracts of EM

Sl. No.	*Tested Organism*	*Biomass (dry weight) in Grams*				
		C	*EA*	*AC*	*HX*	*PE*
1.	*Trichoderma harzianum*	00	00	00	00	00
2.	*Alternaria alternata*	00	00	00	00	00
3.	*Gliocladium*	00	00	00	00	00
4.	*Metarhizium anisopliae*	00	00	00	00	00
5.	*Beauveria bassiana*	00	00	00	00	00

weight was found to be increased in hexane treatment of *Rhizobium leguminosarum* and *Phosphobacterium* which reveals 0.167 g and 0.147 g whereas the respective untreated bacteria organism reveals 0.141 g and 0.120 g. A significant change in dry weight of all the tested bacterial and fungal biocontrol and biofertilizers was not observed in acetone and petroleum ether extracts. In relation to the spore germination of fungal biocontrol agents, inhibition was not recorded in any of the extracts tested (Table 5) Spore germination was not affected in all the treatment of EM extracts as in control (Table The present study clearly reveals the crude metabolites obtained from EM-MA extracts did not show any toxic effect on selected biofertilizers and biocontrol agents. Further study under field trail will helpful to study such a non target effect on the biocontrol agents surely recommend the EM-MA as the major component in organic farming used to improve crop yield and productivity.

REFERENCES

Buerkert (2007). The role of effective microorganisms in the composting of banana (*Musa* spp.) residues. *J. of Plant Nutrition and Soil Science,* 170; 649–656.

Duarte, A., Menendez, J.M. and Trigueiro, N. (1992). Estudio preliminar sobre la compatibilidad de *Metarhizium anisopliae* com algunos plaguicidas quimicos. *Revista Baracoa,* 22; 31–39.

Formowitz, B., Elango, F., Okumoto, S., Müller, T. and Buerkert, A. (2007). The role of effective microorganisms in the composting of banana (Musa ssp.) residues. *J. of Plant Nutrition and Soil Science,* 170; 649–656.

Hirose, E., Neves, P.M.O.J., Zequi, J.A.C., Martins, L.H., Peralta, C.H. and Moino Jr., A. (2001). Effect of biofertilizers and Neem oil on the entomopathogenic fungi *Beauveria bassiana* (Bals.) Vuill and *Metarhizium anisopliae* (Metsch) Sorok. *Brazilian Arch. Biol. Technol,* 44; 419–423.

Javaid, A., Bajwa, R. and Anjum, T. (2008). Effect of heat-sterilization and EM application on wheat (*Triticum aestivum* L.) grown in organic amended sandy loam soil. *Cereal Research Communication*, 36; 489–499.

Namasivayam, S.K.R. and Kirithiga (2010). Effect of formulation of effective microorganism (EM) on post treatment persistence, microbial density and soil macronutrients. *Recent Research in Science and Technology*, 2; 102–106.

Tang, Xiaowei Niu, Qing Sun and Rugang Wang (2009). Bioremediation of petroleum polluted soil by combination of Rye grass with effective microorganisms, *Environmental Science and Information Application*, 2; 51–54.

Effects of Abiotic Factors on the Growth of *T. atroviride* Isolates Obtained from Southern Indian Tea Plantation

S. Anita, P. Ponmurugan* and M. Ramya
Department of Biotechnology, K.S. Rangasamy College of Technology, Tiruchengode, Tamil Nadu
E-mail: anita.sundar@gmail.com

INTRODUCTION

Tea (*Camellia sinensis* (L.) O. Kuntze) is a perennial monoculture crop cultivated in 2.71 million ha covering more than 34 countries across Asia, Africa, Latin America, and Oceania with production of 3.22 million metric tonns of made tea annually. This foliage crop tends to endure severe stress both physically and physiologically owing to the frequent harvest of shoot (Thomas *et al.*, 2009). Several bacterial and fungal antagonists have been successfully used to control wide number of phytopathogen. However, research informations are being inadequate on the biological control of tea diseases (Ponmurugan and Baby, 2006). Biological control is the method or practice through which the phytopathogen is controlled by means of another (beneficial) organism, their genes, and/or products, such as metabolites, that diminish the effects of pathogens and support the growth of plants (Vinale *et al.*, 2007). *Trichoderma* spp. are investigated as an efficient antagonist for over 70 years and it occupies almost 50% of fungal Biocontrol agents (Samuels, 1996). Diverse mechanisms have been recommended for their biocontrol activity, which comprises of antagonism for space and nutrients, secretion of enzymes and antimicrobial compounds. Hence the present study was carried out to screen an efficient

*Corresponding author.

indigenous *Trichoderma* spp. isolated from tea soils of southern India based on the various abiotic factors.

MATERIALS AND METHODS

Biocontrol Agents Used

Trichoderma spp. used for the present study was isolated from the rhizosphere soil samples of southern Indian tea plantation, which includes the districts of Valparai, Coonoor, Munnar, Vandiperiyar, Gudalore and Koppa. The cultures were identified as *Trichoderma atroviride* (MTCC 9641) from Microbial Type Culture Collection (MTCC), Chandigarh.

Assessment of Growth Parameters

Radial growth measurement of *T. atroviride* isolates were observed on PDA by the method demonstrated by Elad *et al.* (1981). Petri dishes (90 mm diameter) containing PDA were centrally inoculated with a 5 mm of agar plugs obtained from peripheral region of *T. atroviride* isolates. Plates were incubated at 27±2°C for three consecutive days and radial mycelial growth was recorded for every 24 hours. Mycelium dry weight of *T. atroviride* isolates was determined by transferring 5 mm of actively growing cultures into 100 mL of PDB and incubated at 27°C. The mycelial mats were harvested at different intervals of time by using Whatman No.1 filter paper and the dry weight of the mycelium was obtained by drying the mat at 80°C for 24 hrs and the dry weight is expressed in mg.

Effect of Various Abiotic Factors on the Growth of *T. atroviride* Isolates

Influence of different abiotic factors such as pH, temperature regimes and light spectra on the growth of *T. atroviride* were studied both in solid and liquid media. The PDA and PDB medium amended with different range of pH form 4.0 to 6.5 were taken for the present study. Similarly different range of temperature (15 to 35°C) was selected for the analysis the radial growth and dry weight of the mycelium. Moreover the *T. atroviride* isolates were made to grow in different light spectra such as UV light, White fluorescent, dark, diffused day light under room condition and direct sun light.

RESULTS AND DISCUSSION

The growth pattern in terms of radial growth and mycelium dry weight of *T. atroviride* isolates are presented in the Table 1. The lowest radial growth rate of

37.67 mm was observed for T MTCC isolate, whereas the highest value of 45 mm was recorded for Tv1 isolate on 5th day of incubation. The mycelium of Tv1 isolate covered the entire plate within five days of incubation with significant difference when compared to other isolates. The mycelium dry weight of *T.atroviride* showed a minimum difference between the 5th and 7th days for all the isolates. The Tv1 isolate had a higher mycelium dry weight of 400.0 mg followed by Tm3 and Tc3 of 370.6 mg and 350.7 mg; respectively on 7th day of incubation. The standard culture (T MTCC isolate) showed the lowest of 280.3% when compared to other isolates on the same day of incubation.

Table 1: Assessment of Growth Parameters of *T. atroviride* Isolates on PDA and PDB Media

T.atroviride isolates	*Radial Growth (mm)*		*Dry Weight of Mycelium (%)*		
	3rd day	*5th day*	*3rd day*	*5th day*	*7th day*
Tv1	40.33^{a}	45.00^{a}	236.8^{a}	370.7^{a}	400.0^{a}
Tc3	33.00^{b}	42.33^{b}	200.6^{a}	320.9^{b}	350.7^{b}
Tm3	31.33^{b}	40.67^{b}	200.3^{b}	350.7^{b}	370.6^{b}
Tp4	32.67^{b}	41.00^{b}	178.2^{c}	300.7c	320.2^{c}
Tg2	32.00^{b}	40.00^{b}	184.6^{c}	315.0^{c}	336.3^{bc}
Tk2	30.00^{b}	40.00^{b}	153.9^{d}	280.0^{c}	300.3^{d}
T MTCC*	30.00^{c}	37.67^{c}	179.6^{c}	254.3^{c}	280.3^{d}
SE ±	1.75	1.98	10.41	12.06	10.93
CD at P=0.05	2.24	2.08	20.1	24.0	20.00

*Standard culture used for comparison purpose.
Similar alphabets followed in a column denote they are statistically insignificant and vice versa at $P < 0.05$.

Effects of Abiotic Factors on the Growth of *T. atroviride* Isolates

Effect of pH

The optimum pH for the growth of *T.atroviride* was found to be 4.5 for all the ingenious isolates (Table 2 and Figure 1. Tv1 isolate registered a high radial growth and covered the 9 cm Petri plate within five days of incubation followed by Tc3 and Tm3 isolates with 42.50 mm and 40.17 mm. Tk2 isolate showed a minimum growth rate of 39.84 mm among the indigenous isolate, however significant difference was obtained when compared with standard MTCC culture. These results were found to be highly correlated with mycelial dry weight with significant difference between indigenous and stranded isolate. pH below 4.5 and above 5.5, the mycelium growth was reduced notably. Moreover, at pH below 4 and above 6.5 the antagonist has failed to grow.

Table 2: Effect of pH Levels on the Radial Growth of *T.atroviride* Isolates on 5th Day of Incubation

Isolates	*pH*					
	4.00	*4.50*	*5.00*	*5.50*	*6.00*	*6.50*
Tv1	36.67^{a}	45.00^{a}	42.50^{a}	41.34^{a}	41.34^{a}	36.34^{a}
Tc3	35.67^{ab}	42.50^{b}	41.67^{b}	38.17^{b}	38.17^{b}	31.17^{ab}
Tm3	35.84^{ab}	40.17^{b}	38.50^{c}	35.50^{c}	35.50^{bc}	30.30^{b}
Tp4	33.84^{ab}	40.34^{c}	38.50^{c}	37.50^{b}	37.50^{b}	28.50^{b}
Tg2	34.34^{ab}	40.34^{c}	40.50^{b}	35.34^{c}	35.34^{c}	28.67^{b}
Tk2	32.84^{bc}	39.84^{c}	34.50^{b}	32.50^{c}	30.50^{c}	27.00^{bc}
T MTCC	28.00^{c}	30.00^{d}	34.67^{d}	31.50^{d}	31.50^{d}	23.17^{c}
SE±	2.74	1.59	1.27	1.30	2.53	2.13
CD at P = 0.05	7.81	4.54	3.62	3.76	5.30	6.07

Similar alphabets followed in a column denote they are statistically insignificant and vice versa at $P < 0.05$.

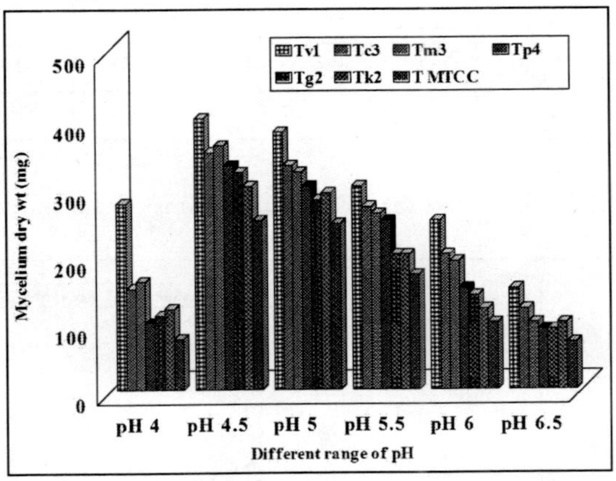

Fig. 1: Effect of pH Levels on the Mycelium Dry Weight of *T. atroviride* Isolate

Effect of Temperature

The optimum temperature regime for the growth of *T. atroviride* isolates was found to be in the range of 25–30°C. On PDA medium, the radial growth of Tv1 isolate recorded the maximum of 45 mm growth at 25°C (Table 3). Temperature below 20°C and above 30°C arrested the growth drastically and at temperature below 10°C and above 35°C the antagonists failed to grow. Tp4 and Tg2 isolates were almost identical with 43.00 mm and 43.33 mm; respectively. However, the radial growth was found to be low when compared to Tv1, Tc3 and Tm3 isolates with significant difference. The results of percentage of mycelial dry weight for the all the six isolates showed in significant difference, expect Tv1isolate which produced higher mycelium percentage of 93.5 at a temperature of 25°C (Figure 2). However, a gradual decline was noticed when there was an increase and decrease of various temperature ranges.

Table 3: Effect of Temperature Regimes on the Radial Growth of *T. atroviride* Isolates on 5th of Incubation

Isolates	*Temperature (°C)*							
	5	*10*	*15*	*20*	*25*	*30*	*35*	*40*
Tv1	NG	28.33^{a}	38.00^{a}	40.67^{a}	45.00^{a}	42.67^{a}	40.67^{a}	19.33^{a}
Tc3	NG	24.33^{a}	31.33^{bc}	36.33^{b}	44.67^{ab}	39.00^{b}	35.67^{c}	NG
Tm3	NG	24.67^{b}	33.00^{b}	35.67^{bc}	45.00^{a}	38.33^{b}	37.67^{b}	NG
Tp4	NG	20.67^{c}	29.33^{c}	33.00^{d}	43.00^{c}	37.33^{b}	30.00^{c}	NG
Tg2	NG	20.33^{bc}	29.67^{d}	34.00^{cd}	43.33^{bc}	37.67^{b}	30.00^{c}	NG
Tk2	NG	17.00^{d}	24.33^{d}	30.67^{e}	41.33^{d}	38.33^{b}	31.67^{c}	NG
T MTCC	NG	NG	21.67^{c}	27.67^{f}	40.33^{d}	31.67^{c}	24.64^{c}	NG
SE ±	–	1.59	1.22	0.91	0.71	1.02	0.59	–
CD at P = 0.05	–	4.54	3.48	2.59	2.03	2.91	1.68	–

NG-No growth was observed;

Similar alphabets followed in a column denote they are statistically insignificant and vice versa at $P < 0.05$.

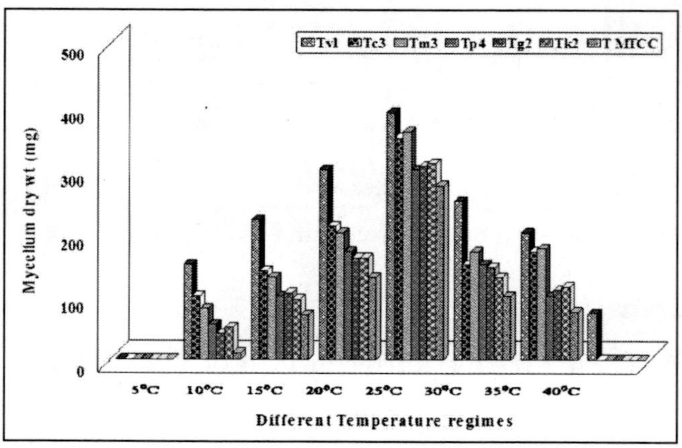

Fig. 2: Effect of Different Range of Temperatures on the Mycelial Dry Weight of *T. atroviride* Isolates

Effect of Light Spectrum

The growth of the antagonists in diffused day light under room condition had the maximum growth followed by fluorescent light, UV light and direct sun light. The dark condition had reduced the growth for all the seven isolates significantly (Table 4). Tv1 isolate exhibited maximum radial growth was 45 mm in diffused day light and minimum growth of 26.33 mm in dark conditions. Exposure of Tv1, Tc3 and Tm3 isolates to different light spectrum showed significant difference except, diffused day light were Tc3 and Tm3 isolates showed radial growth of 43.67 mm and 43.33 mm; respectively. Tp4, Tg2 and Tk2 isolates produced almost similar radial growth of 15.33 mm, 16.33 mm and 15.67 mm; respectively under dark condition. Hence, indigenous isolates showed a significant difference when compared to standard isolate. Percentage of dry weight was found to increased under diffused daylight and white fluorescent light for all the isolates (Figure 3). Radial growth of *T. atroviride* isolates highly correlates with percentage of mycelium dry weight for all the isolates. Hence, T MTCC showed poor growth when compared to indigenous isolates.

Table 4: Effect of Different Light Spectra on the Radial Growth of *T. atroviride* Isolates on 5th of Incubation

Isolates	*Direct Sunlight*	*UV Light*	*White Fluorescent Light*	*Diffused Day Light*	*Dark*
Tv1	30.67^{a}	35.00^{a}	39.67^{a}	45.00^{a}	26.33^{a}
Tc3	26.00^{b}	31.00^{b}	34.33^{b}	43.67^{a}	20.67^{b}
Tm3	25.67^{b}	27.00^{cd}	34.33^{b}	43.33^{a}	21.25^{b}
TP4	19.33^{c}	25.33^{bc}	30.67^{c}	39.67^{b}	15.33^{c}
Tg2	20.00^{cd}	23.00^{bc}	32.00^{c}	39.33^{b}	16.33^{c}
Tk2	17.33^{cd}	22.33^{bc}	28.67^{d}	38.00^{bc}	15.67^{c}
T MTCC	14.00^{d}	20.00^{d}	25.33^{e}	35.00^{c}	12.33^{e}
SE±	1.25	1.50	0.98	0.78	0.80
CD at P = 0.05	3.55	4.28	2.64	2.21	2.50

Similar alphabets followed in a column denote they are statistically insignificant and vice versa at $P < 0.05$.

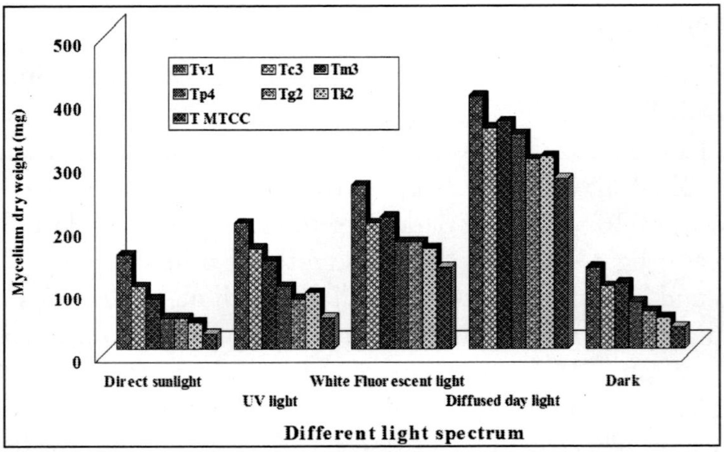

Fig. 3: Effect of Different Light Spectra on the Mycelial Dry Weight of *T. atroviride* Isolates

REFERENCES

Chet, I., Inbar, J. and Hadar, Y. (1996). Fungal antagonists and mycoparasites. In D.T. Wicklow and B.E. Soderstrom (eds) The Mycota, Vol. 5, *Environmental and Microbial Relationships.*

Cook, R.J. and Baker, K.F. (1983). The nature and practice of biological control. *American Phytopathological Soc.* St. Paul. Minnesota, 539 p.

Francesco Vinale, Krishnapillai Sivasithamparam, Emilio L. Ghisalberti, Roberta Marraa, Sheridan L. Woo (2007). Matteo Lorito Trichoderma–plant–pathogen interactions. *Soil Biology and Biochemistry.*

Nioh, I., Osada, M., Yamamura, T. and Maramatsu, K. (1995). Acidophilus and acid tolerant actionomycetes in acid tea field soil, *J. Gen. Appl. Microbiology*, 41: 175–180.

Ponmurugan, P. and Baby, U.I. (2007). Evaluation of fungicides and biocontrol agents against *Phomopsis* canker of tea under field conditions. *Australasian Plant Pathology*, Vol. 36, pp. 68–72.

Samuels, G.J. (1996). *Trichoderma*: a review of biology and systematic of the genus. *Mycol Res* 100: 923–935.

Thomas, J., Mandal, A.K.A., Raj Kumar, R. and Chordia, A. (2009). Role of Biologically Active Amino Acid Formulations on Quality and Crop Productivity of Tea (*Camellia* sp.). *International Journal of Agricultural Research*, Vol. 4, pp. 228–236.

Infectivity and Effectiveness of VAM on Sesame (*Sesameindicum*) and Sorgam (*Sorgambicolor*) Plants

S. Ramya*, K. Nithya and J. Philip Robinson
Department of Biotechnology, K.S. Rangasamy College of Technology, Tiruchengode
E-mail: rampri91@gmail.com

INTRODUCTION

Arbuscular Mycorrhiza *Fungi* is a type of mycorrhiza in which the fungus penetrates the cortical cells of the roots of a vascular plant. Vesicular Arbuscular Mycorrhiza (VAM) is a potential biofertilizer. Mycorrhizal fungi that grow into the root cortex of the host plant and penetrate root cells to form two kinds of specialized structures, arbuscules and vesicles. 95% of the world's presentspecies of vascular plants belong to families that are charcteristically mycorrhizal (Quilambo, 2000). Mycorrhizal Fungi is specifically designed to reduce transplant stress while improving soil hydration and fertility. It has been suggested that VAM plantsacclimatize more efficiently to water stress (Sweat, Davies, 1984). Mycorrhizal fungi serve as a secondary root system to 99% of the earth's plants. It extend themselves far out into the soil to extract nutrients and water for their host plant. For instance, mycorrhizal association is responsible for up to 80% of the total P uptake by plants (Marschner, Dell, 1994). One of the strongest effects of Arbuscular Mycorrhizal Fungi inoculation is an increase in the development of the host plant, which is attributed to an increase in nutrient uptake, particularly those that have low soil mobility and low concentration in the soil solution. Mycorrhizal fungai is fundamental part of the plant, as 95% of all species could not survive without it (Abbott, Robson, 1981). The VAM fungus,

*Corresponding author.

Glomus fasciculatum (Giguspora Species) is known to symbiotically associate with Plants and enhance the nutrient content in the plant. In the present investigation, the growth parameters, mycorrhizal dependency and chlorophyll content in Zeamays were analysed. Mycorrhizal and Non-Mycorrhizal Plants of sesamum and Vesicular Arbuscular Mycorrhiza (VAM) is the most abundant kind of mycorrhiza described as 'a universal plant symbiosis'. They are found in practically every taxonomic group of plants and the list of species not infected is probably far shorter than the infected ones. Lack of host specificity is even more characteristic of this symbiosis than other types known. Studies on VAM fungi conducted during last few decades envisaged their occurrence in a wide variety of hosts, different habitats and variability in quality and quantity.

Sesame Indicum

The major contributing countries (68%) are the world productions of Sesame are India, Sudan, Myanmar and China. Sesame seed is highly nutritive (50% oil and 25% protein) which are traditionally used for direct consumption and as a source of oil of excellent quality because of the presence of natural antioxidants such as sesamin and sesamol. Sesame seeds also contain a high amount of the anti-nutrientphytic acid. The increasing economic importance for food, oil and medicine the yield potential of sesame is not impressive due to its cultivation in sub marginal lands and unavailability of high yielding varieties and inbuilt resistance to biotic and abiotic stress. The hybridization procedures and cross breeding of this cultivar varieties was mostly unsuccessful due to the operation of high degree of crossability barriers. Production of high yielding variety with disease resistance is the present need for the sesame crop improvement. The family Pedaliaceae shows sensitivity to irradiation. The phenotyphical variations among the cultivars are the main source of resistance for these diseases.

Sorgam Bicolor

*Sorghum*is a genus of numerous species of grasses, one of which is raised for grain and many of which are used as fodder plants either cultivated or as part of pasture. The plants are cultivated in warmer climates worldwide. Species are native to tropical and subtropical regions of all continents in addition to the South West Pacific and Australasia. Sorghum is in the subfamily Panicoideae and the tribe Andropogoneae (the tribe of big bluestem and sugar cane). Sorghum has been, for centuries, one of the most important staple foods for millions of poor rural people in the semi-arid tropics of Asia and Africa. For some impoverished regions of the world, sorghum remains a principal source of energy, protein, vitamins and minerals. Sorghum grows even in harsh environments and it can be grown without application of any fertilizers or other inputs by a multitude of small-holder farmers in many countries. In the present study, the percentage of infection and various growth parameters such as the number of shoots, number of roots, leaf length and other parameters were observed and confimed with mycorrhizal and non-mycorrhizal infected plants. Biochemical analysis for the Mycorrhizal plants are analysed to study the effect of Mycorrhizae for the nutrient uptake for cultivated plants.

MATERIALS AND METHODS

The plants Sesame and Sorgam are grown in the green house garden of the Biotechnology Department of K.S. Rangasamy College of Technology, Tiruchengode. Plant samples randomly from a total of 10 patches, which were chosen to cover a range of shoot densities from 100–200 shoots m^{-2}. Plants were dug up with a spade carefully washed free from sediment using a sieve in situ. Plants with undamaged roots were transferred to plastic bags and transported to the laboratory, were they kept in cold storage until the next morning when the roots were examined for VAM. Ten plants were sampled randomly from each of the 10 patches.

Staining for Colanization

Staining of the roots or measure the amount of mycorrhizalcolonization, selected fibrous roots for staining. The roots were cleared in 10% KOH at 90ºC for 15 min. (Philips and Haymen, 1970). After the cleaning, the root samples were stained in 0.05% typanblue in lactoglycerol at 90°C for 5 min. from each of sesame and sorgam plants 1 cm segments was taken for the staining. All the segments were examined for the presence of fungal structures (e.g., Vesicles, arbuscles and hyphae) at 200–400x magnification using Phase contrast microscope (Carl Zeiss, Germany).

Preparation of Inoculum

The pure culture of *Glomus fasiculatum* was maintained and multiplied with Sorgum in shade house. After three months of growth, aerial parts were cut-off from the soil level and discarded. The root portion which are colonized with *G. fasiculatum* was cut down into small pieces and mixed into soil and shade dried.

Seeds of *Sesamum indicum* and Maize were collected and inoculated with VAM fungi. Mycorrizal Plants Growth parameters such as plant length, number of roots and number of leaves were observed and statistically analysed with non mycorrizalplants. Mycorrizal dependency (MD) was calculated as per the formula of plenchette *et al.* (1983).

Biochemical Analysis

Chlorophyll Test

Chlorophyll is a natural pigment found in green plants. It is the primary pigment that absorbs light energy from the sun for photosynthesis. Chlorophyll content was tested by using (Moran and Porath Method 1980). 0.5 gram (W) of fresh leaves were collected and smashed with 80% Acetone. Then 5 ml (V) filtrate was collected and chlorophyll content such as chlorophyll A, chlorophyll B, total chlorophyll and carotenoid pigments were estimated (optical density-OD) under UV spectrometer at various wavelength (480 nm, 647 nm, 666 nm).

Estimation of Chlorophyll

Chlorophyll A = 12.7 (OD 666 nm) – 2.79 (OD 647 nm)

Chlorophyll B = 20.7 (OD 647 nm) – 4.62 (OD 666 nm)

Total chlorophyll = 17.9 (OD 647 nm) + 8.08 (OD 666 nm)

Carotenoid -> Corrected A* = OD 480 nm + (0.114 × OD 666 nm) – (0.638 × OD647 nm)

Carotenoid = Corrected A* (1/100 × V/W)

RESULTS AND DISCUSSION

Effect of growth parameters VAM inoculated plants showed good results when compared with Non-Mycorrizal plants. Although a few studies have been conducted on tropical trees, the results are encouraging as the growth of seedlings and productivity was found to be enhanced. The use of VAM fungi in forestry appears to

Fig. 1: Sorgam Plant Growing in Normal and VAM Applied Soil

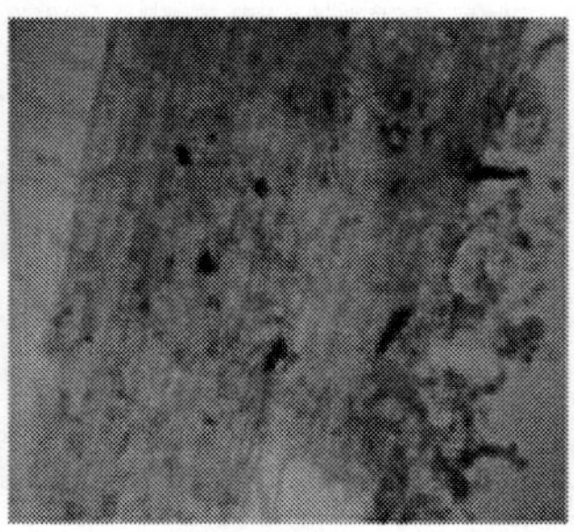

Fig. 2: The Presence of Vesicle Forming in Sesame Plant the Root

Fig. 3: The Presence of Vesicle Forming in Sorgam Plant the Root

be more important than in agriculture because in countries like India no large scale provisions exist to irrigate, fertilize and protect the plantation. The practical use of VAM fungi seems to be more appropriate as they are effective in overcoming the stress conditions like draught, disease incidences and deficiency of nutrients.

Table 1: Effect of Growth Parameters Sesamum Plant in the VAM Treated Soil

S. No.	*Full Length of the Plant*	*Number of Leaves*	*Number of Roots*
1.	4.4±0.35	2±0	1±0
2.	4.96±0.37	2±0	1.2±0.4
3.	4.62±0.23	2.4±0.8	1.2±0.4
4.	4.88±0.37	2.8±0.97	1.4±0.48
5.	4.6±0.38	3.2±0.97	1.6±0.8
6.	5.48±0.24	3.6±0.8	1.8±0.74
7.	7.64±0.36	4±0	2±0.63

Table 2: Effect of Growth Parameters Sesamum Plant in the in the Normal Soil

S. No.	*Full Length of the Plant*	*Number of Leaves*	*Number of Roots*
1.	5.18±0.58	2±0	1.2±0.4
2.	5.2±0.316	2.4±0.8	1.4±0.48
3.	4.82±0.39	2.4±0.8	1.6±0.8
4.	5.04±0.50	2.8±0.9	1.6±0.48
5.	5.3±0.34	2.8±0.97	1.8±0.74
6.	5.72±0.34	3.2±0.97	2±0.89
7.	6.3±0.26	3.6±0.8	2.2±0.74

Table 3: Effect of Growth Parameters of Sorgam Plant in the VAM Treated Soil

S. No.	*Full Length of The Plant*	*Number of Leaves*	*Number of Roots*
1.	15.8±0.70	2.4±0.48	3.4±1.01
2.	19.6±1.00	2.8±0.74	3.6±1.01
3.	26.1±0.66	3±0.63	4.2±1.16
4.	28.3±0.48	3.2±0.4	4±1.14
5.	33.06±0.71	3.2±0.74	4.2±1.46
6.	38.08±0.67	3±0.89	5±1.41
7.	40.36±0.46	3.4±0.48	5.6±1.01

Table 4: Effect of Growth Parameters of Sorgam Plant in the Normal Soil

S. No.	*Full Length of the Plant*	*Number of Leaves*	*Number of Roots*
1.	11.1±0.46	2±0.63	2.8±0.74
2.	21.8±0.37	2.2±0.74	2.6±0.8
3.	24.94±0.7	2.4±0.8	3.4±1.01
4.	23.36±0.58	2.8±0.4	3.6±1.01
5.	25.26±0.69	3±0.63	3.8±0.74
6.	27.97±0.51	3.2±0.74	3.8±1.16
7.	33.04±0.58	2.8±0.4	4.4±1.49

Table 5: Chlorophylltest

Wavelength	*Sesamum*		*Sorgam*	
	Control	*Treated with VAM*	*Control*	*Treated with VAM*
666 nm	0.713	1.240	0.573	0.704
647 nm	0.494	0.892	0.413	0.480
480 nm	1.695	2.947	1.649	1.874

Table 6: Chlorophyll Content

Content	*Sesamum*		*Maize*	
	With VAM	*Without VAM*	*With VAM*	*Without VAM*
Chlorophyll A	13.259	7.678	7.601	6.124
Chlorophyll B	12.735	6.931	6.683	5.694
Total chlorophyll	25.986	14.603	14.280	13.221
Carotenoid	0.352	0.204	0.187	0.201

In the present study VAM colonization was found in the root samples. The length and leaf surface area are considerably increased in the VAM infected plants. From the literature on the interactions between VAM-fungi and terrestrial plants, it is clear that VAM is mainly involved involved in facilitating nutrient uptake (Khan, 1975) although it has been shown that VAM is mainly involved in phosphorous uptake and that its role in uptake of nitrogen or other nutrients is less clear(Smith and Read, 1975).The cultivators of Maize and Sesamum require more amount nutrients in theirearly stage of development of plant system completely based on the development and performance of roots (Cheung *et al.,* 1987). The VAM fungal symbioses were proved to play a vital role in such stress conditions by supplying the nutrients to the host plant. The present study also established that the association of VAM fungi with Sesame as Maize enhanced the growth when compared with control.

REFERENCES

Abbott, L.K. and Robson, A.D. (1981). Infectivity and effictiveness of arbuscular mycorrhical fungi, effect of inoculam type. *Australian. Journal of Agricultural research*, 32: 631–639.

Asmah, A.E. (1995). Effect of phosphorus source and rate of applicationon VAM fungal infection and growth of maize (*Zea mays* L). *Mycorrhiza.* 5: 223–228.

Banik, R.L. (1987). Techniques of bamboo propagation with special reference to prerooted and prehizomed branch cuttings and tissue culture.

Barea, J.M. (1991). Vesicular arbuscular mycorrhiza as modifiers of soil fertility. *Advances in Soil Science*, 15: 1–40.

Chenug, Y.Y., Copper, S.G., Hansken,T.J. and Cheung, Y.C. (1987). Research on the raising of phyllostacys pubesecens seedling. In: Rao, A.N. Dhanarajan, G., Sastyr,C.B. (Eds.). *Proceedings of International Bamboo Workshop. Recent Research on Bamboos, Honghou*, Peoples Republic of China, Oct. 6–14, 1985. The Chinease Academy of Forestry, Peoples Republic of China and IRDC, Canada, 154–159.

Khan, A.G. (1975). Growth efforts of Vasicular-arbuscular mycorrhiza on crops in the field. In.: Sanders, F.E. Mosse, B., Tinker, P.B. (Eds.), *Ectomycorrhizas.* Academic Press London, 419–435.

Marschner and Dell (1994). Nutrient uptake in microbial symbiosis. *Plant and Soil* 159: 89–102.

Moran, P. and Porath, D. (1980). Chlorophyll determination in intact tissues using N', N-dimethylformaldehyde. *Plant Physiol,* 65: 478–479.

Philips, J.M. and Hayman, D.S. (1970). Improved procedures fro clearing roots and staning practecies and vesicular arbuscular mycorrhizal fungi for rapid assessment of infection. *Trans. Br. Mycol. Soc.,* 55: 158–161.

Plenchette, C., Fortin, J.A. and Furlan, V. (1983). Growth response of several plant species to mycorrhizae in a soil of moderate P-fertility. I. Mycorrhizal dependency under field conditions. *Plant Soil,* 70: 199–209.

Quilambo, O.A. (2000). Functioning of peanut (*Arachis hypogaea* L.) under nutrient deficiency and drought stress in relation to symbioticassociations, PhD thesis. University of Groningen, the Netherlands. Van Denderen B.V., Groningen, ISBN 903671284X.

Smith, S.E. and Read, D.J. (1997). Mycorrhizal Symbiosis, academic Press, London.

Sweat, M.R. and Davies, F.T. (1984). Mycorrhiza, water relations, growth and nutrient uptake of Geranium grown under moderately high phosphorus regimes. *J. Am, Soc. Hortic. Sci.,* 109: 210–213.

Cloning, Expression and Purification of Genetically Engineered Membrane Protein Bacteriorhodopsin (Br) in *E. coli*

J. John Peter* and P. Ponmurugan
Department of Biotechnology, K.S. Rangasamy College of Technology, Tiruchengode, Tamil Nadu
E-mail: john24385@gmail.com

INTRODUCTION

Bacteriorhodopsin (BR) is a transmembrane protein that acts as a light-driven proton pump in *Halobacterium salinarum*, converting light energy (500–650 nm) into a proton gradient (Krebs *et al.,* 1993). The ability of BR to convert light energy into chemical energy or sunlight into electricity has been used in different applications mainly optical appliances but also for therapeutic/medical applications and research. BR is the only protein constituent of the purple membrane (PM), a two-dimensional crystal lattice naturally present as part of the plasmic membrane of the bacterium. The cross section of BR with residues important for proton transfer is shown in Figure 1 (Ulrich Haupts *et al.,* 1999). BR contains seven alpha-helices that enclose an *all-trans* retinal chromophore linked via a protonated Schiff base to residue Lys216. Upon light absorption, retinal undergoes an isomerization process that results under normal conditions in the translocation of a proton from the cytoplasmic side to the extracellular side of the membrane.

Escherichia coli is one of the most common systems for heterologous expression of bacteriorhodopsin and its mutant forms (Khorana, 1993). Since formation of the native fold of BR is hampered in E. coli membranes (Karnik *et al.,* 1990; Kamo *et al.,* 2006), recombinant BR was produced in the form of apoprotein-bacterio-opsin

*Corresponding author.

(BO), and then subjected to reconstitution with retinal *in-vitro (Braiman et al., 1987).*The highest level of renatured BR expressed in this way was 17 mg/l (Shand *et al.,* 1991; Miercke *et al.,* 1991). Later a hybrid protein approach was developed, which favoured overexpression of BO in the form of inclusion bodies (Chen and Gouaux, 1996; Pompejus *et al.,* 1993) and thus, alleviated problems connected with target protein instability and toxicity to cells. Maltose-binding protein (MBP) and a large fragment of β-galactosidase (454 aa) were successfully used as N-terminal carrier proteins, and expression levels of about 200 mg/l were achieved for these protein fusions. However, significant losses during purification and reduced renaturation efficiency resulted in relatively low yields of purified and renatured BR (e.g. 6–10 mg/l of BR generated from MBP-BO).

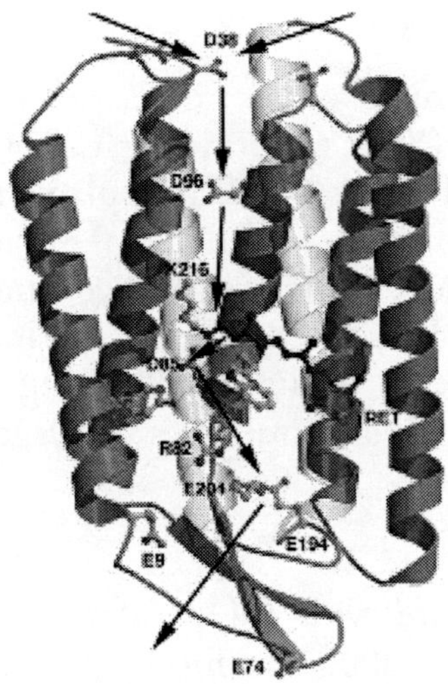

Fig. 1: Cross section of the structural model VI (PDB entry BRR) of bacteriorhodopsin with selected residues important for proton transfer steps shown in ball and stick representation. The possible path of protons is indicated by arrows.

The aim of this work was to achieve a significant level of BO expression in *E. coli* and to work out efficient procedure for BO purification.

MATERIALS AND METHODS

Plasmid Construction and Expression

The codon-optimized gene sequences of BO (GenBank: J02755) was chemically synthesized and inserted into pUC57 vector by GenScript USA Inc. and used for further cloning. BO gene was amplified from pUC57-BO by PCR using BO_F: 5'-TATA**CATATG**CAGGCCCAAATTACGGGTC-3' (NdeI restriction site is shown in bold) and BO_R: 5'-TATA**CTCGAGTCAGTGATGATGATGATGATG**CGA GGTCGCTGCC GCA-3' (XhoI restriction site, Stop codon and His-Tag are shown in bold) and inserted into the expression vector pET41a (+). Insert was confirmed by colony PCR followed by DNA sequencing (Applied Biosystems, 3100-Genetic Analyser).

To express BO from pET28-BO, transformed BL21(DE3) cells (Novagen) were plated on a Luria–Bertani (LB) agar medium (Tryptone: 10 g/L, Yeast extract: 5 g/L, NaCl: 10 g/L and bacterial agar: 15 g/L), containing 25 µg/ml kanamycin, then incubated for 14 hrs at 37°C. The colonies were inoculated into 3 ml of LB broth (tryptone: 10 g/L, yeast extract: 5 g/L and NaCl: 10 g/L), containing 25 µg/ml kanamycin then incubated for 8 hrs at 37°C with shaking at 180 rpm.

The next day, 1 mL of overnight culture of *E. coli* was added to 100 mL of fresh Terrific Broth (TB) medium (Yeast extract: 24 g/L, Tryptone: 12 g/L, K_2HPO_4: 12.5 g/L, KH_2PO_4: 2.3 g/L and Glycerol: 4 ml/L) supplemented with 40 µg/ml kanamycin and grown in 0.75 L flasks at 37°C to an OD_{600} of 0.8. Protein synthesis was induced by addition of 0.05 mM IPTG and cultivation was continued at 18°C for 20 h. Cellular lysates were analyzed by SDS-PAGE according to the method of Laemmli (1970) and detection of the target proteins by immunoblotting was carried out by standard methods using anti-His monoclonal antibodies (Novagen).

Purification of BO

Cells from 100 ml of culture with OD_{560} of about 6.0–8.0 optical units were harvested by centrifugation at 6000rpm for 10 min, suspended in 20 ml of Lysis buffer (50 mM Tris–HCl, pH 8.0, 0.1 M NaCl, 1 mM EDTA), incubated with 0.5 mg/ml of lysozyme and sonicated for 15 min on a Vibra-Cell VCX *500* (Sonics and Materials Inc., USA) at 50% output (20 s pulses with 20 s intervals). Suspension was centrifuged at 25,000 × g for 20 min, the pellet was dissolved in 5 ml of CelLytic B (Sigma) and then diluted into 25 ml of buffer A (50 mM Tris–HCl, pH 8.4, 0.15 M NaCl) containing 0.5% Sarkosyl and incubated for 1 h at ambient temperature with gentle shaking. Suspension was centrifuged as noted above, the obtained supernatant was adjusted to 10 mM concentration of imidazole and applied on a 5 ml column of Ni-NTA agarose (Qiagen) equilibrated with buffer A with 0.25% Sarkosyl and 10 mM imida-zole. The protein–resin complex was washed with a 25 ml of buffer A with 0.25% Sarkosyl and 25 ml of buffer B (0.1 M NaP, pH 8.0, 0.1 M NaCl, 0.2% SDS), 15 ml of buffer C (0.1 M NaP, pH 8.0, 0.2% SDS). The protein was eluted with 15 ml of Elution buffer (0.1 M NaP, pH 8.0, 0.2% SDS, 120 mM imidazole). Solution of bO (1 mg/ml) was dialyzed against 2 × 1 L of buffer D (0.1 M NaP, pH 6.0, 0.2% SDS) for 24 h.

Circular Dichroism Measurements

After dialysis, CD spectra of BO (8 μM) were recorded in buffer D at 25°C using a 0.5 mm path-length quartz cell with a detachable window (Hellma, Germany). Spectra were aquired by a Jasco J-810 spectropolarimeter (JASCO, Japan) in the 190–250 nm range (20 nm/min, 1 nm slit width).

RESULTS AND DISCUSSION

Expression and Purification of BO

Plasmid pET41a-BO coding for a C-terminal His-tagged BO was constructed and expressed in *E.coli* strain BL21 (DE3). High level of expression of the hybrid gene from plasmid pET41a-BO in BL21 (DE3) was obtained when cells were cultivated in TB at 18°C with 0.05 mM IPTG (Figure 2). Increasing the concentrations of inducer did not affect the level of biosynthesis of the hybrid, but significantly lowered the growth rate. To purify BO and to obtain a functionally active recombinant BR, we followed a common approach, which involves extraction of BO from insoluble fraction with organic solvents or detergents and sub-sequent transfer of the target protein to SDS solution (Braiman *et al.*, 1987; Miercke *et al.*, 1991; Pompejus *et al.*, 1993; Chen and Gouaux, 1996).

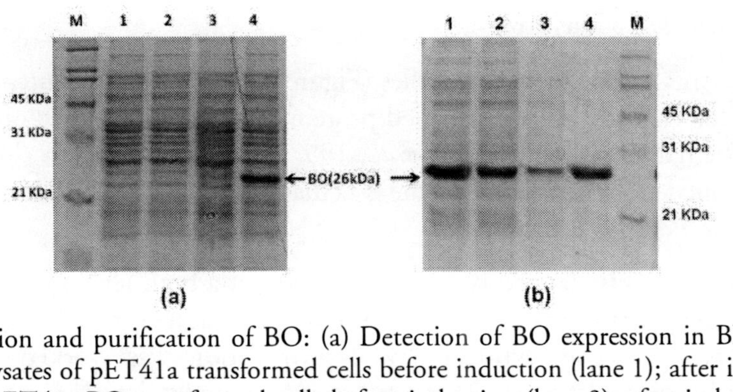

Fig. 2: Expression and purification of BO: (a) Detection of BO expression in BL21 (DE3) by SDS-PAGE. Lysates of pET41a transformed cells before induction (lane 1); after induction (lane 2). Lysates of pET41a-BO transformed cells before induction (lane 3); after induction (lane 4). (b) **Purification of BO**. Solubilised pellet fraction (lane 1), flow through (lane 2), washed (lane 3) and eluted BO (lane 4).

As shown in Figure 2b, the most part of BO protein was found in the pellet obtained by centrifugation of cellular lysate at 25,000 × g for 30 min, and a rather small fraction of the hybrid protein, which retained in the supernatant was further sedimented at 80,000 × g for 1 h. BO could be readily extracted from insoluble fraction with a number of detergents, e.g. SDS, Sarkosyl, Triton X-100 and CelLytic B reagent, less effective with DDM. We chose Sarcosyl for initial solubilisation of BO since this detergent extracts preferentially the cytoplasmic membrane proteins of Gram-negative bacteria (Filip *et al.*, 1973). Sarkosyl at 0.25% concentration is compatible with Ni-NTA purification system.

While His-tagged BO was bound to a nickel chelated resin, Sarcosyl was substituted with SDS by extensive washing of the column with buffer B. SDS (0.2%) was further used for protein elution and dialysis, since SDS is the detergent of choice to promote α-helical formation of BO and to maintain partially folded state of the protein (London and Khorana, 1982; Popot and Engelman, 1990; Booth, 2003). Solution of BO in SDS micelles is usually used as an intermediate in reconstitution experiments (Huang *et al.*, 1981; Miercke *et al.*, 1991; Chen and Gouaux, 1996; Pompejus *et al.*, 1993), and high α-helical content and monomeric state of BO at this stage are especially important for subsequent renaturation.

Properties of the obtained His-tagged BO were characterized by SDS-PAGE method. We show that BO was >95% pure and care was taken to prevent protein aggregation, which markedly increased with lowering the temperature to 5–10°C and/or when BO concentration was kept above 1 mg/ml. The protein at concentrations up to 1 mg/ml was stable for weeks at room temperature.

Circular Dichroism Measurements

CD measurements of BO in SDS micelles (Figure 3) showed high content of α-helices - 66%, suggesting that the affinity-purified protein retained about 90% of the secondary structure of the native protein (Booth *et al.*, 1997). CD spectrum was similar to that of BO/SDS obtained from purple membranes (Huang *et al.*, 1981; London and Khorana, 1982; Booth *et al.*, 1997).

The hybrid protein, His-tagged BO, was expressed at a high level in *E. coli* and it can be used for regeneration process for functionally active recombinant BR. The procedure for its cloning, expression and purification was worked out, which is distinguished by high effectivity of the key stages, namely, an efficient solubilization of a hybrid protein in detergents, high selectivity and high yield of BO. The final yield of recombinant BO produced in E. coli of about 60 mg/l.

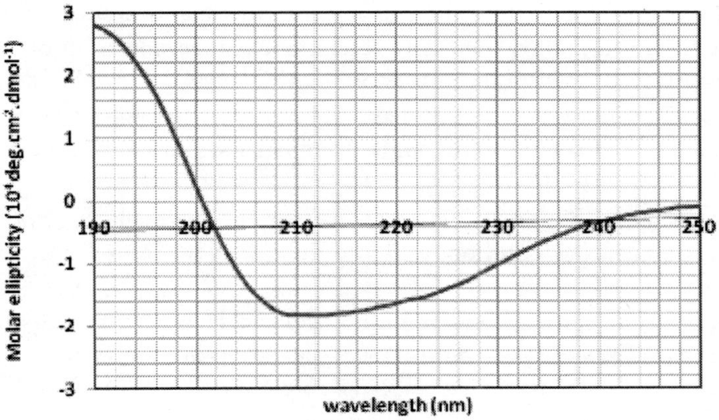

Fig. 3: Circular Dichroism Spectrum of BO in SDS Micelles

ACKNOWLEDGEMENTS

We acknowledge Dr. S. Mahalingam, Professor, Dept. of Biotechnology, IIT Madras, Chennai for allowing to use the facilities in need.

REFERENCES

Alexiev, U., Mollaaghabada, R.., Scherrer, P., Khorana, H.G. and Heyn, M.P. (1995). Rapid long range proton diffusion along the surface of the purple membrane and delayed proton transfer into the bulk. *Proc. Natl. Acad. Sci.* USA 92: 372–76.

Booth, P.J. (2003). The trials and tribulations of membrane protein folding in vitro. Biochim. *Biophys. Acta* 1610, 51–56, PMID: 12586379.

Booth, P.J., Farooq, A. and Flitsch, S.L. (1996). Retinal binding during folding and assembly of the membrane protein bacteriorhodopsin. *Biochemistry* 35: 5902–9.

Booth, P.J., Riley, M.L., Flitsch, S.L., Tem-pler, RH., Farooq, A. *et al.* (1997). Evidence that bilayer formation bending rigidity affects membrane protein folding. *Biochemistry,* 36: 197–203.

Booth, PJ. and Farooq, A. (1997). Intermediatesin the assembly of bacteriorhodopsin in vestigated by time-resolved absorptionspectroscopy. *Eur. J. Biochem*, 246: 674–80.

Booth, PJ., Flitsch, S.L., Stern, L.J., Green-halgh, D.A., Kim, P.S. and Khorana, H.G. (1995). Intermediates in the folding of the membrane protein bacteriorhodopsin. *Struct. Biol.* 2: 139–43.

Braiman, M.S., Stern, L.J., Chao, B.H. and Khorana, H.G. (1987). Structure–function studies on bacteriorhodopsin. IV. Purification and renaturation of bacterio-opsin polypeptide expressed in *Escherichia coli. J. Biol. Chem.* 262, 9271–9276, PMID: 3298254.

Brown, LS., Varo, G., Needleman, R. and Lanyi, J.K. (1995). Functional significance of aprotein conformation change at the cy-toplasmic end of helix F during the bacteriorhodopsin photocycle, *Biophys, J.* 69: 2103–11.

Chen, G.Q. and Gouaux, J.E. (1996). Overexpression of bacterio-opsin in *Escherichia coli* as a water-soluble fusion to maltose binding protein: efficient regeneration of the fusion protein and selective cleavage with trypsin. *Protein Sci.* 5: 456–4567, PMID: 8868482.

Filip, C., Fletcher, G., Wulff, J.L. and Earhart, C.F. (1973). Solubilization of the cytoplasmic membrane of Escherichia coli by the ionic detergent sodium-lauryl sarcosinate. *J. Bacteriol.* 115, 717–722, PMID: 4580564.

Hildebrandt, V., Ramezani Rad, M., Swida, U., Wrede, P., Grzesiek, S. *et al.* (1989). Genetic transfer of the pigment bacteriorhodopsin into the eukaryote Schiz-o Saccharomyces pombe. *FEBS Lett*, 243: 137–40.

Huang, K.S., Bayley, H., Liao, M.J., London, E. and Khorana, H.G. (1981). Refolding of an inte-gral membrane protein. Denaturation, renaturation, and reconstitution of intact bacteriorhodopsin and two proteolytic fragments. *J. Biol. Chem.* 256, 3802–3809, PMID: 7217055.

Keszthelyi, L. and Ormos, P. (1980). Electricsignals associated with the photocycle of bacteriorhodopsin. *FEBS Lett,* 109: 189–93.

Krebs, M.P. and Khorana, H.G. (1993). Mechanism of light-dependent proton translocation by bacteriorhodopsin, *J. Bacteriol.* 175, 1555–1560, PMID: 8383660.

Laemmli, U.K. (1970). Cleavage of structural proteins during the assembly of the head of bacteriophage T4, *Nature* 227, 680–685.

London, E. and Khorana, H.G. (1982). Denaturation and renaturation of bacteriorhodopsin in detergents and lipid-detergent mixtures. *J. Biol. Chem.* 257, 7003–7011, PMID: 7085614.

Miercke, L.J., Betlach, M.C., Mitra, A.K., Shand, R.F., Fong, S.K. and Stroud, R.M. (1991). Wild-type and mutant bacteriorhodopsins D85N, D96N, and R82Q: purification to homogeneity, pH dependence of pumping, and electron diffraction. *Biochemistry* 30, 3088–3098, PMID: 1848786.

Pompejus, M., Friedrich, K., Teufel, M. and Fritz, H.J. (1993). High-yield production of bacteriorhodopsin via expression of a synthetic gene in *Escherichia coli. Eur. J. Biochem.*211, 27–35, PMID: 7916683.

Popot, J.L. and Engelman, D.M. (1990). Membrane protein folding and oligomerization: the two-stage model. *Biochemistry* 29, 4031–4037, PMID: 1694455.

Sonar, S., Marti, T., Rath, P., Fischer, W., Coleman, M. *et al.* (1994). A redirected proton pathway in the bacteriorhodopsin mutant Tyr57! Asp, *J. Biol. Chem.*269:28851–58.

Isolation of PGPRS for the Management of Rhizome Rot Disease in Turmeric Plants

R. Ganesh Babu*, P. Ponmurugan, C. Citra[1] and N. Mathivanan[2]

Department of Biotechnology, K.S. Rangasamy College of Technology, Tiruchengode, Tamil Nadu

[1]Research Scholar, Biocontrol and Microbial Metabolites Lab, Center for CAS in Botany, University of Madras, Chennai, Tamil Nadu

[2]Biocontrol and Microbial Metabolites Lab, Center for CAS in Botany, University of Madras, Chennai, Tamil Nadu

E-mail: gb.babu88@gmail.com

INTRODUCTION

Turmeric (*Curcuma longa* L.) belongs to the family Zingiberaceae is an important commercial spice crop grown in India and it is part of Indian's culture. Indian turmeric is considered the best in the world (Alonso *et al.*, 2004). Turmeric is used as a dye in textile, cosmetics, preparation of medicinal oils, ointments and poultice. It is stomachic, carminative, tonic, blood purifier and an antiseptic. The aqueous extracts have biopesticidal and biofungicidal properties. Moreover, it has been reported that turmeric has antimicrobial, antioxidant and anticancerous properties. It grows in irrigated and rain fed conditions, black clayey loams and red soils having natural drainage. Crop cannot withstand water logging or alkalinity conditions. Turmeric thrives in the tropics and subtropics, where it requires a hot, moist climate, 20 to 30°C optimum temperature with 65–70% relative humidity. Adequate soil moisture is the most significant factor affecting rhizome yield (Reddy *et al.*, 2005). The important turmeric varieties grown in India are, 'Alleppey Finger' (Kerala), 'Erode and Salem turmeric' (Tamil Nadu), 'Raja pore' and 'Sangli turmeric' (Maharashtra)

*Corresponding author.

and 'Nizamabad Bulb' (Andhra Pradesh). In Tamil Nadu, the important varieties cultivated are Erode local, BSR-1 and 2, PTS-10, Roma, Suguna, Sudarsana and Salem local (Ponmurugan *et al.*, 2009).

Rhizome rot caused by a fungal pathogen, *Pythium graminicolum* f.sp. *aphanidermatum* is a serious problem in turmeric cultivation in Tamil Nadu especially Coimbatore, Dharmapuri, Erode, Namakkal, Salem and Villupuram districts where turmeric plants are under intensive cultivation. If the pathogen is infected the crop, the leaves turn yellow and the rhizome become soft, which lead to severe crop losses. Soil drenching of systemic fungicides is the common practice adapted by the farmers in this region for the management of rhizome rot. This practice leads to deleterious effects on beneficial microorganisms and also affects soil health which in turn to pollute the environment and also hazardous to animal and human health. In this scenario, biological control using plant growth promoting rhizobacteria (PGPRs) could be a viable alternative to chemical control for the management of rhizome rot disease of turmeric. The PGPRs especially those belong to the genera, *Bacillus* and *Pseudomonas* have been proved as efficient biocontrol agents in controlling various plant diseases besides, enhancing the plant growth.

The present study envisages identification of efficient strains of PGPRs for the control of rhizome rot disease and enhancing the growth metabolism of turmeric plants. The project is also aims to understand the mechanisms by which the PGPRs suppress turmeric rhizome rot pathogen. The bioformulations are to be developed in the project with PGPRs could be used for the control of rhizome rot disease, which importantly avoid/minimize the use of chemical fungicides so as to ensure the plant health and soil fertility. Biological control of turmeric rhizome rot will be popularized through demonstration trails with farmers' participatory approach.

MATERIALS AND METHODS

Soil Sample Analysis

Soil samples were collected from different agroclimatic zones of turmeric planting districts covering various varieties of turmeric plants in Tamil Nadu. The collected soil samples were subjected to analyze various soil edaphic parameters such as soil reaction as pH, Electrical conductivity, total organic carbon (Walkley and Black, 1934), total nitrogen (AOAC, 1990) and available phosphorous (Jackson, 1973). Moreover, the population density of PGPR consortia was enumerated in the soil samples using selective media like Kings'B medium by following serial dilution technique.

Survey of Rhizome Rot Disease in Turmeric Plants

In order to study the severity of rhizome rot disease in various planting districts of Tamil Nadu, an extensive survey was carried out. A scale ranging from 0–7 has been developed to find out the susceptibility of turmeric cultivars to infection. Disease severity was calculated using a formula, *i.e.*, (Sum of all numerical ratings)/(Total number of plants observed per field × highest grade in the scale) × 100. Per cent disease incidence (PDI) was calculated using the formula PDI = E (n × v)/(N × V) × 100; where, n – number of plants at each infection, v – grade for each group of plant, N – total number of plants observed and V – highest grade in the scale. A total of 3725 ha were covered during the survey in which 13 cultivars were covered to find out their susceptibility to *P. aphanidermatum* infection.

Interaction between Turmeric Rhizomes and PGPRs

Rhizome bits (50 × 25 mm size) of BSR-1, BSR-2, PTS-10, Roma, Suguna and Sudarsana cultivars were taken in different conical flasks containing moist cotton at the bottom and sterilized twice on alternate days. Spore suspensions of *P. aphanidermatum* and *P. fluorescence* (1.5×10^9 spores ml^{-1}) were prepared from PDA cultures and sprayed on the above rhizome bits under aseptic condition. The spore suspension of antagonist was sprayed before inoculating with the pathogen, simultaneously with the pathogen and after the establishment of the pathogen. Interaction was also studied by inoculating *P. fluorescence* on above rhizome bits of different cultivars on 3, 5, 7 and 11 days after the inoculation of the pathogen. The inoculated conical flasks were incubated at room temperature (25±2°C) and observed for the development of the pathogen.

Isolation of *P. aphanidermatum*

P. aphanidermatum was isolated from turmeric rhizomes which were infected with rot disease using *Pythium* selective medium amended with antibiotics like Pentachloronitrobenzene, Rifampicin and Ampicillin Sodium salted. The infected rhizomes were surface sterilized with Tween-20 for 2min, and then washed with sterile distilled water. The rhizomes were cut into small pieces and placed onto Potato Dextrose Agar plates. After incubation for a period of 2–3 days, the development of *P. aphanidermatum* mycelium on PDA plates was observed and purified by repeated sub culturing process. A total of 22 isolates were obtained from infected turmeric rhizomes and these were purified subsequently for dual culture assays.

Isolation of PGPRs from Rhizospheric Samples of Turmeric Plants

An attempt has been made to isolate different PGPRs consortia especially *P. fluorescence* and *Bacillus subtilis* using selective media from soil samples collected from different turmeric soils. The media such as Kings'B medium, *Bacillus thuringiensis* selective media, *Bacillus coagulans* selective media, *Bacillus medium, Bacillus racemilaticus* selective media, *Bacillus polymyxa* selective media, *Bacillus cereus* selective media, *Actinomycetes*-Starch Casein agar, *Streptomyces*-Starch casein KNO_3 agar, *Trichoderma* selective media, and *Streptomyces* selective media were used.

Pathogenicity Test (Koch's Postulates)

The earthen pots containing sterilized soil were inoculated with *P. aphanidermatum* by growing in PDB broth. The healthy rhizomes of turmeric were soaked in water and subsequently put up into the soil. These pots were maintained under greenhouse condition for a period of 2–3 months.

RESULTS

Analysis of Soil Samples

Total number of PGPRs population existing in turmeric soil samples indicated that the population level was found to be 6.7 × 10^3 cfu/gm soil dry weight in Erode and 4.8 × 10^3 cfu/gm soil dry weight in Coimbatore region. Total population was found to be more in Erode and lesser in Namakkal regions. The population density was positively correlated to soil nutrients like total organic carbon, total nitrogen and available phosphorous contents (Table 1).

Table 1: Population Density of PGPRs and Nutrient Status in Turmeric Soils of Tamil Nadu

Name of Planting District	*Population Density of PGPRs**	*Soil Reaction (pH)*	*Organic Carbon (%)*	*Total Nitrogen (ppm)*	*Available Phosphorous (ppm)*
Coimbatore	4.8	7.8	2.5	2734	14.7
Dharmapuri	2.5	6.9	1.8	1777	13.2
Erode	6.7	7.8	2.7	2844	16.8
Namakkal	1.7	6.8	0.8	1146	12.8
Salem	2.8	6.8	1.7	1245	13.0
Villupuram	3.4	7.2	1.6	1305	13.5
CD at P = 0.05	3.55	1.38	1.02	253.37	3.47

**Pseudomonas fluorescence* × 10^3 cfu/gm soil dry wt.

Survey of Rhizome Rot Disease in Turmeric Plants

Rhizome rot disease incidence was relatively high in Erode and Coimbatore followed by Dharmapuri districts. BSR-1, BSR-2, PTS-10, Alleppey Finger and Roma cultivars were found to be highly susceptible to the disease. On the other hand, the cultivars such as Suguna, Sudarsana and Raja pore were recorded as moderate susceptible while Erode local, Salem local, Sangli turmeric, Nizamabad, Bulb-11, Tr-8 and Tr-9 cultivars were highly susceptible to the disease (Table 2).

Table 2: Clonal Susceptibility of Turmeric Cultivars to *P. aphanidermatum* Infection at Different Agroclimatic Zones of Tamil Nadu

Planting Districts in Tamil Nadu	*Clonal Susceptibility*		
	High	*Moderate*	*Low*
Coimbatore (290 ha)	BSR-1	Suguna Sudarsana	Erode local
Dharmapuri (850 ha)	BSR-2	Raja pore	Salem local
Krishnagiri (930 ha)	PTS-10		Sangli turmeric
Erode (1020 ha)	Alleppey Finger		Nizamabad Bulb
Namakkal (180 ha)	Roma		Tr-8
Villupuram (455 ha)			Tr-9

Values in the Parentheses Indicate the Area Surveyed

The results indicated that pathogen failed to develop in treatments where antagonist was inoculated before and simultaneously with the pathogen. When antagonist was sprayed after the establishment of *P. aphanidermatum,* normal development of oospore took place. The spore masses were subsequently colonized by *P. fluorescence* (Table 3). The reproductive stages of the pathogen were arrested to varying levels when the

Table 3: Interaction of *P. fluorescence* with *P. aphanidermatum* on the Rhizomes of Turmeric Plants

Time of inoculation of P. Fluorescence	*Observations*
Post-inoculation*	Produced oospore profusely by *P. aphanidermatum P. fluorescence* completely covered pathogen mycelium including spore mass
Simultaneous**	Pathogen failed to develop
Pre-inoculation***	Pathogen failed to develop
Control (Pathogen alone)	Oospore exuded from the spore mass

* After the establishment of *P. aphanidermatum.*

** *P. aphanidermatum* and *P. fluorescence* inoculated at the same time.

*** Before the establishment of *P. aphanidermatum.*

rhizomes were inoculated with *P. fluorescence* at various time lag on the establishment of the pathogen. There was no significant difference among the different cultivars of rhizome used for the study. The formation of spore mass (zoosporangium) was completely arrested when *P. fluorescence* was inoculated on 3rd day after the inoculation of *P. aphanidermatum*. But when the inoculation was made on 5th, 7th, and 9th day, the normal development of spore mass took place (Table 4).

Table 4: Time of Application of *P. fluorescence* on the Development of *P. aphanidermatum* in Turmeric Rhizomes

*Days after Inoculation of P. Aphanidermatum**	*Observations*
3.	Rhizomes completely colonized by *P. fluorescence*. No oospore Formation by the pathogen.
5.	Oospores exuded from the spore masses by *P. aphanidermatum*. It was completely colonized with profuse sporulation by *P. fluorescence*.
7.	Oospores exuded from the spore masses by *P. aphanidermatum*. It was completely colonized with profuse sporulation by *P. fluorescence*.
9.	Oospores exuded from the spore masses by *P. aphanidermatum*. It was completely colonized with profuse sporulation by *P. fluorescence*.
11.	Normal oospores exuded by *P. aphanidermatum*. It was completely colonized with profuse sporulation of *P. fluorescence* (less in number).
Control**	Old spore masses without oospores exude (spores liberated).

* Time of *P. fluorescence* application.
** *P. aphanidermatum* alone.

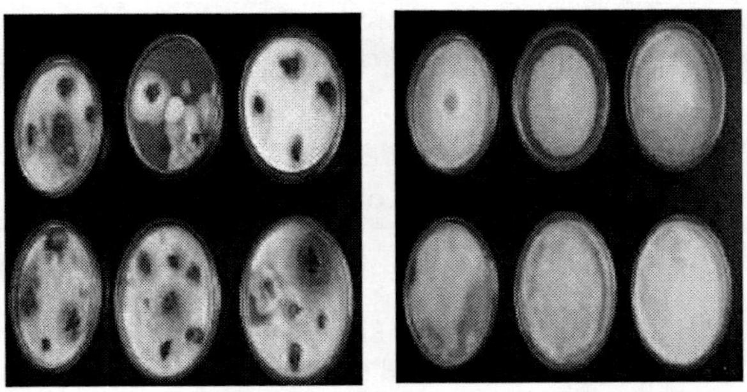

Fig. 1: *P. aphanidermatum* Isolated from the Infected Turmeric Rhizome

Isolation of *P. aphanidermatum*

The culture characteristics of all the isolates were almost same with minor variations in terms of growth pattern. The mycelium was hyaline, branched and septate and the hyphal width ranged between 0.66–1.13 μ. The colony colour was hyaline initially and changed to pale white. The colony margin was entire initially and became irregular. White concentric rings were developed on the colony, which gradually changed to blackish white. Formation of concentric rings is common in many fungi and is obviously due to diurnal periodicity of light and darkness. The size of zoospores were ranged between 6.2–7.4 × 2.3–2.8 μ.

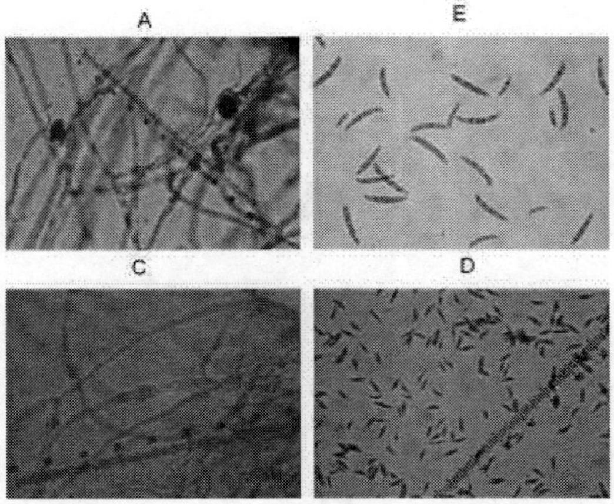

Fig. 2: Culture Characteristics of *P. aphanidermatum*

Isolation of PGPRs from Rhizospheric Samples of Turmeric Plants

A total of 1007 soil samples were collected from which around 642 strains were obtained and these were purified subsequently for further for studies (Table 5). Based on the screening test, there were 103 potential strains identified for further studies. These strains are being stored in 15% glycerol solution in cryo-vials for further use.

Table 5: Details of Sample Collection and Isolation of PGPRs

S. No.	*Name of the Turmeric Planting Districts*	*No. of Soil Samples Collected*	*No. of Strains Obtained*	*No. of Potential Strains*
1.	Erode	189	105	21
2.	Coimbatore	228	120	18
3.	Dharmapuri	157	113	25
4.	Namakkal	208	154	22
5.	Salem	225	150	17
6.	Villupuram	?	?	?
Total		1007	642	103

Screening of Efficient PGPRs

The efficient PGPRs were screened against *P. aphanidermatum* isolated from infected rhizome of turmeric plants by following dual culture techniques. Dual culture medium (1:1) was used for the growth of both bacteria and fungi for dual culture assays (Figure 3). The percentage of inhibition was recorded at every 24 hrs interval. The strain B146 showed a good efficiency (88.09%) in inhibiting the pathogen.

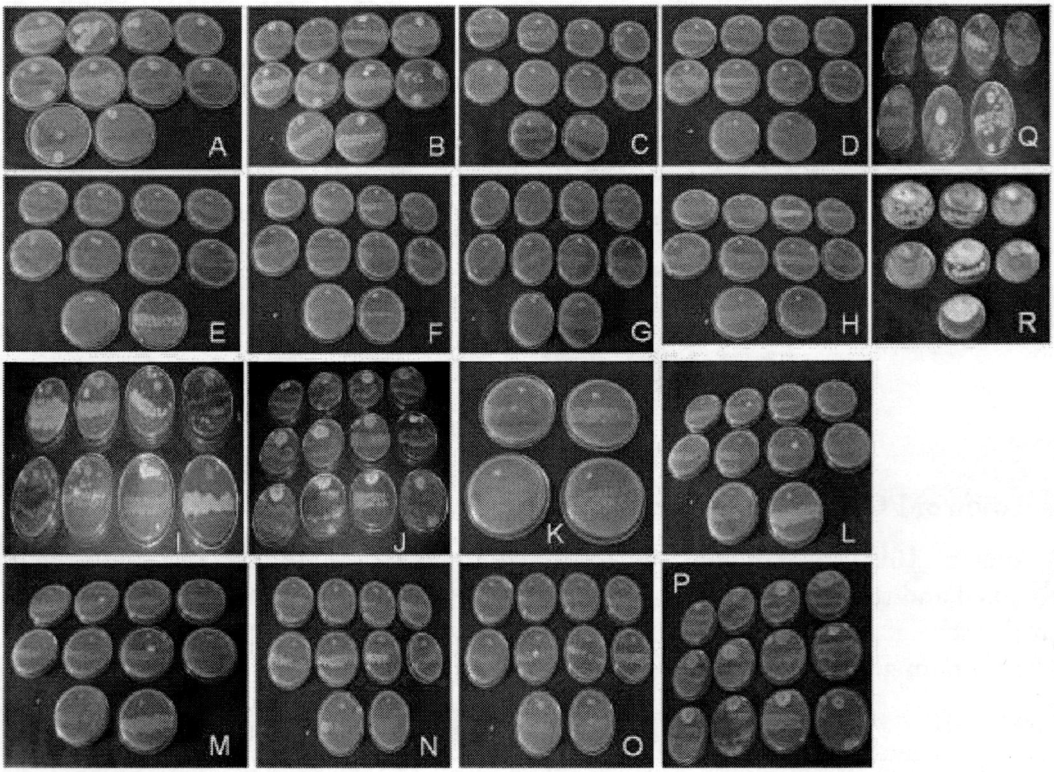

Fig. 3: Dual Culture Assay between *P. aphanidermatum* and *P. fluorescence* (Batch II)

Pathogenicity Test

The plants which were emerged from the rhizomes were observed for leaf curling, chlorosis and other symptoms. Rhizomes were periodically collected from the above pots for re-isolation of the pathogen. The results revealed that the plants were shown leaf curling and chlorosis with stunted growth due to *P. aphanidermatum* infection on 28th day on incubation (Figure 4).

Fig. 4: Study of Pathogenicity Test for *P. aphanidermatum*

DISCUSSION

It was positively correlated with soil nutrients like total organic carbon, total nitrogen and available phosphorous contents (Table 1). Similar observations were reported in tea soils by Baby *et al.* (2002) who observed a correlation between beneficial microorganisms and nutrients in the soil. Similarly, a study was undertaken by Ponmurugan *et al.* (2007) to correlate the population density of actinomycetes with soil nutrients for which soil samples were collected from different agroclimatic zones of southern Indian tea plantations for the production of bioactive secondary metabolites. It has been reported that most of the fungal and bacterial isolates tend to grow in the range of neutral to alkaline soil conditions. On the other hand, competitive saprophytic ability of *Trichoderma* spp. has been reported already by several investigators (Sharma *et al.*, 1999; Naseby *et al.*, 2000; Kim *et al.*, 2002; Dadrick *et al.*, 2003). Higher curcumin content in turmeric with organic amendments was reported by Rao *et al.*, 2005 which coincided with our results. Due to application of biocontrol agents, there was an increase in morphological, biochemical and physiological parameters in *Phomopsis* canker infected tea plants (Ponmurugan and Baby, 2003; Ponmurugan and Baby, 2007). The present study contributed well upon understanding the mechanisms by which the antagonist suppresses *P. aphanidermatum* which in turn improve the plant metabolism by inducing the systemic resistance in plants and also increase the quality of rhizomes. *Pythium graminicolum* has been reported as a cause of severe soft rot of ginger in the field. The fungus has been reported to be carried in seed-pieces. On Kauai this fungus has been isolated from ginger rhizomes showing soft rot by Park, 1935. The leek disc assay is a simple method of determining the Pythium isolates were inoculated with the soil and rhizome were sown in to mixed soils the symptoms were occurred in 28th day from the period of inoculation. Pathogenicity of *Fusarium* isolates recovered from leeks. The test is rapid with results being known within 14 days and *Fusarium* can be identified directly using conidia from sporulating discs rather than isolating from plant material. It is also a good method to produce *Fusarium* inoculum for further experimental work. Testing isolates using leek seedlings is a reliable method to determine Pathogenicity. Two non-pathogenic F. oxysporum isolates, which were morphologically di*ff*erent, significantly reduced the Fusarium root and stem rot incidence of cucumber in glasshouse trials. The number of non-pathogenic isolates, obtained in this study was rather low compared to the previous reports by Nel *et al.* (2006).

REFERENCES

Alonso, Jorge. Tratado de Fitofármacos Nutracéuticos. Barcelona: Corpus, 2004, pp. 395–403 (633.8 ALO).

AOAC (1990). *Official Methods of Analysis of the Association of Official Analytical Chemists.* Helrich, K. (Ed), 15th Edition, Vol. 1 and 2. AOAC Inc. Washington, USA.

Baby, U.I. and Chandramouli, B. (1996). Biological antagonism of *Trichoderma* and *Gliocladium* spp. Against certain primary root pathogens of tea. *Journal of Plantation Crops* 24, 249–255.

Dadrick, C., da Silva, F.G., Shen, Y. and Ronald, P. (2003). Antagonistic interactions between strains of *Xanthomonas oryzae* pv. *oryzae*. *Phytopathology* 93, 705–711.

Jackson, M.L. (1973). *Soil Chemical Analysis.* Prentice Hall of India Pvt. Ltd. New Delhi, pp. 498–516.

Kim, M.J., Choi, J.W., Park, S.M., Cha, B.J., Yang, M.S. and Kim, D.H. (2002). Characterization of a fungal protein kinase from *Cryphonectria Parasitica* and its transcriptional upregulation by hypovirus. *Molecular Microbiology* 45, 933–941.

Naseby, D.C., Pascual, J.A. and Lynch, J.M. (2000). Effect of biocontrol strains of *Trichoderma* on plant growth, *Pythium ultimum* populations, soil microbial communities and soil enzyme activities. *Journal of Applied Microbiology* 88, 161–169.

Nel, B., Steinberg, C., Labuschagne, N. and Viljoen, A. (2006). Isolation and characterization of non-pathogenic Fusarium oxysporum isolates from the rhizosphere of healthy banana plants. *Plant Pathology* 55, 207–216.

Park, M. (1935). Report on the work of the mycological division. Adm. Rept. Dir. Agr., ceylon, 1934, pp. D124-d131 (r*ev. Appl. Mycol.* 15: 136137, 1936).

Ponmurugan, P. and Baby, U.I. (2007). Evaluation of fungicides and biocontrol agents against Phomopsis canker of tea under field conditions. *Australasian Plant Pathology* 36, 68–72.

Ponmurugan, P., Baby, U.I. (2003). *In vitro* interaction of fungal antagonists with *Phomopsis theae*, the causal agent of collar canker disease in tea. *Indian Journal of Microbiology* 43, 41–44.

Ponmurugan, P., Ganesh Babu, R. and Balakrishnan, V. (2009). Evaluation of indigenous Trichoderma atroviride for the biological control of rhizome rot disease in turmeric plants. *J. Phytopath*, 157: In Press.

Reddy, M.N., Sridevi, N.V. and Charitha Devi, M. (2005). Changes in the nitrogen fractions and amino acid m etabolism of turmeric (*Curcuma longa* L.) roots infected with *Fusarium solani. Plant Pathology Bulletin* 14: 221–226.

Sharma, S.K., Verma, B.R. and Sharma, B.K. (1999). Biological control of *Sclerotinia sclerotiorum* causing stem rot of chickpea. *Indian Phytopathology* 52, 44–46.

Walkley, A. and Black, C.A. (1934). An examination of the Degtjareff method for determining soil organic matter and proposed modification of chromic valid titration method. *Soil Science* 37, 29–38.

Diversity and Distribution of Lichen Species at Eastern Ghats of Tamil Nadu

Ayyappadasan G.*, Ponmurugan P., S. Sanjeeva Nayaka[1], D.K. Upreti[1] and K. Karthikeyan

Department of Biotechnology, K.S. Rangasamy College of Technology, Tiruchengode, Tamil Nadu

[1]National Botanical Research Institute, Rana Pratap Marg, Lucknow, U.P.

E-mail: ayyappadasan07@gmail.com

INTRODUCTION

Lichens are fascinating symbiotic associations; actually they represent a biological partnership of fungi (mycobionts) and photoautotrophic algae (phycobionts). Lichens grow under diverse ecological conditions; they occur from arctic habitats to tropical rainforests where they colonize a broad variety of substrates, such as rock, bark, soil and even leaf/bark surfaces (Negi *et al.,* 2000). Lichen metabolites exert a wide variety of biological actions including antibiotic, antimycobacterial, antiviral, anti-inflammatory, analgesic, antipyretic, antiproliferative and cytotoxic effects. Even though these manifold activities of lichen metabolites have now been recognized, their therapeutic potential has not yet been fully explored and thus remains pharmaceutically unexploited (Müller, 2002). The utility of lichens is due of range of secondary compounds produced by them (Boustie and Grube, 2005). The lichens are sensitive to air pollution and microclimatic changes. Hence, widely considered as bioindicators and constitute an important component of biodiversity in a region (Loppi *et al.*, 1995; Myers *et al.*, 2000). Antibiotic properties of the lichens are of special interest to the scientists. According to one estimate, 50% of all lichens have antibiotic properties. Usnic acid isolated from lichens is having wide-spectrum of

*Corresponding author.

antibiotic properties. Usnic acid, inhibited the growth of gram positive bacteria such as *Staphylococcus aureus, Bacillus subtilis* and *Bacillus megaterium*, but the acids had no affect on the gram negative bacteria like *Escherichia coli* and *Pseudomonas aeruginosa*. On the other hand, usnic acid isolated from *Teloschistes chrysophthalmus* and *Ramalina celastri* demonstrated antiviral activity against the arena viruses (Rowe *et al.*, 1999; Turk *et al.*, 2003). Phenolic constituents from the lichen *Parmotrema stuppeum* including methyl orsenillate, orsenillic acid, atranorin and lecanoric acid showed moderate antioxidant activity (Jayapraksha and Rao, 2000).

Present study deals about the effect of antioxidant activity of lichens of different species for biomedical applications.

METHODOLOGY

Sample Collection and Maintenance

Lichen samples were collected from different attitudes covering various agroclimatic zones of Yercaud and Kolli hills belonging to Eastern Ghats of Tamil Nadu for the present study. Different living and non-living substrata were also obtained along with lichen thallus to study the nature of attachment and to check the moisture content, pH and buffering action. Corticolous lichens were peeled-off along with tree bark, but in case of saxicolous lichens collection, small pieces of the rock substrate were collected carefully. Collected lichen samples were deposited in Herbarium sheet without moisture content which would be used for further authentication and voucher specimen for the project.

Survey on Lichen Distribution

Lichens distribution upon different living (Plant bark) and Non-living substrata (brick wall, rock and soil) and subsequently analysed the bark moisture content and pH using ELICO pH meter (model-LI 120). Using the following scale, the percentage of distribution of lichens was calculated.

Grade	*Description*
1	Present in flower pot
2	Present in brick wall
3	Present in rock
4	Present in soil surface
5	Present in tree bark

Percentage of distribution (%) = $(n \times v)/N \times V \times 100$

Where,

n = number of lichens found
v = grade for each lichen group
N = total number of substrata observed
V = highest grade in the scale.

Identification of Lichen Specimens

Lichen identification was carried out by following the standard procedures from genus to species level according to the method of Culberson (1972) and Walker and James (1980). Lichen morphology, anatomy, lichen acid existence, growth forms, powdery appearance on the thallus and nature of fruiting bodies were critically analyzed to identify the lichen communities. Moreover, colour spot tests have been performed by using a few chemical reagents upon applying it on the thallus in which changing the colour of thallus medulla was recorded.

The following tests were performed:

1. *K test (Potassium)*: Prepared 10–25% aqueous solution of potassium hydroxide, applied onto cortex, medulla and part of the apothecium of lichen thallus.
2. *C test (Calcium hypochlorite)*: A freshly prepared aqueous solution of calcium hypochlorite (1:2 w/v) was applied on the thallus.
3. *KC test (Potassium and Calcium hypochlorite)*: At a particular spot of the thallus after removing surface particles potassium hydroxide was applied first and immediately followed by calcium hypochlorite.
4. *Pd test (Paraphenylenediamine)*: One gm of paraphenylenediamine and 10 gm of sodium sulphite were dissolved in 100 ml of distilled water along with 1 ml of liquid detergent (tween 20) which is applied on the thallus medulla.
5. *I test (Iodide)*: 2.5 gm of iodine and 0.5 gm of potassium iodide were dissolved in 100 ml of distilled water and subsequently applied on the lichen thallus.
6. *Other colour spot tests*: A mixture of nitric acid and aqueous solution of ferric chloride (0.5%) was applied on the lichen thallus to identify *Melanelia* and *Buellia* species.
7. *Chromatography test*: A small piece of lichens was taken in which two drops of acetone was added to extract the secondary metabolites and subsequently transferred to chromatographic plates (20 × 20 cm size) containing 30 gm of silica gel G paste to separate the compounds based on molecular weight and Rf values. Silica plates were sprayed by using 10% sulphuric acid gently and then it was kept on hot air oven at 110°C for 3 minutes. The number of spots were detected and compared with the standard for identification of specific secondary metabolites.

Preparation of Lichen Extract

Different groups of lichens (8 gm) were taken and subjected to get the extract powder using different solvent systems (petroleum ether, ethyl acetate, acetone, ethanol and water) using Soxhalet extractor system. Powdered extracts were stored at –20°C in deep freezer for further studies.

Antibacterial Activity of Lichen Extracts by Well Diffusion Method (Bauer and Kirby, 1966)

A lawn was made using standard strain of pathogenic microorganisms covering Gram's positive bacteria such as *Staphylococcus aureus* (MTCC 7443) and *Bacillus subtilis* (MTCC 2422) and Gram's negative bacteria namely *Escherichia coli* (MTCC 1698) and *Klebsiella pneumonia* (MTCC 9509), onto Muller-Hinton agar plates (Himedia, Mumbai) and incubated under controlled condition (37°C). Wells were made by a sterile cork borer (0.85 cm) and 200 µl of lichen extract was added to each well. The experiment was carried out in duplicates and the zone of inhibition was recorded by comparing chloromphenicol and amphicilin antibiotics as the standard.

Antioxidant Activity of Lichen Extracts by DPPH Radical Scavenging Assay

Antioxidant activity of lichen extracts was carried out by following the method of DPPH radical scavenging (Khalaf *et al.,* 2008), Ferric ion reducing/antioxidant power (FRAP) (Oyaizu, 1986) and hydrogen peroxide (Ruch *et al.,* 1989) assays.

RESULTS AND DISCUSSION

Lichen Collection and Analysis of Substratum

Lichen species were collected from different agroclimatic zones of Yercaud and Kolli hills of Tamil Nadu belonging to Eastern Ghats. Attachment of lichen species upon the substrates are being influenced by several factors like pH, nutrients and mineral contents in the available substrates. The results showed that Crustose, Foliose and Terricolous were found abundantly when compared to Fruticose type (Table 1). Existence of lichens on living substrata showed the maximum colonies about 45.7 were observed in palm tree and least in *Mangifera andamnica* about 17.3. It could be correlated with its pH (Table 2). The results indicated that lichens could survive well above pH 4.4 and below 5.8. The optimum pH was between in the ranges of 5.0 to 5.5. pH is also one of the essential factors responsible for existence of lichens on substrata. Existence of lichens on non-living substrata showed the maximum colonies observed in brick wall (36.7) followed by rock (31.3) and least in soil surface (13.0) (Figure 1).

Table 1: Survey of Lichen Flora at Yercaud and Kolli Hills

S. No.	*Family Name*	*Genus Name*	*Number of Species*
1.	Monoblastiaceae	*Anisomeridium*	1
2.	Ramalinaceae	*Bacidia*	1
		*Ramalina**	6
3.	Caliciaceae	*Buellia*	2
		*Pyxine**	2
		Diplotomma	1
		Dirinaria	2
4.	Parmeliaceae	*Bulbothrix*	1
		Canomaculina	1
		*Canoparmelia**	1
		Hypotrachyna	1
		Myelochora	1
		*Parmelinella**	1
		*Parmotrema**	7
		*Usnea**	2
5.	Candelariaceae	*Candelaria*	1
6.	Arthoniaceae	*Cryptothecia*	1
7.	Graphidaceae	*Diorygma*	1
		*Glyphis**	1
		*Graphis**	5
		Thecaria	1
8.	Thelotremataceae	*Diploschistes*	1
9.	Haematommataceae	*Haematomma*	1
10.	Physciaceae	*Heterodermia*	8
		Hyperphyscia	1
		Phaeophyscia	1
11.	Lecanoraceae	*Lecanora**	2
12.	Collemataceae	*Leptogium*	1
13.	Roccellaceae	*Opegrapha*	1
		Sigridea	1
14.	Pertusariaceae	*Pertusaria*	2
15.	Porinaceae	*Porina*	1
16.	Strigulaceae	*Strigula*	1
Total	16	33	61

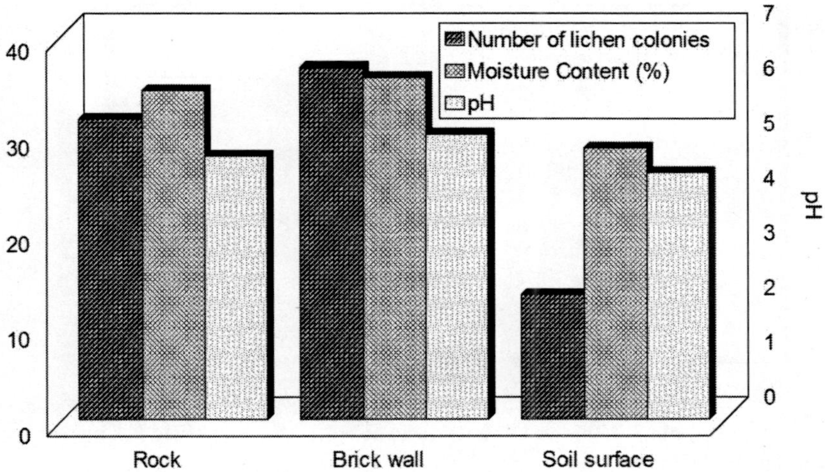

Fig. 1: Distribution of Lichens on Different Non-Living Substrata Irrespective of its pH and Moisture Content

Table 2: List of New Species Recorded to Tamil Nadu Lichen Flora at Different Altitudes

S. No.	*Species Name*	*Altitude (Meter above mean sea level)*
1.	*Bulborthrix isidiza*	800
2.	*Candelaria concolor*	1500
3.	*Diploschistes megalosporous*	1400
4.	*Diplotomma lauricassial*	400
5.	*Graphis intermediella*	1500
6.	*Haematomma puniceum*	1500
7.	*Heterodermia pellucid*	1500
8.	*Heterodermia togochi*	1500
9.	*Heterodermia adglutinata*	300
10.	*Leptogium austroamericanum*	1400
11.	*Myelochora entotheioctora*	1500
12.	*Opegrapha cinerea*	900
13.	*Ramalina pacifica*	1400
14.	*Usnea perplexans*	1500

* Lichen genus also present in Kolli hills.

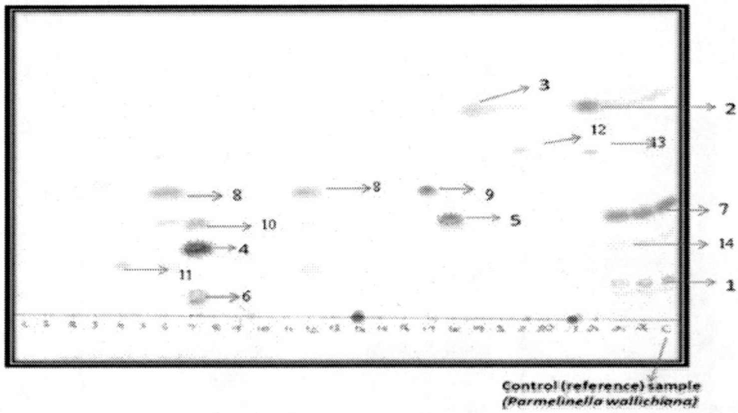

1. Salazinic acid 2. Atraonin 3. Usnic acid 4. Stictic acid 5. Protolichesterinic acid 6. Consalazinic acid 7. Norstictic acid 8–14. Unknown substances

Fig. 2: Identification of Lichen Species by Separating Organic Compounds Using Thin Layer Chromatography Method

The least population of lichens on soil surface might be due to poor adherence and uneven surface.

Lichen Flora at Yercaud and Kolli Hills

Lichen species were identified based on the morphology, anatomy and colouring tests including thin layer chromatography methods (Figure 3). The results indicated that there were about 61 different species covering 13 families with 33 genera of lichens. Interestingly, it has been recorded nearly about 50 and 22 new species of lichens at Yercaud and Kolli hills; respectively. Totally (Including previous results) Yercaud

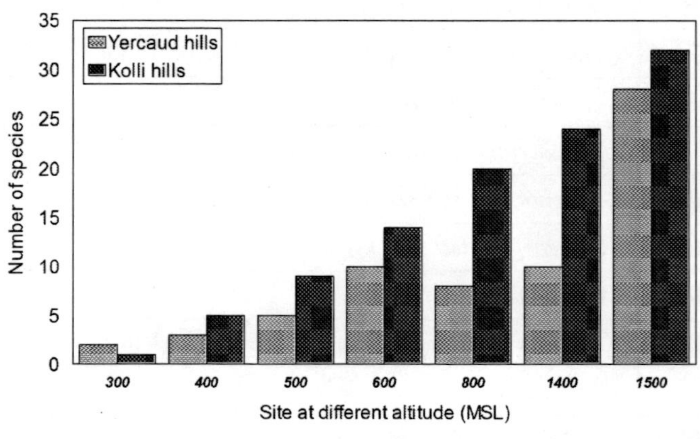

Fig. 3: Diversity of Lichen Species at Different Altitudes of Yercaud and Kolli Hills

Table 3: Total Antioxidant Capacity of Different Extracts Containing Lichen Compounds and its Rate of Inhibitory Concentration (IC50)

IC_{50} Values (μg/ml)								
Species	*Benzene*	*Isopropanol*	*Methanol*	*Water*	*Petroleum Ether*	*Ethanol*	*Ethyl Acetate*	*Acetone*
Parmotrema tinctorum	252	52.15	–	–	57.98	128.13	18.42	–
parmotrema hababianum	169.69	–	–	–	480	–	450.45	–
Parmotrema reticulatum	–	–	19.2	54.36	16.47	20.95	34.43	20.95

hills have 230 species of lichens which covering 70 genera and 35 families. Lichen species were identified and authenticated by Dr. D.K. Upreti, Head and Senior Scientist, Lichenology laboratory, CSIR-National Botanical Research Institute (NBRI), Lucknow, and also voucher specimen of identified lichen samples were deposited on NBRI Herbarium Centre and K.S. Rangasamy College of Technology Herbarium Collection Centre for future authentication purpose.

Lichen diversity was found to be high at higher altitude areas (1400–1500 MSL) compared to low altitude regions. Species diversity was increased with the increase to the height of the hills which may be due to pure atmospheric air, lack of air pollution, low temperature, low sunshine with high RH. Among all the species surveyed, a total of 14 new species were recorded in Tamil Nadu lichen flora diversity (Table 2). In Kolli hills, lichen diversity was gradually increased with increase in altitude similar to that

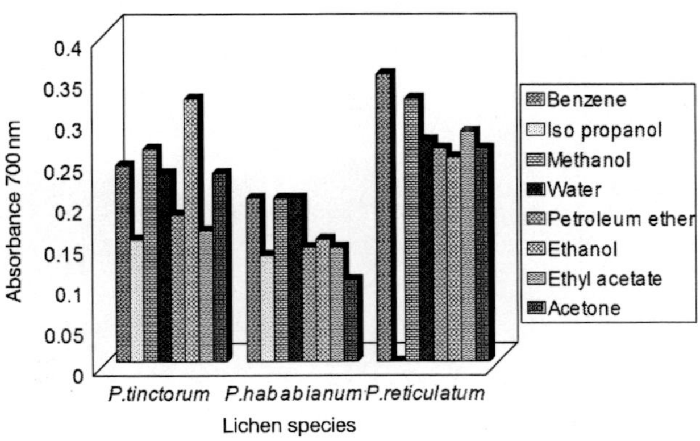

Fig. 4: Effects of Various Solvents on Ferric Acid Reducing Power for Different *Parmotrema* Species

of Yercaud hills. Lichen population was found to be high at 1500 M above mean sea level. Mostly it was noted as *Ramalinaceae* and *Parmeliaceae* families of lichens. Low altitude areas were mostly covered with *Caliciaceae* and *Graphidaceae* families (Figure 4).

REFERENCES

Boustie, J. and Grube, M. (2005). Lichens—a promising source of bioactive secondary metabolites. *Plant Genetic Resources* 3: 273–278.

Culberson, C.F. (1972). Improved conditions and new data for the identification of lichen products by a standardized thin-layer chromatographic method. *J. Chrom.* 72: 113–125.

Jayapraksha, G.K. andd Rao, L.J. (2000). Phenolic constituents from the lichen *Parmotrema stuppeum* (Nyl.) Hale and their antioxidant activity. *Zeitschrift für Naturforschung* 55 (11–12): 1018–22.

Loppi, S., Putorti, E., Dominics, V. and Barbaro, A. (1995). Lichens as bioindicators of air quality near a municipal solid waste incinerator (Central Italy). *Allionia* 33: 121–129.

Müller, K. (2002). Pharmaceutically relevant metabolites from lichens. *Appl. Microbiol. Biotechnol.* 56 (1/2): 9–16.

Myers, N., Mittermeier, R.A., Mittermeier, C.G., da Fonseca, G.A. and Kents, J. (2000). Biodiversity hotspots for conservation priorities. *Nature* 403: 853–858.

Negi, H.R. and Upreti, D.K. (2000). Species diversity and relative abundance of lichens in Rambak catchement of Hemi National Park in Ladakh. *Curr. Sci.* 78: 1105–1111

Turk, A.O., Meral, Y., Merish, K. and Turk, H. (2003). The antimicrobial activity of extracts of the lichen *Cetraria aculeata* and its protolichesterinic acid constituent. *Zeitschrift für Naturforschung* 58: 850–854.

Molecular Characterization of Enterotoxigenic *Bacillus cereus* in Tropical Fisheries Environment of Cochin, India

Sanjoy Das* **and K.V. Lalitha**

Central Institute of Fisheries Technology, Matsyapuri, Cochin, Kerala

E-mail: sanjoy125@yahoo.co.in

INTRODUCTION

Bacillus cereus is a Gram positive spore-forming organism, which causes two different types food poisoning viz. diarrhoeal type and emetic type. Apart from these, *B. cereus* has also been reported to be associated with opthalmitis, respiratory tract infection and central nervous system disorder (Bekemeyer and Zimmerman, 1985; Beer *et al.*, 1990; Barrie *et al.*, 1992). In *B. cereus, hbla, bceT* and *entFM* are the important genes, which are responsible for diarrhoeal enterotoxin production (Heinrichs *et al.*, 1993; Agata *et al.*, 1995; Asano *et al.*, 1997). Being a spore-former organism, it can survive heat-treated foods. Some strains of *B. cereus* are psychrotrophic in nature and thus can grow even in refrigerated food items (TeGiffel *et al.*, 1997). This organism is also capable of producing biofilm on different food-contact surfaces (Oosthuizen *et al.*, 2002) and the possibility of transmission from those surfaces to different food products can't be ruled out. According to Centers for Disease Control, United States of America, approximately 2% of the total food-borne illnesses in USA are due to enterotoxigenic *B. cereus* (Centers for Disease Control and Prevention, 1994). In European countries, food poisoning caused by enterotoxigenic *B. cereus* is very common and attributes to 47, 33 and 22% of the total food poisoning cases reported in Iceland, Norway and Finland, respectively (Granum and Lund, 1997). Although in India, no major outbreak due to this

*Corresponding author.

pathogen has been reported so far, enterotoxigenic *B. cereus* has been frequently isolated from different types of food items (Kamat *et al.*, 1989; Agarwal *et al.*, 1997; Das *et al.*, 2009). In one study carried out among diarrhoea patients in and around Kolkata, India, *B. cereus* was isolated from 3.5% of the stool samples of the patients (Banerjee *et al.*, 2011). As per the tropical seafood is concerned, it was observed that 24.91% of the samples collected from Cochin and surrounding area were positive to enterotoxigenic *B. cereus* (Das *et al.*, 2009). So, this is very much required to study the occurrence of this pathogen in the tropical fisheries environment. Keeping this in view, the present study was designed to assess the occurrence and molecular characterization of *B. cereus* from tropical fisheries environmental samples of Cochin and surrounding areas.

MATERIALS AND METHODS

Samples and Isolation of the Organism

A total of 51 samples collected during 2009–10 from fish markets and landing centres of Cochin and surrounding areas comprising of 30 ice, 18 mud and sand and 3 sea water were screened for the presence of *B. cereus. S*elective plating technique on Polymixin-pyruvate-egg yolk-mannitol-bromocresol purple agar (PEMPA) was done for Isolation of *B. cereus* (Szabo *et al.*, 1984). Twenty five grams of sample was blended in 225 ml of sterile normal saline in a stomacher blender (Seward Ltd, U.K.) and serial 10 fold dilutions were made on normal saline. From each dilution, 0.5 ml was spread-plated on PEMPA plates. After incubating at 37°C for 24 h, suspected *B. cereus* colonies showing lecithinase reaction were subcultured and subjected to identification using Gram's staining and standard biochemical reactions as mentioned in Bacteriological analytical manual of United States Food and Drug Administration (Rhodehamel and Harmon, 1998). The starch hydrolytic ability of the organisms was checked on nutrient agar containing 1% starch.

Screening of *B. cereus* Isolates for Haemolytic Activity and Production of Enterotoxin

The haemolytic activity of the suspected isolates was checked on spot inoculation on blood agar viz. Columbia agar base with 5% human blood. The blood agar plates were incubated at 37°C for 24 h. The ability to produce diarrhoeal enterotoxin was detected by Reverse passive latex agglutination (RPLA) test using BCET-RPLA kit (Oxoid) as per manufacturer's instruction.

Preparation of Crude PCR Template

Crude PCR template from the isolates was prepared by boiling method. Overnight grown culture of the isolate was centrifuged at 7000 × g for 10 minutes at 4°C. The pellet was washed once with normal saline and suspended in autoclaved MilliQ distilled water. The suspended bacterial cells were lysed on boiling water bath for 10 min. Then it was immediately frozen at –20°C. After freezing, the lysate was thawed and centrifuged at 4000 × g to pellet the debris. 5 µl of the supernatant was used as template in the PCR reaction.

Polymerase Chain Reaction

Total 4 sets of primers were used for this study. For *B. cereus* group specific PCR, BalF and BalR primers were used. All the isolates were screened for the presence of three virulent genes viz. *hbla*, *entFM* and *bceT* genes, which are generally responsible for enterotoxin production. The primer pairs HblA1/HblA2, BceT1/BceT2 and ENTA/ENTB were used to detect the presence of *hbla*, *bceT* and *entFM* genes, respectively. The details of primers including sequences have been depicted in Table 1. PCRs were carried out in 25 µl of reaction mixture, which consisted of 5 µl of template, 10 mM Tris-HCl (pH 9), 50 mM KCl, 1.5 mM $MgCl_2$, 200 µM of each dNTP (Fermentas), 1 U *Taq* DNA polymerase (Fermentas) and 10 p mol of respective primers. For each PCR assay at the beginning of the reaction, the initial

Table 1: PCR Primers and Cycling Condition Details

Primer	*Sequence*	*Target and Reference*	*PCR Cycling Condition*
BalF BalR	5'- TGCAACTGTATTAGCACAAGC T -3' 5'- TACCACGAAGTTTGTTCACTACT -3'	*Bacillus cereus* Group (Chang *et al.*, 2003)	94°C for 45 seconds (denaturation), 55°C for 45 seconds (annealing) and 72°C for 45 seconds (extension).
HblA1 HblA2	5'- CTAATGTAGTTTCACCTGTAGCAAC- 3' 5'- AATCATGCCACTGCGTGGACATATAA- 3'	*hbla* gene (Mantynen and Lindstrom, 1998)	94°C for 30 seconds, 58°C for 45 seconds and 72°C for 1 minute.
ENTA ENTB	5'- ATGAAAAAAGTAATTTGCAGG- 3' 5'- TTAGTATGCTTTTGTGTAACC- 3'	*entFM* gene (Asano *et al.*, 1997)	94°C for 45 seconds, 52°C for 45 seconds and 72°C for 1 minute.
BceT1 BceT2	5'- GAATTCCTAAACTTGCACCATCTC G-3' 5'- CTGCGTAATCGTGAATGTAGTCAAT- 3'	*bceT* gene (Mantynen and Lindstrom, 1998)	94°C for 45 seconds, 55/65°C for 45 seconds and 72°C for 45 second

denaturation was done at 95°C for 5 min and at the end of the cycle, the final extension was carried out at 72°C for 5 min. The other details of PCR cycling condition for each PCR have been mentioned in Table 1. *bceT* gene specific PCR was run at two different annealing temperatures viz. 55 and 65°C. Following PCR, the amplified products were resolved in 1.5% agarose gel prepared in 0.5 × TBE buffer (45 mM Tris, 45 mM boric acid and 2 mM EDTA, pH 8.0) containing ethidium bromide (0.3 μg/ml). Electrophoresis was done at 7-volts/cm for 1 hour and the gel was photographed by an UV gel documentation system (Alpha Innotech Corporation, USA).

RESULTS AND DISCUSSION

Occurrence of *B. cereus* and Enterotoxin Production

Among the 51 samples screened, the presence of *B. cereus* was detected in 16 samples (4 ice and 12 mud and sand). Total 41 isolates have been recovered from the positive samples. All the isolates were haemolytic on blood agar and 38 isolates showed starch hydrolytic property. By Reverse passive latex agglutination (RPLA) test, 36 isolates were found to produce diarrhoeal enterotoxin. Those 36 isolates were sourced from 4 ice and 11 mud and sand samples. All the enterotoxin producing isolates were found to hydrolyze starch (Table 2).

Detection by *B. cereus* Group Specific PCR

In PCR using *B. cereus* group specific primers BalF/BalR, all the 41 isolates yielded 533 bp amplified product. Standard strains of *Bacillus subtilis* NCIM 2545 and *Staphylococcus aureus* NCIM 2079 were also employed in this PCR as negative controls, but none of them yielded any amplified product in this PCR.

Screening for Virulence Genes

The isolates were screened for the presence of three enterotoxin producing genes viz. *hbla*, *entFM* and *bceT* genes. *hbla* gene specific PCR using primers HblA1/HblA2 yielded 834 bp amplified product in case of 36 enterotoxigenic isolates. The isolates, which were negative to enterotoxin production by RPLA method, didn't yield any amplified product in this PCR. In case of *entFM* gene specific PCR using ENTA/ENTB primers, 1300 bp amplified product was observed in case of 35 isolates. On the other hand, *bceT* gene specific PCR didn't yield any amplified product in any of the isolates tested. The results of each PCR, enterotoxin production by RPLA test and starch hydrolysis test for each isolate has been depicted in Table 2.

Table 2: Details of *B. cereus* Isolates

Isolates	*Starch Hydrolysis*	*RPLA**	*BCG***	*Hemolysis*	*hbla PCR*	*entFM PCR*	*bceT PCR*
BC-env-1	+	+	+	+	+	+	–
BC-env-2	+	+	+	+	+	+	-
BC-env-3	+	+	+	+	+	+	–
BC-env-4	+	+	+	+	+	–	–
BC-env-5	+	+	+	+	+	+	–
BC-env-6	+	+	+	+	+	+	–
BC-env-7	+	+	+	+	+	+	–
BC-env-8	+	+	+	+	+	+	–
BC-env-9	+	+	+	+	+	+	–
BC-env-10	+	–	+	+	–	+	–
BC-env-11	+	+	+	+	+	+	–
BC-env-12	+	+	+	+	+	+	–
BC-env-13	+	+	+	+	+	+	–
BC-env-14	+	+	+	+	+	+	–
BC-env-15	+	+	+	+	+	+	–
BC-env-16	+	+	+	+	+	+	–
BC-env-17	–	–	+	+	–	–	–
BC-env-18	+	+	+	+	+	+	–
BC-env-19	+	+	+	+	+	+	–
BC-env-20	+	+	+	+	+	+	–
BC-env-21	+	+	+	+	+	+	–
BC-env-22	+	+	+	+	+	+	–
BC-env-23	+	+	+	+	+	+	–
BC-env-24	+	+	+	+	+	+	–
BC-env-25	+	+	+	+	+	+	–
BC-env-26	+	+	+	+	+	–	–
BC-env-27	+	+	+	+	+	–	–
BC-env-28	+	+	+	+	+	+	–
BC-env-29	+	+	+	+	+	+	–
BC-env-30	–	–	+	+	–	+	–
BC-env-31	–	–	+	+	–	+	–
BC-env-32	+	+	+	+	+	+	–
BC-env-33	+	+	+	+	+	+	–
BC-env-34	+	+	+	+	+	+	–
BC-env-35	+	+	+	+	+	+	–
BC-env-36	+	+	+	+	+	+	–
BC-env-37	+	+	+	+	+	+	–
BC-env-38	+	+	+	+	+	+	–
BC-env-39	+	+	+	+	+	+	–
BC-env-40	+	–	+	+	–	–	–
Bc-env-41	+	+	+	+	+	–	–

* BCG = Bacillus cereus group specific PCR

** RPLA = Enterotoxin production by RPLA test

The aim of the this study was to study the occurrence of enterotoxigenic *B. cereus* in fishery environmental samples collected from fish markets and landing centers and also to study the presence of different virulence genes in the isolates of *B. cereus.* In this study, 31.37% (16 out of 51) of the samples were found to possess *B. cereus* and enterotoxigenic *B. cereus* was isolated from 15 samples (Occurrence rate 29.41%). The occurrence of enterotoxigenic *B. cereus* was found 13.33% (4 out of 30) and 61.11% (11 out of 18) in case of ice and mud/sand, respectively. On other hand, all the three sea water samples were found negative for this organism. The comparatively high occurrence was found in case of mud and sand collected from the fishery environment. In previous study, it was observed that 29.41% of fishes of tropical area of Cochin possessed enterotoxigenic *B. cereus* (Das *et al.*, 2009). As per this study, it can be stated that occurrence of enterotoxigenic *B. cereus* is moderately high in tropical fisheries environment and the fisheries environment may be possible source of transmission of *B. cereus* to fish.

On screening for virulence genes, it was observed that all the enterotoxigenic isolates possessed *hbla* gene while this gene was absent in all non-enterotoxigenic isolates (Table 2). Similar type of finding was reported by Das *et al.* (2009) in case of *B. cereus* isolates from fish. *entFM* gene was detected in three non-enterotoxigenic isolates, while this gene was not detected in four enterotoxigenic isolates (Table 2). So, it can be stated that no relationship can be established between presence of *entFM* genes and enterotoxin production. *bceT* gene was not detected in any of the isolates. So, this gene can be considered rare in case of tropical fishery environmental isolates of Cochin. So based on this study, it can be concluded that *hbla* gene can be considered as marker for diarrhoeal enterotoxin production in case of *B. cereus* isolates from tropical fishery environment.

ACKNOWLEDGEMENTS

The authors are thankful to the Director, Central Institute of Fisheries Technology for providing necessary facilities for conducting this research work.

REFERENCES

Agarwal, R.K., Kapoor, K.N., Bachhil, V.N., Singh, B.R., Kumar, A. and Bhilegaonkar, K.N. (1997). Studies of toxigenicity of *Bacillus cereus* strains isolated from food. *Ind J Comp Microbiol Immunol Infect Dis,* 18: 162–165.

Agata, N., Ohta, M., Arakawa, Y. and Mori, M. (1995). The *bceT* gene of *Bacillus cereus* encodes an enterotoxic protein. *Microbiology,* 141: 983–988.

Asano, S.I., Nukumizu, Y., Bando, H., Hzuka, T. and Yamamoto, T. (1997). Cloning of novel enterotoxin gene of *Bacillus cereus* and *Bacillus thuringiensis*. *Appl Environ Microbiol,* 63: 1054–1057.

Banerjee, M., Nair, G.B. and Ramamurthy, T. (2011). Phenotypic and genotypic characterization of *Bacillus cereus* isolated from acute diarrheal patients. *Ind J Med Res,* 133: 88–95.

Barrie, D., Wilson, J.A., Hoffman, P.N. and Kramer, J.M. (1992). *Bacillus cereus* meningitis in two neurosurgical patients: an investigation into the source of the organism. *J Infect,* 25: 291–297.

Beer, P.M., Ludwig, I.H. and Packer, A.J. (1990). Complete visual recovery after *Bacillus cereus* endophthalmitis in a child. *Am J Ophthalmol,* 110: 212–213.

Bekemeyer, W.B. and Zimmerman, G.A. (1985). Life-threatening complications associated with *Bacillus cereus* pneumonia. *Am Rev Respir Dis,* 131: 466–469.

Centers for Disease Control and Prevetion (1994). *Bacillus cereus* food poisoning associated with fried rice at two day care centers-Virginia, 1993. *Morbidity and Mortality weekly report,* 43: 177–178.

Das, S., Surendran, P.K. and Thampuran, N. (2009). PCR-based identification of enterotoxigenic isolates of *Bacillus cereus* from tropical seafood. *Ind J Med Res,* 129: 316–320.

Granum, P.E. and Lund, T. (1997). *Bacillus cereus* and its food poisoning toxins. *FEMS Microbiol Lett,* 157: 223–228.

Heinrichs, J.H., Beecher, D.J., Macmillan, J.D. and Zilinskas, B.A. (1993). Molecular cloning and characterization of *hbla* gene encoding the B component of hemolysin BL from *Bacillus cereus*. *J Bacteriol,* 175: 6760–6766.

Kamat, A.S, Nerkar, D.P. and Nair, P.M. (1989). *Bacillus cereus* in some Indian foods: incidence and antibiotic, heat and radiation resistance. *J Food Safety,* 10: 31–41.

Mantynen, V. and Lindstrom, K. (1998). A rapid PCR-based DNA lest for enterotoxigenic *Bacillus cereus*. *Appl Environ Microbiol,* 64: 1634–1631.

Oosthuizen, M.C., Steyn, B., Theron, J., Cosette, P., Lindsay, D., Holy, A.V. and Brozel, V.S. (2002). Proteomic analysis reveals differential protein expression by *Bacillus cereus* during biofilm formation. *Appl Environ Microbiol,* 68: 2770–2780.

Rhodehamel, E.J. and Harmon, S.M. (1998). *Bacillus cereus*. In Bacteriological Analytical Manual, 8th Ed., Revision A, Chap. 14, U.S. Food and Drug Administration.

Szabo, R.A., Todd, E.C.D. and Rayman, M. (1984). Twenty-four hour isolation and confirmation of *Bacillus cereus* in foods. *J Food Prot,* 47: 856–862.

Te Giffel, M.C., Beumer, R.R., Granum, P.E. and Rombouts, F.M. (1997). Isolation and characterization of *Bacillus cereus* from pasteurized milk in household refrigerators in Netherlands. *Int J Food Microbiol,* 34: 307–318.

Phenazine Antibiotic Producing *Pseudomonas* Spp. VSMKU 4047 against Sheath Blight of Rice

V. Shanmugaiah*, H. Harikrishnan and K. Nithya

Department of Microbial Technology, School of Biological Sciences, Madurai Kamaraj University, Madurai, Tamil Nadu

E-mail: vshanmugaiah@yahoo.co.in

INTRODUCTION

Sheath blight disease caused by *Rhizoctonia solani* (Kuhn) was considered as important production constraints of rice in South and South East Asia and other parts of world (Ou, 1985). Effective fungicides are existing to deal with these but not considered as long term solution because of concerns about exposure risks, health and environmental hazards, residue persistence and development of tolerance soil-borne bacteria antagonistic to plant pathogens might contribute substantially to prevent plant diseases and represent an alternative to the use of chemical pesticides in agriculture. *Pseudomonas* spp. have been studied mainly because of their widespread distribution in soil, their ability to colonize the rhizosphere of host plants and ability to produce a wide range of compounds inhibitory to a number of serious plant pathogens (Weller 1988; Copper and Higgins 1993; Mathivanan *et al.*, 2005). *Pseudomonas fluorescens* was reported to provide 79–82% control of rice blast and sheath blight and thus increase grain yield in rice (Gnanamanickam *et al.,* 1998, Shanmugaiah, 2007). The ability of *P. fluorescens* to suppress soilborne fungal pathogens depends on their ability to produce antibiotic metabolites such as pyoluteorin, pyrrolnitrin, Phenazine-1-carboxylic acid and 2, 4-diacetylphloroglucinol (Sullivan and Gara 1992, Shanmugaiah *et al.*, 2010). One of these metabolites Phenazine is a major factor in the biological control of a range of plant pathogens mainly sheath blight of rice (Shanmugaiah, 2007, Shanmugaiah *et al.*, 2010). A few reports are available

*Corresponding author.

on the characterization of phenazine and its antifunagal activity against plant pathogens (Shanmugaiah *et al.*, 2010). In this investigation attempts were made to identify and characterize antifungal compounds in *P. fluorescens* isolated from rhizosphere region of rice crop. The present study investigate were made to identification of phenazine gene from fluorescent pseudomonades isolated from rice rhizosphere and their biological action sheath blight of rice from selected potential isolate *Pseudomonas* sp VSMKU4047.

MATERIALS AND METHODS

Isolation of Fluorescent Pseudomonades (FPs) from Rice Rhizosphere

10 g of soil samples collected from rice rhizosphere was taken in a conical flask. To which, 90 ml of sterile distilled water was added and the flasks were placed on a rotary shaker at 120 rpm for 30 min. The resulted suspension was serially diluted and spread plated on to King's B agar (KBA) (King *et al.*, 1954). The plates were incubated at room temperature (28 ± 2°C) for 48 h and observed under UV light. Distinct single colonies of fluorescent bacteria were picked, sub-cultured to purity and maintained in KBA slants and also in 30% glycerol vials.

Antifungal Activity in Dual Plate Assay

The selected isolate VSMKU4047 was tested towards fungal pathogens for its antifungal activity against *Rhizoctonia solani, Macrophomina phaseolina, Fusarium udum, Alternaria alternata, Fusarium oxysporum* by dual plate assay (Shanmugaiah *et al.*, 2010).

Identification of Phenazine Gene by Polymerase Chain Reaction (PCR)

The Identification of metabolite, phenazine gene from fluorescent pseudomonadesby heat lysis method, phenazine gene was determined by PCR amplification using the specific primer of the Phenazine forward and reverses primers Phz1a 5'-GGCG ACATGGCTCGCTCAAC-3' and PHz2a 5'-CGCGTGGCGGCGGCGTATAT-3' PCR amplification was performed using forward primer *Phz* and reverse primer *Phz2* (Medox, India), which were developed from the *Phz* gene sequence of *Pseudomonas fluorescens* Q2-87 (Raaijmakers *et al.,* 1997). The conditions used in PCR reaction are given below: PCR amplification was carried out in 12 μl of reaction was used an then the amplifications were performed with a Cyber lab (Smart PCR, USA) using the following PCR conditions: Initial denaturation for 2 min at 94°C followed by 25 PCR cycles of amplification, and each cycle had 1 min denaturation at 94°C, 45 sec annealing at 56°C, extension at 72°C for 1.75 min and a final extension at 75°C for 1min (Delaney *et al.*, 2001).

Gel Electrophoresis

Samples of the PCR products (6 µl) were separated on a 1.5% agarose gel in 1 X TAE buffer at constant voltage of 100 V for 2 h. The gel was stained with ethidium bromide for 20 min and destained in distilled water for 30 min. The gel was visualized under UV transilluminator and photographed.

Identification of Selected Strain

The selected isolate VSMKU4047 was identified as *Pseudomonas* sp based on 16S rRNA, physiological and morphological characters exhibited in the Bergery's Manual of determinative bacteriology (Williams *et al.,* 1994).

Lytic Enzyme Production

The selected strain *Pseudomonas* sp VSMKU4047 was streaked on the respective substrate (0.5%) amended agar medium and then the plates were incubated at room temperature. After 24 h and 48h, Protease, Amylase and Pectinase, Chitinase and Cellulase production was observed. By flooding Gram's iodine over the starch agar plate amylase activity was tested. Mercuric chloride was flooded over the gelatin and casein agar plates to check the zone of clearance for protease, similarly flooding 1% Congo red over the colloidal pectin, chitin and cellulose agar plates (Shanmugaiah 2007).

Screening of Phosphate Solubilization

To detect the phosphate solubilizing from selected isolate *Pseudomonas* sp VSMKU4047 were streaked onto Pikovskaya's agar medium, which contains (per liter): 0.5 g yeast extract, 10 g Dextrose, 5 g $Ca_3(PO_4)_2$, 0.5 g $(NH_4)_2SO_4$, 0.2 g KCl, 0.1 g $MgSO_4.7H_2O$, 0.0001 g $MnSO_4.H_2O$, 0.0001 g $FeSO_4.7H_2O$ and 15 g agar. After 3 days of incubation at 28°C, strains that induced clear zone around the colonies were considered as positive (Pikovskaya,1948) .

HCN and Siderophore Production

Production of hydrocyanic acid (HCN) was determined by method of (Lorck, 1948). Bacterial strains (24 h old) were separately streaked on nutrient sucrose agar with 4.4 g glycine with simultaneous addition of a filter paper soaked in 0.5% picric acid in 1% Na_2CO_3 in the upper lid of Petri plate. The plates were sealed with parafilm. The control plates did not receive inoculum. After incubation at 28±1°C changes in colour from yellow to reddish brown were examined. Siderophore production by different strains FPs *was* tested by chromo azural S (CAS) assay (Schwyn and Neilands, 1987). The strains were streaked on citramide agar and incubated for 48h at 30°C. After incubation a thin layer of CAS reagent in 0.7% agar was spread on the bacterial growth and plates were again incubated for 24h at 30^{oC} formation of yellow orange zone around the colonies indicates siderophore production.

IAA Production

It was done as per the protocols of (Patten and Glick, 1996). FPs was grown in Luria broth supplemented with L-tryptophan (1 µg/ml) for 72 h. At the end of the incubation, cultures were centrifuged at 10,000*g* for 10 min and the supernatants collected. One ml of this culture filtrate was allowed to react with 2 ml of Salkowsky reagent (1 ml of 0.5 M $FeCl_3$ in 50 ml of 35% $HCIO_4$) at 28±2°C for 30 min. At the end of the incubation, pink color developed which indicates the presence of IAA. Quantification of IAA was done by measuring the absorbance in a spectrophotometer at 530 nm.

Detached Leaf Assay

Healthy rice leaves were cut into 10 cm length segments and their surface sterilized with 2% Sodium hypochlorite, after three minutes washed with sterile distilled water. Rice leaves were dipped in culture, culture filtrate of *Pseudomonas* sp.VSMKU4047, carbendazim and water for three minutes, and then allowed for air dried. After air dry the uniform size of sclerotium was placed on the above treated rice leaves. Then leaves were placed in the moist chamber for 7 days. After seven days Sheath blight lesion development in rice leaf segments was observed. The lesion length around the sclerotium was measured and ShB severity was rated by Relative Lesion Height (RLH) method with the following formula:

% RHL = 100 × Total height of lesions/Total leaf height.

Detection of Phenazine by TLC

Selected strain *Pseudomonas* sp VSMKU4047 were grown with shaking in King's B broth for 48 h at 27°C, and phenazines were extracted with ethyl acetate as described previously (Mavrodi *et al.,* 1998) and slight modification. Commercial Phenazine compound purchased from (Sigma Aldrich, Mumbai) was used as a standard. The extracts were dried, suspended in methanol, and spotted on silica gel plates (Merck, Germany). Chromatography was carried out using benzene—acetic acid (95:5), and chromatograms were examined at 254 nm.

Antifungal activity of Phenazine against *R. solani*

The antifungal activity of phenazine against *R.solani* was studied by well diffusion assay test using Potato Dextrose Agar (PDA). Phenazine compounds were scrapped from TLC plate. Mycelial discs of *R.solani* were cut from 3 days old cultures and placed in the center of PDA plates. Four wells were made in each plate using sterile cork borer and 25 µl, of suspension of ethyl acetate extract phenazine from TLC was used in the respective well. 25 µl of carbendazim and ethyle acetate was placed in the control well. The mycelial growth of fungal pathogens was observed after 3 days and the zone of inhibition was measured.Same methodology were used for other concentration of crude Phenazine compound againt *R. solani* such as 50 µl 75 µl and 100 µl respectively.

RESULTS

Isolation and Screening of Antagonistic Fluorescent Pseudomonads

Out of 63 Flourescent pseudomonads were isolated from rice rhizosphere soils and screened for antifungal activity against *R.solani.* Among them, 36 Flourescent pseudomonads were showed inhibitory activity against *R.solani.* The zone of inhibition of mycelia growth of the above fungal pathogens ranged from 10 mm to 32.6 mm. Further secondary screening of selected isolated VSMKU4047 towards other fungal pathogens such as *Macrophomina phaseolina, Fusarium oxysporum, Fusarium udum and Alternaria alternata.* (Table 1).

Table 1: Antagonistic Activity of Rice Rhizosphere FPs against Fungal Pathogen

Fungal Pathogens	*Zone of Inhibition (mm)*
Rhizoctonia solani	28.9
Macrophomina phaseolina	29.0
Fusarium oxysporum	32.6
Fusarium udum	32.6
Alternaria alternata	34.5

Screening of Phenazine Gene by Polymerase Chain Reaction (PCR)

Phenazine gene from antagonistic bacteria was identified by method with using forward primer phz1a and reverse phz2b primers. Out of 36 antagonistic bacteria, 15 isolate showed Phenazine positive with characteristic future of 1Kb DNA fragments amplification compare with that of Phenazine positive bacteria *Pseudomonas aeruginoisa*

MML2212. The isolates of PCR bands not visualized considered as negative isolates of Phenazine gene (Table 2).

Identification of VSMKU4047

The selected strain VSMKU-4047 was identified as *Pseudomonas* sp. based on the morphological, physiological tests and 16s rRNA sequence analysis (Table 2).

Production of Lytic Enzymes, Phosphate Solubilization, HCN, Siderophore and IAA

Production of lytic enzymes was investigated from selected strain VSMKU4047 among five lytic enzymes, the isolate VSMKU4047 showed the production of protease and amylase. Phosphate solubilization and Indole acetic acid was produced by isolate VSMKU4047 where as Siderophore and Hydrogen cyanide production as absent of the same isolate (Table 2).

Table 2: Physiological and Biochemical Traits of Lytic Enzymes and Secondary Metabolites Production by Rice Rhizosphere Isolate *Pseudomonas* sp VSMKU4047

Biochemical Lytic Enzyme and Metabolite Production	*Characters*
Gram strains	+
Fluorescent Character	+
Motility	+
Oxidase	+
Catalase	+
Secondary metabolites	*Characters*
Phosphate solubilization	+
HCN	–
Siderophore	–
Phenazine	+
Lytic enzyme and growth hormone	*Character*
Chitinase	–
Protease	+
Amylase	+
Cellulase	–
Pectinase	–
IAA	+

+: Positive –: Negative

Detached Leaf Assay

Detached leaf assay was established that treatment with culture and culture filtrate of *Pseudomonas* sp. VSMKU4047 significantly inhibited the sclerotial viability. More over the sclerotia failed to establish Sheath blight symptoms on detached leaves, whereas other treatments showed different levels of lesion development (Table 3).

Table 3: Evaluation of *Pseudomonas* sp.VSMKU4046 against Rice Sheath Blight

Treatments	*Sheath Blight Severity (%)*
R.solani (sclerotium)	87
R.solani (sclerotium) + carbendazim	55
Pseudomonas sp.VSMKU4047+ *R.solani* (sclerotium)	15
Culture filtrate of VSMKU4047+ *R.solani* (sclerotium)	20

Antifungal Activity of Crude Phenazine

25µl, 50 µl 75 µl and 100 µl of ethyle acetate extracted from *Pseudomonas* spVSMKU4047 crude Phenazine significantly inhibited the mycelial growth of *R.solani* compared to commercial Phenazine, Carbendazim and ethyle acetate control (Table 4).

Table 4: Effect of Crude Phenazine against *R.solani*

Concentration (µl)	*Phenzine (Authentic)*	*Crude Phenazine*	*Carbendazim*	*Ethylacetate*
25	17	16	12	6
50	21	20	15	10
75	27	24	22	10
100	35	35	28	10

Phenazines are N-containing heterocyclic molecules. These molecules have broad spectral antagonistic activity against both fungi and bacteria, they also play a significant role in colonization of microbes in an around plant rhizosphere. The present study indicate, the selected isolate *Pseudomonas* sp. VSMKU4047 have broad spectral activity against various fungal pathogens such as *R. solani, M.phaseolina, F.oxysporum, F.udum* and *A. alternate* with zone inhibition range 29 mm to 35 mm. The use of plant growth promoting fluorescent pseudomonades for control of phyto pathogens and their mechanism of biocontrol strategy have been reported recently (Compant *et al.*,

2005, Shanmugaiah *et al.,* 2010). The selected strain was identified VSMKU 4047 was identified as *Pseudomonas* sp based on the morphology, biochemical and 16s rRNA sequence analysis. Phenazine genes were identified using universal primer of Phenazine by PCR with Phenazine positive strains. Similar result was obtained from Fluorescent Pseudomomonades Broad spectrum and action of Phenazine against active against fungal pathogens and root knot nematode (Kavitha *et al.* 2005). In this juncture the selected isolate VSMKU4047 have already demonstrate the production of Phenazine, lytic enzymes and plant growth hormones. The broad-spectrum antifungal strain, *P. aeruginosa* PUPa3 and *P.aeruginosa* MML2212 can be used as an effective biological control candidate against devastating fungal pathogens that attack tobacco, rice, groundnut, banana, cotton, tea, chili, sugarcane and mango. Other functional traits such as siderophores, IAA and phosphatase may enhance its potential as biofertilizer. Based on the results obtained from NMR and MS the active metabolite that mediates the broad-spectrum antifungal property has been identified as PCN (Kumar *et al.*, 2005, Shanmugaiah *et al.,* 2010). The remarkable reduction of sheath blight was recorded with culture and culture filterate of *Pseudomonas* sp. VSMKU4047 and Carbendazim with 15% , 20% and 55% compare to water shocked sclerotium control. The significant antagonistic strain VSMKU4047 is now being screened for their effects on development of sheath blight disease in the greenhouse. Further purification and characterization of Phenazine are in progress.

ACKNOWLEDGMENT

The authors gratefully acknowledge University Grant Commission (UGC), New Delhi, provides financial support for this work through Major Research Project sanction *(F.No.39–214/2010) (SR)).*

REFERENCES

Compant, S., Duffy, B., Nowak, J., Clément, C. and Barka, E.A. (2005). Use of growth-promoting bacteria for biocontrol of plant diseases: principles, mechanisms of action, and future prospects. *Appl. Environm. Microbiol.* 71(9): 4951–4959.

Copper, A.L. and Higgins, K.P. (1993). Application of Pseudomonas fluorescens isolates to wheat as potential biological control agents against take-all. *Plant Pathol.* 42: 560–567.

Delaney, S.M., Mavrodi,D.V., Bonsal, R.F and Thomashow, L.S. (2001). PhzO, a Gene for Biosysnthesis of 2-Hydroxylated Phenazine compounds in Pseudomomnas aureofaciens 30–84. *J. Bacteriol.* 183: 318–327.

Gnanamaickam, S.S., Valasubramanian, K., Thara, V. and Chaterjee A.K. (1998). Microbial antagonists for rice diseases control: molecular approaches in microbes: for health, wealth and sustainable environment. Ed Ajit varma pp. 371–388.

Kavitha, K., Mathiyazhgan, S., Sendhilvel, M., Nakkeran, S., Chandrasekar, A. and Dilantha Fernando, W.G. (2005a). Broad spectrum action of phenazine against active and dormant structures of fungal pathogens and root knot Nematode. *Archiv. Phytopathol. Plant Protect.*, 38(1): 69–76.

King, E.O., Ward, M.K. and Raney, D.E. (1954). Two simple media for the demonstration of pyocyanin and flourescein. *J. Lab. Clin. Med.,* 44: 301–307.

Kumar, R.S., Ayyadurai, N., Pandiaraja, P., Reddy, A.V., Venkateswarlu, Y., Prakash, O. and Sakthivel, N. (2005). Characterization of antifungal metabolite produced by a new strain Pseudomonas aeruginosa PUPa3 that exhibits broad-spectrum antifungal activity and biofertilizing traits. *J. App.Microbiol,* 98(1): 145–54.

Lorck, H. (1948). Production of hydrocyanic acid by bacteria. *Physiol. Plant.* 1: 142–146.

Mathivanan, N., Prabavathy, V.R. and Vijayanandraj, V.R. (2005). Application of talc formulations of Pseudomonas fluorescens Migula and Trichoderma viride Pers. Ex. S.F. Gray decrease the sheath blight disease and enhance the plant growth and yield in rice. *Journal of Phytopathology*, 153: 697–701.

Mavrodi, D.V., Ksenzenko, V.N., Bonsall, R.F., Cook R.J., Boronin, A.M. and Thomashow, L.S. (1998). A seven-gene locus for synthesis of phenazine-1-carboxylic acid by Pseudomonas fluorescens 2–79, *J. Bacteriol.,* 180: 2541–2548.

Ou, S.H. (1985). Rice diseases. Great Britain (UK): Commonwealth Agricultural Bureau 380p Kloepper, J.W., Leong, J., Teintze, M. and Schroth, M.N. (1980). Pseudomonas siderophores: a mechanism for explaining disease suppressive soils. *Curr. Microbiol.* 4: 317–320.

Patten, C. and Glick, C.R. (1996). Bacterial biosynthesis of indole-3-acetic acid. *Can J Microbiol* 42: 207–220.

Pikovskaya, R.I. (1948). Mobilization of phosphorus in soil connection with the vital activity of some microbial species. *Microbiologia,* 17: 362–370.

Raaijmakers, G.M., Weller, D.M. and Thomashow, L.S. (1997). Frequency of antibiotic producing Pseudomonas spp. in the natural environments. Appl. Environ. Microbiol. 63: 881–887.

Schwyn, B. and Neilands, J.B. (1987) Universal chemical assay for the detection and determination of siderophores. *Anal Biochem* 160(1): 47–56.

Shanmugaiah, V. (2007). Biocontrol potential of phenazine-1-carboxamide producing plant growth promoting rhizobacterium, Pseudomonas aeruginosa MML2212 against sheath blight disease of rice. PhD Thesis, University of Madras, Chennai, India.

Shanmugaiah, V. Mathivanan, N and Varghese, B. (2010). Purification, crystal structure and antimicrobial activity of phenazine-1-carboxamide produced by a growth-promoting biocontrol bacterium, Pseudomonas aeruginosa MML2212. *Journal of Applied Microbiology,* 108: 703–711.

Sullivan, D. and Gara, F. (1992). Traits of fluorescent pseudomonas spp. involved in suppression of plant root pathogens. *Microbiol,* 40: 1064–1066.

Weller, D.M. (1988). Biological control of soil borne plant pathogens in the rhizosphere with bacteria. *Annul Rev Phytopathol* 26: 379–407.

Williams, S.T. (eds) (1994). Bergey's Manual of Determinative Bacteriology, 9th edition Williams and Wilkins Co, Baltimore, USA.

A Study on Antimicrobial Activity of *Camellia sinensis* and *Lowsonia inermis* against MRSA

Laishram Sahela Devi*, Suchitra Hijam and S. Arumugam[1]

Department of Biotechnology, Nandha Arts and Science College, Erode

[1]Department of Botany, Government Arts and Science College, Salem

E-mail: sahela2@gmail.com

INTRODUCTION

Methicillin *staphylococcus aureus,* highly virulent gram positive bacteria that occurs in microscopic clusters resembling grapes. It is a predominantly extra cellular pathogen that causes a variety of community acquired a nosocomial infection that range from harmless superficial skin infections to serious invasive disease such as endocarditis, osteomyelitis and pyarthritis.The global emergence and increase of methicillin resistant *staphylococcus aureus* (MRSA) caused a shortage of effective beta-lactam antibiotics to MRSA based infections, resistance to methicillin and a variety of the other antibiotics with mechanism of action similar to the beta-lactam antibiotics.

Plants play an important role in our traditional medicine as the produce secondary metabolites during stress. These secondary metabolites can be used as medicine. In the present study two plants are taken for finding their antimicrobial activity against MRSA. *Camellia sinensis*, the tea plant is the species of whose leafs and leaf buds are used to produce tea. It is an evergreen shrub or small tree that is usually trimmed below two meters (six feet) when cultivated for its leaves. It has lots of medicinal values and shows antibacterial activity. *Lowsonia inermis* (henna) is a small tree or large shrub, growing to six meters high. It has lateral branch with leaves that grow in pair, to four centimetres long. Henna extracts show antibacterial, antifungal and ultraviolet light screening activity.

*Corresponding author.

MATERIALS AND METHODS

Isolation of Methicillin Resistant *Staphylococcus aureus*

MRSA was isolated from pus samples collected from infected wound patient of government hospital, Erode, by serial dilution technique and spread on mannitol salt agar. It was confirmed by gram staining, haemolysis test, coagulase test, catalase test, IMVC test, sugar fermentation test etc.

Antimicrobial Sensitivity Test

Extraction Methods

The effective of an antimicrobial sensitivity testing is based on the size of the zone of inhibition that surrounds a disc that has been impregnated with a specific concentration of the agent. Two methods were used for the extraction of plant sample; 1) crude extraction method 2) soxlets methods and four solvents were used viz chloroform, acetone, ethanol and methanol.

In crude extraction method, the selected plant leaves were dried and grinded to obtain the powder of the leaves. About 20 g leaf powders were weighed and suspended to 100 ml of selected solvents (ethanol, acetone, methanol and chloroform) and kept on rotatory shaker for 24 hours at room temperature. The extract was then filtered through whatmann's no. 1 filter paper and filtrate was allowed to evaporate at room temperature.

In soxlets method, 6 g of the selected dried leaves powder leaves were transferred to 30ml soxhlation unit and extracts were filtered and stored at 4°C for further use.

Plants Extract Discs

100 µl, 200 µl and 300 µl of the plant extract were impregnated on the sterile discs.

Antibacterial Activity

The antimicrobial activity of selected plants were assayed using the disc diffusion techniques by Kirby-Bauer method, 50 ml of the sterile nutrient broth were aseptically inoculated with stock culture and incubated at 37°C for 24 hours. The test cultures were swabbed on air dried nutrient agar plates using sterile cotton swabs. The disc loaded with plant extracts were placed on the surface of nutrient agar plate swabbed with bacterial cultures. Then the plates were incubated at 37°C for 24 hours. The zone of inhibition was measured including the disc and recorded.

RESULTS

MRSA was isolated from the pus of wound infected patients. It was identified and characterized by different confirmation methods and the results were recorded on Table 1.

Table 1: Confirmation of Methicillin Resistant *Staphylococcus aureus*

Sl. No.	*Test*	*Results*
1.	Gram staining	Positive,cocci
2.	Beta-Hemolysis	positive
3.	Catalase	Positive
4.	Coagulase	Positive
5.	Mannitol fermentation test	Positive
6.	Biochemical test:	
	Indole test	Negative
	Methyl red test	Positive
	VP test	Negative
	Citrate utilization test	Negative

The gram staining revealed that the isolated colony has gram positive bacteria in cluster form.The Beta-hemolysis and coagulase test showed positive results. Mannitol fermentation test showed positive results but for only acid formation and not the gas.The colony when subjected to biochemical test, revealed positive response for Methyl red test and negative response for indole, VP and citrate utilization test. The antibacterial sensitivity test was carried out with two plant extracts in accordance with the Kirby-Bauer method (Figures 1–8).

Camellia Sinensis

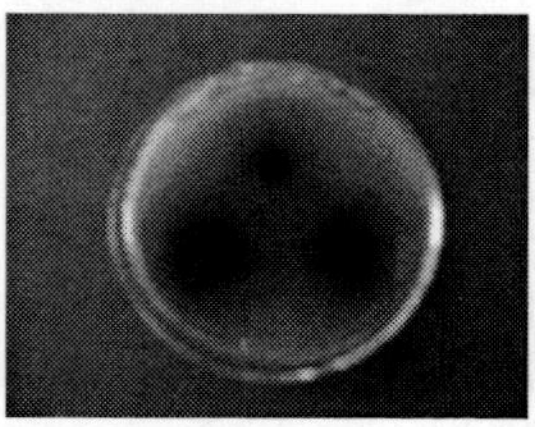

Fig. 1

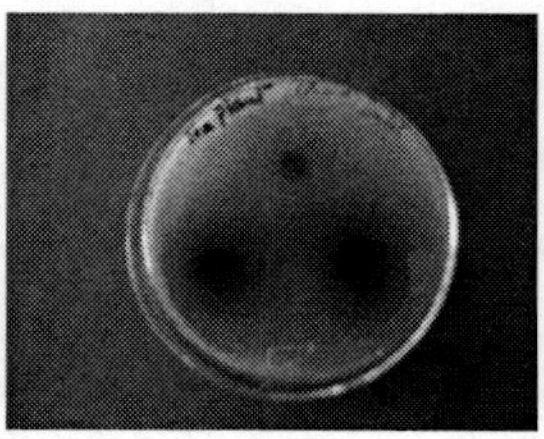

Fig. 2

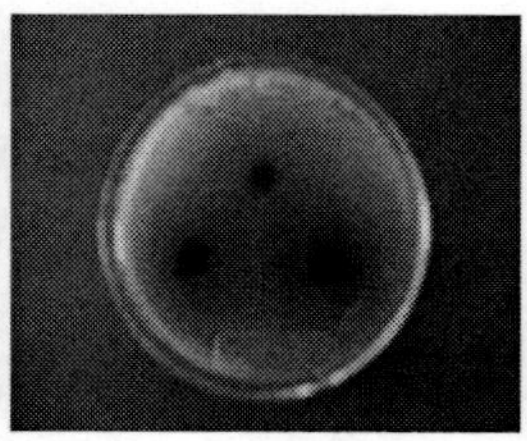

Fig. 3

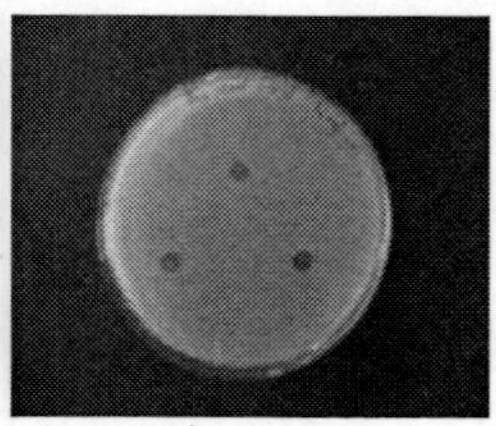

Fig. 4

Lowsonia Inermis

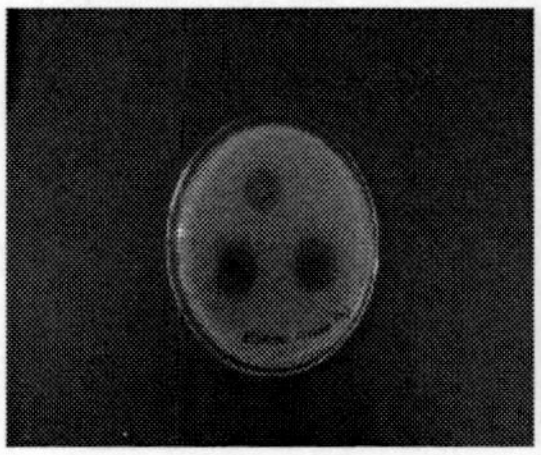

Fig. 5

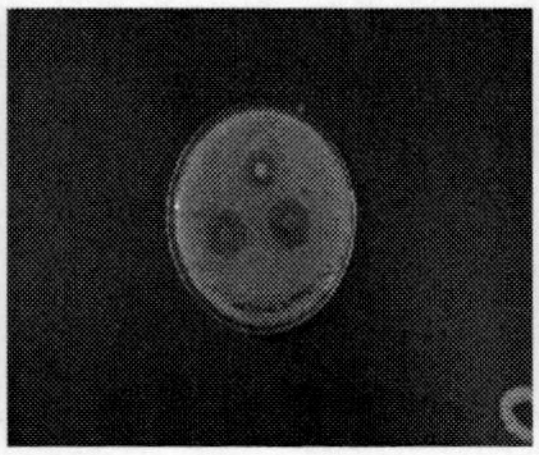

Fig. 6

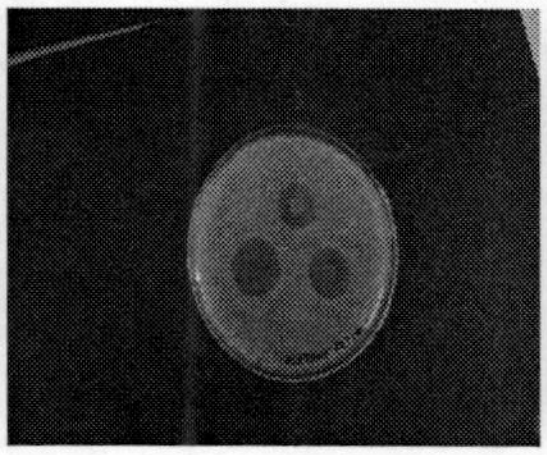

Fig. 7

Fig. 8

All the plants showed no zone of inhibition when chloroform was used for the extraction in both crude and soxlet method.

Camellia sinensis: The zone of inhibition varied from 10–13 mm in crude method and 11–15 mm in soxlet method (Table 2).

Table 2: Crude Method and Soxlet Method

Extract	*Crude Method*			*Soxlet Method*		
	Zone of Inhibition in Diameter (mm)			*Zone of Inhibition in Diameter (mm)*		
	100 µl	*200 µl*	*300 µl*	*100 µl*	*200 µl*	*300 µl*
Acetone	10	11	12	11	12	14
Chloroform	–	–	–	–	–	–
Ethanol	11	11	12	11	12	14
Methanol	11	12	13	14	15	16

Lawsonia inermis: The zone of inhibition varied from 12–19 mm in crude method and 16–22 mm in soxlet method (Table 3).

Table 3: Crude Method and Soxlet Method

Extract	*Crude Method*			*Soxlet Method*		
	Zone of Inhibition in Diameter (mm)			*Zone of Inhibition in Diameter (mm)*		
	100 µl	*200 µl*	*300 µl*	*100 µl*	*200 µl*	*300 µl*
Acetone	16	17	19	17	19	22
Chloroform	–	–	–	–	–	–
Ethanol	12	13	18	16	19	20
Methanol	16	16	18	17	19	20

CONCLUSION

The MRSA was isolated and identified successfully from the infected wound. The antibacterial sensitivity test was achieved with two plants viz; *Camellia sinensis* and *Lawsonia inermis*. Among these two *Lawsonia inermis* showed more sensitivity than *Camellia sinensis* against MRSA.

REFERENCES

Alexander Ogston A9 (1880). Report upon micro organisms in disease, BMJ; 1:369. Ambler RP: (1980) the amino acid sequence of the *Staphylococcus aureus* l *Pencillinase. Journal of Biochem* 151:197.

Aquuil, F., Khan, M.S., Owais, M. and Ahmed, I. (2002). Department of Agricultural microbiology, Faculty of Agriculture Science, Aligard Muslim University, Aligard-20.

Gurran, C. (2002). Freemen hospital. Journal of Medicine, Volume 91, 148–151. Didenberg *et al.* (1990) comparison of the sequence of the sequence of class A, ß-lactamase and of the secondary structure elements of penicillin recognizing protiens. Antimicrob Agents Chemother, 35: 294, 191.

Hiramatsu, K., Hanaki, H., Ino, T., Yabuta, K., Oguri, T. and Tenover, F.C. (1997). "PDF Methicilin resistant Staphylococcus aureus clinical strain with reduced vancomycin susceptibility". *Journal of Antimirob Chemother,* 40(1): 135–136.

Horiba, N., Maekawa, Y., Ito, M., Matsumoto, T. and Nakamura, H. (1991). A pilot study of Japanese green tea as medicament: antibacterial and bacterial effects; 17: 122–24.

Schliefer, K.H. and Kloos, W.E. (1975). Isolation and characterisation of staphylococci from Human Skin. *Int Journal of Syst Bacteriol,* 25: 50.

Shivananda Nayak, B. and *et al.* (2007). The evidence based wound healing activity of Lowsonia inermis Linn., Phytother.

Wang, R. *et al.* (2007). "Identification of novel cytolytic peptides as Key virulence determinants of community-associated MRSA". *Nature Medicine DOI:* 10: 1038 nm, 1656.

Molecular Analysis of van A Gene – From Vancomycin Resistant *Staphylococcus aureus* Isolates

Ganesh Manikandan S.* and Job Gopinath[1]
Department of Biotechnology,V.M.K.V. Engineering College,
Vinayaka Missions University, Salem, Tamil Nadu
[1]Department of Zoology, Govt. Thirumagal Mills College,Gudiyatham, Vellore
E-mail: bioganesh2k2@gmail.com

INTRODUCTION

Staphylococcus aureus are facultative anaerobic Gram-positive cocci that are catalase positive and oxidase negative. Under the microscope they usually appear as grape-like clusters. *S. aureus* are fairly heat sensitive; the heat stable toxin is produced in the food by the bacterium that causes the illness. Serious infections produced by *S. aureus* includes bacteraemia, pneumonia, osteomyelitis, acute endocarditis, myocarditis, pericarditis, encephalitis, meningitis, choriamnionitis, mastitis, scalded skin syndrome and abscesses of the muscle, urinogenital tract and various intra-abdominal organs (Murray *et al.*, 2003).

Antibiotic resistance is the ability of a microorganism to withstand the effects of an antibiotic. It is a specific type of drug resistance. Antibiotic resistance evolves naturally via natural selection through random mutation, but it could also be engineered by applying an evolutionary stress on a population. Once such a gene is generated, bacteria can then transfer the genetic information in a horizontal fashion (between individuals) by plasmid exchange. If a bacterium carries several resistance genes, it is called multi-resistant.

*Corresponding author.

Vancomycin Intyermediate S.aureus (VISA) and Vancomycine Resistant S.aureus (VRSA) are specific types of antimicrobial-resistant staph bacteria. While most staph bacteria are susceptible to the antimicrobial agent vancomycin some have developed resistance. VISA and VRSA cannot be successfully treated with vancomycin because these organisms are no longer susceptible to vancomycin. However, to date, all VISA and VRSA isolates have been susceptible to other Food and Drug Administration (FDA) approved drugs (CDC).

Vancomycin-resistant *Staphylococcus aureus* (VRSA) is a strain of *Staphylococcus aureus* that has become resistant to the glycopeptide antibiotic vancomycin. With the increase of staphylococcal resistance to methicillin, vancomycin (or teicoplanin) is often a treatment of choice in infections with methicillin-resistant *Staphylococcus aureus* (MRSA).

Vancomycin resistance is still a rare occurrence. Unfortunately, VRSA may also be resistant to meropenem and imipenem, two other antibiotics that can be used in sensitive staphylococcus strains.

At present, high-level resistances to both glycopeptide and β-lactam antibiotics in *S. aureus* seem to be mutually exclusive, in that both resistances are not seen at once in the same strain of bacterium. However, this is not due to a fundamental biochemical incompatibility. Theoretically, a superbug displaying high-level resistances to both classes of antibiotics could evolve given the current selective environment (Chang *et al.*, 2003).

The isolate was tested for antimicrobial susceptibility by using Microscan® overnight panels (Dade Behring, Deerfield, Illinois); vancomycin MIC was 4 μg/mL. Further testing by Etest® (AB Biodisk North America, Inc., Piscataway, New Jersey) indicated that the isolate was resistant to vancomycin (MIC >256 μg/mL). After notification and subsequent analysis by the New York State Department of Health (NYSDOH), the isolate was forwarded to CDC, where it was confirmed to be VRSA (vancomycin MIC = 64 μg/mL, using the National Committee for Clinical Laboratory Standards broth microdilution reference method). The isolate contained both the *mecA* and *vanA* genes mediating oxacillin and vancomycin resistance, respectively. The isolate was susceptible to chloramphenicol, linezolid, minocycline, quinupristin-dalfopristin, rifampin, and trimethoprim-sulfamethoxazole (CDC 2002). In tha present study the samples were collected ramdomly from the clinical diagnostic centre in salem city for identification, isolation and amplification of *vanA* gene for futher research was carried out.

MATERIALS AND METHODS

Various clinical laboratories in Salem and Vellore, Tamil Nadu, India were opted for sample collection. Pus samples were collected from respective districts and transported to the laboratory after bedside inoculation using standard microbiological methods. Pus was removed according to the WHO (1999). Swab intact pustules with 70% ethyl alcohol were allowed to dry. Either piercing with a sterile needle or absorbing the contents with a sterile swab or aspirating contents was done. A sterile swab to sample areas of superficial pyoderma, preferably sampling from the margin of the lesion was used.

All clinical samples were carefully analyzed for the identification of *S.aureus* straibs using standardized methods like Gram staining, hemolytic activity, the coagulase test, and gelatinase test. Various obtained from High-Media Laboratory, India were used for the antibiotic assay. The oligonucleotide primers designed to amplify the above gene were as follows: Van A, (Forward 5′-GGGAAAACGACAATTGC-3′ and 5'ATTGCCGTAACATG-3'Reverse primer) The PCR reaction consisted of denaturation at 92°C for 3 min, followed by 30 cycles of heating to 92°C for 1 min, cooling to 56°C for 1 min, then heating to 72°C for l min. Vials were then kept at 72°C for 3 min. A sample of 10 micro liters from each reaction was analyzed by gel electrophoresis in a 0.8% agarose gel with ethidium bromide, and results were observed.

RESULTS AND DISCUSSION

VRSA culture is Gram-Positive and is grown at 35 to 37°C on nutrient agar media. Isolate of VRSA were highly resistant to methicillin (5 μg), penicillin G (10U), tetracycline (30 μg), gentamycin (10 μg) and vancomycin (30 μg) (Figure 1). Isolated colonies produced β-hemolysis (Figure 2) isolates of VRSA were coagulase positive; it forms the colloidal nature of plasma. When colonies were transferred to gelatin medium, it become liquid in nature at 4°C. Isolated colonies were produces golden yellow colour on the selective media (Mannitol-Salt Agar media) (Figure 3) and they are non motile. Chromosomal DNA isolated from vancomycin resistant *S.aureus* , Prominent band was observed on Agarose gel electrophoresis compared with standard marker. The present study got *VanA* gene amplified fragment at the region of 120bp on AGE (Figure 4).

A VRSA culture is Gram-Positive and is grown at 35 to 37°C on nutrient agar media. Most of the colony appeared as yellow in color. VRSA colonies were smooth, circular and convex. Some of the cultures are appeared as white in color. Most of the isolates well grown on 7–15% of NaCl containing nutrient agar media. Isolate of

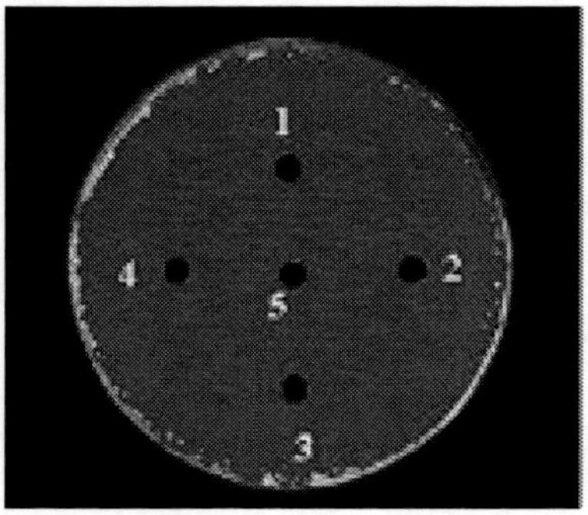

Fig. 1: 1. Methicillin (5 μg) 2. Penicillin G (10U),
3. Tetracycline (30 μg) , 4. Gentamycin (10 μg) and 5. Vancomycin (30 μg)

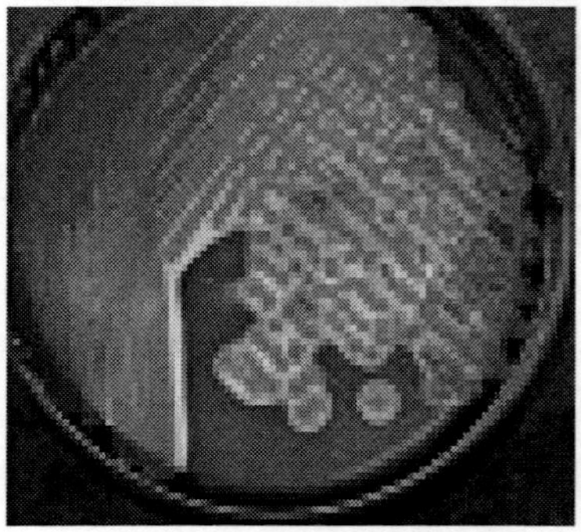

Fig. 2: Isolates Produced β-haemolysis on Blood Agar Medium

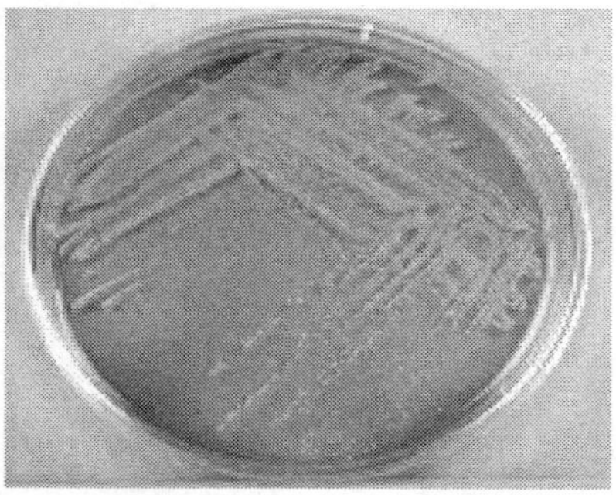

Fig. 3: Golden Yellow Colour Formation on MSA Medium

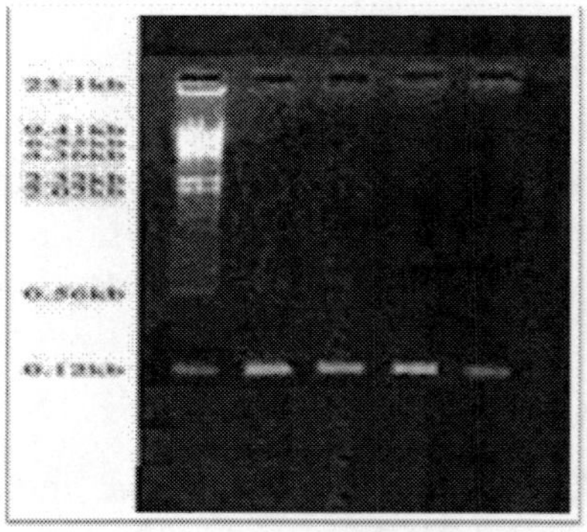

Fig. 4: 1. Marker, 2. S1, S2, S3 and S4 are VRSA DNA samples

VRSA were highly resistant to methicillin (5 μg), penicillin G (10U), tetracycline (30 μg), gentamycin (10 μg) and vancomycin (30 μg) (Figure 1) Disc diffusion test revealed that all vancomycin resistant staphylococcal isolates have shown antimicrobial resistance to most of the commonly used antimicrobials.

Isolated colonies produced β-hemolysis (Figure 2) isolates of VRSA were coagulase positive; it forms the colloidal nature of plasma. When colonies were transferred to gelatin medium, it become liquid in nature at 4°C. Tiwari HK and Sen MR (2006) reported that out of 1681 staphylococcal isolates, 761 isolates were slide coagulase positive, 772 were slidex staph plus positive, and 783 were tube coagulase positive. Thus, 783 strains were confirmed as *S. aureus* and remaining 898 strains as CoNS. Isolated colonies were produces golden yellow colour on the selective media (Mannitol-Salt Agar media) (Figure 3) and they are non motile.

In the present study showed 100% similar to their study in the field of Isolation and Identification of VRSA. However the present isolated strains well demonstrated the presence of *vanA* gene (Figure 4).

An extensive search for VRSA in all the patient's contacts was undertaken, and no VRSA was identified. However, a close friend of the patient did carry a vancomycin-susceptible, methicillin-resistant strain of *S. aureus* that was indistinguishable from the VRSA on pulsed-field gel electrophoresis. This finding underscores the importance of extending efforts to prevent and reduce the spread of methicillin-resistant *S. aureus* beyond inpatient facilities. The situation reinforces the recommendation of the Centers for Disease Control and others to test all strains of staphylococci for resistance to vancomycin, and to use vancomycin prudently. Further work on the population analysis and mechanisms of resistance in these strains is proceeding (Hiramatsu, 1997). The current study also showed 100% similar vancomycin resistance and biochemical characters.

The presence of the *vanA* gene was confirmed by PCR and was located on a 60 kb plasmid. The DNA sequence of the VRSA *vanA* gene was identical to that of a vancomycin-resistant strain of *Enterococcus faecalis* recovered from the same catheter tip culture. (Charles MD 2002).

Saha B *et al.,* (2008) submitted the following aspects of results. A pathogenic vancomycin-resistant *Staphylococcus aureus* (VRSA) isolate (MIC >/=64 mug ml (–1)) was obtained from a Kolkata hospital in June 2005. Species identification was confirmed by Gram staining, standard biochemical tests and PCR amplification of the nuc gene, which encodes the thermostable nuclease that is highly specific for *S. aureus.* The VRSA isolate was also resistant to beta-lactams (amoxicillin, ampicillin, cefepime, cefotaxime, cefuroxime, cephalexin and meticillin), chloramphenicol, streptomycin, macrolides (erythromycin and roxithromycin), clindamycin, rifampicin and trimethoprim-sulfamethoxazole. However, the isolate was susceptible to gentamicin (an aminoglycoside) and ciprofloxacin (a fluoroquinolone). The resistance to vancomycin was inducible *in vitro,* because the MIC of vancomycin increased from 64 mug ml(–1) initially to 1024 mug ml(–1) during culture of this VRSA strain in the presence of vancomycin. The VRSA isolate contained a large plasmid (approximately 53.4 kb) and four small plasmids of approximately 6, 5.5, 5.1 and 1.5 kb. The large plasmid of approximately 53.4 kb harboured the vancomycin-resistance genes *vanHAX,* which was confirmed by PCR amplification using the same plasmid as template and, separately, primers specific for the 2.61 kb *vanHAX* gene cluster, *vanH* (969 bp), *vanA* (1032 bp) and *vanX* (609 bp). The VRSA isolate was also positive for *mecA*. Vancomycin resistance was successfully transferred from this VRSA donor to a vancomycin-sensitive recipient *S. aureus* clinical isolate by a broth mating procedure. The MIC of vancomycin for the transconjugant was 32 mug ml(–1), as against 2 mug ml(–1) for the parent strain. Nucleotide sequencing of the PCR product showed partial homology with van genes of an enterococcal transposon Tn1546-like element. This is believed to be the first Indian *S. aureus* isolate that has been shown to be phenotypically vancomycin-resistant, presumably due to a *vanHAX* analogue. In the current study, VRSA isolates were showed 100% similar antibiogram pattern to their study but *vanA* gene found at 120bp on AGE . The PCR result slightly varied from their results.

REFERENCES

Centre for Disease Control and Prevention (CDC) (2002). *Staphylococcus aureus* resistant to vancomycin—United states, 2002.*MMWR*.51:565–567.

Chang, S., Sievert, D.M., Hageman, J.C., Boulton, M.L. and Tenover, F.C., Downes, F.P., Shah, S., Rudrik, J.T., Pupp, G.R., Brown, W.J., Cardo, D. and Fridkin, S.K. (2003). Infection with vancomycin—resistant Staphylococcus aureus containing the vanA gene. *N Engl J Med.*, 348:1342–1347.

Hiramatsu, K., Hanaki, H., Yubuta, Oguri T. and Tenover, F.C. (1997). Methicillin-Resistant Staphylococcus aureus clinical strains with reduced vancomycin susceptibility. *J. Antimicrobial Chemother,* 40: 135–136.

Mackie, T.J. and McCartney, J.E. (1989). Microbial Infections. Medical Microbiology. 13th Edition Longman Group Limited, London, pp. 236–245.

Maniatis, T., Fritsch, E.F. and Sambrook, J. (1982). Molecular Cloning: A Laboratory Manual.Cold Spring Harbor Laboratory, Cold spring Harbor, New York, p. 545.

Murray, M.G. and Thompson, W.F. (1980). Rapid isolation of high-molecular-weight plant DNA. *Nucl. Acids Res.* 8:4321–4325.

Murray, P.R., Baron, E.J., Jorgensen, J.H., Pfaller, M.A. and Yolken, R.H. (2003). Manual of Clinical Microbiology, 8th edn. Washington, *DC: American Society for Microbiology.*

SECTION–II

Shri K.N. Narasimhaiah Award

Apoptosis Inducing Activity of *Euphorbia hirta* Leaf Extract in Breast Cancer Cells

K. Kavithaa*, P.R. Padma and S. Sumathi
Department of Biochemistry, Biotechnology and Bioinformatics
Avinashilingam Institute of Home Science and
Higher Education for Women University, Coimbatore
E-mail: kavi.kavithu@gmail.com

INTRODUCTION

Cancer is a class of diseases in which the body cells become abnormal and divide indiscriminately. Cancer cells may become invasive and transform normal adjacent cells into malignant cells. They may also spread through the blood stream and lymphatic systems to other parts of the body to form metastatic tumors in distant organs. Cancer is a major public health problem worldwide with millions of new cancer patients diagnosed each year and many deaths resulting from this disease (Engel *et al.,* 2011).

Chemotherapy is one of the important treatment methods, but the side effects are difficult to tolerate, and as a result, people have being paying more attention to searching for new antitumor agents from natural products. Chemotherapy remains the principal mode of treatment for various cancers (Dai *et al.,* 2011). Chemotherapy and radiotherapy, the conventional cancer treatments used nowadays, are expensive and cause many side effects, including such minor ones as vomiting, alopecia, diarrhea, constipation, and major ones such as myelo suppression, neurological, cardiac, pulmonary and renal toxicity. All such side effects reduce the quality of life and discourage patients to observe medication protocols which then lead to the progression of cancer and associated complications (Castro *et al.,* 2011). A major

*Corresponding author.

problem of cancer therapy is the side effects of chemotherapy. Another key problem in cancer treatment is the reducing sensitivity of tumor cells to cytotoxic drugs. The majority of chemotherapy medicines have serious adverse effects in addition to their clinical effects. Patients find these side effects hard to tolerate, which often causes chemotherapy failure (Ko *et al.,* 2011).

Minimizing side effects and maximizing efficacy is a major goal in the development of cancer treatment. Now days the usage of combination of naturally derived active compounds with existing anticancer drug has been dramatically increased, because of its selectivity of target cell, reduced level of side effects, enhanced activity to controlling the cancer cell proliferation when compared with standard drug alone. Complementary medicine is a group of medicines and practices that may be used in addition to the standard treatments for enhanced cancer therapy. Furthermore the synergetic effect of complementary medicines is biocompatible, satisfactory stability, low toxicity and target-orientation (Kuete and Efferth, 2011).

Euphorbia hirta is of the family Euphorbiaceae, commonly known as "Snake weed" is a medicinal, rhizomatous herb distributed in South Western Ghats of India and North East Coast of Tamil Nadu. The plant is native to India but it is a pan tropical weed. The leaves are opposite, elliptical, oblong with a faintly toothed margin and darker on the upper surface. The flowers are small, numerous and crowded together in dense cymes of about 1 cm in diameter. Previous studies in our laboratory have shown that, the leaves of *Euphorbia hirta* do possess radical scavenging effect. Among the different organic extracts tested, methanolic extract showed better scavenging and rich in phytochemicals (Alkaloids, phenolics, flavanoids, saponins, glycosides, tannins, steroids and terpenoids). Hence the present study is aimed to develop a standard anticancer agent to control the breast cancer proliferation using extract of leaves of *Euphorbia hirta*.

MATERIALS AND METHODS

Maintenance of Cell Line

Breast cancer MCF-7 cell lines was obtained from National Centre for Cell Science (Pune, India) and revived in 20% Dulbeccos Minimal Essential Medium (DMEM) containing 10% foetal bovine serum (FBS). The culture medium was supplemented with antibiotic 1 µl of Penstrep. The cultures were maintained in 25 cm^2 culture flasks with the growth condition maintained at 37°C and 5% CO_2 in an air jacketed CO_2 incubator.

MTT Assay

The cell proliferation ability was assessed by MTT assay. The viability of the cells in the presence and absence of tamoxifen, with different concentrations was examined by MTT assay. This assay was performed using different concentrations namely 0.2, 0.4, 0.8, 1.6 and 3.2 mg of the methanolic leaf extract *Euphorbia hirta* for different time intervals 6, 12, 24 and 36 hours to test the toxicity and IC_{50} value was calculated.

Acridine Orange/Ethidium Bromide Staining

Ao/EtBr staining technique is used to differentiate between quiescent and actively proliferating cells. Apoptotic cells were identified by Ao/EtBr staining Parks *et al.* (1979) as explained in Appendix IV.

Nuclear Changes during Apoptosis

The nuclear changes such as chromatin condensation and fragmentation in MCF-7 cells were investigated in the presence and the absence of leaf extract and with or without tamoxifen was observed by Propidium iodide, DAPI and Hoechst staining.

Propidium Iodide Staining

Propidium iodide (PI) staining is based on the principle that apoptotic cells, among other typical features, are characterized by DNA fragmentation and consequently, loss of nuclear DNA content which can be analyzed using a flourochrome such as PI that interacts with nucleic acids and can stain the nuclear changes of apoptotic cells. The detail of this procedure developed by *Sarker et al.* (2000) is explained in Appendix V.

DAPI Staining

Nuclear condensation and fragmentation of apoptotic cells can be visualized by staining the nucleus and its fragments with a fluorescent dye such as DAPI that binds the DNA. Intense staining of apoptotic nucleus may result from the increased permeability of the dye to altered chromatin configuration.

Hoechst Staining

Chromatin condensation is a late apoptotic event. Hoechst staining determines the chromatin condensation. Hoechst dyes are intercalating, Benzimidazole derivatives that bind preferentially to AT base region and emit blue fluorescence when excited by UV light at about 360 nm. The nuclear morphology of cells will be checked under a fluorescence microscope (360 nm). Apoptotic cells were judged based on the nuclear morphology changes.

DNA Fragmentation

DNA fragmentation which is the hall mark of apoptosis occurs at the late phase of apoptosis. The DNA isolated from plant extract treated as well as untreated cells were subjected to agarose gel electrophoresis and observed for the electrophoretic pattern occurring during apoptosis.

Flowcytometric Analysis

The ability of plant extracts to influence cell cycle arrest was analysed by flow cytometry. Flow cytometer is an instrument that illuminates cells (or other particles) as they flow individually in front of a light source and then detects and correlates the signals from those cells that results from the illumination. The measurements are made as the cells or particles pass through the flow cytometer, in a fluid stream. Here, the fluorescent cells are illuminated with the light from a laser. Analysis and differentiation of cell is based on size and whether cell is carrying fluorescence molecule either in the form of antibodies or dyes.

RESULTS AND DISCUSSION

Anti Proliferative Activity

The cells were treated with various concentrations namely 0.2, 0.4, 0.8, 1.6 and 3.2 mg and different time intervals namely 6, 12, 24 and 48 hours. The results revealed that treatment with different concentrations resulted in loss of cell viability of MCF-7 cells in a concentration and time dependent manner. The IC_{50} was estimated to be 1.6 mg in MCF-7 cells. Once IC_{50} was determined of the leaf extract was tested for its

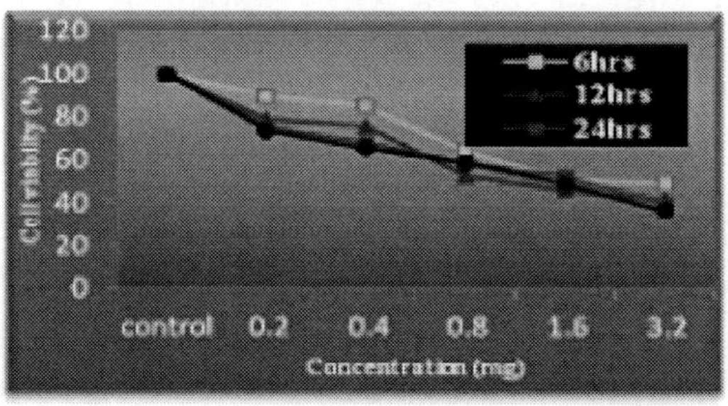

Fig. 1: The Values are Mean of Triplicates. The Control was Fixed as having 100% Cell Survival and the Percent Survival in the Other Groups was Calculated

effect on cell viability along with a standard chemotherapeutic drug tamoxifen. The results showed that extract along with tamoxifen has higher cytotoxicity. Treatment with plant extract and tamoxifen alone showed comparatively lower cytotoxic activity.

Acridine Orange/Ethidium Bromide Staining

Proapoptotic activity of methanolic leaf extract of *Euphorbia hirta* was investigated using fluorescence microscopy. Viable cells have uniform bright green nuclei. Early apoptotic cells have green nuclei but perinuclear chromatin condensation is visible as bright green patches or fragments. Late apoptotic cells have orange to red nuclei with condensed structure.

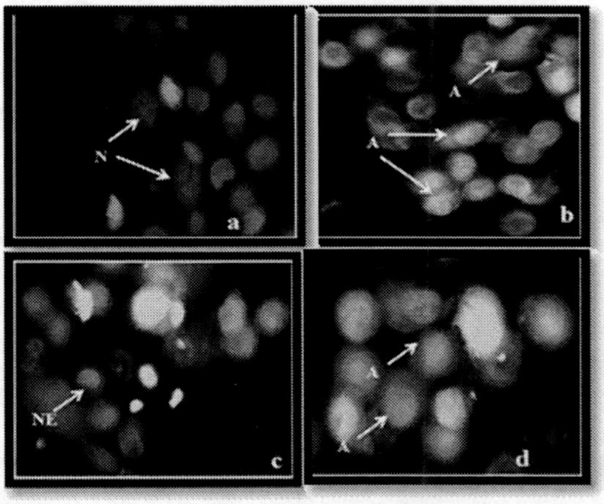

Fig. 2: N – Normal cells, A – Apoptotic cells, NE – Necrotic Cells (a) Control, (b) Cells + Methanolic Leaf Extract of *Euphorbia hirta*, (c) Cells + Tamoxifen, (d) Cells + Methanolic Leaf Extract of *Euphorbia hirta* + Tamoxifen

Propidium Iodide Staining

Cells treated with methanolic leaf extract of *Euphorbia hirta* showed moderate increase in the population of cells undergoing apoptosis. The other group of tamoxifen and methanolic leaf extract of *Euphorbia hirta* showed an increased population of apoptotic cells. The present study explains that the combinatorial effect of drug and leaf extract caused effective control of MCF-7 cell proliferation and it is due to synergetic effect. In the case of leaf extract alone group it showed moderate effect in staining of nucleus. These findings suggest that, the methanolic leaf extract of *Euphorbia hirta* along with tamoxifen induced membrane damage mediated cell death.

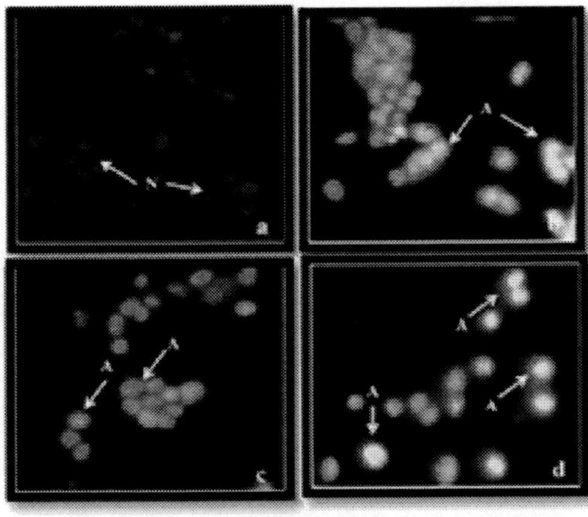

Fig. 3: N – Normal cells, A – Apoptotic cells (a) Control, (b) Cells + Methanolic Leaf Extract of Euphorbia Hirta, (c) Cells + Tamoxifen, (d) Cells + Methanolic leaf Extract of *Euphorbia hirta* + Tamoxifen.

DAPI Staining

Cells treated with methanolic leaf extract of *Euphorbia hirta* showed moderate increase in the population of cells undergoing nuclear fragmentation. The other group of tamoxifen and methanolic leaf extract of *Euphorbia hirta* showed an increased population of nuclear fragmentation.

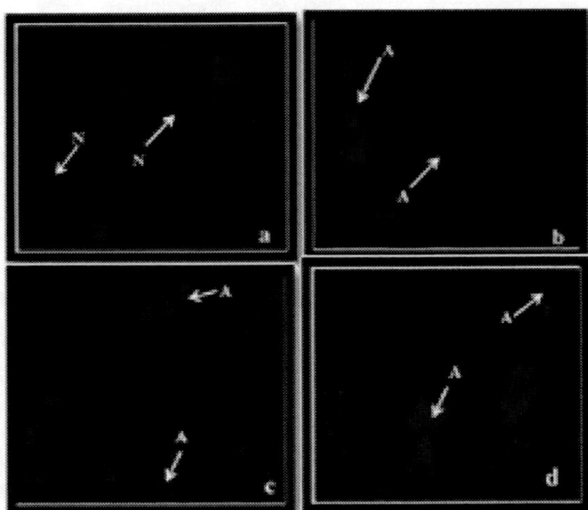

Fig. 4: N – Normal Cells, A – Apoptotic Cells (a) Control, (b) Cells + Methanolic Leaf Extract of *Euphorbia hirta*, (c) Cells + Tamoxifen, (d) Cells + Methanolic Leaf Extract of Euphorbia Hirta + Tamoxifen

Hoechst Staining

Cells treated with methanolic leaf extract of *Euphorbia hirta* showed moderate increase in the population of cells undergoing nuclear fragmentation. The other group of tamoxifen and methanolic leaf extract of *Euphorbia hirta* showed an increased population of nuclear fragmentation.

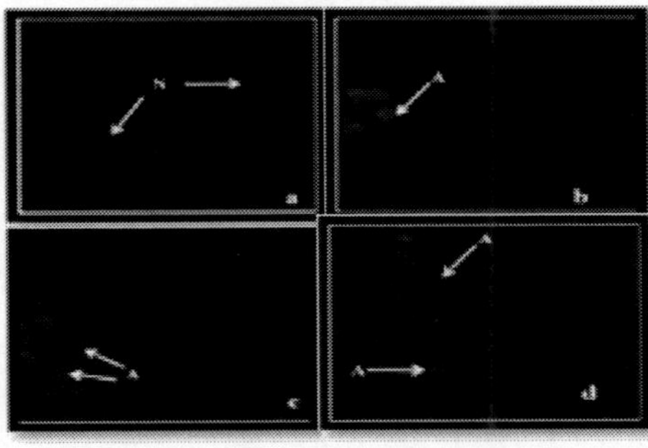

Fig. 5: N – Normal Cells, A – Apoptotic Cells (a) Control, (b) Cells+ Methanolic Leaf Extract of *Euphorbia hirta*, (c) Cells + Tamoxifen, (d) Cells+ Methanolic Leaf Extract of *Euphorbia hirta* + Tamoxifen

DNA Fragmentation

In the present study it was evident that methanolic leaf extract of *Euphorbia hirta* caused DNA fragmentation in MCF-7 cells as indicated by decrease in the intensity of the bands in Lane 2. In Lane 4 the cells treated with tamoxifen and methanolic leaf extract of *Euphorbia hirta* it was noticed that prominent fragmentation of DNA has occurred and this effect was more pronounced and evident by DNA laddering. In the other groups only decrease in the intensity of the band was observed.

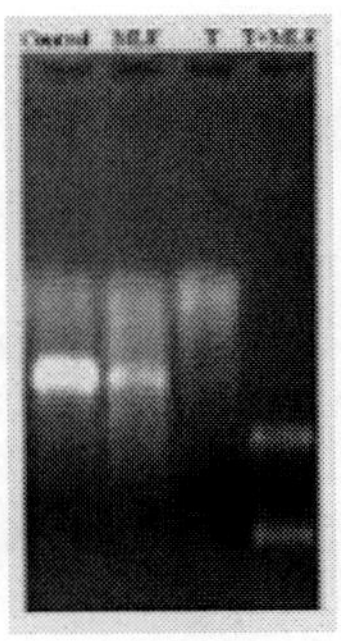

Fig. 6: Control-Cells alone, MLE-Methanolic Leaf Extract of *Euphorbia hirta*, T-Tamoxifen, T+MLE-Tamoxifen with Methanolic Leaf Extract of *Euphorbia hirta*

Flowcytometric Analysis

Flow cytometric analysis confirmed the apoptotic pathway of mechanism of growth inhibition by the methanolic leaf extract of *Euphorbia hirta* in MCF-7 cells. Cells treated with the extract combined with tamoxifen induced apoptosis and the intensity of the cells was more in late apoptotic phase.

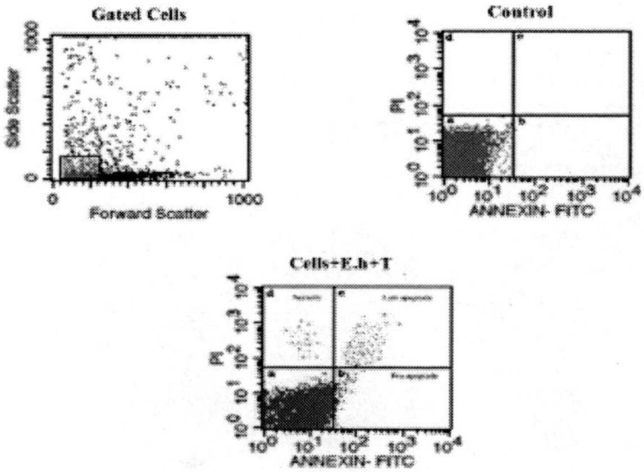

Fig. 7

Thus the leaves of *Euphorbia hirta* have potential apoptotic activity in MCF-7 cells. These extracts can be used to treat the breast cancer as a phytotherapeutic agent. It also justifies and validates the combination of standard anticancer chemotherapeutic drug with plant based extracts to improve anticancer therapy.

REFERENCES

Castro, A.J.A., Villarreal, M.L., Olivo, L.A.S., Sanchez, M.G., Dominquez, F. and Carranca, A.J. (2011). Mexican medicinal plants used for cancer treatment: Pharmacological, phytochemical and ethnobotanical studies, *J. Ethnopharmacol*, 133, 945–972.

Dai, Z., Gao, J., Li, Z., Ji, Z., Kang, H., Guan, H., Diao, Y., Wang, B. and Wang, X. (2011). *In vitro* and *In vivo* antitumor activity of *Scutellaria Barbate* extract on murine liver cancer, Molecules, 16, 4389–4400.

Engel, N., Oppermann, C., Falodun, A. and Kragl, U. (2011). Proliferative effects of five traditional Nigerian medicinal plant extracts on human breast and bone cancer cell lines, *J. Ethanopharmacol*, 137, 1003–1010.

Ko, Y.M., Wub, T.Y., Wub, Y.C., Chang, F.R., Guh, J.Y. and Chuang, L.Y. (2011). Annonacin induces cell cycle-dependent growth arrest and apoptosis in estrogen receptor related pathways in MCF-7 cells, *J. Ethnopharmacol*, 137, 1283– 1290.

Kuete, V. and Efferth, T. (2011). Pharmacogenomics of Cameroonian traditional herbal medicine for cancer therapy. *J. Ethnopharmacol*, 137, 752–766.

Grape Compound Piceatannol Induces Mitochondria Mediated Apoptosis in Breast Cancer Cells

Karthick V., Shruthi S., Kadalmani B.[1] and Mohamed Ali S.*

Department of Biotechnologe, Rajalakshmi Engineering College, Thandalam, Chennai, Tamil Nadu

[1]Department of Animal Sciences, Bharathidasan University, Tiruchirapalli, Tamil Nadu

E-mail: mohamedali.s@rajalakshmi.edu.in

INTRODUCTION

Over the recent years, there has been growing interest in naturally occurring phytochemical compounds with anticancer potential owing to their non toxicity, inexpensiveness and availability in ingestive forms. Piceatannol (3, 4, 3', 5'-tetrahydroxy-trans-stilbene) is a natural stilbene, is an analog and a metabolite of Resveratrol belongs to the large group of biologically active substances found in plants. This compound is classified as phytoestrogen because of its ability to interact with estrogen receptor. Numerous beneficial effects of Piceatannol described in the literature involves cardio protective, (Woo *et al.,* 2010); anti-cancer, anti-inflammatory (Kim *et al., 2008),* and antioxidant action (Kita *et al.*, 2012; Vo *et al.,* 2010; Kim *et al.*, 2009; Woolter *et al.*, 2002) an anti-atherosclerotic effect and beneficial effects in the treatment of atherosclerosis (hardening of the arteries) (Choi *et al.*, 2010; Lee *et al.*, 2009). The present study investigated the toxicity, cell cycle, morphological changes, level of anti-oxidant enzymes, mitochondrial trans-membrane potential drop induced by Piceatannol and gene expression pattern of both pro and anti-apoptotic in breast cancer cell line MCF-7.

---

*Corresponding author.

MATERIALS AND METHODS

Both human breast MCF-7 and normal fibroblast cells (NHLC) were cultured in RPMI medium at 37°C in a humidified 5% CO_2 atmosphere. The medium was supplemented with 10% foetal bovine serum, (Gibco), 100 units/ml penicillin, 100 µg/ml streptomycin and 2 mM glutamine. The cells were maintained in an exponential growth in monolayer cultures with regular medium changes. Routine methods and the following instruments were used to study the apoptotic hallmarks and cellular mechanisms such as Confocal Laser Scanning Microscope (CLSM), flowcytometry for morphological and cell cycle analysis, Spectofluorimetry was used for enzyme analysis. mRNA expression study was conducted by quantitative real-time PCR.

RESULTS AND DISCUSSION

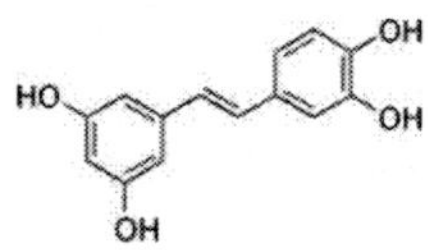

Fig. 1: Chemical Structure of Piceatannol

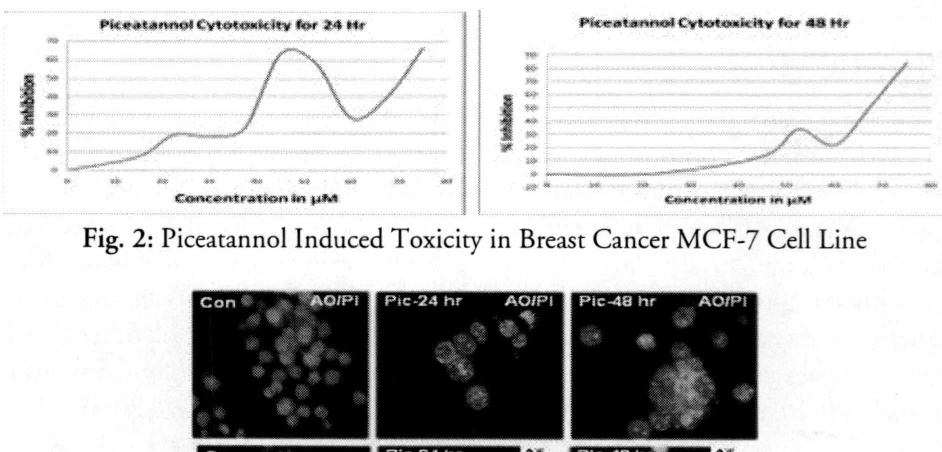

Fig. 2: Piceatannol Induced Toxicity in Breast Cancer MCF-7 Cell Line

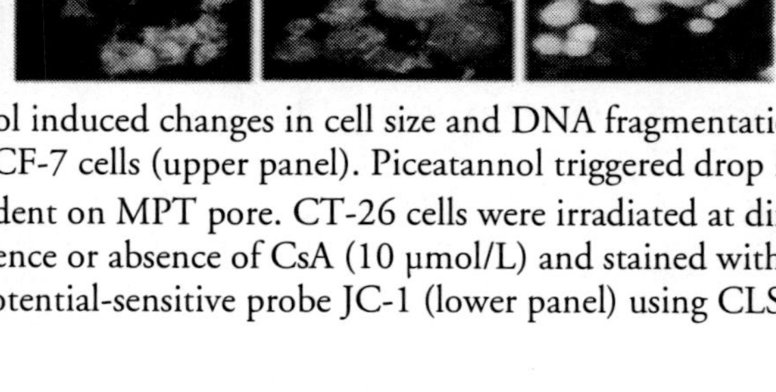

Fig. 3: Piceatannol induced changes in cell size and DNA fragmentation as compared to drug-treated MCF-7 cells (upper panel). Piceatannol triggered drop in mitochondrial $\Delta\psi_m$ is dependent on MPT pore. CT-26 cells were irradiated at different periods in the presence or absence of CsA (10 µmol/L) and stained with membrane potential-sensitive probe JC-1 (lower panel) using CLSM

Breast cancer became a major health problem world-wide among women and affects as many as one in eight women in their lifetime (Jemal *et al.,* 2006). It is the most common cancer among women and the second leading cause of death in women is lung cancer. Defective apoptosis (programmed cell death) is one of the causative factors responsible for tumorigenicity and is implicated in multiple stages of cancer development and progression (Burns and Deiry, 2003). Additionally, the ability of tumor cells to evade apoptosis plays a significant role in promoting resistance to conventional chemotherapy and radiation therapy (Daniel *et al.,* 2001).

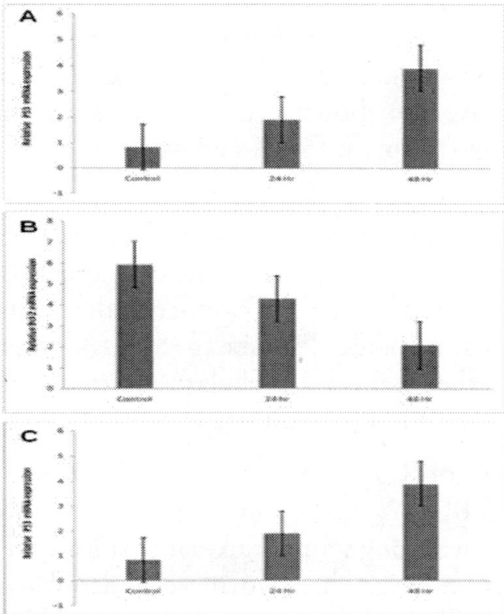

Fig. 4: Piceatannol treated cells induced pro-apoptotic Bax mRNA was gradually increased as early as from 24 hrs to 48 hrs (A) whereas the anti-apoptotic Bcl-2 mRNA is indeed decreased (B). Note there is a dramatic increase of p53 mRNA expression confirming the apoptotic mediated cell death in Piceatannol treated cells compared to non-control cells.

The present study focused on the anti cancer potential of the Picaetannol in breast cancer cell line in which the compound significantly inhibits the cell proliferation and induced cell death. Apoptosis is essential during development and in the maintenance of tissue homeostasis, and serves to delete damaged or redundant cells. Viable and apoptotic cells were identified by fluorescence microscopy using Acridine Orange and Ethidium Bromide uptake, the number of apoptotic cells and the total numbers of cells were counted using 40X magnification. Only the condensed or fragmented nuclei were considered as apoptotic cells. At least 150 cells were counted per area and each apoptotic cell was evident with hallmarks like shrunken cytoplasm, condensed and hyperchromatic chromatin and fragmented nuclei. Biochemical features of Apoptotic cells exhibit several biochemical modifications such as protein cleavage, protein cross-linking, DNA breakdown, and phagocytic recognition that together result in the distinctive structural pathology described elsewhere (Ali *et al.,* 2002).

DNA fragmentation has been considered as gold standard method of apoptosis. By using Hoechst 33258 staining, the present study has revealed some of the molecular changes subsequent to the treatment of the cells with Picaetannol, the extent of DNA damage was increased proportionately with the treatment period. It is well established that p53 expression is up regulated in apoptosis as this controls the expression of many genes implicated in apoptosis (BCL-2 and BAX) or cell cycle regulation. Our results have also shown that p53 expression was increased in MCF-7 cells upon the Piceatannol treatment for 24 and 48 hr.

Glutathione Peroxidase (GPx) is a classical intracellular enzyme that catalyzes the reduction of hydrogen peroxide (H_2O_2) to water and that of a variety of organic hydro-peroxides to the corresponding alcohols, with a specific requirement for the reductant substrate, glutathione (GSH). Together with Catalase (CAT), which also reduces H_2O_2, as well as superoxide dismutases (SODs), which scavenge superoxide radicals, GPx provides the main intracellular enzymatic defense system against reactive oxygen species (ROS) in the mammalian body.

Recently, the involvement of ROS in apoptotic cell death was reported (Possel *et al.,* 1997; Li *et al.,* 2012). BCL-2, a protein that inhibits apoptosis, is reported to suppress lipid peroxidation through an antioxidant pathway. Proto-oncogene BCL-2 appeared to function by inhibiting the mitochondria depolarization. Conversely, Bax induced mitochondria depolarization. The ratio of Bcl-2 to Bax has been reported to be correlated with apoptosis in cancer cells (Bras *et al.,* 2005 Kowaltowski *et al.,* 2009). Indeed our results confirmed that Bcl-2 ratio decreases with an increase in Bax. In addition to that Piceatannol treated cells compared to the control have shown a consistent increase in p53 level confirming that p53 dependent apoptotic cell death has occurred.

A major mechanism responsible for detoxification of radicals in mammalian cells is the breakdown of ROS by scavengers such as GPx, and SOD. The former could specifically catalyse the GSH to reduce the hydroperoxide (Yano *et al.,* 1995; Jing *et al.,* 1999), and the latter could clear the superoxide anion free radical. Previous studies have shown that Piceatannol has both pro and anti-apoptotic activity (anti-oxidant) depending on the concentration used, whereas, in this study it was found to be pro-oxidative. Results from our study showed that Piceatannol enhanced the sensitivity of breast cancer cells by increasing the level of ROS and decreasing GSH level as reported earlier by Dai *et al.* (1999). To summarize, Piceatannol induces apoptosis by depleting the intra cellular GSH, whereas GPx potentiates ROS production but not Catalase.

The increase in ROS production enhances the loss of mitochondrial membrane depolarization and cell cycle arrest, thus potentiating apoptosis in MCF-7 cells.

ACKNOWLEDGEMENTS

Authors thank Dr. Kadalmani, Bharathidasan for providing laboratory space to carry out this entire work reported here.

REFERENCES

Ali, S.M. and Olivo, M. (2002). Bio-distribution and subcellular localization of Hypericin and its role in PDT induces apoptosis in cancer cells. *Intl J Oncol.* 21: 531–540.

Bras, M., Queenan, B. and Susin, S.A. (2005). Programmed Cell Death via Mitochondria: Different Modes of Dying Biochemistry (Moscow), 70: 231–239.

Burns, T.F. and El-Deiry, W.S. (2003). Microarray analysis of p53 target gene expression patterns in the spleen and thymus in response to ionizing radiation. *Cancer Biol Ther.* 2: 431–443.

Choi, K.H., Kim, J.E., Song, N.R., Son, J.E., Hwang, M.K., Byun, S., Kim, J.H., Lee, K.W. and Lee, H.J. (2010). Phosphoinositide 3-kinase is a novel target of piceatannol for inhibiting PDGF-BB-induced proliferation and migration in human aortic smooth muscle cells. *Cardiovasc Res.* 85: 836–44.

Dai J., Weinberg, R.S., Waxman, S. and Jing, Y. (1999). Malignant cells can be sensitized to undergo growth inhibition and apoptosis by arsenic trioxide through modulation of the glutathione redox system. *Blood.* 93: 268–277

Daniel, B. (2002). Treatment options after failure of radiation therapy. *A Review Rev Urol 4* (Suppl): S12–S17.

Jemal, A., Siegel, R., Xu, J. and Ward, E. (2010). Cancer statistics. *CA Cancer J Clin.* 60: 277–300.

Kim, E.J., Park, H., Park, S.Y., Jun, J.G., Park, J.H.( )2009. The grape component Piceatannol induces apoptosis in DU145 human prostate cancer cells via the activation of extrinsic and intrinsic pathways. *J Med Food.* 12: 943–51.

Kim, Y.H., Kwon, H.S., Kim, D.H., Cho, H.J., Lee, H.S., Jun, J.G., Park, J.H. and Kim, J.K. (2008). Piceatannol, a stilbene present in grapes, attenuates dextran sulfate sodium-induced colitis. *Int Immunopharmacol.* 8: 1695–702.

Kita, Y., Miura, Y. and Yagasaki, K. (2012). Antiproliferative and Anti-Invasive Effect of Piceatannol, a Polyphenol Present in Grapes and Wine, against Hepatoma AH109A Cells. *J Biomed Biotechnol.*

Kowaltowski, A.J., de Souza–Pinto, N.C., Castilho, R.F. and Vercesi, A.E. (2009). Mitochondria and reactive oxygen species. *Free Radical Biol Med.* 47: 333–343.

Lee, B., Lee, E.J., Kim, D.I., Park, S.K., Kim, W.J. and Moon, S.K. (2009). Inhibition of proliferation and migration by Piceatannol in vascular smooth muscle cells. *Toxicol in Vitro.* 23:1284–91.

Li, L., Ishdorj, G,C. and Gibson, S.B. (2012). Reactive oxygen species regulation of autophagy in cancer: Implications for cancer treatment. *Free Radic Biol Med.* Jul 19.

Possel, H., Noack, N., Augustin, W., Keilhoff, G. and Wolf, G. (1997). 2, 7–Dihydrodichlorofluorescein diacetate as a fluorescent marker for peroxynitrite formation. *FEBS Lett.* 416: 175 –178.

Vo, N.T., Madlener, S., Bago-Horvath, Z., Herbacek, I., Stark, N., Gridling, M., Probst, P., Giessrigl, B., Bauer, S., Vonach, C., Saiko, P., Grusch, M., Szekeres, T., Fritzer-Szekeres, M., Jäger, W., Krupitza, G. and Soleiman, A. (2010). Pro- and anticarcinogenic mechanisms of piceatannol are activated dose dependently in MCF-7 breast cancer cells. *Carcinogenesis.* 31: 2074–81.

Wolter, F., Clausnitzer, A., Akoglu, B., Stein, J. (2002). Piceatannol, a natural analog of resveratrol, inhibits progression through the S phase of the cell cycle in colorectal cancer cell lines. *J Nutr.* 132: 298–302.

Woo, A., Min, B. and Ryoo, S. (2010). Piceatannol-3'-O-beta-D-glucopyranoside as an active component of rhubarb activates endothelial nitric oxide synthase through inhibition of arginase activity. *Exp Mol Med.* 42: 524–32.

Yano, T., Takahashi, S. and Ichikawa, T. (1995). Active oxygen generated in the process of carcinogen metabolism can induce oxidative damage in nuclei. *Res Commun Mol Pathol Pharmacol,* 87: 367–370.

Yongkui, Jing; Jie, Dai; Ruth, M.E., Chalmers-edman; Willam, G., Tatton and Samuel, Waxman (1999). Arsenic trioxide selectively induces acute promyelocytic leukemia cell apoptosis via a peroxide dependent pathway, *Blood,* 94: 2102–2111.

Biomedical Potential of Parmeloid Lichens Obtained from Local Shop of Erode and Tripur Districts of Tamil Nadu

R. Nanthini*, M. Mithirai, G. Ayyappadasan, P. Ponmurugan, S. Sanjeeva Nayaka[1] and T. Madhu Preetha

Department of Biotechnology, K.S.R. College of Technology, Tiruchengode

[1]Lichenology Laboratory, National Botanical Research Laboratory, Rana Pratap Marg, Lucknow

E-mail: ayyappadasan07@gmail.com

INTRODUCTION

Antioxidant is a molecule capable of inhibiting the oxidation of other molecules. Oxidation is a chemical reaction that transfers electron from a substance to an oxidizing agents. Oxidation reactions can produce free radicals. In turn these radicals can start chain reaction. The chain reaction occurs in a cell. It can cause damage or death to the cell. Antioxidants are widely used as ingredients in dietary supplements and have been investigated for the prevention of diseases such as cancer, coronary. Lichens are organisms formed by symbiotic association and integration of an alga and a fungus. Lichen symbiosis is very successful one and thallus shows no resemblance to either partner. It is slow growing associations of fungi (mycobionts) and photosynthetic partners (photobionts). That may be unicellular green algae or cyanobacteria. It is present in a wide range habitats throughout the world and dominates terrestrial ecosystem (80%). 20000 species of lichens have been reported globally. The Indian is sub continent has 2450 species of lichens of which India alone has about 2040 species (Aswathi, 2000). It extends from Orissa to Tamil Nadu. They occupy a vast range of habitats and substrates are from yercaud hills.o lichens produce a wide range

*Presenting author.

of an organic compounds that can be divided into two groups called primary metabolites and secondary metabolites (Elix., 1996). Secondary metabolites are produced by the fungus aloneand secreted onto the surface of lichens hypae either in amorophous forms or as crystals. The chemistry of about one third of all lichens species as has been studied upto now 350 secondary metabolites are known from lichens.

Phenolic constituents from the lichen *Parmotrema stuppeum* including methyl orsenillate, orsenillic acid, atranorin and lecanoric acid showed moderate antioxidant activity (Jayapraksha and Rao, 2000). An animal study reported antioxidant activity of lichen *Cetraria islandica* which revealed that the results were encouraging significantly (Gülçin *et al.*, 2002). Stictic acid derivatives from the lichen *Usnea articulata* were reported to have significant antibiotic properties (Lohèzic-Le Dèvèhat *et al.*, 2007). Pannarin obtained from lichens inhibited cell growth and induces cell death in human prostate carcinoma DU-145 cells (Maier *et al.*, 1999).

The orcinol derivatives, tenuiorin and methyl orsellinate present in extract of *Peltigera leucophlebia* exhibited *in vitro* inhibitory activity. A correlation has been observed between 5-lipoxygenase inhibition and antiproliferative effects for related lichen metabolites. On this account, tenuiorin and methyl orsellinate were further tested for antiproliferative activity on cultured human breast, pancreatic and colon cancer cell lines. Methyl orsellinate lacked antiproliferative activity but tenuiorin depicted moderate activity (Ingolfsdottir, 2002). Bianthraquinone glycosides, colleflaccinosides isolated from *Collema flaccidum* collected in Israel and Russia, were reported to have antitumor activity (Rezanka and Dembitsky, 2006).

MATERIALS AND METHODS

Sample Collection and Maintenance

Lichens were purchased from Local market of Erode and Tirupur district. Purchases of lichens were obtained from four areas of Erode main bazaar, Tirupur main and Villages. It was maintained in refrigerator (4° Celsius) for further use.

Lichen Identification

Lichen samples were identified based on the chemical and morphology appearance. Samples were authenticated by National Botanical research Institute, Lucknow.

K Test

Lichens were placed on a dish and a drop of Potassium Hydroxide was applied to the lichen specimen. The lichen was then observed for color changes using a microscope at time intervals of fifteen seconds, one minute and two days.

C Test

Samples designated to receive the C test were placed on a white filter paper, and then each sample (while still on the filter paper) was positioned individually on a dish. A drop of Chlorine bleaches solution (household bleach) to the specimen. The specimens were observed with a microscope at intervals of fifteen seconds, thirty seconds and one minute for color changes.

Pd Test

The Pd test was performed by placing the lichen, each on a separate sheet of filter paper, on a dish and the para-phenylenediamine was dropped onto the specimens by squeezing the tube to drop approximately 2 ml of para-phenylenediamine on the specimen. The specimen was then observed using a microscope at time intervals of fifteen seconds, one minute and two days for changes in color.

Preparation of Lichen Extract

The solid material (lichens) was packed in the filter paper and placed inside the soxhalet extractor flask. The solvent was taken in the round bottom flask. The Condenser was placed on top of the soxhalet by connecting inlet and outlet. The round bottom flask is placed in heating mantle. The temperature was set based on the boiling point of the solvent. The solvent was used based on their polarity in increase order like choloroform, acetone, ethanol, ethyl acetate, water respectively. Then the extracts were obtained from the Rotatory evaporator at low pressure point.

Total Phenolic Content

The total phenolic content was determined using Folin-Ciocalteu method. Gallic acid monohydrate was used as standard. The total phenolic content was expressed as grammes of gallic acid equivalents (GAE) per 100 g extract.

Antimicrobial Activity (Well diffusion method)

Antimicrobial activity of different region of lichen extracts were tested against pathogenic yeast such as *Candida albicans, Candida tripicas* and *Candida krusi.* The potato dextrose agar was prepared and poured in petriplates. Then the fungal species such as *Candida albicans, Candida tripicas, Candida kursi* is swapped.The antifungal disc such as conozole, itraconozole, tetracycline was used for the measurement of zone of inhibition.

DPPH Free Radical Scavenging Activity

The free radical scavenging activity of the samples were measured by using 1, 1-diphenyl-2-picryl-hydrazyl (DPPH).

$$\text{DPPH scavenging effect (\%)} = [(A_0 - A_1)/A_1] \times 100$$

Where,

A_0 – absorbance of the control reaction.

A_1 – absorbance in the presence of the sample.

The extract concentration providing 50% inhibition (EC50) was calculated from the graph of DPPH scavenging effect against extract concentration. Methanol was used as control.

Hydrogen Peroxide Radical Scavenging Activity

The hydrogen peroxide radical scavenging activity was assessed by Ruch *et al.* (1989) method using methanolic extract of lichens sample of 200–1000 µg/ml were added to a 0.6 ml hydrogen peroxide (40 mM) with the already prepared phosphate buffer (pH 7.4). The reaction mixtures were incubated at room temperature for 10 mins. After incubation, the reaction mixture read at 230 nm against the blank solution with phosphate buffer (pH 7.4). The percentage of inhibition calculated based on the formula.

Cell Viability Assay

MTT assay was conducted to assay cytotoxicity and cell viability, dependent on the conversion of yellow tetrazolium salt to purple formazan product. Cells (1×10^5 cells/mL) were grown on a 96-well plate. Cells were treated with LMG (0–200 µg/mL) or LMG (0–150 (g/mL) combined with 200 (M H_2O_2 for 24 h, after which an MTT solution (1 mg/mL in PBS) was added, and cells were incubated for

2 h. After the supernatant was discarded and DMSO was added to the well, the absorbance at 570 nm was measured using a spectrophotometer.

RESULTS

Lichen samples of different location were studied and it was estimated for the potential value for biological applications like antimicrobial and anticancer potential. Lichen samples of different places has mixer of Parmeloid lichens such as *Parmotrema tinctorum, Pamotrema retticulatum, Parmotrema hababianum parmotrema austrosinense* and some of the fruticose lichens such as Ramaalina sps. The photometric evaluation of the antioxidant capacity of the ethanolic extract of Erode city variety showed good antioxidant capacity than the tirupur city. Significant decreases were observed in the DPPH radical activity due to the scavenging ability of the extracts. The IC_{50} value of the ethanolic extract of erode and tirupur city and standard antioxidant (ascorbic acid) such as 90 µg/ml, 160 µg/ml and 110 µg/ml respectively. A lower IC_{50} value indicates a higher free radical scavenging activity. The scavenging ability of the both extracts were obtained L-ascorbic acid was used as standard antioxidant and found the IC_{50} value 300 µg/ml. Tirupur village variety posses scavenging activity in high compared to erode village and IC_{50} values were 350 µg/ml and 450 µg/ml respectively. Anticancer activity has showed better activity in cell line.

Fig. 1: Lichen Sample at Tree Bark

Table 1: Lichen Sample at Market Place

S. No.	*Variety*	*Species*
1.	Erode city	Parmotrema retticulatum and parmotrema tinctorum
2.	Erode village	Parmotrema austrosinense and parmotrema retticulatum, some of the Ramaliana sps
3.	Tirupur Variety	Parmelinella wallichiana and parmotrema hababianum
4.	Tirupur village	Parmotrema retticulatum and Parmotrema tinctorum and unidentified sps.

Table 2: Extractive Values of Lichen Samples

S. No.	*Solvent Name*	*Total Gram*	*Extractive Value (gm)*	*Extractive Value (gm)*	*Extractive Value (gm)*	*Extractive Value (gm)*
1	Chloroform	7	0.119	0.10	0.21	0.09
2	Ethyl acetate	7	0.051	0.14	0.19	0.12
3	Acetone	7	0.366	0.18	0.32	0.17
4	Ethanol	7	0.090	0.11	0.18	0.2
5	Water	7	0.187	0.09	0.2	0.10

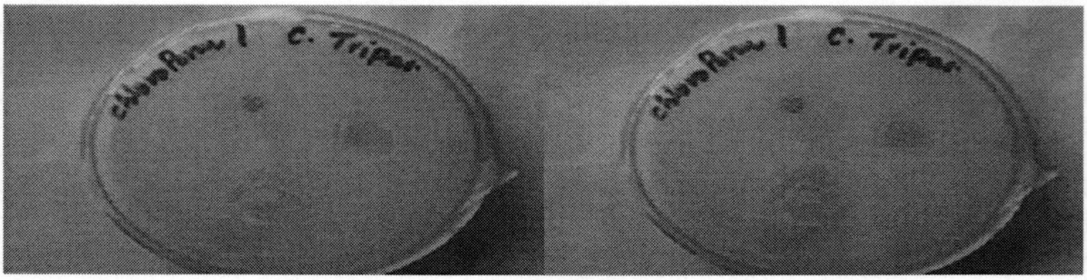

Fig. 2: Zone of Inhibition of Lichen Extracts

Table 3: Zone of Inhibition of Different Variety of Lichen Samples

S. No.	*Extract*	*Organism*	*Zone of Inhibition (mm) Erode Town*	*Zone of Inhibition (mm) Erode Village*	*Zone of Inhibition (mm) Tirupur Town*	*Zone of Inhibition (mm) Tirupur Village*
1.	Ethanol	*Candida tripicus*	14	12	14	16
2.	Ethanol	*Candida kursi*	12	11	13	18
3.	Ethanol	*Candida albicans*	13	15	08	17
4.	Chloroform	*Candida tripicus*	16	13	10	15
5.	Ethyl acetate	*Candida tripicus*	11	12	11	13

Table 4: DPPH Assay of Erode Variety of Lichen Sample

S. No.	*Solvant Extract*	*30 min %*	*40 min %*	*50 min %*
1.	Chlorofor m	0.97	1.74	–ve
2.	Ethyl acetate	70	47	42
3.	Acetone	–ve	–ve	–ve
4.	Ethanol	84.01	70.65	62.92
5.	Water	15.3	3.13	–ve

Table 5: Effect of Different Extracts of *Parmotrema reticulatum*

Concentration	*18.75 µg*	*37.5 µg*	*75 µg*	*150 µg*	*300 µg*	*Cont*
ABS	0.64	0.453	0.229	0.018	0	0.624
	0.646	0.485	0.204	0.022	0	0.65
	0.611	0.468	0.215	0.012	0	0.642
Avg	0.632333	0.468667	0.216	0.017333	0	0.638667
Conc (µg)	% Cell Inhibition					
18.75	0.991649			IC 50	56.62 µg/ml	
37.5	26.61795					
75	66.17954			R^2	0.9951	
150	97.28601					
300	100					

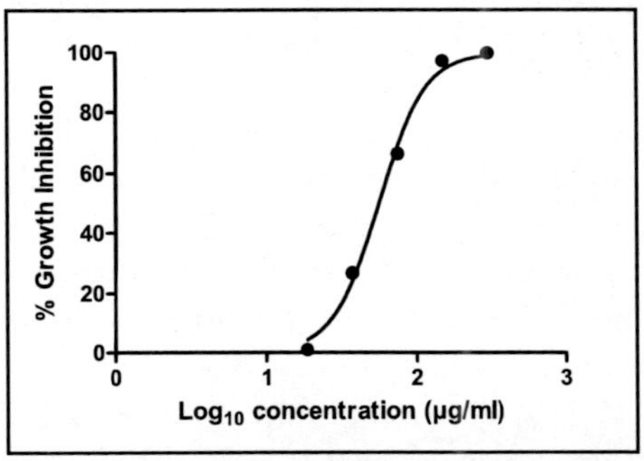

Fig. 3: Effect of Inhibition of Hela Cell Lines Using *P. retticulatum* Extracts

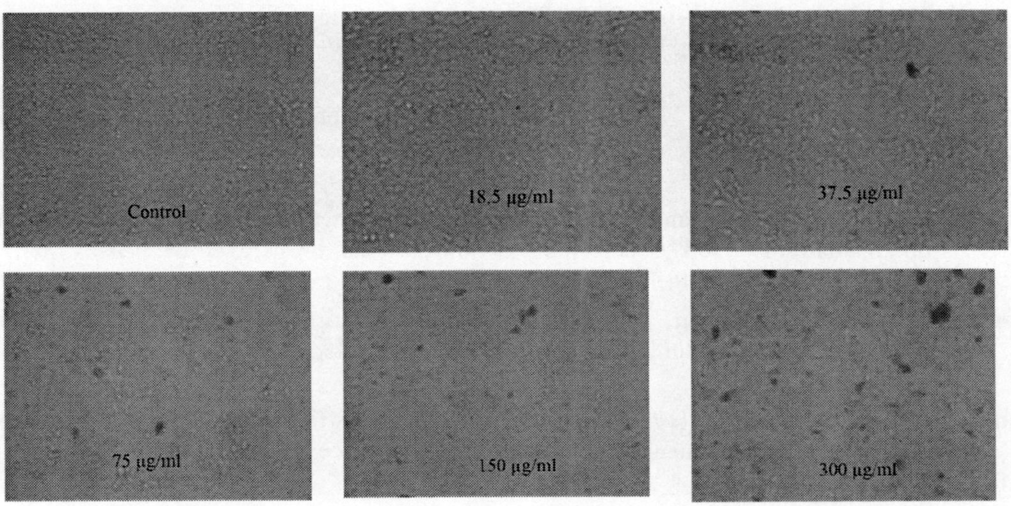

Fig. 4: Effect of Different Concentration of Extracts of *Parmotrema reticulatum*

ACKNOWLEDGEMENTS

This work was supported by Indian Council of Medical Research, New Delhi for providing financial Supports and also the Management of K.S. Rangasamy College of Technology, Tiruchenogde for providing the lab facility. The author is very grateful to National Botanical Research Institute for their Guidance and Constant support towards this project.

REFERENCES

Crittenden, P.D. and Porter, N. (1999). Lichen forming fungi: Potentialsources of novel metabolites. *Trends Biotechnol,* 9: 409–414.

Desai, I.D. (1984). Vitamin E analysis methods for animal tissues. *Methods Enzymol,* 105: 138–147.

Fyson, P.F. (1997). The flora of the South Indian hill stations: Ootacamund, Coonoor, Kotagiri, Kodaikanal, Yercaud, and the country round. Today and Tomorrow's Printers and Publishers, New Delhi.

Gülçin I.. Boustie, J. and Grube, M. (2002). Determination of antioxidant activity of lichen *Cetraria islandica* (L) Ach. *Journal of Ethnopharmacol,* 79: 325–329.

Gupta, V.K., Darokar, M.P., Saikia, D., Pal, A., Fatima, A. and Khanuja, P.S. (2007). Antimycobacterial activity of lichens. *Pharmaceutical Biol,* 45: 200–204.

Habig, W.H, Pabst, M.J. and Jakoby, W.B. (1974). Glutathione-S-transferases. The first enzymatic step in mercapturic acid formation. *J. Biol. Chem,* 249: 7130–7139.

Hill, D.J. and Ahmadjian, V. (1973). Relationship between carbohydrate movement and the symbiosis in lichens with green algae. *Planta* 103: 267–277.

Ingolfsdottir, K. (2002). Usnic acid. *Phytochem,* 61 (7): 729–736.

Jayapraksha, G.K. and Rao, L.J. (2000). Phenolic constituents from the lichen *Parmotrema stuppeum* (Nyl.) Hale and their antioxidant activity. *Zeitschrift für Naturforschung* 55 (11–12): 1018–22.

Kadavul, K. and Parthasarathy, N. (1999). Plant biodiversity and conservation of tropical semi-evergreen forest in the Shervarayan hills of Eastern Ghats, *India Biodiversity and Conservation* 8: 419–437.

Karthikaidevi, G., Thirumurugan, G., Manivannan, P., Anantharaman, K., Kathiresan, K. and Balasubramanian, T. (2009). Screening of antibacterial properteies of lichens *Roccella belangeriana* (Awasthi) from Pichavaram Mangrove (*Rhizophora* Spp). *Adv. Biol. Res.* 3: 127–131.

Kristmundsdóttir, T., Jónsdóttir, E., Helga M. Ögmundsdóttir and Ingólfsdóttir, K. (2006). Solubilization of poorly soluble lichen metabolites for biological testing on cell lines. *Anti-Cancer Drugs,* 17: 510–512.

LeBlanc, F. and DeSloover, J. (1970). Relation between industrialization and the distributions and growth of epiphytic lichens and mosses in Montreal, *Can. J. Bot.* 48: 1485–1496.

Lohèzic-Le Dèvèhat, F. Fridovich, I. and Mannervik, B. (2007). Stictic acid derivatives from the lichen *Usnea articulata* and their antioxidant activities. *J. Natural Products* 70: 1218–1220.

Loppi, S.; Putorti, E.; Dominics, V. and Barbaro, A. (1995). Lichens as bioindicators of air quality near a municipal solid waste incinerator (Central Italy). *Allionia* 33: 121–129.

Mayer, M.; O'Neill, M.A.; Murray, K.E.; Santos-Magalhaes, N.S.; Carneiro-Leao, A.; Thompson, A.M. and Appleyard, V.C.L. (2005). Usnic acid: a non-genotoxic compound with anti-cancer properties. *Anti-Cancer Drugs*. 16: 805–809.

Misra, H.P. and Fridovich, I. (1972). The role of superoxide anion in the autoxidation of epinephrine and a simple assay for superoxide dismutase. *J. Biol. Chem*. 247: 3170–3175.

Moron, M.S., Depierre, J.W. and Mannervik, B. (1979). Levels of glutathione, glutathione reductase and glutathione-S-transferase activities in rat lung and liver. *Biochem. Biophys. Acta.* 582: 67–68.

Müller, K. (2002). Pharmaceutically relevant metabolites from lichens. *Appl. Microbiol. Biotechnol.* 56 (1/2): 9–16.

Myers, N.; Mittermeier, R.A.; Mittermeier, C.G.; da Fonseca, G.A. and Kents, J. (2000). Biodiversity hotspots for conservation priorities. *Nature* 403: 853–858.

Nanjo, F., Goto, K., Seto, R., Suzuki, M., Sakai, M. and Hara, Y. (1996). Scavenging effects of tea catechins and their derivatives on 1, 1,-diphenyl-2-picrylhydrazyl radical. *Free Radic. Biol. Med.* 21: 895–902.

Biological Synthesis of Gold Nano Particle from *Chryseobacterium* Spp. and its Medical Application in the Cancer Diagnosis

R. Kaviyadhivya* and R. Subbaiya

Department of Biotechnology, K.S. Rangasamy College of Technology, Thiruchengode

E-mail: ravijeya.dhivya680@gmail.com

INTRODUCTION

Nanotechnology is a emerging field that is developing in many areas, including the field of medicine, lifescience and in new technologies. Due to their unique properties they are the leading phenomenon in present world. The metal gold has some distinctive features like less oxidation compared to other metals. Gold nanoparticles possesses number of potential applications for instance, an effective drug delivery against microorganism. Recent studies had showed that the synthesis of nanoparticles using microorganisms had led to exciting area of research in the cancer diagnosis. Cancer is a disorder of cells and although it usually appears as a tumour (a swelling) made up of a mass of cells, the visible tumour is the end result of a whose series of changes which may have taken many years to develop (Grzelczak *et al.*, 2008). Cancer is a important cause of mortality worldwide people and the number of people who are affecting is increasing. The origin of cancer can be related to metabolic alteration, such as mitochondrial increase of glycolysis, which largely depends on this metabolic pathway needed to convert glucose to pyruvate for the generation of ATP to meet the cancer cell energy need (Sperling *et al.*, 2008). The present study demonstrates the efficacy of biologically synthesised silver nanoparticles (AgNPs) as a anti-tumour agent using lung cancer cell lines *invitro* and *invivo.*

*Corresponding author.

Gold nanoparticles (*Au*NPs) are currently playing a significant role for human welfare in the field of clinical diagnosis as well as several biomedical applications. More and more research shows that *Au*NPs-based technologies are becoming promising approaches in cancer research and AIDS treatment. Gold (*Au*) is unique compared to other metals because of its resistance to tarnishing. According to the earliest records use of *Au* for medical purposes can be traced back to the Chinese civilization in 2500 BC, and after that, several ancient cultures have utilized *Au*-based materials for medicinal purpose for the treatment of a variety of diseases such as smallpox, skin ulcers, measles, and syphilis (Shankar *et al.,* 2008). In today's era of nanotechnology, gold nanoparticles (*Au*NPs) have been used for the treatment of diseases like rheumatoid arthritis, and so forth, while considerable research is currently going on for unveiling potential anticancer and antimicrobial and biodiagnostic applications of *Au*-based materials and compounds for clinical applications (McGilvray *et al.,* 2006).

Diagnosis and treatment of cancer in nascent stages are of great importance because of the widespread occurrence of the disease, high death rate, and the frequency of reoccurrence even after treatment. According to cancer statistics 2010 done by American Cancer Society, estimated new cancer cases were 1,529,560 and 569,490 death in both males and females in the US.

The gold nanoparticles was produced by both chemical and biological synthesis using tri sodum citrate as reducing agent and *using chryseobacterium* in biological synthesis. GNPs are easy of synthesis, functionalization and biocompatibility (Shree R.singh *et al.,* 2011). Actinomycetes have the ability to produce antibiotics and *Chryseobacterium* which comes under actinomycetes are gram negative and widely distributed in soil has eminent properties in the synthesis of gold nanoparticles. The synthesized GNPs for drug delivery is the preliminary application in a particular target. The main objective of this study is that synthesizing the gold nanoparticle and with that nanoparticle the cancer will be diagnoised by the targeted drug delivery system. The cancer activity will be tested using the animal cell line and further implemented on the animal model (mouse) for checking the activity of gold nanoparticle having the activity against the cancer.

MATERIALS AND METHODS

Collection of the Sample

The soil sample was collected from the costal region of Rameshwaram, Ramanathapuram district, Tamil Nadu. From that soil sample the *Chryseobacterium* spp was isolated.

Synthesis of Gold Nanoparticles

Chemical Reduction Method

The chloroauric acid [0.02M] was reduced to gold nanoparticles by using trisodium citrate [0.01M] which acts as reducing agent when boiled at 80 degree Celsius for 30 minutes at water bath where a color change occurred to purple. This method is also called as citrare synthesis method. This shows that chloroauric acid is reduced and gold ions were produced. The solution was filtered with the whattman filter paper and lyophilized at a temperature of-70 degree celcius for 48 hours and further it was characterized at various methods in order to confirm that synthesized particle is nanoparticle or not and its size was characterized.

Biological Synthesis

The biological synthesis method was done by a *Actionomycete* spp called *Chryseobacterium* which produce enzymes namely reductase and dehydrogenase which has the ability to reduce chloroauric acid. The synthesis was done with an ISP2 media. 0.1 M stock solution was prepared and taken at different concentrations (0 mM, 0.5 mM, 0.8 mM, 1 mM, 1.5 mM, 1.8 mM and 2 mM). Two control were taken with and without culture. The samples were made up to 1ml using ISP2 media, and then *Chryseobacterium* was inoculated in all the test tubes and incubated for two days at 37 degree Celsius.

After incubation the color change was obtained from green to yellow.then this medium was inoculated to 100 ml of ISP2 medium with different concentration (0.2 M, 1.5 M and 2 M). The wavelength obtained for concentrations was 533 nm, 528 nm, 541 nm and 325 nm respectively by using UV-Spectrophotometer.there was no color change in both the controls.

The *Chryseobacterium* spp is a gram negative bacterium.It is marine bacteria which was collected from Rameshwaram marine soil. The *Chryseobacterium* has the ability to reduce the chloroauric acid into gold nanoparticle.

Fig. 1: Chryseobacterium

CHARACTERIZATION

The synthesized gold nanoparticles were lyophilized at temperature of-70 degree celcius for 2 days.

UV-Spectrophotometer

Initially the gold nanoparticle was conformed by UV-Spec and the wavelength obtained was at 525 nm with absorbance of 0.488.

SEM Analysis

The synthesized nanoparticles were lyophilized for 3 days at and the sample was conformed as nanoparticles.

X-Ray Diffraction

The presence nanoparticles were conformed in X-Ray diffractionby calculating the devdy scherr equation.

Flurosence Analysis

The flurosence was measured and a nanoparticle was conformed.

RESULT AND DISSCUSION

UV-Spectrophotometer is used to find out the simple molecules in a reaction mixtures. UV visible spec for GNPs from *Chryseobacterium* spp is 525 nm. The UV-Spectrophotometer is used for the initial conformation of the synthesized gold nanoparticle. For the GNPs the wavelength was observed at 528.0 nm and the absorbance range about 0.361. This wavelength initially confirms the produced particle was gold nanoparticle. After the UV Spec readings the nanoparticle was confirmed with other characterization method which includes Scanning Electron Microscopy, Flourimetry, and FTIR methods. The Scanning Electron Microscopy is used to find out the size of the gold nanoparticle produced.

The flourimetry is used to find the fluorescence capacity of the produced gold nanoparticle. The fluorescence was calculated with the fluorescence emission and fluorescence expiation. For the good fluorescence capacity the particle should have the fluorescence value of 1. The FTIR (Fourier Transform Infra Red) method is used to characterize the produced gold nanoparticle. In the FTIR the chemical bonds present in the synthesized particle was identified. Thus with these information the produced particle was confirmed as gold nanoparticle.

The produced gold nanoparticle was applied in the human non small cell lung carcinoma cell line for the identification of the activity of the nanoparticle. The gold nanoparticle was bounded with the drug molecule which was targeted at the site of the cancer. Thus this is technically called as targeted drug delivery system. The gold nanoparticle kills the cancer cells in the cell line studies thus my further work was studying the activity of the gold nanoparticle in the animal model studies.

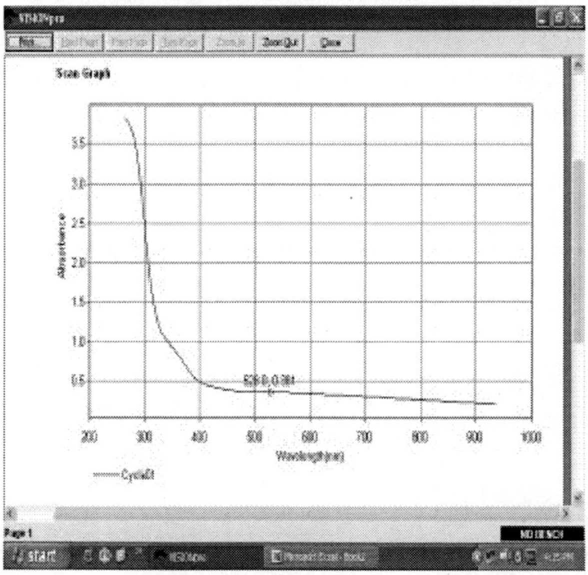

Fig. 2: UV Spectrophotometer

This graph show that the nanoparticles were produced at the nanometer of 528 and absorbance was found at 0.0361. This is the preliminary stage to confirm the nanoparticles by its wavelength.

Flourimetry

This is used to find out the emission and excitation of the GNPs. The result showed by SEM analysis.

The morphology (size and shape) of GNPs was amalysedusing SEM. The size of the GNPs may be elliptical or spherical.

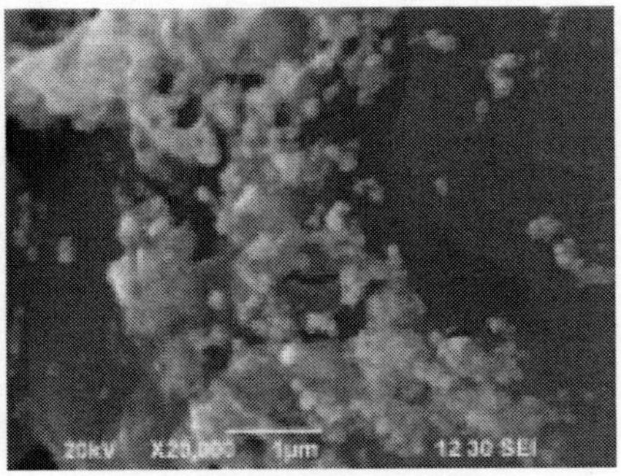

Fig. 3: SEM Analysis

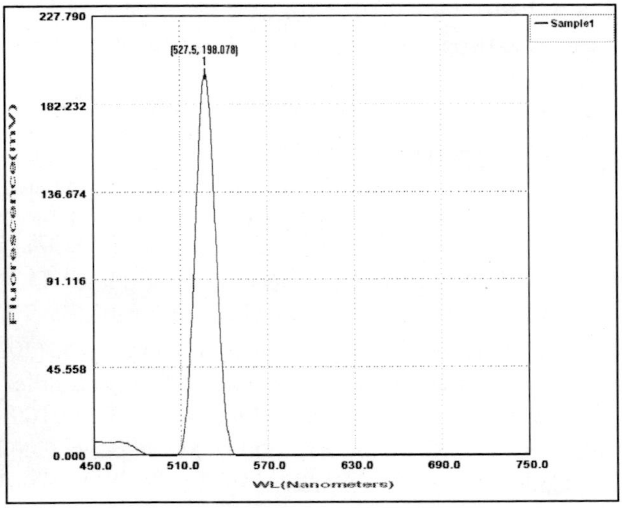

Fig. 4: Fluroscence Excitation

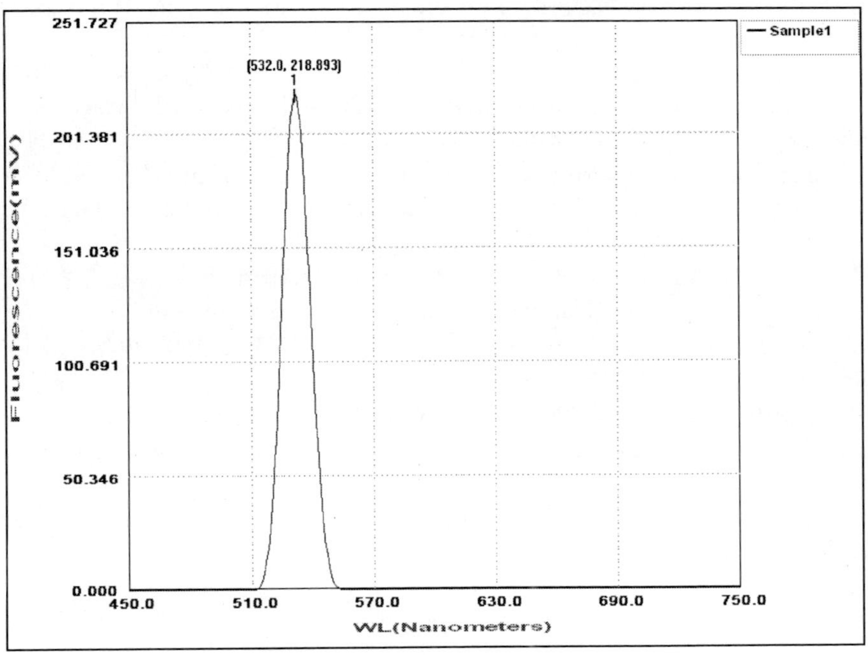

Fig. 5: Fluorescence Emission

The fluorescence can be calculated by using the formula emission/excitation=quantum yield.

The value obtained by using this formula should be less than 1. The fluorescence can be used to estimate fluorescence capacity of gold nano particle the value of fluorescence is less than 1.

REFERENCES

Grzelczak, M., Pérez-Juste, J., Mulvaney, P. and Liz-Marzán, L.M. (2008). Shape Control in Gold Nanoparticle Synthesis. *Chem. Soc. Rev.*, 37: 1783–1791.

Jadzinsky, P.D., Calero, G., Ackerson, C.J., Bushnell, D.A. and Kornberg, R.D. (2007). Structure of a Thiol Monolayer-Protected Gold Nanoparticle at 1.1 Å Resolution. *Science,* 318: 430–433.

Liao, H., Jiang, Y., Zhou, Z., Chen, S. and Sun, S. (2008). Shape-Controlled Synthesis of Gold Nanoparticles in Deep Eutectic Solvents for Studies of Structure-Functionality Relationships in Electrocatalysis. *Angew. Chem. Int. Ed.,* 47: 9100–9103.

*L*ofton, C. and Sigmund, W. (2005). Mechanisms Controlling crystal Habits of Gold and Silver Colloids. *Adv. Funct. Mater.*, 15: 1197–1208.

McGilvray, K.L., Decan, M.R., Wang, D. and Scaiano, J.C. (2006). Facile Photochemical Synthesis of Unprotected Aqueous Gold Nanoparticles. *J. Am. Chem. Soc.,* 128: 15980–15981.

Nie, S. and Emory, S.R. (1997). Probing Single Molecules and Single Nanoparticles by SurfaceEnhanced Raman Scattering. *Science,* 275: 1102–1106.

Pellegrino, T., Kudera, S., Liedl, T., Javier, A.M., Manna, L. and Parak, W.J. (2005). On the Development of Colloidal Nanoparticles towards Multifunctional Structures and their Possible Use for Biological Applications, *Small,* 1: 48–63.

Rosi, N.L., Giljohann, D.A., Thaxton, C.S., Lytton-Jean, A.K.R., Han, M.S. and Mirkin, C. A. (2006). Oligonucleotide-Modified Gold Nanoparticles for Intracellular Gene Regulation. *Science,* 312: 1027–1030.

Shankar, S.S., Rai, A., Ankamwar, B., Singh, A., Ahmad, A. and Sastry, M. (2004). Biological Synthesis of Triangular Gold Nanoprisms. *Nature Materials* 3: 482–488.

Sperling, R.A., Gil, P.R., Zhang, F., Zanella, M. and Parak, W.J. (2008). Biological applications of gold nanoparticles. *Chem. Soc. Rev.,* 37: 1896–1908.

Sperling, R.A., Pellegrino, T., Li, J.K., Chang, W.H. and Parak, W.J. (2006). Electrophoretic Separation of Nanoparticles with a Discrete Number of Functional Groups. *Adv. Funct. Mater,* 1: 943–948.

Templeton, A.C., Wuelfing, W.P. and Murray, R.W. (2000). Monolayer-Protected Cluster Molecules. Acc. *Chem. Res.,* 33: 27–36.

Differential Effect of *Majorana hortensis* Leaf Extract on Etoposide-Induced Apoptosis in Primary Chick Embryo Fibroblasts and HepG2 Cells

Radha Palaniswamy* and P.R. Padma

Department of Biochemistry, Biotechnology and Bioinformatics, Avinashilingam Institute for Home Science and Higher Education for Women, Coimbatore

E-mail: radhapalaniswamy@yahoo.com; prpadma@yahoo.co.in

INTRODUCTION

Cellular oxidative stress is involved in many physiological and pathophysiological processes, including proliferation and senescence or death of normal and cancer cells (Archana, A., 2009). Apoptosis is the hallmark of cancer that is highly regulated and organized cell death process, controlling the development and homeostasis of multicellular organisms. It is characterized by well-defined features like cellular morphological change, chromatin condensation and oligonucleosomal DNA cleavage (Chih, H.W. *et al.*, 2001). These oxidative damages could be retarded by endogenous defense systems such as antioxidant enzymes, but these systems are not completely efficient. Hence, an exogenous source of antioxidants is required. The natural antioxidant reserves in the daily food we eat have gained considerable attention in this regard (Han, P. *et al.,* 2009).

An antioxidant (free radical scavenger) is a compound that inhibits or delays the oxidation of substrates and help in preventing various devastating diseases and cancer (Ji, H. *et al.*, 2012). Local herbalists have been treating various cancers and cancer-related conditions with many plants have been reported to be useful (Kalpana R. *et al.*,

*Corresponding author.

2011). With this backdrop, in the present study, the leaves of the candidate plant, *Majorana hortensis* (*M. hortensis*) have been subjected to a series of analyses to determine the anti-apoptotic and anticancer potential.

METHODOLOGY

Fresh leaves of *M. hortensis* were collected and 1g was homogenized thoroughly in 10ml of methanol, and dried at 60°C, weighed and dissolved in DMSO to obtain a final concentration of 200 μg/5μl of DMSO. Untransformed (primary chick embryo) and transformed cells (HepG2, human liver carcinoma cell line) were used to understand the oxidative stress induced apoptosis, and the effect of the leaf extract on them. The chick embryo fibroblasts were prepared from 8-day-old chick embryo and cultured in (DMEM) as per the procedure described in http://homepages.gac.edu/ ~cellab/chpts/chp12/ex12-10.html. HepG2 cells were procured from NCCS, Pune, India and maintained under standard conditions. Once the cells had attained confluent growth, the cells were trypsinized using Trypsin–EDTA and 10^5 cells were used for various assays. The treatment groups were untreated (negative) control, H_2O_2 treated (positive) control, methanolic extract of *M.hortensis* treated group, H_2O_2 + methanolic extract treated group. The concentration of hydrogen peroxide used for all the assays was 200 μM. The oxidant and plant extract exposure was given for 24 hours. The apoptotic events that occurred after the oxidant treatment were analysed using various parameters-giemsa (Mercille, S. and Massie, B., 1994), ethidium bromide (EtBr) (Montecucco, F. *et al.*, 2009), propidium iodide (PI) (Priyadarsini, R.V. *et al.*, 2010) and DAPI staining (Rashmi, R. *et al.*, 2003) and viewed under a fluorescent microscope.

RESULTS

Effect of *M. hortensis* Leaf Extract on the Morphological Changes in Etoposide Induce Stress in Primary Chick Embryo Fibroblasts and Hep2 Cells: The morphological changes observed in primary chick embryo fibroblasts and Hep2 cells stained with giemsa are depicted in Figure 1(a) and (b). respectively. Etoposide caused a steep increase in the number of cells (cancerous) showing apoptotic morphology in both chick embryo fibroblasts and HepG2 cells; however, the effect of *M. hortensis* leaf extract in the two types of cells, showed a markedly differential response. In the chick embryo fibroblasts, the presence of the extract, along with the oxidant showed a recovery in survival, with a decrease in the apoptotic cells. Whereas, in the case of HepG2 cells, the administration of the leaf extract alone increased the number of apoptotic cells compared to control, indicating anticancer activity of the leaf extract. Co-exposure of the Hep2 cells with leaf extract and etoposide caused a further increase in the number of apoptotic cells. This observation indicates that the *M.hortensis* leaf extract augments the cytotoxicity of the chemotherapeutic agent (etoposide) only in the cancer cells, while protecting the non-cancerous cells from its cytotoxicity. Effect of *M. hortensis* Leaf Extract on the Nuclear Changes of Apoptosis Observed Primary Chick Embryo Fibroblasts and Hep2 Cells: Similar results as in giemsa were observed in EtBr (Figure 2(a) and (b)) in chick embryo and HepG2 cells. The PI staining results were also in accordance with the above for chick embryo and for HepG2 cells). DAPI staining also proved the differential effect as very significant, since the *M. hortensis* leaf extract protects non-cancerous cells from oxidative death at the same time rendering the cancerous cells more susceptible to the chemotherapeutic agent-induced oxidative death.

Control Etoposide

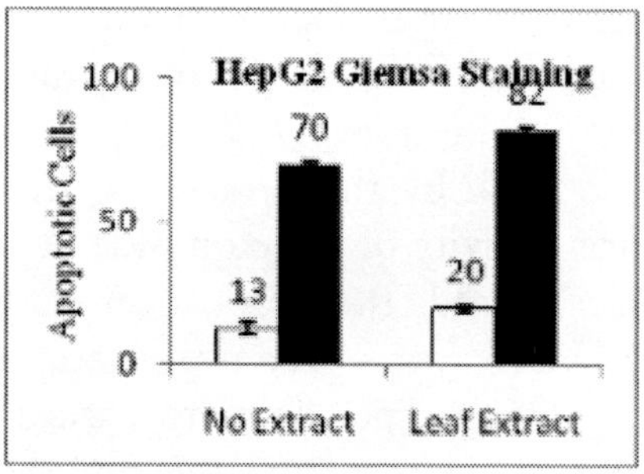

Fig. 1(a)

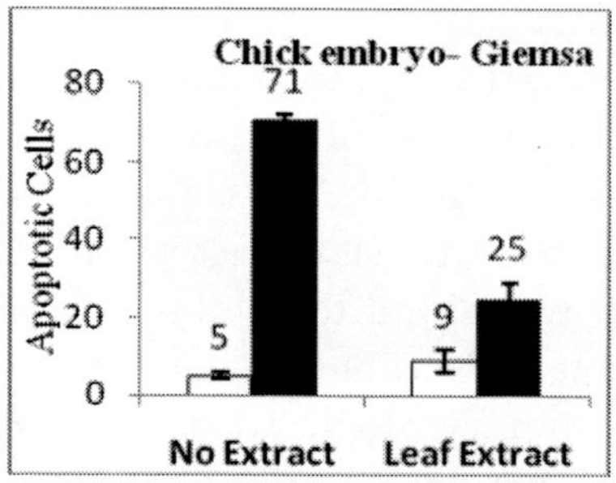

Fig. 1(b)

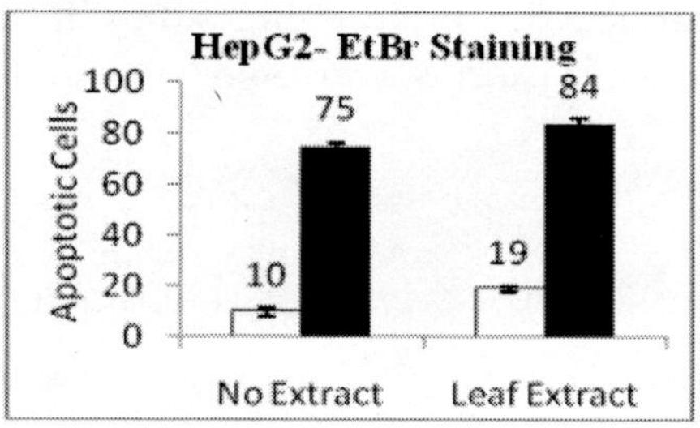

Fig. 2(a)

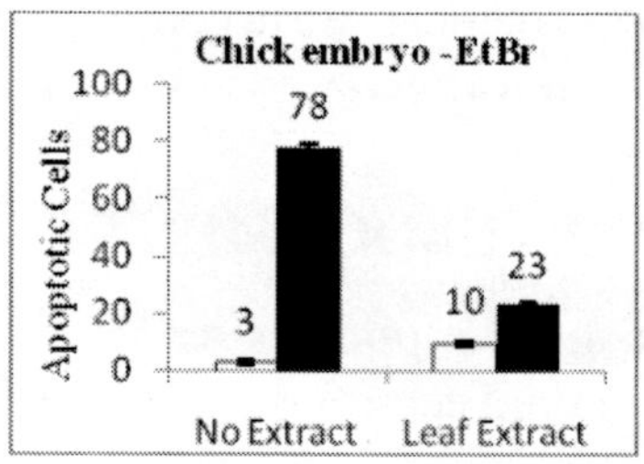

Fig. 2(b)

DISCUSSION

Etoposide caused morphological changes in chick embryo fibroblasts and the *Triticum aestivum* leaves were very effective in decreasing the etoposide-induced apoptosis in chick embryo fibroblasts as determined by giemsa staining (Sarker, K.P. *et al.*, 2000). The effects of tamoxifen were investigated on the growth and apoptosis in the human bile duct carcinoma cell line QBC939 using giemsa staining to characterize the morphological changes (Sharma, P.K. *et al.,* 2011). The apoptosis inducing effect of essential oil of *Ocimum viride* in human colorectal adenocarcinoma (COLO 205) cell line with ethidium bromide (EtBr) for nuclear changes showed apoptotic nuclear changes (Sowemimo, A. *et al.*, 2011). The oxaprozin-induced apoptosis in CD 40L-treated human monocytes was encountered and evaluated by PI staining (Storz, P., 2011). The effect of quercetin on HeLa cells and the changes in the apoptotic nuclei before and after treatment with quercetin was studied in chick embryo fibroblasts with EtBr and DAPI staining (Sun, L. *et al.*, 2011).

SUMMARY

In the chick embryo fibroblasts, etoposide caused an increase in the number of apoptotic cells, but with the co-administration of the leaf extract, this was brought down drastically, indicating the protection rendered by the leaf to the untransformed cells. In the case of Hep2 cells, the steep increase in the number of apoptotic cells due to the presence of etoposide was not reverted back by the presence of the leaf extract. But the leaf extract augmented the anticancer activity of the etoposide. The leaf extract was found to exhibit a differential response towards the non-cancerous (anti-apoptotic effect) and cancerous (anticancer effect) cells in the presence of etoposide. It was not cytotoxic to non-cancerous cells, showing its specificity to cancer cells alone. Therefore, the leaf extract (or its components) are targeting the cancer cells, while sparing the normal cells. This effect of the leaf extract has a very important bearing in cancer treatment, as they can minimize the undesirable side effects of cancer chemotherapy, while rendering the cancer cells more susceptible.

ACKNOWLEDGEMENT

The financial support is rendered by the Women Scientist Scheme-A, Department of Science and Technology, New Delhi, India.

REFERENCES

Archana, A. (2009). Influence of *Triticum Aestivum* leaf extract on etoposide-induced apoptosis in primary cultured chick embryo fibroblasts.A thesis submitted to Avinashilingam University for Women, Coimbatore in partial fulfillment of the requirement for the degree of Master of Philosophy in Biochemistry.

Chih, H.W., Chill, H.F., Tang, K.S., Chang, F.R. and Wu, Y.C. (2001). Bullatacin a potent antitumour annonaceous acetogenin, inhibits proliferation of human hepatocarcinoma cell line by apoptosis induction, *Life Sci.*, 69, 1321–1331.

Han, P., Kang, J., Li, H., Hu, S., Lian, S., Qiu, P., Zhang, J., Li, W. and Chen, Q. (2009). Antiproliferation and apoptosis induced by tamoxifen in human bile duct carcinoma QBC939 cells via upregulated p53 expression, *Biochem. Bioph. Res. Co.*, 385, 251–256.

Ji, H., Zhang, L., Zhang, Y. Jin, H. and Wang, L. (2012). Antioxidant properties of Laoying tea (Litsea Coreana L.) extracts, *Adv. Mat. Res.*, 340, 209–214.

Kalpana, R., Mital, R. and Sumitra (2011). Vegetable and fruit peels as a novel source of antioxidants, *J. Med. Plants Res.*, 5, 63–71

Mercille, S. and Massie, B. (1994). Induction of apoptosis in nutrient-deprived cultures of hybridoma and myeloma cells, *Biotech, Bioengin*, 44, 1140–1154.

Montecucco, F., Bertolotto, M., Ottonella, L., Quercioli, A., Mach, F. and Dalleqri, F. (2009). Oxaprozin-induced apoptosis on CD40 ligand-treated human primary monocytes is associated with the modulation of defined intracellular pathways, *J. Biomed. Biotechnol.*, 48, 11–22.

Priyadarsini, R.V., Murugan, R.S., Maitreyi, S., Ramalingam, K., Karunagaran, D. and Nagini, S. (2010). The flavonoid quercetin induces cell cycle arrest and mitochondria-mediated apoptosis in human cervical cancer (HeLa) cells through p. 53 induction and NF-κB inhibition, 649, *Eur. J. Pharmacol*, 84–91.

Rashmi, R., Santhosh Kumar, T.R. and Karunagaran, D. (2003). Human colon cancer cells differ in their sensitivity to curcumin-induced apoptosis and heat shock protects them by inhibiting the release of apoptosis-inducing caspases, *FEBS Lett*, 538, 19–24.

Sarker, K.P., Obara, S., Nakata, M., Kitajima, I. and Maruyama, I. (2000). Anandamide induces apoptosis of PC-2 cells: involvement of superoxide and caspases-3, *FEBS Lett.*, 472, 39–44.

Sharma, P.K., Singh, S.S., Singh, N.P. and Porwal, M. (2011). *In vitro* antioxidant activity of Gossypium herbaceum Linn., *Int. Res. J. Pharm.*, 2, 166–170.

Sowemimo, A., Venter, M.V., Baatjies, L. and Koekemoer, T. (2011). Cytotoxicity evaluation of selected Nigerian plants used in traditional cancer treatment, *J. Med. Plants Res.*, 5, 2442–2444.

Storz, P. (2011). Forkhead homeobox Type O tanscription factors in the responses to oxidative stress, *Antioxid. Redox Sign.* 14, 593–605.

Sun, L., Chen, T., Wang, X., Chen, Y. and Wei, X. (2011). Bufalin induces reactive oxygen species dependent bax translocation and apoptosis in ASTC-a-1 cells, eCAM, Article ID 249090, 12 pages, doi:10.1093/ecam/nep082.

Biosensors: Towards Point-of-Care Cancer Diagnostics

D. Pradeep*
Department of Biotechnology, Adhiyamaan College of Engineering, Hosur
E-mail: jessin333@gmail.com

INTRODUCTION

Cancer: An Overview

Cancer (Medical term: malignant neoplasm) is a class of diseases in which a group of cells display uncontrolled growth (division beyond the normal limits), invasion (intrusion on and destruction of adjace nt tissues), and sometimes metastasis (spread to other locations in the body via lymph or blood). These three malignant properties of cancers differentiate them from benign tumors, which are self-limited, do not invade or metastasize.

Cancer may affect people at all ages, even fetuses, but the risk for most varieties increases with age. Cancer causes about 13% of all deaths. According to the American Cancer Society, 7.6 million people died from cancer in the world during 2007. Cancers can affect all animals.

GENESIS OF A CANCER CELL

Nearly all cancers are caused by abnormalities in the genetic material of the transformed cells. These abnormalities may be due to the effects of carcinogens, such as tobacco smoke, radiation, chemicals, or infectious agents. Other cancer-promoting genetic abnormalities may be randomly acquired through errors in DNA replication, or are inherited, and thus present in all cells from birth. The heritability of cancers are

*Corresponding author.

usually affected by complex interactions between carcinogens and the host's genome. Genetic abnormalities found in cancer typically affect two general classes of genes.

- Activation of proto-oncogene to oncogene
- Inactivation of tumor suppressor gene.

Proto-oncogenes are normal genes which codes for several proteins that act as growth factors, growth factor receptors, signal transducing proteins or transcription factors. Proto-oncogenes normally control cell division and differentiation but ultimately can be converted to oncogenes as a result of point mutation, chromosomal translocation or viral infection. The activated oncogenes induce cellular proliferation and therefore tumour development.

Tumour suppressor genes are responsible for switching the cell division process off, whereas oncogenes turn on cell division. There are two types of tumour suppressor genes:

- *Caretaker genes*—These genes check uninduce cell prliferation and if DNA is damaged they induce the repair of damaged genes. Eg: BRCA1, BRCA2, MLH1.
- *Gatekeeper genes*—If damaged DNA can not be repaired then they induce the cell death (apoptosis). Eg: p. 53, p. 21, Rb1, BaX, APC.

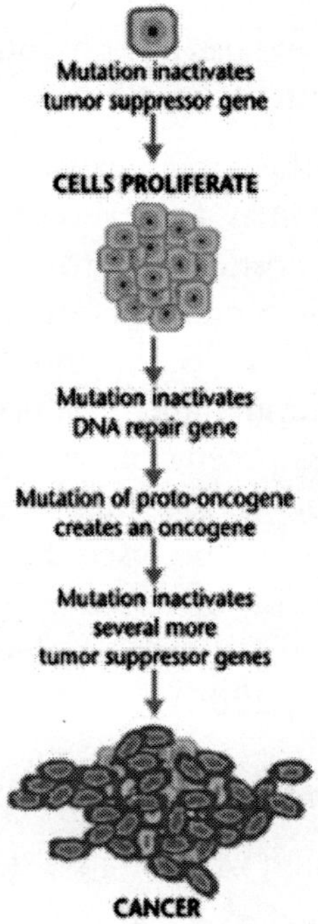

Fig. 1: Genesis of Cancer Cells

If these tumour suppressor genes are lost or become inactive then cells are liberated from the constraints imposed by these genes, yielding the uncontrolled growth and division.

CANCER DETECTION AND DIAGNOSTICS

Methods currently used to detect cancer include medical imaging, tissue biopsy, and bioanalytic assays of body fluids. None of these are sensitive and/or specific enough to detect early stage cancers. Lack of diagnostic tests for early stages of specific cancers contributes to high death rates. In fact, most solid tumors cannot be detected until they have reached a diameter of 1cm or larger. A tumor this size contains millions of cancer cells and may indicate that the cancer has metastasized or spread to other areas of the body. Once a malignant tumor is identified, pathologists look at biopsy specimens under a microscope to visually differentiate stages of cancer. Although skilled, a pathologist has obvious limitations in making judgments based solely on appearance (smith *et al.*, 2006).

So, the earlier cancer detection and accurate diagnosis are critical for improving the outcomes for cancer patients. Currently, the most important cancer diagnostic and prognostic indicators are morphological and histological characteristics of tumors or single biomarkers, such as prostate specific antigen (PSA). However, there is increasing use of molecular tools, both genomic and proteomic, to profile molecular alterations in tumors and produce "molecular signatures" based on many variables. Since tumor development involves many biological changes, the signatures can be very complex. Correlation of these molecular signatures with clinical parameters may yield information that can be used by physicians to aid in patient management decisions such as selection of effective cancer treatment. The challenge to the research community is to measure these complex molecular signatures in the clinical setting (Avraham Rasooly *et al.*, 2006).

CANCER BIOMARKERS

The FDA defines a biomarker as "a characteristic that is objectively measured and evaluated as an indicator of normal biologic processes, pathogenic processes, or pharmacologic responses to a therapeutic intervention." Cancer biomarkers are molecular changes that can be detected in the tumor or in the blood, urine, or other body fluids of cancer patients. These biomarkers or molecular signatures can be produced either by the tumor itself or by the body in response to the presence of cancer. They may include DNA, DNA modifications, RNA, proteins or protein modifications or other biological molecules. There are five types of biomarkers (Avraham Rasooly *et al.*, 2006).

Risk, Screening and Detection Markers

These biomarkers can be used to screen asymptomatic individuals who are at risk of cancer such as testing for mutations in BRCA1.

Diagnostic and Disease Monitoring Markers

These markers can be used for identification, characterization and monitoring progression of cancer. It also used to identify diseased individuals among those presenting with vague symptoms, or to distinguish among subclasses of disease. This includes PSA for detection and monitoring for recurrence of prostate cancer or CA-125 monitoring for recurrence of ovarian cancer.

Prognostic Markers

These markers allow the natural course of an individual cancer to be predicted, distinguishing 'good outcome' tumors from 'poor outcome' tumors, and they guide the decision of whom to treat (or how aggressively to treat). Notable recent examples include breast cancer gene expression signatures that estimate the probability of the original breast cancer recurring after it has been resected (that is, surgically removed). These multigene-expression tests can now be used to decide who should receive systemic therapy to eliminate any remaining tumor cells (that is, adjuvant therapy) after surgery, to reduce the risk of relapse (Sawyers L, 2008).

Predictive Biomarkers

These biomarkers are used to assess the probability that a patient will benefit from a particular treatment. Patients with breast cancer in which the gene ERBB2 (also known as HER2 or NEU) is amplified benefit from treatment with trastuzumab (Herceptin), whereas when the gene encoding the estrogen receptor is expressed by the tumor, the patients respond to treatment with tamoxlfen instead (Sawyers L, 2008).

Pharmacodynamic Biomarkers

It measures the near-term treatment effects of a drug on the tumor (or on the host) and also it is used to guide dose selection in the early stages of clinical development of a new anticancer drug.

Technologies for Biomarker Development (John W. Park *et al.*, 2004)

Recent technological advances now provide an unprecedented opportunity for the development of biomarkers relevant to cancer mechanisms. Prominent among these are:

Molecular Profiling Approaches

It based on genomics or proteomics to define the molecular signature of cancer, patterns of gene and protein expression and DNA alterations (*e.g.*, gene hyper-methylation, single nucleotide polymorphisms, microsatellite instability, and gross chromosomal aberrations) associated with different cancers.

Mass Spectrometry

Mass spectrometry in conjunction with bioinformatics algorithms has identified serum protein profiles that can be used to diagnose ovarian cancer, even though many of the proteins are unknown.

Microarray Technology

Expression microarrays are now being widely used to characterize the transcriptional profiles associated with various tumor types, stages, and/or phenotypes, for example in colon carcinogenesis.

Comparative Genomic Hybridization

It can be used to detect chromosomal alterations associated with certain cancers, such as hematological malignancies.

Cell Based Technologies

These technologies relevant to biomarker development include those permitting detection and quantification of circulating tumor or endothelial and endothelial progenitor cells. Methods to sensitively detect and enumerate rare circulating tumor cells have shown promise as a marker of disease status or treatment response, including studies in breast and prostate cancers. Isolation of such cells offers the further possibility of applying molecular profiling techniques to cells that may be surrogates of minimal residual disease ("micrometastasis") or metastatic cancer. Assays of circulating endothelial cells or endothelial progenitors may also be informative about cancer progression and/or tumor angiogenesis.

Novel Imaging Technologies

These molecular and functional imaging approaches include technologies to assess cellular metabolism (*e.g.,* ^{18}F-fluorodeoxyglucose positron emission tomography), cell proliferation and apoptosis (e.g., ^{18}F-fluoro-L-thymidine and ^{99}mTc-annexin imaging), resistance to chemotherapy (^{99}mTc-sestamibi imaging), angiogenesis and vascular dynamics (*e.g.,* dynamic contrast enhanced computed tomography and magnetic resonance imaging). Other technical advances include simultaneous combined modality imaging (*e.g.,* positron emission tomography/computed tomography or positron emission tomography/magnetic resonance imaging), which offers great promise in integrating functional and anatomical information. Optical technologies under development using fluorescence and bioluminescence may also prove useful in early drug development.

However, Proteomic and genomic analyses generate vast amounts of data but these approaches are primarily research tools that may not be practical for routine cancer clinical testing because of their complexity. The cumbersome and lengthy protocols introduce issues of reproducibility and must be carried out by highly skilled technicians under almost ideal conditions. Nevertheless, the actual trend in clinical practice is to monitor the health–state of the patient is directed towards point-of-care screening analysis, and household equipment. Moreover, most underserved population in developing countries does not access to state-of-the art diagnostic methods. The emergence of technology that changes sophisticated diagnostic tools by low-cost, accurate and portable systems seems to be an urgent need, making possible to reach even rural people who hardly move to Health Centers. Nowadays, the development of such technology seems reachable. Research on bio sensors is expected to greatly help to achieve this goal (Maria Soledad Belluzo *et al.*, 2008).

BIOSENSORS

Biosensors have been defined as compact analytical devices that bring together the use of a biological, a biologically-derived or a biomimic element to recognize the analyte. They are closely associated with, or integrated to, a signal conversion unit, the so-called physicochemical transducer, which eventually leads to a read-out. The use of biological materials as recognizing elements gives biosensors a remarkable ability to specifically react with the analyte of interest, distinguishing it from structurally similar compounds. (Maria Soledad Belluzo etal., 2008).

It has several potential advantages over other clinical analysis methods including increased assay speed and flexibility, capability for multi-target analyses, automation, reduced costs of diagnostic testing and a potential to bring molecular diagnostic assays to community health care systems and to underserved populations (Avraham Rasooly *et al.*, 2004).

Components of Biosensors

- ***Ligands*** (The sensitive biological element, e.g. Tissues, Microorganisms, Organelles, Cell receptors, Enzymes, Antibodies, Nucleic acids, Synthetic receptors, Sensing organs etc).
- ***Transducer*** (Acts as an interface, measuring the physical change that occurs with the reaction at the bioreceptor then transforming that energy into measurable electrical output)
- ***Detector Element*** (Signals from the transducer are passed to a microprocessor where they are amplified and analyzed, the data is then converted to concentration units and transferred to a display or/and data storage device).

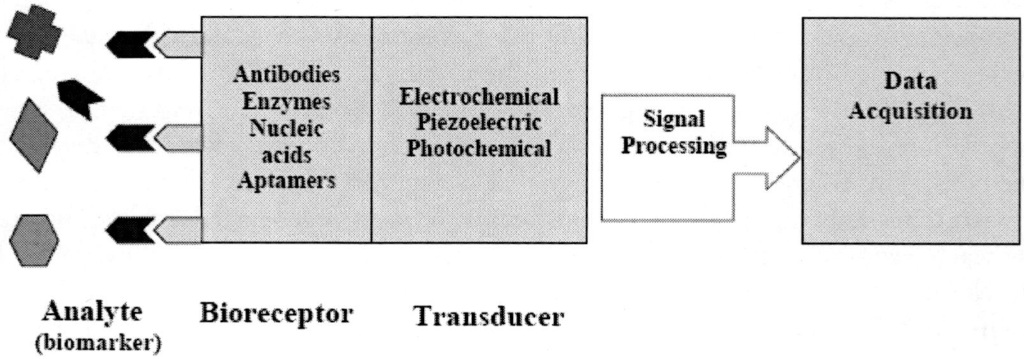

Fig. 2: Schematic of a Typical Biosensor (Heather Watson, 2008)

Ligands for Cancer Diagnostics Biosensor

The most important characteristics for ligands are affinity and specificity. Biosensors that use antibodies as recognition elements (immunosensors) are common, because antibodies are highly specific, versatile, and bind strongly and stably to the antigen. However, the complexity of cancer detection based on molecular signatures will require multiple ligands, and it is difficult to produce large quantities of uniform antibodies for numerous antigens. The lack of easily synthesized high affinity ligands is a major limiting factor for development of high throughput biosensor-based detection systems for cancer diagnostics. Therefore, researchers are attempting to develop new types of ligands more suited for high-throughput screening and chemical synthesis including synthetic peptides and aptamers, protein binding nuclic acid (Avraham Rasooly, 2006).

For example Cy5-labeled IgG antibodies for detection of anti-toxin human serum by an array biosensor (Frances Ligler), anti-HER2 antibodies for fluorescent nanoparticle-based detection of breast cancer (Zeev Rosenzweig) and monoclonal anti-cytokeratin antibodies labeled with fluorescent dye to recognize human cytokeratins in circulating tumor cells in blood (Richard Bruce). Ellington describes work by his lab in which conjugated 70-mer modified RNA aptamers were identified that bind prostate-specific membrane antigen (PSMA) to luminescent CdSe and CdTe nanocrystals for detection.

Prostate specific antigen (PSA) is an established biomarker currently used in the diagnosis and monitoring of prostate cancer. PSA and PSA monoclonal antibody (PSA-mAb) interactions are specific and bind tightly with affinities (Kd) of 1.1 + or- 0.2 nM. PSA is a 237 residue glycoprotein produced by prostate cells and its levels are significantly elevated in men with prostate cancer. It was the first biomarker to see widespread diagnostic use, primarily through blood tests of PSA. Methods currently used to detect PSA often give false positives due to poor specificity. These methods often include immunoassays that have radio-labeled compounds and are time consuming, hazardous, and expensive (Heather Watson, 2008).

Transducers for Cancer Diagnostics

The transducer converts the biochemical interactions into a measurable electronic signal. Optical, calorimetric, Electrochemical, Piezoelectric, acoustical and mechanical, transducers are among the many types found in biosensors. Electrochemical biosensors often detect the exchange of electrons between electrodes and redox proteins. Piezoelectric biosensors produce an electric signal in response to an applied mechanical stress. These are typically based on cantilevers, surface acoustic waves (SAWs) or quartz crystal microbalances (QCMs) where changes in mass produce a change in electric potential. The transducer works either directly or indirectly (Avraham Rasooly, 2006).

Direct Detection Biosensor

Direct recognition sensors, in which the biological interaction is directly measured typically, using non-catalytic ligands such as cell receptors or antibodies. Interferometric biosensors, which use self-referencing surface-normal interferometry to detect molecules bound to the surface of a spinning disc, is discussed by Nolte and co-workers. The biological compact disc (BioCD) (Varma *et al.*, 2004) allows high-speed sensing of sub-nanometer surface displacements, and has capacity for more than a million spots on a single disc using antibodies as ligands. Another type of direct recognition sensor is a quartz resonator transducer that measures changes in acoustic resonance. Mascini and co-workers (Tombelli *et al.*, 2005) used this technology for the detection of the TP53 mutation common in many cancers using synthetic oligonucleotides to analyze PCR amplicons.

Indirect Detection Biosensor

The second class of transducers, indirect detection sensors, relies on secondary labeled elements that are often fluorescently tagged antibodies or catalytic elements such as enzymes. Some examples of secondary elements are the enzyme alkaline phosphatase and fluorescently tagged antibodies that enhance detection of a sandwich complex.

Electrochemical transducers are also commonly used for indirect biosensor detection. Articles from the Gau and Wang labs describe the use of electrochemical transducers to measure the oxidation or reduction of an electroactive compound on the secondary ligand in one common type of indirect detection sensor. Wang and co-workers have combined electrochemical transducers with nanoscale materials to produce a unique coding capability for simultaneous measurement of multiple cancer markers. Gau *et al.* (2005) focus on the manufacturing of electrochemical biosensors using plastic rather than silicon substrates, and then using this platform for analysis of IL-8 protein in saliva. IL-8 in saliva is being investigated as a biomarker for head and neck cancers.

NANOTECHNOLOGY APPLICATIONS FOR CANCER DIAGNOSTICS

Recent research efforts in the area of biosensing are focused on nanoscale assemblies which can be used for detection and may change their mechanical, optical or electric properties in response to changes on surface. Some of the nanoscale assemblies under development include micro-cantilevers which measure bending because of changes in mass surface, particles like Quantum Dots and metal nanoparticles that can broaden the spectrum of fluorescent dyes and may change their optical properties because of changes on their surface. Nanotransducers offer new potential for miniaturization to nanoscale size, ability to detect large number of molecules simultaneously, reduced cost of production, high throughput and an enormous number of channels. Brown and co-workers utilized Quantum Dot technology for labeling peptide ligands labeling, while Ellington and co-workers used them to label aptamers.

POINT OF CARE TESTING

One of the most hopeful applications of biosensor is for this point-of-care testing; diagnostic testing that is performed on site. Point-of-care- testing offers the potential for faster and cheaper diagnostics. Biosensor- based diagnostics might facilitate cancer screening and improve the rates of earlier detection and attendant improved prognosis. Such technology may be extremely useful for enhancing health care delivery in the community (Priya Pathak *et al.*, 2007).

As biosensor technology advances, the range of applications broadens. Several biosensor applications for cancer diagnostics are described in this issue. Point of care testing, which is diagnostic testing performed on site, might significantly improve healthcare and healthcare delivery. As discussed above, POCT may permit rapid and less costly cancer diagnostics, thereby permitting increased testing in underserved populations. Successful development of biosensor based clinical testing will require continued

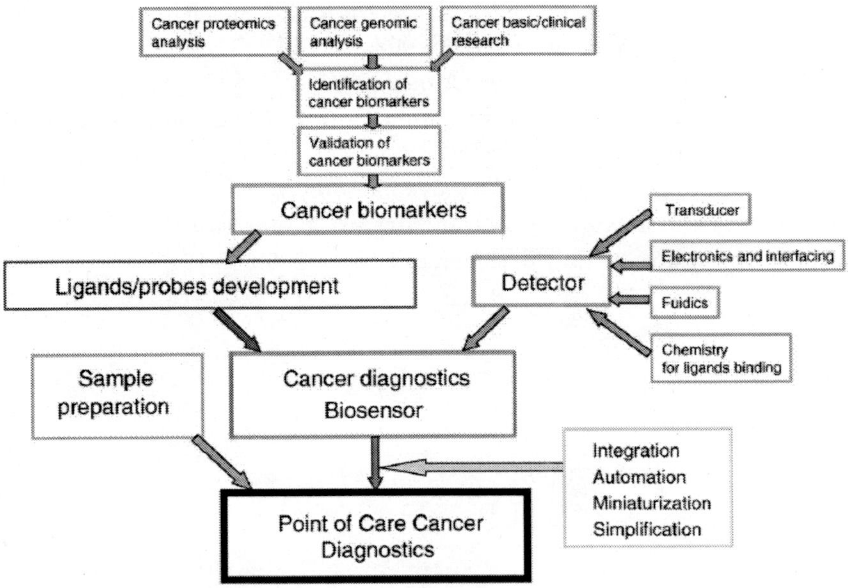

Fig. 3: The Main Elements Required to Adopt for Cancer Clinical Testing (Avraham Rasooly *et al.*, 2006)

development and validation of biomarkers and development of ligands for those biomarkers, more sensitive transducers, and multi-channel biosensors, miniaturization, and integration. Finally, cost reduction is crucial to make this technology widely available and useful.

REFERENCES

Avraham, Rasooly (2006). Guest Editorial, Moving Biosensor to point of care cancer diagnostics. *Biosensors and Bioelectronics,* 21, 1847–1850.

Avraham, Rasooly and James, Jacobson (2006). Development of Biosensors for Cancer Clinical Testing. *Biossensors and Bioelectronics,* 21, 1851–1858.

Brown, K.C., *Current Opinions in Chemical Biology*, 2000, 4, 16–21.

Ellington, A.D. and Szostak, J.W., *Nature,* 1992, 355, 850–852.

Gau, V., Ma, S.C., Wang, H., Tsukuda, J., Kibler, J. and Haake, D.A., *Methods,* 2005, 7, 73–83.

John W. Park, Robert S. Kerbel, Gary J. Kelloff, J. Carl Barrett, Bruce A. Chabner, David R. Parkinson, Jonathan Peck, Raymond W. Ruddon, Caroline C. Sigman, Dennis J. Slamon, Rationale for Biomarkers and Surrogate End Points in Mechanism-Driven Oncology Drug Development. *Clinical Cancer Research,* 2004, 10, 3885–3896.

María Soledad Belluzo, María Élida Ribone, Claudia Marina Lagier, Assembling amperometric Biosensors for Cancer Clinical Diagnostics. *Sensors,* 2008, 8, 1366–1399.

Priya Pathak, Katiyar V.K., shibashish Giri, Cancer research-Nano Particles and Their Use in Cancer Research. *Journal of Nanotechnology*, 2007, 3, 1–14.

Sawyers, L., The Cancer Biomarker Problem. *Nature,* 2008, 452, 542–552.

Smith, Dave, Nie, True, Gao, Multicolor Quantum Dots for Molecular Diagnostics of Cancer. Future Drugs, 2006, 6, 231–244.

Tombelli, S., Minunni, M., Mascini, M., *Biosensors and Bioelectronics*, 2005, 20, 2424–2434.

Varma, M.M., Nolte, D.D., Inerowicz, H.D. and Regnier, F.E., *Opt. Lett,* 2004, 29, 950–952.

Watson, Heather, Sensors for the early detection of Cancer, 2008, 1–12.

SECTION–III

Kunnath Pharmaceuticals Award

In vitro Antimicrobial Activity of *Ficus religiosa* (Linn.) Plant Leaves and Barks Extracts

G. Ramakrishnaiah*[*] and V. Prasanna Gandhi[1]

Department of Biotechnology Sree Vidyanikethan Engineering College,
A. Rangampet, Tirupati, Andhra Pradesh

[1]Department of Microbiology, Sree Vidyanikethan Degree College,
A. Rangampet, Tirupati, Andhra Pradesh

E-mail: ramakrishnaiah@ymail.com

INTRODUCTION

Medicinal plants are naturally gifted with invaluable bioactive compounds which form the backbone of traditional medicines. The use of plants as medicines antedates history. There has been an increasing interest worldwide on therapeutic values of natural products from plants due to disenchantment with modern synthetic drugs. Approximately 80% of the 4,000 inhabitants of the earth rely on herbal medicines for their primary health (Mukherjee *et al.,* 2006). Interest in medicinal plants with antibacterial properties has revived due to several problems associated with the use of antibiotics (Abu-Shanab *et al.,* 2004). Plants are rich source of bioactive compounds and thus serve as important raw materials for drug production (Chetty; 2008). Many works have been done which aim at knowing the different antimicrobial and phytochemical constituents of medicinal plants using them for the treatment of microbial infections as possible alternative to chemically synthetic drugs to which many infectious microorganisms have become resistant.

Ficus religiosan is commonly known as people tree in English. In India it is known by several vernacular names, the most commonly used ones being Asvatthah (Sanskrit), Sacred fig (Bengali), Peepal (Hindi), Arayal (Malayalam), Ravi (Telgu) and Arasu

*Corresponding author.

(Tamil) (Warrier, 1996). It belongs to Moraceae family. Peepal tree is one of the most familiar trees in India. It is one of the longest living trees of the world. *Ficus religiosa* is used in traditional medicine for about 50 types of disorders including asthma, diabetes, diarrhea, epilepsy, gastric problems, and inflammatory disorders, infectious and sexual disorders (Singh *et al.*, 2011; Sivarajan *et al.,* 1994). The barks of *F. religiosa* is an important ingredient in many Ayurvedic formulations, such as *Nalpamaradi tailam, Chandanasavam, Nyagrodhadi churna* and *Saribadyasavam* (Simha *et al.,* 2007; Newman *et al.,* 2003). In this context we selected historically important ancient plant of India, i.e. *Ficus religiosa* to study its antimicrobial activity against some selected microbes.

MATERIALS AND METHODS

Collection of Plant Material

Fresh leaves of *Ficus religiosa* were collected from the Sree Vidyanikethan Engineering College campus, A. Rangampet, Tirupati, Andhra Pradesh, India. The leaves and bark were washed under running tap water, air dried and then ground into powder by using electric grinder.

Methanol Extraction of Leaves and Bark of *Ficus religiosa*

Both the dried leaves and bark were powdered (each 10 g) and extracted separately with ethanol in a Soxhlet extractor. The extracts obtained were stored in refrigerator at 4°C for further use.

Diethyl Ether Extraction of Leaves and Bark of *Ficus Religiosa*

Both the dried leaves and bark were powdered (each 10 g) and extracted separately with diethyl ether in a Soxhlet extractor. The extracts obtained were stored in refrigerator at 4°C for further use.

Ampicillin (20 mg) was used as a positive control.

Microbes Used

Two Gram negative bacteria (*E.coli* and *Pseudomonas aeruginosa*), one Gram positive bacteria (*Staphylococcus aureus*) and one fungus (*Aspergillus niger*) were collected from the Department of Microbiology, S.V. University, Tirupati, A.P. India and were used in the present study. All the bacterial samples were maintained in nutrient agar slants. The fungus sample was maintained in PDA medium plates.

Screening of Antibacterial and Anti Fungal Activity

In vitro antibacterial and antifungal studies were conducted against selected bacteria and fungus using disc diffusion method (9). Two types of solvents (methanol and diethyl ether) were used for bark and leaves extractions. The different concentrations of methanol and diethyl ether extractions (100, 200, 300 and 400 mg/ml) of both bark and leaves of people tree were prepared. The Whatman no: 1 filter paper discs size of 30 mm was used for disc diffusion method. Paper discs were incubated in plant extracts for overnight at room temperature and air dried before used in the experiments.

RESULTS AND DISCUSSION

The antimicrobial activity was measured as zone of inhibition of two types of solvent extracts on three bacteria and one fungus. Tables 1 and 2 showed the antimicrobial activity of methanol and diethyl ether extracts of both leaves and bark of *Ficus religiosa* on selected microbes.

Table 1: Antimicrobial Activity of Methanol Extracts of Leaves and Bark of *Ficus religiosa*

S. No.	*Tested Microbes*	*Methanol Extracts*							
		Leaves Extract Zone of Inhibitions (mm)				*Bark Extract Zone of Inhibitions (mm)*			
		10 mg/ml	*20 mg/ml*	*30 mg/ml*	*40 mg/ml*	*10 mg/ml*	*20 mg/ml*	*30 mg/ml*	*40 mg/ml*
1.	*E.coli*	0.7	0.9	1.1	2.4	0.6	0.7	0.9	1.8
2.	*Pseudomonas aurignosa*	0.6	0.8	1.0	2.2	–	–	0.8	1.1
3.	*Bacillus subtilis*	0.9	1.2	1.5	2.8	0.8	1.0	1.2	2.2
4.	*Aspergillus niger*	–	–	–	1.6	–	–	1.0	1.2

Values were the mean of five replicates.

In table one the methanol extracts of leaves and bark were showed antimicrobial activity against three bacteria. At lower concentrations methanol extracts showed less antimicrobial activity and showed higher activity at 400 mg/ml concentration three tested bacteria. Both *Bacillus subtilis* (2.8 and 2.2 mm) and *E. coli* (2.4 and 1.8 mm) bacteria showed higher inhibitions than *P. aurignosa* (2.2 and 1.1 mm) in methanol extractions of leaves and bark. But at lower concentrations no activity and at higher concentrations (40 mg/ml) very less activity was observed against fungi (*A. niger*).

Table 2: Antimicrobial Activity of Diethyl Ether Extracts of Leaves and Bark of *Ficus religiosa*

S. No.	*Tested Microbes*	*Diethyl Ether Extracts*							
		Leaves Extract Zone of Inhibitions (mm)				*Bark Extract Zone of Inhibitions (mm)*			
		10 mg/ml	*20 mg/ml*	*30 mg/ml*	*40 mg/ml*	*10 mg/ml*	*20 mg/ml*	*30 mg/ml*	*40 mg/ml*
1.	*E.coli*	0.5	0.7	0.9	1.0	0.4	0.6	0.7	0.9
2.	*Pseudomonas aurignosa*	0.4	0.6	0.7	0.9	–	–	–	–
3.	*Bacillus subtilis*	0.8	1.0	1.2	1.4	0.7	0.9	0.9	1.2
4.	*Aspergillus niger*	–	–	–	–	–	–	–	–

Values were the mean of five replicates.

It is in line with the studies of phytochemistry and pharmacological properties of *Ficus religiosa* by Inder Kumar Makhija *et al.* (10).

In table two diethyl ether extracts of leaves and bark were showed varied antimicrobial activity against tested three bacteria and no activity against on *A. niger* in all concentrations. In case of diethyl ether extract of bark E. coli (1.0 and 0.9 mm) and *B. subtilis* (1.4 and 1.2 mm) bacteria showed inhibition and no activity on *Pseudomonas aurignosa* and *A. niger.* Diethyl ether extracts of leaves showed inhibitory activity on three rested bacteria and no activity on *A. niger*. It indicates that the diethyl ether extract of leaves and barks had no activity on *A. niger* in all tested concentrations. The results showed that the methanol leaves and barks extracts of *Ficus religiosa* had considerable inhibition activity on tested bacteria than diethyl ether extracts of leaves and bark of the *Ficus religiosa*.

REFERENCES

Abu-Shanab, B., Adwan, G., Abu-Safiya, D., Jarrar, N. and Adwan, K. (2004). Antibacterial activities of some plant extracts utilized in popular medicine in Palestine. *Turkish J. Biol.,* 28: 99–102.

Chetty, K.M. (2008). Flowering Plants of Chittoor. 1st Edn., Andhra Pradesh, India, 132.

Inder Kumar Makhija; Indra Prakash Sharma and Devang Khamar (2010). Phytochemistry and Pharmacological properties of F*icus religiosa*: An overview. *Annals Biol. Res.*, 1: 171–180.

Mukherjee, P.K. and Wahil, A. (2006). Integrated approaches towards drug development from Ayurveda and other system of medicine. *J. Ethnopharmacol,* 103: 25–35.

Newman, D.J., Cragg, G.M. and Snader, K.M. (2003). Natural products as sources of new drugs over the period 1981–2002, *J. Nat. Prod.,* 66: 1022–1037.

Simha, K.R.G. and Laxminarayana, V. (2007). *Indian J. Trad., Know*, 6: 648–652.

Singh, D., Singh, B. and Goel, R.K. (2011). Traditional uses, phytochemistry and pharmacology of *Ficus religiosa:* A review. *J. Ethnopharmacol,* 01–05.

Sivarajan, V.V. and Balachandran, I. (1994). Ayurvedic drugs and their sources, Oxford and IBH Publishing Co. Pvt. Ltd., New Delhi, 374–376.

Warrier, P.K. (1996). Indian medicinal plants—A compendium of 500 species, Orient Longman Ltd. Chennai, Vol. III, 38–39.

Screening of South Indian Herbs for Anti-Quorum Sensing Activity

K. Anitha and T. Mahalakshmi*

Department of Life Science MASS College of Arts and Science, Kumbakonam, Tamil Nadu

E-mail: mahatamilarasan@gmail.com

INTRODUCTION

Bacteria monitor their environment for the presence of other bacteria by producing and responding to certain chemicals signaling molecules. The concentration of these molecules is in direct proportion to the load or density of the bacteria; in other words, there is a direct relationship between the concentrations of chemical signaling molecules and that of the bacteria present in a given environment. This is termed as Quorum Sensing (QS).

It allows bacteria to communicate with each other by synchronizing/regulating the expression of several genes that are necessary for bacteria to behave in unison as a group (Anjali *et al.,* 2008).

Phytochemicals have been used for thousands of years for the treatment and management of diseases (Rasmuseen and Givskov, 2006). The secretion of enzymes that destroy the auto inducers and the production of auto inducers antagonists are used by competitor bacteria and susceptible eukaryotic host to render QS bacteria mute and deaf respectively. Analogues synthetic strategies are now being explored for the development of novel antimicrobial therapies (Stephan and Bonnie, 2011).

For many years, control of bacterial infections by inhibiting microbial growth has been a primary approach of antimicrobial chemotherapy. An emergency problem associated with continual indiscriminant are of this therapeutic strategy is the selection of resistant bacteria with higher level of tolerance against broad-spectrum

*Corresponding author.

antibiotics. It has been recognized that there is a need for a strategy that can block very basic mechanisms of bacterial communication that appear to control bacterial virulence factors leading to pathogencity (Vattem *et al.*, 2008).

The anti-pathogenic activities of the extracts were also explored using to Quorum Sensing inhibition assay. The mutant strain *Chromobacterium violaceum* CV026 was used as a biosensor to detect anti pathogenic activity (MC Lean *et al.,* 2004).

It has been reported that many gram-negative bacteria. including pathogenic species (e.g. Pseudomonas aeruginosa) use small organic molecules known a N-acylated homoserine lactones AHLs for Cell-Cell communication. (Water and Bassler, 2005). Pseudomonasa an increasingly prevelant opportunistic human pathogens, is the most common gram-negative bacterium found in nosocomial infections. The capacity of pseudomonas aeruginosa to produce such diverse, often over whelming infections is due to an arsenal of virulence factors. Many extra cellular virulence factors secreted by Pseudomonas aeruginosa have been shown to be controlled by a complex regulatory circuit involving cell-to-cell signaling system that allow the bacteria to produce these factors in a coordinated, cell-density dependent manner (Passador *et al.,* 1995).

Pseudomonas aeruginosa produces several extracellulars products that after colonization can cause extensive tissue damage, blood stream invasion, and dissemination. In vivo studies have shown that mutants defective is the production of exotoxin A, exoenzymes, elastase or alkaline protease are essential for maximum virulence of Pseudomonas aeruginosa however the relative contribution of a given factors may vary with the type of infection (Nicas *et al.*, 1985).

Antibiotic resistance provides a great therapeutic and economic burden in the treatment of infections disease and it may threaten the success of antimicrobial chemotherapy. It is estimated that antibiotic resistance double hospital stay and morbidity single antibiotic resistance itself is a great problem, however the appearance of multiple antibiotic resistant (MDR) strains came a more pronounced obstacle for patients and health care propossionals (Levy, S.B. *et al.,* 2004; Smith, M.A., 2005). Bacteria by reversal of resistance, effect inhibition, inhibition of biofilm formations, interference in bacterial Quorum sensing, etc.

MATERIALS AND METHODS

Plant Material and Extract Preparation

The leaf of *Terminalia Arjuna, Tragia involucrata, Clerodendrum Phlomids, Solanum trilobatum, Solanum surattense, Solanum nigrum, Solanum torvum* and *Clerodendrum inerme*. The seed of *Lawsonia inermis, Myristica fragrans, Terminalia chebula, Piper longum* and *Solanum melongena*. The root of *Glycyrrhiza Glabra, Withania somnifera*, used in this study were obtained from the local market and were identified based on its physical characteristics.

Air dried and powered plant materials approximately from 100 to 300 g were submitted to sequential maceration with 500 ml of water at room temperature. After each step the extracts were filtered and the solvents were removed under vacuum at 30°C until dry aqueous extracts were obtained.

Strains and Media

Pseudomonas aeruginosa MTCC 2453 and Chromobacterium violaceum 2656 were collected from Microbial type culture collection center. IMTECH, Chandiragh. The purchased culture was propagated in the nutrient broth and incubated at 37°C for 24 hours of incubation. The grown culture was sub cultured at regular intervals of 30 days and stored at 4°C for future use. The Luria Bertani (LB) medium was used throughout the study.

Culture Condition

For all the assay except those for biofilm formation the culture condition were as follows. Overnight culture of Pseudomonas aeruginosa were grown in LB medium at 37°C with shaking. The cultures were then diluted 100-fold in fresh LB medium and allowed to grow to an optical density at 600 nm (OD600) 1.7 (early stationary phase). At this point, the culture was divided into 10ml aliquots and additional 1ml of fresh medium containing plant extract or media control was added to a final concentration of 1 mg/1ml extract. Cultures were recovered at late stationary phase (approximately 72 h after addition). The cells were separated from the growth medium by centrifugation at 10,000x g for 10 min (Turbo *et al.,* 2003).

AHL Bioassay

The extracted plant material was placed directly on to LB plates spread with *Chromobacterium violaceum* supplemented with AHL were incubated at 30°C and QS inhibition was detected by a ring of colorless. Purified halogenated furanone (100 μg) was used as a positive control for QS inhibition and ethanol (20 μg) as a negative control. The ethanol was allowed to evaporate from the control and sample discs before testing to eliminate toxicity.

RESULTS AND DISCUSSIONS

Screening results of 15 different medicinal plant extract for their ability to inhibit QS regulated violacein production against *C. violaceum.* The plant extracts namely Terminalia Chebula showed the greater inhibition, where as *Lawsonia trilobactum, solanum melongena, Terminalia arjuna* are the good sources of producing AHL antagonists (Table 1).

Table 1

S.No.	*Botanical Name*	*Family*	*Parts Used*	*Folk Name*	*Medicinal Uses*
1.	*Terminalia chebula*	Combretaceae	Seed	Kadukai	Flatulence, Constipation, Piles, Cough and Colds
2.	*Solanum torvum*	Solanaceae	Leaves	Sundakai	Tolerate diseases
3.	*Solanum surattense*	Solanaceae	Leaves	Karimulli	Toothache, cough, Asthma, Catarrh
4.	*Myristica fragrans*	Myristicaceae	Seed	Jathikai	Skin diseases, leprosy, colds, cough, pimples
5.	*Piper longum*	Piperaceae	Seed	Arisi Thippili	Throat irritation
6.	*Terminalia arjuna*	Combretaceae	Leaves	Maruthu	Heart disease, hemorrhages and ulcer
7.	*Solanum melongena*	Solanaceae	Seed	Kandakathri	Venereal disease, fever, dysentery
8.	*Tragia involucrata*	Euphorbiaceae	Leaves	Kanjsori	Curing eczema, fevers, wheezing and diabetes
9.	*Clerodendrum phlomides*	Lamiaceae	Leaves	Thaluthalai	Swelling
10.	*Solanum trilobactum*	Solanales	Leaves	Thuthuvalai	Cooking, Cough and mildly toxic
11.	*Lawsonia inermis*	Lythraceae	Leaves	Maruthani	Skin dyeing, reducing Heat
12.	*Clerodendrum inerme*	Verbenaceae	Leaves	Bechangu	Hypertension dysentery, cough, Skin rashes and heart related disorders
13.	*Withania somnifera*	Solanales	Leaves	Ashwagantha	Tumors, tubercular glands, Carbuncles, and ulcers
14.	*Solanum nigrum*	Solanaceae	Leaves	Manathathakkali	Mouth ulcers, pepticulcers
15.	*Glycrrhiza Glabra*	Fabaceae	Root	Athimathuram	Mouth ulcers, pepticulcers

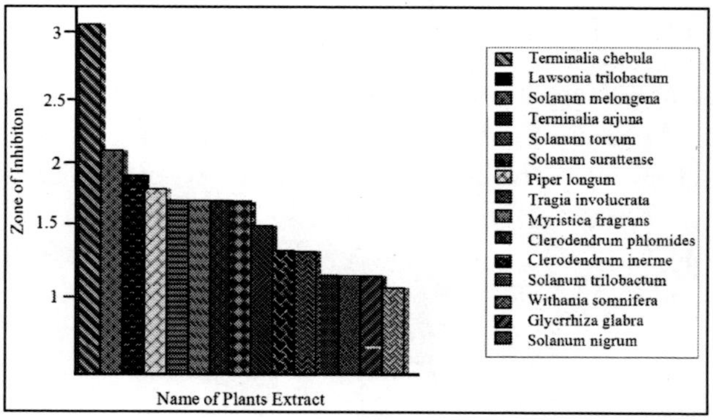

Fig. 1: Effect of Extract on Quorum Sensing Activity

The present study proves the potentiality of the Indian medicinal plants as a source of anti QS compounds. Ethnomedicinal uses of these plants are focused on the development of molecules that are structurally analogues to the AHL or Autoinducers. Such molecules are potential use as an antimicrobial drug aimed at bacteria as anti QS molecules. Further researches are undergoing to view as plants are potential targets for novel antimicrobial drug.

REFERENCES

Anjali, Kumari; Patrizia, Pasini and Sylivia, Daunert (2008). Detection of bacterial quorum sensing N-acyl homoserine lactones in clinical samples. *Annal Bioanal. Chem.* 391: 1619–1627.

Levy, S.B. and Marshall, B. (2004). Infectious Disease Informatics. *Nat Med.* 10: 122–129.

MC Lean, R.J.C., Pierson, L.S. and Fugua, C. (2004). A Simple Screening Protocol for the indentification of Quorum signal antagonists. *J. Microbiol. Methods*, 58: 351.

Nicas, T. and Iglewski, B.H. (1985). The contribution of exoproducts to virulence of Pseudomonas aeruginosa. *Can. J. Microbial,* 31: 387–392.

Passador, L. and Iglewski, B.H. (1995). Quorum sensing and virulence gene regulation in Pseudomonas aeruginosa In: Roth JA, editor, virulence mechanisms of bacterial pathogens. 2nd ed. Washington. *American Society for Microbiology,* 65–78.

Rasmuseen, T.B., Givskov, M. (2006). Quorum-sensing inhibitors as anti-pathogenic drugs. *Int. J. Med. Microbial,* 296: 149–161.

Smith, M.A. (2005). Antibiotic resistance. *Nurs. Clin. North. Am.*, 40: 63–75.

Stephan, Schauder and Bonnie Bassler, L. (2001). The languages of bacteria. *Genes and Development,* 15: 1468–1480.

Vattem, D.A, Huerta, V., Mihalik, K. and Crixell, S.H. (2008). Herbs Spices and Medicinal can Decrease Quorum Sensing dependent virulence in Pseudomonas aeruginosa. *Int. J. Appl. Res. Nat. Prod.,* 1: 9–15.

Waters, C.M. and Bassler, B.L. (2005). Cell-Cell Communication in Bacteria. *Annu. Rev. Cell Develop. Biol.* 21: 319–346.

Protein Pattern of Temperature Tolerant Proteins Obtained from Foliose Lichens for Medical Applications

Deepak Kumar P., Shankar R., Ayyapadassan G.* and P. Ponmurugan

Department of Biotechnology, K.S. Rangasamy College of Technology, Tiruchengode

E-mail: ayyapadassan07@gmail.com

INTRODUCTION

Lichens are fascinating symbiotic associations; actually they represent a biological partnership of fungi (mycobionts) and photoautotrophic algae (phycobionts). Lichens grow under diverse ecological conditions; they occur from arctic habitats to tropical rainforests where they colonize a broad variety of substrates, such as rock, bark, soil and even leaf/bark surfaces. Some lichens are claimed to be good for coughs, jaundice, rabies and restoring lost hair. Herbal medicine texts included account of several species of lichens including *Cladonia, Evernia, Lobaria, Parmelia, Peltigera, Pertusaria, Physica, Rocella, Usnea* and *Xanthoria*. Lichens like *Lobaria pulmonaria* and *Parmelia sulcata* have been used in the treatment of pulmonary and cranial diseases, respectively. Similarly, *Xanthoria parietina* is used to cure jaundice. *Cetraria islandica* commonly used as a remedy for lung diseases, kidney and bladder complaints as well as externally for poorly healed wounds (Müller, 2002). Traditionally it is used mild inflammation of the oral and pharyngeal mucosa. It is also valued in dyspepsia and loss of appetite. In European folk medicine, it is used predominantly in cancer treatment (Müller, 2002).

Lichens are a peculiar group of plant varieties whose traditional usage in various medical applications remains conclusive and its mystery that many properties were unexplored. Information on the edible and medicinal uses of the lichens is scattered

*Corresponding author.

(Crittenden, 1999). Many lichens are known to have potent antibiotic properties, and many are edible. The lichens have been household items of Indians since ancient times as medicines and in various cultural events. One way that they have been utilized is in the extraction of blue, red, brown or yellow dyes in the garment industry. Also, the indicator pigments used in litmus paper is also derived from lichens.

The lichens are sensitive to air pollution and microclimatic changes. Hence, widely considered as bioindicators and constitute an important component of biodiversity in a region (Loppi *et al.,* 1995, Myers *et al.,* 2000). The Proteins which are having the ability to withstand even at high temperature without Denaturation is called Temperature tolerant one. As they are tolerant to that temperature, they remain stable. Such proteins exhibit a wide range of applications. Some of that Proteins are Protease, Amylase, cellulase and chitinase.

Present work deals about the lichen proteins for food industry. Isolated enzymes are having the highest activity in the range of 55–60° Celsius. We have isolated and partially purified the enzymes like Proteases, Amylases and Cellulase from the *Parmotrema retticulata* and *Parmotrema tinctorum* species. We have applied the enzymes for liquefaction process and also clarification of fruit juices at high temperature. Starch liquefy was important in the food process. Here the thermal amylase was used as thinning agents to improve the process efficiency.

MATERIALS AND METHODS

Sample Collection and Maintenance

Lichen samples were collected from different attitudes covering various agroclimatic zones of Yercaud hills belonging to Eastern Ghats of Tamil Nadu for the present study. Different living and non-living substrata were also obtained along with lichen thallus to study the nature of attachment and to check the moisture content, pH and buffering action. Corticolous lichens were peeled-off along with tree bark, but in case of saxicolous lichens collection, small pieces of the rock substrate were collected carefully. Collected lichen samples were deposited in Herbarium sheet without moisture content which would be used for further authentication and voucher specimen for the project.

Identification of Lichens Species

Identification of *Parmotrema tinctorum* and *Parmotrema retticulata* were done on the basis of colour test. Species were authenticated by Dr. S. Sanjeeva Nayaka, Scientist, National Botanical Research Institute, Lucknow.

Fig. 1: *P. reticullatum*

Extraction of Crude Enzymes from Lichen

The crude samples were obtained by grinding the Lichen species *P. retticulatum, P. tinctorum with 1 M Tris Hcl (pH 8).* Then it was maintained at 4 degree Celsius for further studies.

Test Sample

The Crushed samples were subjected to various assay's. The test Sample T_1 is taken as the *P. retticulatum,* and the Test Sample T_2 is *P. tinctorum.*

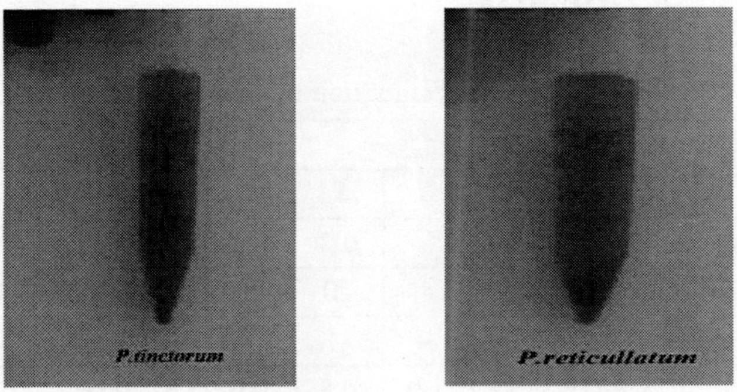

Fig. 2: Crude Enzymes of Parmeloid Lichens

Amylase Assay

Procedure

Add 0.1 ml of different enzyme solution to different test tubes. Add 0.5 ml of starch solution. Makeup the volume to 1 ml. Then incubate the reaction mixture for 10 minutes at room temperature. After 10 minutes stop the reaction by adding 2 ml of

DNS. Keep the test tube in boiling water bath for 5 min and then add 1ml of potassium sodium tartarate. Add 2 ml of water to the above mixture. Note down the absorbance at 540 nm.

Cellulase Assay

Add 0.1 ml of different enzyme solution to different test tubes. Add 0.5 ml of cellulose solution. Makeup the volume to 1ml. Then incubate the reaction mixture for 10 minutes at room temperature. After 10 minutes stop the reaction by adding 2 ml of dns. Keep the test tube in boiling water bath for 5 min, then add 1ml of potassium sodium tartarate. Add 2 ml of water to the above mixture. Note down the absorbance at 540 nm.

Lowry (Folin-Ciocalteau) Method

The Lowry method is reasonably sensitive, detecting down to 10 μg cm^{-3} of protein, and the sensitivity is moderately constant from one protein to another. When the Folin reagent (a mixture of sodium tungstate, molybdate and phosphate), together with a copper sulphate solution, is mixed with a protein solution, a blue-purple color is produced which can be quantified by its absorbance at 660 nm.

RESULTS AND DISCUSSIONS

Table 1: Protein Estimation by Lowrys Method

S. No	*Content*	*Standard (ml)*						*Unknown*	
		B	S_1	S_2	S_3	S_4	S_5	T_2	T_1
1.	Volume of working std (ml)	–	0.2	0.4	0.6	0.8	1	–	–
2.	Concentration (ml)	–	20	40	60	80	100	–	–
3.	Volume of Lichen sample (ml)	–	–	–	–	–	–	0.1	0.1
4.	Volume of distilled water (ml)	1.0	0.8	0.6	0.4	0.2	–	0.9	0.9
5.	Volume of reagent alkaline copper sulfate (ml)	5	5	5	5	5	5	5	5
Incubate for 10 minutes									
6.	Volume of Folin-Ciocaffer reagent (ml)	.5	.5	.5	.5	.5	.5	.5	.5
Incubate for 30 minutes									
	O.D at 660 nm	0.00	0.13	0.22	0.31	0.43	0.52	3.61	1.666

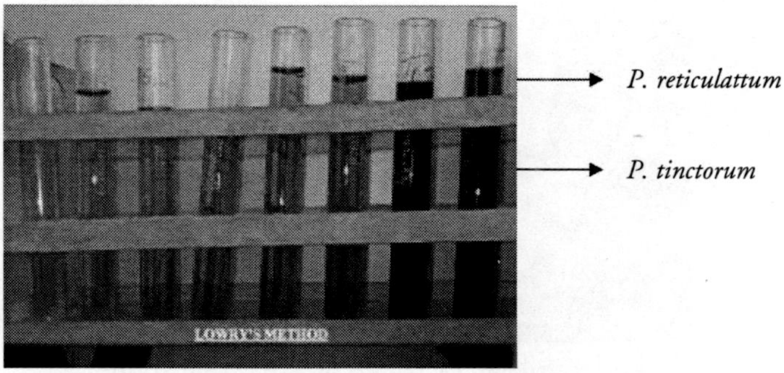

Fig. 3: Estimation of Proteins by Lowrys Method

Table 2: Amylase Assay of Different Lichen Species

S. No.	*Vol of Starch*	*Vol of Enzyme*	*Vol of Distilled Water*	10 min incubation 37°C	*Vol of DNS*	5 min incubation in boiling H_2O bath	*Vol of Sodium Potasium tartarate*	*Vol of DW*	*O.d 540 nm*
1	0.5	0	1		2		1	2	0
2	0.5	0.1 (T_1)	1		2		1	2	0.90
3	0.5	0.1(T_2)	1		2		1	2	0.46

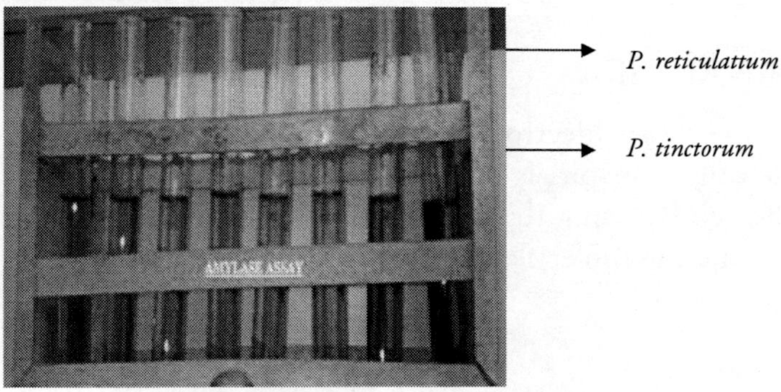

Fig. 4: Enzymatic Assay of Amalase Enzyme

Table 3: Cellulase Assay of Different Lichen Species

S. No.	*Vol of CMC*	*Vol of Enzyme*	*Vol of Distilled Water*	10 min incubation 37°C	*Vol of DNS*	5 min incubation in boiling H_2O bath	*Vol of Sodium Potasium tartarate*	*Vol of DW*	*O.d 540 nm*
1	0.5	0	1		2		1	2	0
2	0.5	0.1(T_1)	0.4		2		1	2	1.0
3	0.5	0.1(T_2)	0.4		2		1	2	0.6

Fig. 5: Enzymatic Assay of Amalase Enzyme

Table 4: Protease Assay of Different Lichen Species

Sample	*Volume of BSA (ml)*	*Volume of Sample (ml)*	*Volume of Water (ml)*	*Reagent C (ml)*	*FC (ml)*	*OD*
Std 1	0.2	–	0.8	0.5	0.5	0.15
Std 1	0.4	–	0.6	0.5	0.5	0.28
Std 1	0.6	–	0.4	0.5	0.5	0.42
Std 1	0.8	–	0.2	0.5	0.5	0.55
Std 1	1	–	0	0.5	0.5	0.62
T_1	–	0.1	0.9	0.5	0.5	0.61
T_2	–	0.1	0.9	0.5	0.5	3.58

SDS PAGE ANALYSIS

SDS-Polyacrylamide Gel Electrophoresis is carried out with a discontinuous Buffer System in which the reservoirs is of a different pH and ionic strength from the buffer used to cast the gel. By using the markers of Known molecular weight it is therefore Possible to estimate the molecular weight of the molecular weight of the Polypeptide chains.

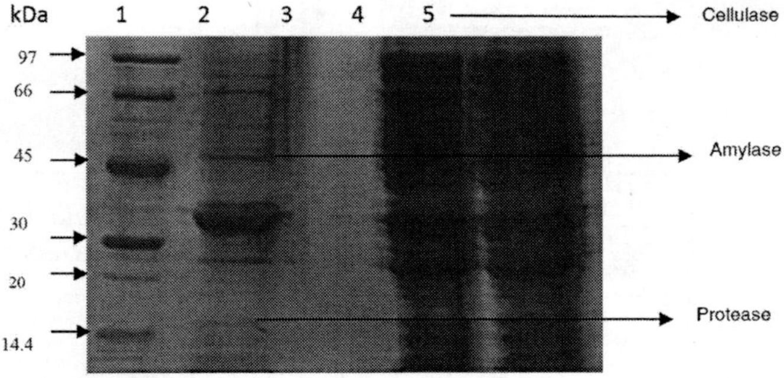

Fig. 6: SDS-PAGE: Lane 1. Protein Marker (97, 66, 45, 30, 20, 14.4 kDa—GE Healthcare), Lane 2. Protein Sample of *P. tinctorum*, lane 4. *P. reticullatum* Lane 5-*P. reticullatum* (control)

Table 5: Enzymatic Activity of Different Tolerant Proteins

Samples	*Enzymes*	*Specific Enzyme Activity (units/mg)*	*Enzyme Activity (units/ml)*
T_1	Amylase	35.05	2.66
	Cellulase	45.81	3.48
	Protease	123.75	3.96
T_2	Amylase	163.37	5.22
	Cellulase	182.93	5.85
	Protease	30.28	2.302

From the results of various assay's the Proteins are seems to be stable and there is no such sort of activity even though there is a High Temperature To analyse its activity an experiment was conducted with apple juice samples.

Clarification of Fruit Juice

An apple of about 150 gms was crushed to make juice with sucrose. It was treated with the Enzymes obtained from the lichen Sample. It was kept undisturbed for about 10 days.

Fig. 7: Protease Applications for Juice Clarification

After 10 days of inhibition, the sample get cleared due to the action of Enzymes.

From the above results it is come to know that Isolated enzymes are having the highest activity in the range of 55–60° Celsius. They liquefy the starch and also clarify the fruit juices at high temperature. Hence these enzymes can be utilized in the beverage industry as an effective clearing agent.

REFERENCES

Crittenden, P.D. and Porter, N. (1999). Lichen forming fungi: Potentialsources of novel metabolites. *Trends Biotechnol*, 9: 409–414.

Loppi, S., Putorti, E., Dominics, V. and Barbaro, A. (1995). Lichens as bioindicators of air quality near a municipal solid waste incinerator (Central Italy). *Allionia* 33: 121–129.

Müller, K. (2002). Pharmaceutically relevant metabolites from lichens. *Appl. Microbiol. Biotechnol.* 56 (1/2): 9–16.

Myers, N., Mittermeier, R.A., Mittermeier, C.G., da Fonseca, G.A. and Kents, J. (2000). Biodiversity hotspots for conservation priorities, *Nature* 403: 853–858.

Characterization of *Nomuraea rileyi* against *Panchaetothrips indicus* Infecting Leaves of *Curcuma longa* L.

R. Arutselvi* and T. Balasaravanan[1]

Research and Development Centre, Bharathiar University, Coimbatore
K.S. Rangasamy College of Technology, Tiruchengode
[1]Nehru Arts and Science, Coimbatore
E-mail: arutselviram@gmail.com

INTRODUCTION

India, "the land of spices" is the largest producer, consumer and exporter of turmeric in the world. Turmeric (*Curcuma longa* L.) stands next to black pepper and cardamom in earning foreign exchange. India is the major turmeric producer in the world accounting for about 80 per cent of the world production and 60 per cent of world in exports. In India, turmeric is raised in an area of 1,34,000 ha with an annual production of 5,34,000 tonnes (Ravikumar, 2002). The major turmeric growing areas are concentrated in Andhra Pradesh, Tamil Nadu, Maharashtra, Orissa and West Bengal. In Tamil Nadu it is cultivated over an area of 33,000 ha. with a production of 1,58,643 tonnes and productivity of 4807 kg ha^{-1}.

Turmeric (*Curcuma longa;* Zingiberaceae) is a medicinal plant extensively used in Indian traditional medical systems of Ayurvedha, Unani, and Siddha medicine as a home remedy for several diseases. The antibacterial effect of turmeric leaves and rhizomes has been proved earlier (Arutselvi *et al.*, 2012). Though turmeric is well known for its medicinal value, its cultivation is affected by several pests. Among them, thrips is the most threatening one which infects the leaves of turmeric plants.

*Corresponding author.

Panchaetothrips indicus was reported as a pest of turmeric, *Curcuma domestica* in India (Bagnall, 1912). *P. indicus* commonly called as thrips is a sucking pest of leaves. It is a small reddish brown fragile insect, though specific pests of this crop are found only occasionally. Its occurrence starts in the month of August and lasts until November. If the thrips infestation occurs in early stages of the crop growth, it would result in total loss of leaves of affected plants. There is a considerable crop loss because following severe infestation, young plants die and the rhizome yield is proved to get decreased.

Over the years, control of thrips has been almost exclusively based on use of chemical insecticides. There are long-term concerns about insecticide exposure risks, health and environmental hazards, residue persistence, development of resistance and elimination of natural enemies. Besides, some of the insecticides reported effective earlier such as carbofuron and quinalphos are now being reported to cause leaf folder resurgence (Devanesan *et al.*, 1995). Therefore, effective alternative control measures are necessary if long term suppression of thrips is to be attained. Biological control, including the use of entomopathogenic fungi, offers a sound alternative management strategy against several pests. Entomopathogens are promising biocontrol agent for a number of crop pests. Several species belonging to order Lepidoptera, Coleoptera, Homoptera, Hymenenoptera and Diptera are susceptible to various fungal infections. Among the entomopathogens, fungi are important as they are virulent, infect by contact, persists in environment for long time and many of them can be mass produced in liquid or solid media (Bhattacharya *et al.*, 2004). Keeping this in view, the present experiments were conducted to evaluate the isolates of entomopathogenic fungi in order to select the most virulent isolates for the control of *P. indicus.* Standard cultures of *Metarhizium anisophilae* (MTCC No: 4100), *Beauverria bassiana* (MTCC No. 6095) and *Nomurayae rileyi* (MTCC No: 4171) were procured from Microbial Type Culture Collection, Chandigarh, India and used for the experiments.

MATERIALS AND METHODS

Plant Materials

C. longa plants were grown in the Botanical garden of K.S. Rangasamy College of Technology, Tiruchengode, Tamil Nadu, India. A part of it was maintained in disease free conditions without any pest attack.

Preparation of Fungal Extracts

For *Beauverria bassiana*, and *Metarhizium anisophilae* Potato Dextrose Agar medium and *Nomuraea rileyi*, Sabarouds Yeast Maltose Peptone (SYMP) medium was used for multiplication. A loopfull of fungal culture was inoculated to 500 ml of broth in a conical flask and maintained in a shaking incubator at 28°C at 120 rpm for 10 days. Spore suspensions of each pathogenic fungus were prepared by using 80–100 ml of sterile distilled water containing 0.05% Tween-80 solution. The conidial count of cultures were then determined by using a Neubaur hemocytometer and adjusted to 1×10^7/ml and stored in 4°C until use.

Laboratory Bioassays

The *P.indicus* was grown in plastic containers covered by a muslin cloth for aeration. Each container consists of 30 larvae and three replicates were maintained. Two ml of spore suspension of the fungi were taken and sprayed over the insects. The control insects were sprayed with 0.02% Tween 80 alone. The containers with insects were maintained at 26 ±1°C temperature; relative humidity 70 ± 10% and photoperiod of 16: 8 L: D. Insect mortality was recorded at every 6 hours interval for two days after treatment and the data was analyzed statistically.

Field Studies

To evaluate the effectiveness, the fungal formulations were studied by the field application. The turmeric field in Karungalpalayam, Erode was chosen. The experimental plots were laid in randomized complete block design with eleven treatments with a plot size of 10 sq.m with fifty plants in each plot and three replicates were maintained for each treatment. The infected plants were treated with a concentration of 1×10^7 conidia/ml of the different entomopathogenic fungus namely, *Metarhizium anisopliae, Beauveria bassiana and Nomuraea rileyi.* Another set of the samples treated with commercial pesticides was used for the comparative studies. One set of samples were untreated with fungus and commercial pesticides and was used as the control. The fungal suspension was mixed with a diluted amount of starch solution for the proper attachment of the fungal suspension with the plants. By using a snapshot sprayer the fungus starch mixture was spared on to the turmeric plants and leaves of ten plants were observed randomly for mortality. The observation was tabulated and represented graphically.

Secondary Metabolite and GC MS Analysis of *Nomuraea rileyi*

Secondary metabolites from *Nomuraea rileyi* was extracted following Dubey *et al.*, 2011. The *N. rileyi* culture was grown in SYMP broth in a flat bottomed flask for 12 days at 26 ± 1°C. The culture filtrate was obtained by straining through muslin cloth. Metabolites were extracted through solvent extraction method into ethyl acetate at the ratio of 4:1(v/v). The upper layer of the solvent was collected through separating funnel into conical flasks. The solvent was evaporated using rotary evaporator. The fungal biomass was air dried for five days and cut into pieces. The pieces were covered in blotting paper and packed in a Soxhlet apparatus and the metabolites were extracted by water bath distillation using methanol for 3 days after which the solvent was removed by evaporating and the metabolites were subjected to GC MS analysis. GC MS analysis of *N. rileyi* in Ethyl acetate was performed to identify the compounds present in the fungus which might be responsible for the mortality of the pest. The analysis was carried with GC Clarus 500 Perkin Elmer equipment. Compounds were separated on Elite-1 cappillary column (100% Dimethylpolysiloxane). Samples were injected with a split ratio of 10:1 with a flow rate of helium 1 ml/min. (carrier gas). Mass detector-Turbo Mass gold-Perkin Elmer Software-Turbomass 5.2 was used as a detector. Other conditions are oven temperature up to 110°C—2 minutes hold; up to 280°C at the rate of 5 deg/min-9 minutes hold. Injector temperature was maintained at 250°C. The constituents were identified after comparison with those available in the Computer Library (NIST ver. 2.1) attached to the GC-MS instrument and reported.

Statistical Analysis

The experimental design for all trials was completely randomized. The data were analyzed using the analysis of variance (ANOVA) and the mean values were compared.

RESULTS AND DISCUSSION

Management of plant diseases by chemical pesticides has lead to the development of pesticide resistant strains of plant pathogens. Hence, biological control of disease is the distinct possibility and can be exploited within the framework of integrated disease management system.The usage of different entomophagous fungus was used in pest management (Bhattacharya *et al.*, 2004). They explained about the usage of various fungal pathogens particularly *Beauveria bassiana, Metarhizium anisopliae, Verticillium lecanii*, and *Nomuraea rileyi* have found to be promising in the control of several agricultural pests. They also explained about the mode of action of these entomopathogenic fungi. The entomophagous fungi are one of the important organisms as there are several advantages of using these fungal pathogens as pest control agent. Most of these are host specific and less toxic to mammal having a wide host range. But in most of the cases the success has been limited where the ambient environmental conditions can be very accurately controlled.

The mortality rates of *P. indicus* in lab conditions after fungal applications are presented in Figure 1. The observations were made at six hours interval, after 6 hrs, 12 hrs, 18 hrs, 24 hrs, 30 hrs, 36 hrs, 42 hrs and 48 hrs after treating with fungus. The mortality of *P. indicus* showed a gradual increase at 12 to 30 hrs. After 36 hrs, there was no specific increase in pest mortality. *Nomuraea rileyi* and *Metarhizium anisopliae* showed significant mortality among the tested fungi. The field experiments are also on par with the lab assays (Figure 2). The death of the pest started on day 1 and gradually increased upto 4 days after which no effect was recorded. Superiority of *N. rileyi* and *M. anisopliae* was evident right from 2nd day and continued

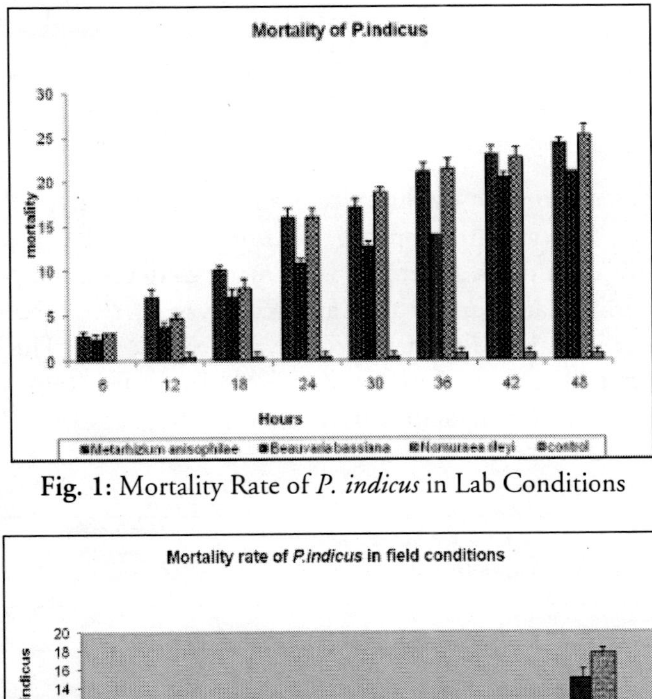

Fig. 1: Mortality Rate of *P. indicus* in Lab Conditions

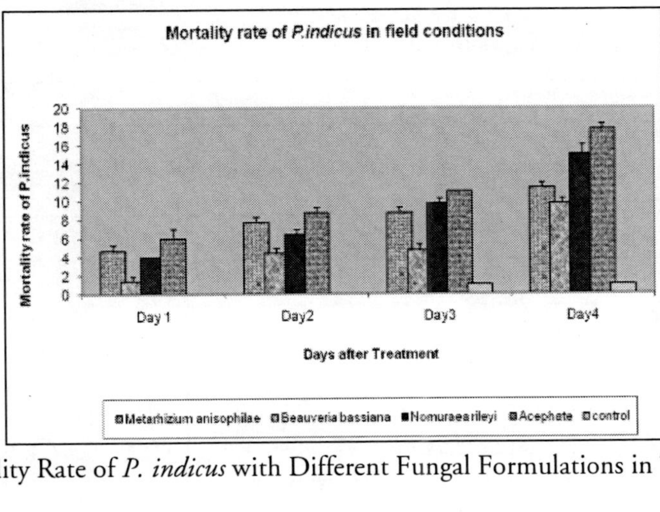

Fig. 2: Mortality Rate of *P. indicus* with Different Fungal Formulations in Turmeric Field

throughout the period of experimentation (15 ± 1 and 11.33 ± 0.58 respectively). On the other hand, *Beauveria bassiana* showed the least activity of 9.67 ± 0.58 among the tested fungi. The fungi *N. rileyi* which showed better result was used for further studies and subjected to GC MS analysis for identification of compounds responsible for the mortality of *P. indicus* and the results are given in Figure 3.

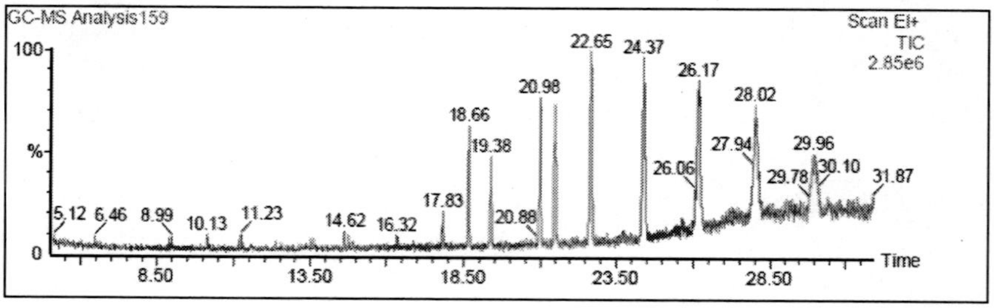

Fig. 3: GC MS Analysis of *Nomuraea rileyi*

Nomuraea rileyi is a dimorphic hyphomycete having invasive hyphal stages and yeast like vegetative stage.It is an entamopatogenic fungus which has a potential use as a biocontrol agent for crop pests. Attempts were made to develop biopesticide based on *N. rileyi.* It is an important fungus that attacks larvae of rice insects. It attacks the larvae of stems borers, leaf folders, armyworms and caseworms. The GC MS analysis revealed the presence of various compounds (Table 1 and Figure 4) which was found to be 1, 2-Benzenedicarboxylic acid, diisooctyl ester ($C_{24}H_{38}O_4$), 2-Propen-1-amine,

Table 1: Compounds Identified in GC MS Analysis of *Nomuraea rileyi* Ethyl Acetate Extract

No.	RT	Name of the compound	Molecular formula	MW	Peak Area %
1.	18.66	2-Phenazinol, 6-amino-	$C_{12}H_9N_3O$	211	4.05
2.	19.38	2,2-Dimethylglutaric anhydride	$C_7H_{10}O_3$	142	4.73
3.	20.98	2-Methylcyclohexylamine	$C_7H_{15}N$	113	8.78
4.	21.48	1,2-Benzenedicarboxylic acid, diisooctyl ester	$C_{24}H_{38}O_4$	390	7.43
5.	22.65	1H-Azepin-1-amine, hexahydro-	$C_6H_{14}N_2$	114	14.86
6.	24.37	1-Piperazineethanamine	$C_6H_{15}N_3$	129	16.89
7.	26.17	dl-Citrulline	$C_6H_{13}N_3O_3$	175	18.92
8.	28.02	2-Propen-1-amine, N-2-propenyl-	$C_6H_{11}N$	97	18.92
9.	29.96	Undecane, 1,2-dibromo-2-methyl-	$C_{12}H_{24}Br_2$	326	4.05
10.	32.09	2H-Pyran, tetrahydro-2-(12-pentadecynyloxy)-	$C_{20}H_{36}O_2$	308	1.35

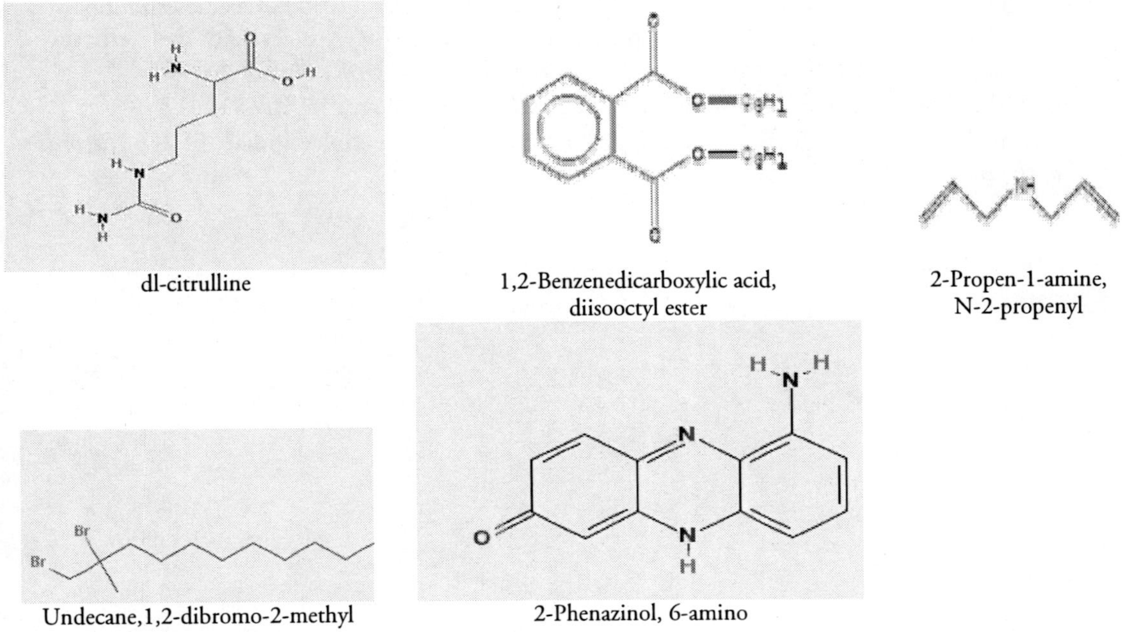

Fig. 4: Structure of Compounds Detected in GC MS Analysis of *Nomuraea rileyi*

N-2-propenyl ($C_6H_{11}N$), Undecane, 1, 2-dibromo-2-methyl ($C_{12}H_{24}Br_2$), 2-Phenazinol, 6-amino ($C_{12}H_9N_3O$), dl-Citrulline ($C_6H_{13}N_3O_3$), etc. Among these the dl-citrulline is found to be in present in larger quantities with a peak area of 18.92%, might be the compound which has a lethal effect on the pest. Further purification and analysis has to be carried out.

REFERENCES

Arutselvi, R., Balasaravanan, T., Ponmurugan, P., Suresh, P. and Ramachandran, N. (2012). Comparative Studies of Anti-microbial Activity of Turmeric and Selected Medicinal Plant Leaf Extracts Used in Indian Traditional Medicine, *Journal of Herbs, Spices and Medicinal Plants,* 18: 3, 231–239.

Bagnall, R.S. (1912). A new genus of Indian thrips (Thysanoptera) Injuries to turmeric. *Record of the Indian Museum,* 7: 257–260.

Bhattacharya, A., Samual, A.C. and Kar, S. (2004). "Entomophogus fungus in pest management. Department of Environmental Science Newsletter, Vol. 5, pp. 1–3.

Devanesan, S., Kumar, V., Ramachandran Nair, V., Bijumathew, T., Ravi, S. and Visalakshi, A. (1995). Leaf folder resurgence, a side effect of insecticide application in rice field, *Entomon* 20, pp. 79–81.

Dubey, S.C., Aradhika Tirupathi, Dureja P. and Anita Grover (2011). Characterization of secondary metabolites and enzymes produced by *Trichoderma* species and their efficacy against plant pathogenic fungi. *Indian Journal of Agricultural Sciences* 81(5): 455–61.

Rathaiah, Y. (1982). Rhizome rot of turmeric. Indian Phytopathology, 35(3): 451–417.

Ravikumar, P. (2002). Production technology for organic turmeric, Spice India, 15: 2–6, *Record of the Indian Museum* 7: 257–260.

Ocimum species a Potential Natural Source with Antimicrobial and Antioxidant Properties

Prasad M.P.*, Sushant Shekhar and Ritesh Singh
Sangenomics Research Lab, Bangalore
E-mail: Prasad_m_p@hotmail.com

INTRODUCTION

Traditional medical system has great value and also many medicinal plants have been identified from indigenous pharmacopoeias, because of which plants are still making imperative contribution to healthcare regardless of modern medicines which has many advances (Afolabi *et al.,* 2007). The medicinal values of plants lie in their Phytochemicals, which produce definite physiological actions on the human body. Medicinal plants have been found to be helpful in curing many diseases and have always promoted the search for different extracts from plants which could act as a potential source of new antimicrobial agent (Beegum *et al.,* 2003 and Lee *et al.,* 2004). Thousands of medicinal species found in different parts of the India are being used from the ancient times and are of great medicinal and economical value (Ramesh *et al.,* 2010).

In recent years, it has been seen that because of indiscriminate use of commercial antimicrobial drugs, rate of acquiring multiple resistances in human pathogenic microorganisms has increased (Javanmardi *et al.,* 2003). This fact has led the scientists to search for novel antimicrobial substances and screen medicinal plants regularly for new antibacterial agents. The genus *Ocimum* (Tulasi) comes under Lamiaceae family and is found in many part of the world like tropical and

*Corresponding author.

subtropical regions of Asia, Africa, and Central and South America. It is a source of essential oils and aroma compounds, a culinary herb and an attractive, fragrant ornamental plant. Tulasi is considered as a sacred plant and its various medicinal properties have been mentioned in ancient medicinal text "*Ayurveda*". Different parts of this plant are used for treatment of various ailments. *Ocimum* is believed to decrease lipid peroxidation and increase the activity of superoxide dismutase (Politeo *et al.,* 2007). The constituents of *Ocimum* species have antibacterial, antifungal, antioxidant and radio protective activity (Kahkonen *et al.,* 1999 and Blois *et al.,* 1958). Studies show that many *Ocimum* species are useful for the treatment of disorders in central nervous system (CNS) and also as antidepressant (Maryam Zahin *et al.,* 2009).

Oxygen derived free radicals and reactive oxygen species (ROS) plays role in many diseases and in food deterioration and spoilage (Ramesh *et al.,* 2010), by inhibiting the initiation or propagation of oxidative chain reactions, Antioxidants can delay or inhibit the oxidation of lipids or other molecules (Hakkim *et al.,* 2008). Synthetic antioxidants used in foods are very unstable and thought to be carcinogenic, because of which interest in finding out new natural and non toxic antioxidants is increasing (Farrukh *et al.,* 2006). Though much antibacterial and phytochemical studies has been conducted on *Ocimum species,* keeping in view the pharmacology, the present work has been designed to screen antibacterial agents by using various solvent extracts and their action on various pathogens and also to evaluate the antioxidant potential of the extracts to explore the basis for its traditional use.

MATERIALS AND METHODS

Plant Material

The plant material of *Ocimum species* were obtained from the local area in and around Ernakulam District, Kerala and University of agriculture, Bangalore India. The plants were indentified based on their description given in the literature. The plant leaves were air dried under shade and powdered.

ANTIBACTERIAL ACTIVITY

Collection of Microorganisms for Antibacterial Activity

The pathogenic microorganism were isolated from clinical samples and indentified on the basis of morphological, biochemical and physiological characteristics according to Bergey's manual of determinative bacteriology. The isolated microorganisms were found to be *K. pneumoniae, Proteus sp., E. faecalis, S. typhi, S. aureus* and *B. subtilis.*

Solvent Extraction

Plant leaves were dried under shade and dried leaves were crushed using mortar and pestle. Then 5 g of crushed plant material was kept on rotary shaker along with 50 ml of different solvents for 2 days. The extract was concentrated by solvent evaporation and used for antibacterial activity.

Determination of Antibacterial Activity

The antibacterial activity was determined by the agar well-diffusion method. Overnight grown bacterial culture was spread on to sterile Petri plate with Mueller Hinton agar medium (Hi Media Laboratories Limited, Mumbai, India). Wells of 6 mm were punched into the previously seeded MH agar plates using sterile cork borer. About 80 μl of the different *Ocimum* extract was placed in the wells and allowed to diffuse for 2 h at 4°C and the plates were incubated at 37°C for 24 hrs. The activity was determined by measuring the diameter of the inhibition zones for each well and expressed in millimeter.

Determination of Minimal Inhibitory Concentration (MIC)

The extracts that exhibited considerable activity were used for MIC determination. The extracts of the test samples were tested in four dose levels of 20 µl, 40 µl, 60 µl and 80 µl. The overnight grown bacterial culture was transferred on MH agar plate and wells were punched out using a sterile 6 mm cork borer. Different concentration (20–80 µl) of the extract was placed in separate wells, allowed to diffuse for 2 h at 4°C and then the plates were incubated at 37°C for 24 hrs. The zone of inhibition was observed and the lowest concentration of the test sample showing zone of inhibition was recorded as the MIC.

ANTIOXIDANT ASSAY

Extract Preparation for Antioxidant Activity

For extraction, 5 g powder of *Ocimum* was mixed with 50 ml of methanol. Extraction continued until the extraction solvents became colorless. The obtained extracts were filtered over Whatmann No. 1 paper and the filtrate was collected, then methanol was removed by evaporation.

DPPH (1, 1-diphenyl-2-picrylhydrazyl) free radical scavenging assay)

Scavenging activity on DPPH was assessed according to the method reported by (Adiguzel *et al.,* 2005) with a slight modification. Briefly, 500 µl of extracts (0.2 to 1 mg/ml) were mixed with 3 ml of methanolic solution of 0.1 mM DPPH. The mixture was shaken well and incubated at room temperature for 30 min and absorbance was measured at 517 nm in a spectrophotometer. Experiment was performed in triplicate and average was taken for determination of percentage inhibition.

Reducing Power Assay

The reducing power of extracts was determined by following method 0.5 ml of extracts (0.2 to 1 mg/ml) was mixed with 0.5 ml of 0.2 M phosphate buffer (pH 6.6) and 0.5 ml potassium ferrocyanide (1%). After incubating the mixture at 50°C for 20 min., 0.5 ml of 10% trichloroacetic acid was added, and then mixture was centrifuged at 3000 rpm for 10 min. 1 ml of supernatant was mixed with 1 ml of distilled water and 0.2 ml $FeCl_3$ (0.1%) and the absorbance was measured at 700.

Thiobarbituric Acid (TBA) Method

The method of [24] was referred; 1 ml of 20% trichloroacetic acid and 1 ml of 0.67% 2-thiobarbituric acid was added to 0.5 ml of sample solution prepared in ethanol. The mixture was placed in a boiling water bath. After cooling, absorbance of supernatant was measured at 552 nm.

RESULTS AND DISCUSSION

The antioxidant activity of *Ocimum* species was determined by three different assays such as reducing power assay, 1-1-diphenyl-2-picrylhydrazyl (DPPH) assay and Thiobarbituric acid (TBA) assay. *Camphor basil* showed maximum activity in reducing power assay at different concentrations as shown in Figure 1. In the DPPH method the absorbance was measured and the percent inhibition of the DPPH radical by *Ocimum* was calculated based on the measured absorbance. Antioxidant capacities in series of concentrations for *Ocimum* was used for calculating the percentage of antioxidant activity as showed in Figure 2. *Camphor basil* showed maximum antioxidant activity compared to other *Ocimum* species. The DPPH radical scavenging activities of *O. gratissimum* was higher compared to other species at lower concentration: whereas, antioxidant activity of *Camphor basil* was higher in other assays. In the TBA assay *Camphor basil* was most active as an antioxidant followed by *O. sanctum purple* at various concentrations as shown in Figure 3.

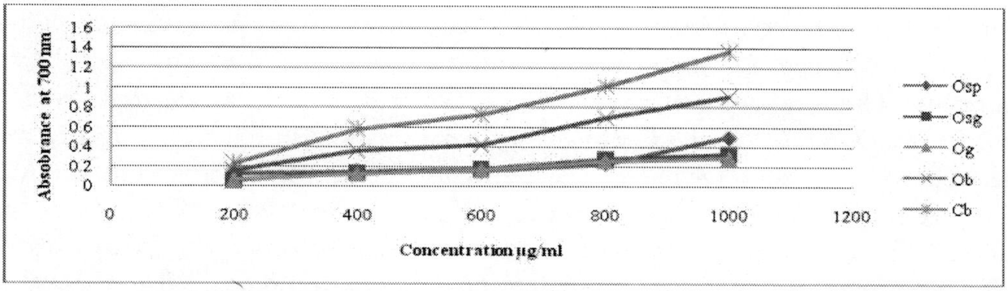

Osp- O. sanctum purple, **Osg**- O. sanctum green, **Og**- O. gratissimum, **Ob**- O. basilicum and **Cb**- Camphor basil

Fig. 1: Reducing Power Assay

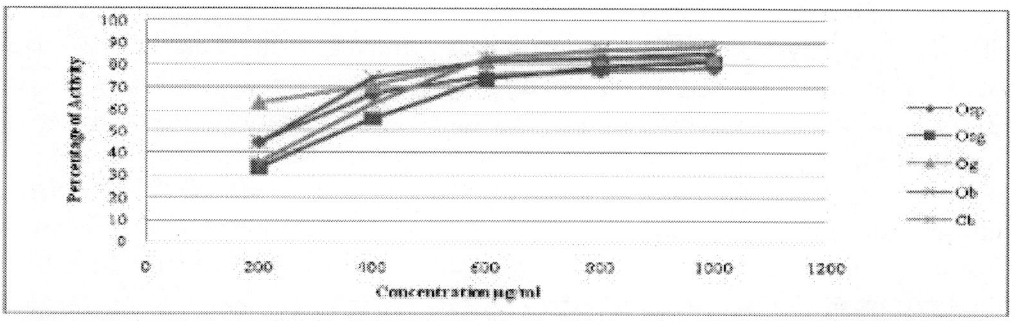

Osp- O. sanctum purple, **Osg**- O. sanctum green, **Og**- O. gratissimum, **Ob**- O. basilicum and **Cb**- Camphor basil

Fig. 2: Antioxidant Activity by DPPH Assay

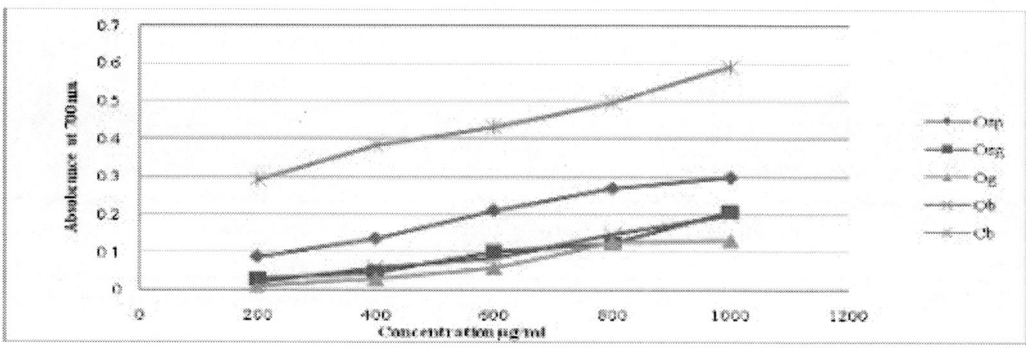

Osp- O. sanctum purple, **Osg**- O. sanctum green, **Og**- O. gratissimum, **Ob**- O. basilicum and **Cb**- Camphor basil

Fig. 3: Thiobarbituric Acid Assay

The antibacterial activities of different extracts of *Ocimum* were evaluated qualitatively and quantitatively against the pathogenic microorganisms by the presence or absence of inhibition zones and zone diameter. To determine the antibacterial activity, various solvent extracts like ethanol, Methanol, Propanol, chloroform, petroleum ether and Isoamyl alcohol was evaluated and compared using agar diffusion method. All the test extracts of *Ocimum* showed significant antibacterial activity against tested pathogens. The antibacterial activity of *O. sanctum purple, O. sanctum green, O. gratissimum, O. basilicum and Camphor basil are* presented in figures 4-8 respectively. The isoamyl extract of *Ocimum* was found to be the most active and produced highest zone of inhibition. Isoamyl extract of *O. sanctum purple* produced 24 mm zone and *O. sanctum green produced* 32 mm zone against *B. subtilis.* Isoamyl extract of *O. gratissimum* produced 26 mm zone, *O. basilicum* produced 28 mm zone *and Camphor basil* produced 22 mm zone against *S. typhi.* Among the tested pathogens *B. subtilis* and *S. typhi* were most sensitive for *Ocimum* extract and no inhibition activity observed against *E. faecalis.* In the MIC assay all the species exhibited diverse antibacterial activity at varied concentrations as summarized in Table 1. The isoamyl alcohol extract was most active compared to other solvent extracts and *B. subtilis* and *S. typhi* were most sensitive among the tested microorganism. Our results are different from [12], in our study zone of inhibition was seen for Ethanol and Methanol extract of *O. basilicum* on *Proteus sp.* and *S. aureus.* The minimal inhibitory concentration was determined at different concentrations and varied for different species. Among the tested microorganism most were sensitive for isoamyl alcohol extract of all *Ocimum* species.

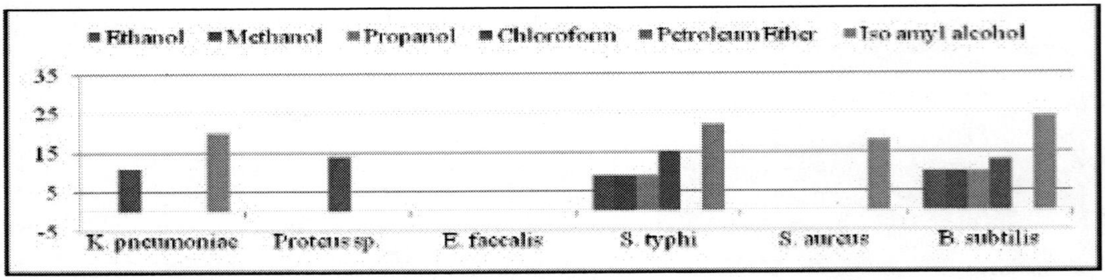

Fig. 4: Antibacterial Activity of *Ocimum sanctum purple*

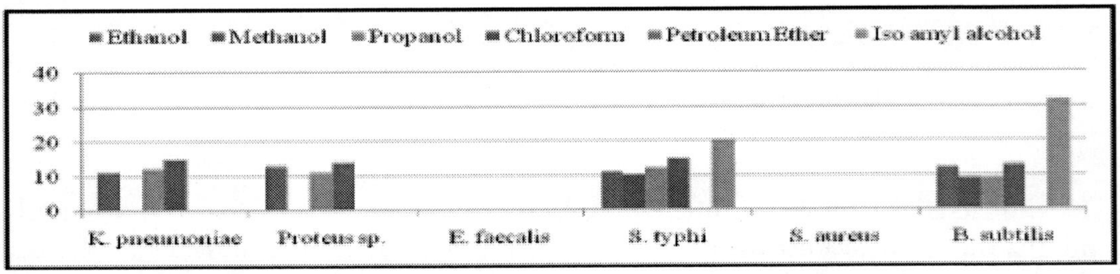

Fig. 5: Antibacterial Activity of *Ocimum sanctum green*

Table 1: Minimal Inhibitory Concentration of *Ocimum species*

Plant Sample and Extraction Solvent	*Pathogens*					
	K. pneumoniae	*S. typhii*	*S. aureus*	*E. faecalis*	*Proteus sp.*	*B. subtilis*
Ocimum sanctum purple						
Ethanol Extract	–	80	–	–	–	60
Methanol Extract	60	80	–	–	–	80
Propanol Extract	–	80	–	–	–	60
Chloroform Extract		20	–	–	40	20
Petroleum Ether	–	0	–	–	–	–
Isoamyl alcohol Extract	20	20	20	–	–	20
Ocimum sanctum green						
Ethanol Extract	40	60	–	–	40	40
Methanol Extract	–	40	–	–	–	60
Propanol Extract	20	40	–	–	40	60
Chloroform Extract	40	20	–	–	20	40
Petroleum Ether	–	–	–	–	–	–
Isoamyl alcohol Extract	–	20	–	–	–	20
Ocimum gratissimum						
Ethanol Extract	–	40	60	–	–	60
Methanol Extract	80	40	80	–	60	40
Propanol Extract	–	60	–	–	60	60
Chloroform Extract	60	80	–	–	–	60
Petroleum Ether	–	–	–	–	–	40
Isoamyl alcohol Extract	20	20	20	–	–	20
Ocimum basilicum						
Ethanol Extract	60	20	80	–	–	–
Methanol Extract	60	40	80	–	–	60
Propanol Extract	40	60	80	–	–	60
Chloroform Extract	40	–	40	–	40	20
Petroleum Ether	–	–	–	–	–	–
Isoamyl alcohol Extract	20	20	20	–	–	20
Camphor basil						
Ethanol Extract	80	80	–	–	–	80
Methanol Extract	80	40	–	–	–	80
Propanol Extract	80	–	–	–	–	60
Chloroform Extract	–	–	60	–	20	–
Petroleum Ether	–	–	–	–	–	–
Isoamyl alcohol Extract	20	20	60	–	–	20

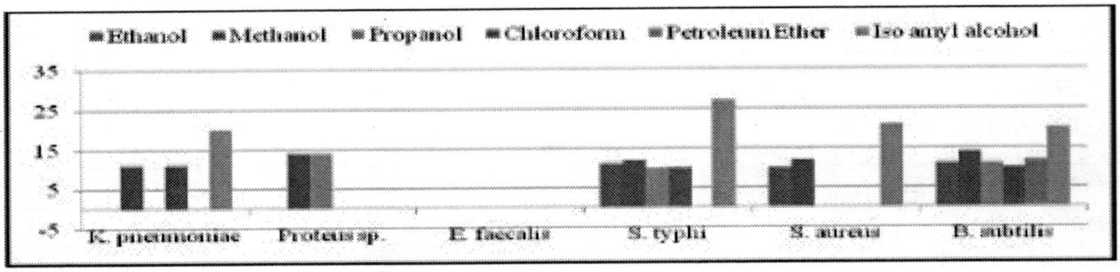

Fig. 6: Antibacterial Activity of *Ocimum gratissimum*

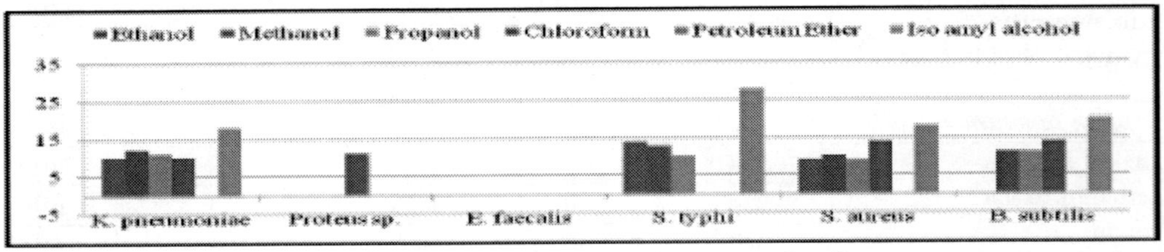

Fig. 7: Antibacterial Activity of *Ocimum basilicum*

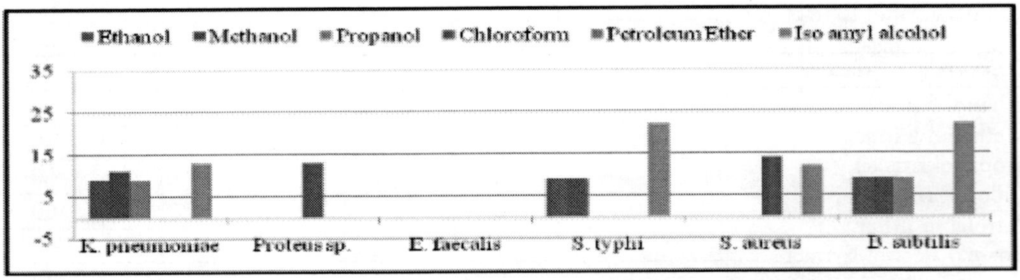

Fig. 8: Antibacterial Activity of Camphor basil (*Ocimum kilimandscharicum*)

The fluctuation in antioxidant potential might be due to the different principles of antioxidant assays. The isoamyl extract showed stronger and broader spectrum of antibacterial activity compared to other solvent extracts. These results predict that isoamyl alcohol is a better solvent for consistent extraction of antibacterial substances from *Ocimum species*. As *Ocimum* is widespread in India, it can be recommended as an easily available and renewal source for screening of antibacterial agent instead of synthetic chemicals. The present findings indicate that *Ocimum* possesses compounds with antibacterial properties against pathogenic microorganisms.

REFERENCES

Adiguzel, A., Gulluce, M., Sengul, M., Ogutcu, Sahin, F. and Karaman, I. (2005). Antimicrobial effects of Ocimum basilicum (Labiatae) Extract. *Turk. J. Biol*: 29: 155–156.

Afolabi, C.A., Ibukun, E.O., Emmanuel, A., Obuotor, E.M. and Farombi, E.O. (2007). Phytochemical constituent and antioxidant activity of extract from the leaves of Ocimum gratissimum. *Sci. Res. Essay.*: 2(5): 163–166.

Beegum, N.R.B. and Devi, T.G. (2003). Antibacterial activity of selected seaweeds from Kovalam south west coast of India. *Asian JR. of Microbiol. Biotech. Env. Sc.* 5(3): 319–322.

Blois, M.S. (1958). Antioxidant determinations by the use of a stable free radical. *Nature*: 181: 1199–1200.

Farrukh, A., Iqbal, A. and Zafar, M. (2006). Antioxidant and Free Radical Scavenging Properties of Twelve Traditionally Used Indian Medicinal Plants. *Turk J. Biol*: 30: 177–183.

Hakkim, F.L., Arivazhagan, G. and Boopathy, R. (2008). Antioxidant property of selected Ocimum species and their secondary metabolite content. *Journal of Medicinal Plants Research*: 2: 250–257.

Javanmardi, J., Stushnoff, C., Locke, E. and Vivanco, J.M. (2003). Antioxidant activity and total phenolic content of Iranian Ocimum accessions. *Food Chemistry*, 83(4): 547–550.

Kahkonen, M.P., Hopia, A.I., Vuorela, H.J., Rauha, J., Pihlaja, K., Kujala, ST. and Heinonen, M., Antioxidant activity of plant extracts containing phenolic compounds. *J. Agric. Food Chem.* 1999: 47: 3954–3962.

Lee, K.W., Everts, H. and Beynen, A.C. (2004). Essential oils in broiler nutrition. *Int. J. Poult. Sci.* 3: 738–752.

Maryam Zahin, Farrukh Aqi. and Iqbal Ahmad (2009). The invitro antioxidant activity and total and total phenolic content of four Indian medicinal plants. *International Journal of Pharmacy and Pharmaceutical Sciences*, 1(1): 88–95.

Politeo, O., Jukic, M. and Milos, M. (2007).Chemical composition and antioxidant capacity of free volatile aglycones from basil (Ocimum basilicum L.) compared with its essential oil. *Food Chemistry*, 101: 379–385.

Ramesh, B. and Satkopan, V.N. (2010). Antioxidant Activities of Hydroalcoholic Extract of Ocimum sanctum Against Cadmium Induced Toxicity in Rats. *Ind. J. Clin. Biochem.*, 25(3): 307–310.

Role of Curcumin as a Nutritional Supplement to Enhance Cell Proliferation and Survival in Hypoxia Induced Hepatocyte Injury – *in vitro* Model

Roshni Baby Thomas, T.R. Anju, Shilpa Joy and C.S. Paulose*

Molecular Neurobiology and Cell Biology Unit, Centre for Neuroscience, Department of Biotechnology, Cochin University of Science and Technology, Cochin, Kerala

E-mail: cspaulose@cusat.ac.in; paulosecs@yahoo.co.in

INTRODUCTION

Liver is a multifunctional vital organ that plays essential roles in metabolism, biosynthesis, excretion, secretion, detoxification and glycogen storage. These processes require energy, making the liver a highly aerobic, oxygen-dependent tissue. These processes also impart vulnerability of the liver to hypoxia/anoxia. Liver hypoxia occurs in a number of clinical conditions such as cirrhosis, portal vein thrombosis, low cardiac output and generalized hypoxemia and can contribute to liver atrophy and/or acute liver failure (Zeenat, F.Z. (2010). Hypoxic conditions can occur during episodes of hepatic inflammation or established chronic disease. Hepatocytes exposed to a hypoxic microenvironment would therefore be potentially sensitized to cell death (Ricky, H.B. *et al.,* 2010). The primary cellular target during Ischemia and reperfusion (I/R) injury is the hepatocyte. Hepatocyte death seen during the hypoxic and reperfusion phases of IRI occurs within a relatively hypoxic environment, and hepatic IRI can be divided into 2 phases. The early phase is thought to involve the activation of Kupffer cells (KCs), which release proinflammatory cytokines and reactive oxygen species (ROS) (Jaeschke, H.; Farhood, A., 1991).

*Corresponding author.

(Jaeschke, H., 2003) The late phase is characterized by increased expression of chemokines and adhesion molecules and hepatic recruitment of effector cells, which amplify the tissue damage. The latter phenomenon is a feature common to many other hepatic diseases. There is evidence suggesting that hepatocyte injury occurs during the relatively hypoxic early phase of IRI (Zhang, W. *et al.*, 2007). ROS release is one of the earliest and most important components of tissue injury after the reperfusion of ischemic organs and is a major contributor to hepatocyte death during reperfusion (Allan, T. *et al.,* 2005). IRI is a pathophysiologic process whereby hypoxic organ damage is accentuated following return of blood flow and oxygen delivery to the compromised tissue (Decker, K., 1990). Because aerobic metabolic rates decrease in hypoxia-sensitive cells under oxygen-limiting conditions, the demand for glucose or glycogen for anaerobic glycolysis may rise drastically as a means of making up for the energetic shortfall (Hochachka, P.W., 1986).

Cellular response to hypoxia involves multiple and complex events. Several studies indicate that the preserving ionic homeostasis. An inadequate supply of ATP impairs the cell's ability to perform critical functions, and may therefore contribute to cell death. Thus when metabolism and membrane functions are decoupled, the cells become sensitive to hypoxia (Hochachka, P.W. (1986). However, the naturally occurring acidosis of ischemia and hypoxia delays onset of necrotic cell death by as much as hours (Gores, G.J., 1998, 1999). Reduced ATP levels in the liver are one major factor that contributes to liver cell death (Cunningham, C.C. and Van Horn, C.G., 2003). The sequence of events from ATP depletion to cell death may involve altered membrane permeability and disruption of ion homeostasis, i.e. increased free intracellular Ca2+, concentrations and increased hydrolysis of membrane phospholipids leading to cell damage or even to cell death (Hochachka, P.W., 1986). It has been shown that fructose can protect against hypoxic damage in the liver (Anundi, I., 1987) and after simple respiratory inhibition, glycolytic substrates prevent cell death by providing an alternate source of cellular ATP (Lemasters, J.J., 1997). Comparing the effect of hypoxia on well fed rats and fasted rats, this was suggested that nutritional status is an important determinant of damage to hepatocytes due to hypoxia (Israel, Y., 1975 and Bradford, B.U., 1986). Hypoxia has already been shown to decrease amino acid and protein synthesis in isolated hepatocytes prior to ATP loss (Chan, T.S., 2008). Inhibition of protein synthesis is an early event and consequently a sensitive marker for chemical-mediated hepatocyte damage (Goothals, F., 1984 and Buc-Calderon, P., 1991).

Curcumin is a natural phenolic compound found in the rhizomes of Curcuma longa and endowed with beneficial biological activities including antioxidant, anticarcinogenic and hepatoprotective effects. In this study curcumin was tested for its potential ability as a nutritional supplement to induce cell division and enhance cell survival in a model system represented by primary cultures of rat hepatocytes subjected to oxygen limiting conditions. Earlier results suggest that curcumin is able to modulate in vitro both expression and function of hepatic Pgp and support the hypothesis that curcumin, a chemopreventive phytochemical, could reveal itself also as a compound endowed with chemosensitizing properties (Hochachka, P.W., 1986).

MATERIALS AND METHODS

Biochemicals used in the present study were purchased from Sigma Chemical Co., St. Louis, USA. [^{3}H] Thymidine and [^{3}H] leucine were purchased from Amersham Life Science, UK.

Animals and Experimental Design

Male Wistar rats of 90–110 g body weight were used for hepatocyte isolation for invitro studies. They were housed in separate cages under 12 h light and 12 h dark periods. Neonatal rats of four days old were used for invivo studies. Neonatal rat were maintained with their mothers under optimal conditions 12 hour light and 12 hour dark periods and were fed standard food and water ad libitum.Rats had free access to standard food and water *ad libitum*. All animal care procedures were done in accordance with the Institutional and National Institute of Health guidelines. All efforts were made to minimize the number of animals used and their suffering. Neonatal Wistar rats were purchased from Amrita Institute of Medical Sciences, Kochi.

Hepatocyte Culture and *in vitro* Hypoxic Induction

Liver perfusion was done using PBS and collagenase buffer in anaesthetised rats. Hepatocytes were isolated by mincing in calcium free buffer and collagenase buffer followed by centrifugation steps and the cells were cultured in William's E medium. Hypoxia was induced in vitro by by placing the hepatocytes in a hypoxic chamber provided with 2.6% oxygen for 30 minutes.

In vivo Hypoxia Induction

Control neonatal rats were given atmospheric air (20.9% oxygen) for 30 minutes (C); (ii) Hypoxia was induced by placing the neonatal rats in a hypoxic chamber provided with 2.6% oxygen for 30 minutes

Thymidine and Leucine Incorporation Studies of Pancreatic Beta Cells

The isolated cells were re-suspended in tissue culture media and cell viability was determined. Cell suspension was diluted and 10^6 cells were added to two different sets of culture plates. [^{3}H] thymidine of specific activity 18 Ci/mmol was added to one set of culture plates for all the three experimental groups to determine DNA synthesis and [^{3}H] leucine of specific activity 63 Ci/mmol to the next set to determine the protein synthesis. After 24 hr. incubation under appropriate conditions, cells were scrapped off from the culture plates and centrifuged at 2000 xg for 20 minutes. The supernatant was discarded and the pellet was re-suspended in 1M NaOH and kept for overnight incubation. Bound radioactivity was counted with cocktail-T in a liquid scintillation counter.

Analysis of Gene Expression by Real-Time PCR

RNA was isolated from the pancreas of experimental rats using the Tri-reagent according to the procedure of (Chomczynski and Sacchi, 1987). Total cDNA synthesis was performed using ABI PRISM cDNA archive kit in 0.2 ml microfuge tubes. The reaction mixture of 20 μl contained 0.2 μg total RNA, 10 × RT buffer, 25 × dNTP mixture, 10 × random primers, MultiScribe RT (50 U/μl) and RNase free water. The cDNA synthesis reactions were carried out at 25°C for 10 min and 37°C for 2 h using an Eppendorf Personal Cycler. Real-time-PCR assays were performed in 96-well plates in an ABI 7300 Real-time-PCR instrument (Applied Biosystems). The specific primers and probes were purchased from Applied Biosystems. The TaqMan reaction mixture of 20 μl contained 25 ng of total RNA-derived cDNAs, 200 nM each of the forward primer, reverse primer and TaqMan probe for assay on demand and endogenous control β-actin and 12.5 μl of Taqman 2 × Universal PCR Master Mix (Applied Biosystems) and the volume was made up with RNAse free water. The following thermal cycling profile was used (40 cycles): 50°C for 2 min, 95°C for 10 min, 95°C for 15 s and 60°C for 1 min. Fluorescence signals measured during amplification were considered positive if the fluorescence intensity was 20-fold greater than the standard deviation of the baseline fluorescence. The $^{\Delta\Delta}$CT method of relative quantification was used to determine the fold change in expression. This was done by normalizing the resulting threshold cycle (CT) values of the target mRNAs to the CT values of the internal control β-actin in the same samples ($^{\Delta}$CT = CT Target−CTβ-actin). It was further normalized with the control ($^{\Delta\Delta}$CT = $^{\Delta}$CT−CT Control). The fold change in expression was then obtained as (2−$^{\Delta\Delta}$CT) and the graph was plotted using log 2−$^{\Delta\Delta}$CT.

Statistics

Data are presented as means with their standard errors. Statistical evaluations were done by ANOVA using In Stat (version 2.04a; GraphPad Software Inc.). The Student–Newman–Keuls post hoc test was used to compare data among the groups. Relative Quantification Software (Relative Quantification Software) was used for analysing real-time PCR results. In all cases, $P < 0{\cdot}05$ was taken to indicate statistical significance.

RESULTS

Effect of Curcumin on DNA Synthesis in Hepatocytes Exposed to Hypoxic Condition *in vitro*

In vitro studies on the effect of curcumin on the hepatocytes exposed to hypoxic condition rats showed a significant increase ($p < 0.01$) in the DNA synthesis of the control and hypoxic groups treated with curcumin compared to control. This reflects a drastic reduction in the cell differentiation in hypoxic condition which is reversed to near control levels when supplemented with curcumin (Figure 1).

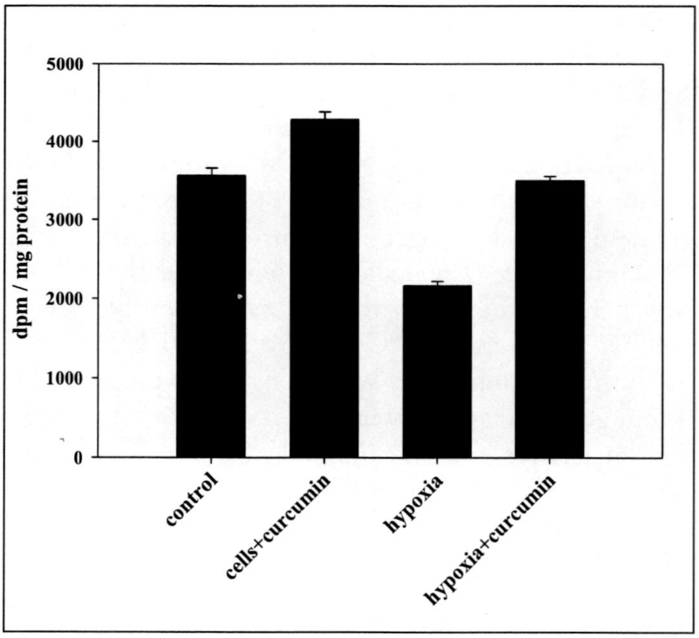

(a) Effect of curcumin on DNA synthesis in hepatocytes exposed to hypoxic condition *in vitro*

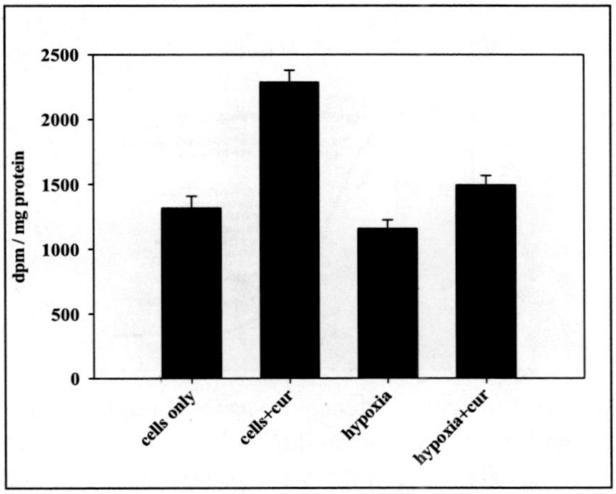

(b) Effect of curcumin on protein synthesis in hepatocytes exposed to hypoxic condition *in vitro*

Experimental groups	Body weight (g)	
	Initial day of Experiment	Final day of Experiment
C	8.56 ± 0.29	12.16 ± 0.35
Hypoxia	7.35 ± 0.54	13.22 ± 0.55

(c) Body weight (g) of hypoxic neonatal rats

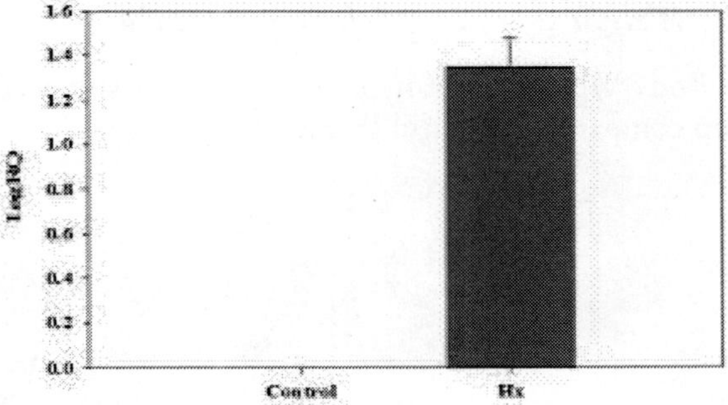

(d) Real Time PCR analysis of NFκB mRNA in the brain stem of hypoxic neonatal rats

Fig. 1: The $^{\Delta\Delta}$CT method of relative quantification was used to determine the fold change in expression. The relative ratios of mRNA levels were calculated using the $^{\Delta\Delta}$CT method normalized with β-actin CT value as the internal control and Control CT value as the calibrator. PCR analyses were conducted in the cerebral cortex with gene-specific primers and fluorescently labeled Taq probe Rn01502266_m1. Values are mean± S.E.M. of 4–6 separate experiments. Each group consist 6–8 rats. C- Control, H- Hypoxia.

Effect of Curcumin on Protein Synthesis in Hepatocytes Exposed to Hypoxic Condition *in vitro*

In hypoxic group, protein synthesis was significantly down regulated ($p < 0.001$) compared to control. The altered parameters are reversed to near control in Curcumin treated groups (Figure 2).

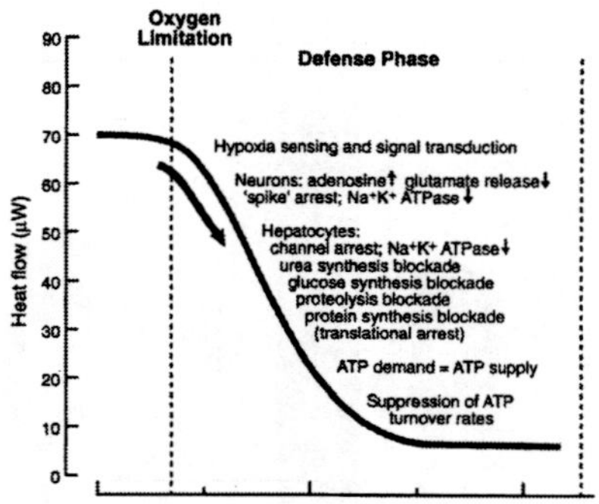

Fig. 2: Summary model of molecular defense processes thought to account for the extreme tolerance to oxygen limitation expressed by cells from hypoxia-tolerant species. Probable sequences are ordered from top to bottom (direction of arrow); vertical axis is heat output of turtle hepatocytes (24).

Body Weight (g) of Hypoxic Neonatal Rats

The Body Weight (g) of hypoxic neonatal rats showed a no significant regulation in the hypoxic group compared to control (Figure 3).

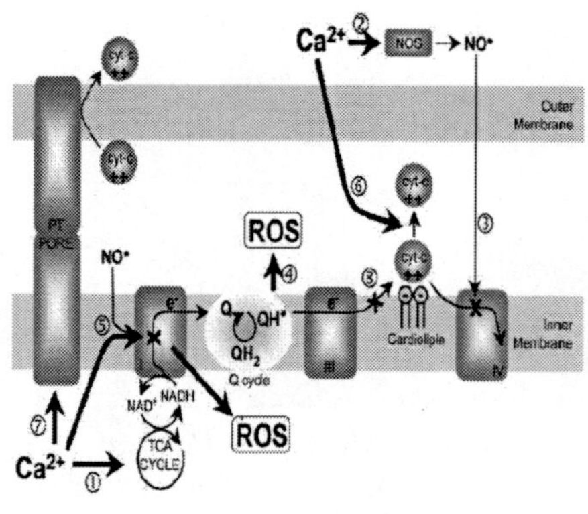

Fig. 3: Mechanisms for Ca^{2+} stimulation of mitochondrial ROS generation. Ca^{2+} stimulation of the TCA cycle (*1*) will enhance electron flow into the respiratory chain, and Ca^{2+} stimulation of nitric oxide synthase (NOS) and subsequent nitric oxide (NO^{-}) generation (*2*) would inhibit respiration at complex IV (*3*). These events would enhance ROS generation from the Q cycle (*4*). In addition, NO^{-} and Ca^{2+} can inhibit complex I, possibly enhancing ROS generation from this complex (*5*). Ca^{2+} also dissociates cytochrome *c* (cyt-c) from the inner membrane cardiolipin (*6*) and at high concentrations triggers PTP opening and cytochrome *c* release across the outer membrane (*7*). The subsequent inhibition at complex III (*8*) would enhance ROS generation at the Q cycle (*4*) Complex II is omitted from this diagram for clarity. (33)

Real Time PCR Analysis of NFκB mRNA in the Brain Stem of Hypoxic Neonatal would Rats

The Real Time PCR analysis of NFκB mRNA in the brain stem of hypoxic neonatal rats shows a drastic increase ($p < 0.001$) compared to control.

DISCUSSION

Tissue hypoxia and ischemia causes anoxia, ATP depletion, glycolytic substrate exhaustion, and acidosis (21). The first lines of defense against hypoxia include a balanced suppression of ATP-demand and ATP-supply pathways; this regulation stabilizes (adenylates) at new steady-state levels even while ATP turnover rates greatly decline. Hypoxic ATP demands of protein synthesis are down-regulated by translational arrest. In hypoxia sensitive cells this translational arrest seems irreversible. Under normoxic conditions, the main energy sinks are: (i) protein synthesis, (ii) protein degradation, (iii) Na+/K+ pumping, (iv) urea biosynthesis, and (v) glucose biosynthesis (22). In effect, the ATP-consuming processes are arranged in a hierarchy, with protein synthesis and RNA/DNA synthesis falling off rather sharply as energy becomes limiting (23). The present study established the effect of hypoxia on the DNA and protein synthesis in hepatocytes *in vitro* and evaluated the effectiveness of curcumin in increasing the cell survival by removing the transcriptional block and enhancing the cell differentiation.

In the present study, we observed a decline in the DNA and protein synthesis in hypoxic hepatocytes which might have been a cellular response to Reactive Oxygen Species formation and subsequent ATP depletion. As we also observed a down regulation of regulatory transcription factor NFκB in the respiratory centre of the brain, the brain stem, this might have augmented the regulation of cell proliferation and differentiation under hypoxic stress.

One mechanism prominent in ischemia/reperfusion injury is an onset of the mitochondrial permeability transition (MPT) (25). Opening of high-conductance permeability transition (PT) pores in the mitochondrial inner membrane causes an onset of the MPT, leading to ATP depletion and necrotic cell death (26, 27). After the MPT, mitochondria uncouples and lose their ability to synthesize ATP. With uncoupling, futile hydrolysis of ATP occurs, catalyzed by the mitochondrial ATP synthase operating in reverse. Thus cellular Ca^{2+} overloading after MPT onset is likely the result of ATP depletion and inhibition of ATP-driven Ca^{2+} pumps and secondary ion exchangers that serve to translocate Ca^{2+} out of the cytosol across the plasma membrane and into the sarcoplasmic reticulum. Ca^{2+} overloading is also suggested to play a key role in ischemia-reperfusion (28, 29). Mitochondria function as O_2 sensors and signal hypoxia by releasing ROS to the cytosol (30). Superoxides are generated at the Qi or Qo sites in complex III, resulting in ROS release to the matrix and/or the inter membrane space/cytosol (31) These ROS can damage cellular components such as proteins, lipids, and DNA (32).

We put forth the possible mechanisms by which curcumin restores the protein and DNA synthesis in the hypoxia affected hepatocytes. Scavenging of intracellular reactive oxygen species (ROS) is one of the potential mechanisms contributing to the protective effects of many antioxidants. Curcumin, a natural product, is an effective ROS scavenger. Curcumin is an effective antioxidant in different *in vitro* assays including reducing power; DPPH, $ABTS^+$, O^{2-} and $DMPD^+$ radical scavenging; hydrogen peroxide (H_2O_2) scavenging; and metal chelating activities (Gulcin, T.I., 2008). It also inhibits UV irradiation- and methylglyoxal-induced oxidative stress and apoptosis in A431 cells, embryonic stem cells, and blastocysts (Chan, W.H.; Wu, C.C., Yu, J.S. (2003 Hsuuw, Y.D., 2005); protects mitochondria from oxidative damage; and attenuates apoptosis (Zhu, Y.G., 2004).

The other possible pathway is by the NF-κB pathway in the regulation of cell proliferation and cell differentiation in both neonatal (Zhu, Y.G., 2004) and adult (Marc, H.G., 2003 and Mark, P.M., 2005) brain structures. Exposure of the brain to oxidative stress results in neuronal damage which starts in the brain stem and became more widespread in other areas of the brain at higher exposure levels (Marc, H.G.; Aase, A., 2003). NF-kB is activated in neurons and glial cells in acute neurodegenerative conditions which involves ROS species (Mark, P.M., 2005) thus playing a major role in modifying neuronal survival and recovery from injury. Brainstem parasympathetic circuits modulate the digestive functions of the stomach, liver and proximal duodenum (Alberto, R.T., 2006).

We observed an increased DNA and protein synthesis in Curcumin treated hypoxic cells. Treatment of the hepatocytes exposed to hypoxic insult with Curcumin enhanced cell survival and differentiation effectively thus reversing the negative effects of ischemia and reperfusion (I/R) injury in liver. This might have in part been contributed by the antioxidant properties of Curcumin. Our study thus points to the possibility of Curcumin as a potential nutritional supplement to counterbalance physical stress like hypoxia.

ACKNOWLEDGEMENT

This work was supported by research grants from DBT, DST, ICMR, Govt. of India and KSCSTE, Govt. of Kerala to Dr. C.S. Paulose.

REFERENCES

Alberto, R.T., Gerlinda, E.H., Kirsteen, N.B. and Richard, C.R. (2006). Brainstem Circuits Regulating Gastric Function. *Annu Rev Physiol.,* 68: 279–305.

Allan, T., Rosemary, A.H., Kunihiko, I., Nathan, D.C., Atsunori, N., Meagan, H.C., Michael, T.L., David, A.G. and Timothy, R.B. (2005). Hepatic Ischemia/Reperfusion Injury Involves Functional TLR4 Signaling in Nonparenchymal Cells. *J Immunol.,* 175: 7661–7668.

Anundi, I., King, J., Owen, D.A., Schneider, H., Lemasters, J.J. and Thurman, R.G. (1987). Fructose prevents hypoxic cell death in liver. *Am. J. Physiol.,* 253: G390–G396.

Bagchi, D., Wetscher, G.J., Bagchi, M., Hinder, P.R., Perdikis, G., Stohs, S.J., Hinder, R.A. and Das, D.K. (1997). Interrelationship between cellular calcium homeostasis and free radical generation in myocardial reperfusion injury. *Chem Biol Interact.,* 104: 65–85.

Boutilier, R.G. (2001). Mechanisms of cell survival in hypoxia and hypothermia. *The Journal of Experimental Biology.* 204, 3171–3181.

Bradford, B.U., Marotto, M., Lemasters, J.J. and Thurman, R.G. (1986). New, simple models to evaluate zone-specific damage due to hypoxia in the perfused rat liver: time course and effect of nutritional state. *J. Pharmacol Exp Ther.,* 236: 263– 268.

Buc-Calderon, P., Latour, I. and Roberfrold, M. (1991). Biochemical changes in isolated hepatocytes exposed to tert-butyl hydroperoxide. Implication for its cytotoxicity. *Cell Biol. Toxicol.,* 7: 129–143.

Chan, T.S., Gottschalk, S., Zwingmann, C., O'Brien, P., Mazer, D., Darby, P. and Bilodeau, M. (2008). Liver hypoxia leads to the utilization of aminoacids by the citric acid cycle to maintain e nergy stores. www. pulsus.com/cddw2008/abs/057.htm.

Chan, W.H., Wu, C.C. and Yu, J.S. (2003). Curcumin inhibits UV irradiation-induced oxidative stress and apoptotic biochemical changes in human epidermoid carcinoma A431 cells. *J. Cell Biochem,* 90, 327–338.

Chandel, N.S., McClintock, D.S., Feliciano, C.E, Wood, T.M, Melendez, J.A., Rodriguez, A.M. and Schumacker, P.T. (2000). Reactive oxygen species generated at mitochondrial complex III stabilize hypoxia-inducible factor-1a during hypoxia: a mechanism of O2 sensing. *J. Biol. Chem.,* 275, 25130–25138.

Crompton, M., Ellinger, H. and Costi, A. (1988). Inhibition by cyclosporin A of a Ca^{2+}-dependent pore in heart mitochondria activated by inorganic phosphate and oxidative stress. *Biochem J.,* 255: 357–360.

Cunningham, C.C. and Van Horn, C.G. (2003). Energy availability and alcohol-related liver pathology. *Alcohol Res Health.,* 27: 291–299.

Decker, K. (1990). Biologically active products of stimulated liver macrophages (Kupffer cells). *Eur. J. Biochem.* 192: 245–261.

Goothals, F., Krack, G., Deboyser, D., Vossen, P. and Roberfrold, M. (1984). Critical biochemical functions of isolated hepatocytes as sensitive indicators of chemical toxicity. *Fundam. Appl. Toxicol.,* 4: 441–450.

Gores, G.J., Nieminen, A.L., Fleishman, K.E., Dawson, T.L., Herman, B., Lemasters, J.J. (1988). Extracellular acidosis delays onset of cell death in ATP-depleted hepatocytes. *Am. J. Physiol.* 255(1): C315–C322.

Gores, G.J., Nieminen, A.L., Wray, B.E., Herman, B. and Lemasters, J.J. (1989). Intracellular pH during 'chemical hypoxia' in cultured hepatocytes. *J. Clin Invest.*, 83: 386–396.

Gulcin, T.I. (2008). Antioxidant and radical scavenging properties of curcumin. *Chem Biol Interact.,* 174, 27–37.

Hochachka, P.W. (1986). Defense strategies against hypoxia and hypothermia. *Science,* 17; 231 (4735): 234–41.

Hochachka, P.W., Buck, L.T., Dollt, C.J. and Landt, S.C. (1996). Unifying theory of hypoxia tolerance: Molecular/metabolic defense and rescue mechanisms for surviving oxygen lack. *Proc. Natl. Acad. Sci.* USA. 9493– 9498.

Hsuuw, Y.D., Chang, C.K., Chan, W.H. and Yu, J.S. (2005). Curcumin prevents methylglyoxal-induced oxidative stress and apoptosis in mouse embryonic stem cells and blastocysts. *J. Cell Physiol.* 205, 379–386.

Israel, Y., Kalant, H., Orregot, H., Khanna, J.M., Videla, L. and Phillipst, J.M. (1975). Experimental Alcohol-Induced Hepatic Necrosis: Suppression by Propylthiouracil (liver damage/thyroid hormones/hypermetabolic state/hypoxia/alcoholism) *Proc. Nat. Acad. Sci.,* 72(3): 1137–1141.

Jaeschke, H. (2003). Molecular mechanisms of hepatic ischemia reperfusion injury and preconditioning. *Am. J. Physiol Gastrointest Liver Physiol.* 284: G15–G26.

Jaeschke, H., Farhood, A. (1991). Neutrophil and Kupffer cell induced oxidant stress and ischemia-reperfusion injury in the rat liver. *Am. J. Physiol.* 260(1): G355–G362.

Lemasters, J.J. and Nieminen, A.L. (1997). Mitochondrial Oxygen Radical Formation during Reductive and Oxidative Stress to Intact Hepatocytes. *Bioscience Report* 17(3): 281–291.

Marc, H.G., Aase, A. and Richard, D.H. (2003). Chronic dietary mercury exposure causes oxidative stress, brain lesions, and altered behaviour in Atlantic salmon (Salmo salar) parr. *Aquatic Toxicology,* 65, 1:8, 55–72.

Mark, P.M. (2005). Survival and Plasticity of Neurons. *Neurochemical research,* 30, 6–7, 883–893.

Nadia, R., Roberto, T., Fabrizio, C. and Elisabetta, C. (1998). Effects of curcumin on P-glycoprotein in primary cultures of rat hepatocytes. Life Sciences, 62: 25, 2349–2358.

Nayler, W.G. (1983). Calcium and cell death. *Eur. Heart. J. 4, Suppl.,* C: 33–41.

Nazareth, W., Yafei, N. and Crompton, M. (1991). Inhibition of anoxia-induced injury in heart myocytes by cyclosporin A. *J. Mol. Cell Cardiol.,* 23: 1351–1354.

Paul, S.B., Yisang, Y., James, L.R., Anders, M.W. and Shey-Shing S. (2004). Calcium, ATP, and ROS: A mitochondrial love-hate triangle. *Am. J. Physiol. Cell Physiol.*, 287: C817–C833.

Qian, T., Nieminen, A.L., Herman, B. and Lemasters, J.J. (1997). Mitochondrial permeability transition in pH-dependent reperfusion injury to rat hepatocytes. *Am. J. Physiol. Cell Physiol 273*: C1783–C1792.

Qian, T., Nieminen, A.L., Herman, B., Lemasters, J.J. (1997). Mitochondrial permeability transition in pH-dependent reperfusion injury to rat hepatocytes. *Am. J. Physiol. Cell Physiol.,* 273: C1783–C1792.

Ricky, H.B., Stuart, M.C., Christopher, J.W., David, H.A. and Simon, C.A. (2010). Reactive Oxygen Species Mediate Human Hepatocyte Injury during Hypoxia/Reoxygenation. *Liver Transplantation.* 16:1303–1313.

Robert, D.G., Beatrice, H., Emmanuel, R., Hong, C., Liping, L., Kyle, D.M., Celeste, M.S., Ulrich, H. and Paul, T.S. (2005). Mitochondrial complex III is required for hypoxia-induced ROS production and cellular oxygen sensing. *Cell Metabolism*. 401–408.

Tomomi, I., Hiroyuki, T., Shunji, H., Dongchon, K., Nobuhiro, S., Keiichiro, N., Hideo U, Naotaka, H., Akira, T. (2001). Mitochondrial DNA Damage and Dysfunction Associated With Oxidative Stress in Failing Hearts After Myocardial Infarction. *Circ. Res.*, 88: 529–535.

Zeenat, F.Z. (2010). Periportal Necrosis in Rat Liver Exposed to Sodium Nitrite-induced Hypoxia. *Research Journal of Animal and Veterinary Sciences*. 5(2): 111–116.

Detection of Extended Spectrum β-lactamase Producing Bacteria from Urinary Tract Infections from Chennai

Ramesh Kumar M.R.*, Archana M., Vijaykanth N., Manikandan N. and Arunagirinathan N.

Department of Microbiology and Biotechnology,
Presidency College (Autonomous), Chennai, Tamil Nadu
E-mail: rameshmicrobiologist@gmail.com

INTRODUCTION

Urinary Tract Infections (UTIs) are one of the most prevalent extra-intestinal bacterial infections. Nowadays, it represents one of the most common diseases encountered in medical practice affecting people of all ages from the neonate to the geriatric age group (Kunin, 1994). Worldwide, about 150 million people are diagnosed with UTI each year (Gupta, 2001). Most infections are caused by retrograde ascent of bacteria from the faecal flora via the urethra to the bladder and kidney especially in the females who have a shorter and wider urethra and is more readily transferred by microorganisms (Jones *et al.*, 2006).The structure of the female's urethra and vagina makes it susceptible to trauma during sexual intercourse as well as bacteria been massaged up the urethra and into the bladder during pregnancy or child birth (El-Sweih *et al.*, 2008; Kolawale *et al.*, 2009). Majority of UTIs are not life threatening and do not cause any irreversible damage. However, when the kidneys are involved, there is a risk of irreparable tissue damage with an increased risk of bacteremia (Hvidberg *et al.*, 2000). Nowadays, drug resistance is a huge growing problem in treating infectious diseases like Malaria, Tuberculosis (TB), Diarrheal diseases, Urinary Tract Infections (UTIs), etc. As suggested by Goldman

*Corresponding author.

and Huskins (1997) the improper and uncontrolled use of many antibiotics resulted in the occurrence of antimicrobial resistance, which became a major health problem worldwide. Drug resistance of pathogens is a serious medical problem, because of very fast arise and spread of mutant strains that are insusceptible to medical treatment.

ESBL

In Gram negative pathogens, β-lactamase production remains the most important contributing factor to β-lactamase resistance. The four major groups of β-lactamsase antibiotics penicillin, cephalosporins, monobactams and carbapenems have a β-lactam ring which can be hydrolysed by β-lactamases resulting in microbiologically ineffective compounds (Bush and Mobashery, 2009). The persistent exposure of bacterial strains to a multitude of β-lactamase has led to overproduction and mutation of β-lactamases. These β-lactamases are now capable of hydrolyzing penicillins, broad-spectrum cephalosporins and monobactams. Thus these are new β-lactamases and are called as extended spectrum β-lactamases—ESBLs (Bush, 2001). In Gram negative bacteria these enzymes remain in the periplasmic space, where they attack the antibiotic before it can reach its receptor site (Stratton, 1996).

MATERIALS AND METHODS

Isolation and Identification of UTI Isolates

Urine samples from UTI patients were collected from different Hospitals from Chennai. For the isolation of UTI causing bacterial strains, loop full of urine samples was streaked on to nutrient agar plate and incubated at 37°C for 24 hours. Isolated colonies were selected and identified on the basis of morphological, cultural and biochemical characteristics (Cheesbrough, 2000). To check morphological characteristics: gram-staining, capsule staining and motility test were performed. To check the growth pattern, different media including MacConkey's agar, Eosine Methylene Blue agar were used. For biochemical characteristics, sugar fermentation (Lactose, glucose, mannitol, maltose, sucrose and xylose), TSI, IMVIC (Indole, MR, VP and Citrate) and nitrate test were performed (Cheesbrough, 2000). Stock cultures were maintained in vials by growing the UTI isolates in 3 ml nutrient broth and next day overlaying with 3 ml 40% glycerol (Sambrook *et al.*, 1989). Vials were then freezed at –70°C.

Combination Disc Method

The Combination disc method using both cefotaxime and ceftazidime, alone and in combination with clavulanic acid, was performed for the detection of ESBL according to the CLSI guidelines. In this test, an overnight culture suspension of the test isolates was adjusted to 0.5 McFarland's standard. Lawn culture was made on the surface of Mueller Hinton Agar (MHA) plate. The Cefotaxime (30 μg) and cefotaxime-clavulanic acid (30 μg/10 μg) discs were placed 20 mm apart on the agar surface. Similarly, the ceftazidime (30 μg) and ceftazidime-clavulanic acid (30 μg/10 μg) discs were also placed. After incubating overnight at 37°C, a ≥ 5 mm increase in the zone diameter was interpreted as positive for ESBL production (NCCLS, 2000).

Antibiotic Sensitivity Test

The susceptibility of the ESBL producing bacteria were detected using the following antibiotics namely Amikacin (30 mcg), Ampicillin (10 mcg), Ciprofloxacin (5 mcg), Doxycycline (30 mcg), Gentamycin (10 mcg) and Tetracycline (30 mcg). Antibiotic sensitivity tests were done on MHA plates by the Kirby-Bauer disc diffusion method according to the CLSI guidelines (NCCLS, 2000).

RESULTS

Out of 178 urine samples, 111 (62.35%) showed significant growth of uropathogens. Among 111 urine isolates 39 (35.13%) were from males and 72 (64.86%) from females. Considering age distribution of culture-positive cases, 18 (16.21%), 47 (42.34%), 31 (27.92), 15 (13.51%) were between age groups 1–20, 21–40, 41–60 and 61–80 respectively. Out of 111 urine isolates, 69 (62.16%) have been identified as *E.coli,* 30 (27.02%) as *K. pneumoniae,* 8 (7.2%) as *P.aeruginosa* and 4 (3.6%) as *Proteus* spp. (Tables 1 and 2). A total of 111 isolates from urine samples were tested for ESBL production and it was found that 62 (54.86%) were positive (Figure 1). Among these 49 (71.01%) were *E.coli,* 11 (36.66%) *K. pneumoniae,* and 2 (25%) *P. aeruginosa* (Table 2 and Figure 2).

Table 1: Total Number of Microorganisms Isolated from UTI Patients and Their Distribution According to Age Groups and Gender

S.No.	*Age*	*Clinical Isolates*				*Total %*
		Male	*%*	*Female*	*%*	
1	01– 20	6	5.4	12	10.81	16.21
2	21–40	14	12.61	33	29.72	42.34
3	41–60	12	10.81	19	17.11	27.92
4	61–80	7	6.3	8	7.20	13.51
Total		39	35.13	72	64.86	100

Table 2: The Different Clinical Isolates and Their ESBL Positive Percentage

S. No	*Clinical Isolates*	*Total*	*%*	*ESBL Positive*	*%*
1	*E. coli*	69/111	62.16	49	71.01
2	*K. pneumoniae*	30/111	27.02	11	36.66
3	*P. aeruginosa*	8/111	07.20	2	25.00
4	*Proteus spp.*	4/111	03.60	0	–

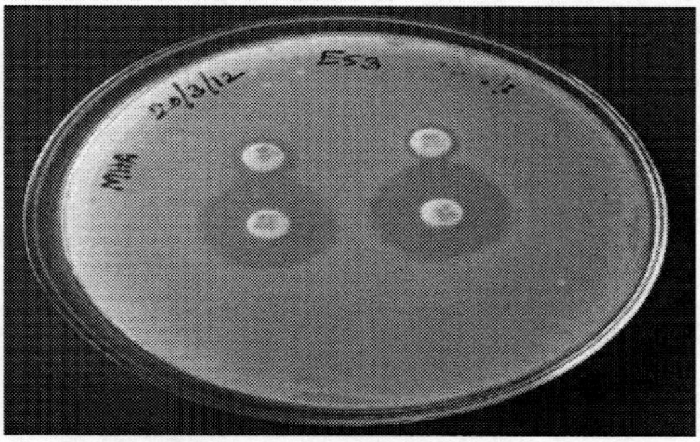

Fig. 1: ESBL Producing Isolates from the Urine Samples on MHA Plates

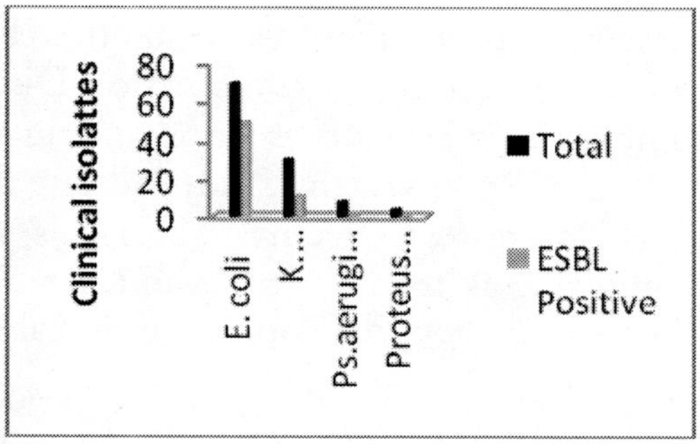

Fig. 2: Different Clinical Isolates and Their ESBL Positively

DISCUSSION

UTI is one of the most common infectious diseases diagnosed in outpatients and in hospitalized patients. The increase in nosocomial UTI is a serious concern because of the longer hospitalization and higher health care losts added to this due to the underlying diseases.The availability of new antimicrobial agents and improved pharmaceutical management of the urinary tract disorders has recently improved. Among the pathogens which cause UTI, *E. coli* is the most frequent bacterium isolated from urine sample (Francesco *et al.*, 2007). Similar to above study, our results also showed that out of 178 samples processed 69 found to be positive for *E. coli*. The prevalence of UTI was found to be more in females than males. The age wise distribution of the hospitalized patients revealed that the middle aged females (21–40) had more UTI. These results are corroborated with the results of Akram *et al.* (2007).

Out of 111 isolates belonging to 4 different genus, *E.coli* showed high resistance to common antibiotics viz. Ampicillin (30 mcg) 85.50%, Ciprofloxacin (5 mcg) 62.31%, Tetracyline (30 mcg) 43.47%, Doxycycline (30 mcg) 37.68%, Gentamycin (10 mcg) 28.98% and Amikacin (30 mcg) 2.89%. In this study all the isolates showed high resistance to Ampicillin and low resistant to Amikacin. Extended spectrum cephalosporins are important therapeutic agents in modern medicine and are often used as first line agents for invasive gram negative bacterium. Over the past decade there have been a number of new β-lactamases in clinical isolates of members of family Enteobacteriaceae (Chlebicki and Oh, 2004). Emergence of MDR bacteria and transfer of resistance genes among organisms against 3rd generation Cephalosporins include cefotoxime, ceftriazone, ceftazidime and cefoxitin. In this study 49 *E. coli*, 11 *K. pneumoniae* and 2 *P. aeruginosa* isolates were found to be resistant to more than two antibiotics. These antibiotic resistant isolates were subject to ESBL screening by combination Disc Method. Gram negative bacteria were the commonest organisms isolated among them *E.coli* was the principal urinary tract pathogen. Results of this study indicate that antibiotics commonly used for treatment of UTIs are less effective. Since this was a cross-sectional study, further regular monitoring is required to establish reliable information about resistant pattern of urinary pathogens for optimal empirical therapy of patients with UTI.

ESBLs have become a widespread serious problem and several aspects of them are worrying. These enzymes are becoming increasingly expressed by many strains of pathogenic bacteria with a potential for dissemination. They compromise the activity of wide-spectrum antibiotics creating major therapeutic difficulties with a significant impact on the outcome of patients. The continued emergence of ESBLs presents diagnostic challenges to the clinical microbiology laboratories, which need to be more aware of the need for their detection. ESBLs occurrence and spread need to be controlled. Appropriate antimicrobial selection, surveillance systems and effective infection control procedures are the key partners in their control.

REFERENCES

Akram, M., Shahis, M. and Khan, A.U. (2007). Etiology and antibiotic resistance patterns of community-acquired urinary tract infections in J N M C Hospital Aligarh, India. *Annals of Clinical Microbiology and Antimicrobials*, 6: 1–7.

Bush, K. (2001). New beta-lactamases in Gram-negative bacteria: diversity and impact on the selection of antimicrobial therapy. *Clinical Infectious Diseases*, 32: 1085–1099.

Bush, K. and Mobashery, S. (1998). How beta-lactamases have driven pharmaceutical drug discovery from mechanistic knowledge to clinical circumvention. *Advanced Experimental Medical Biology*, 456: 71–98.

Cheesbrough, M. (2000). *District Laboratory Practice in Tropical Countries*, part 2, 2nd edn. London: Cambridge University Press.

Chlebicki, M.P. and Oh, H.M.L. (2004). Extended–spectrum β–Lactamases in Clinical Isolates of *Escherichia coli* and *Klebsiella spp.* in a Singapore Hospital: Clinical Spectrum. *Annals Academy of Medicine*, 33: 302–306.

El-Sweih, N., Jamal, W. and Rotimi, V.O. (2008). Spectrum and antibiotic resistance of uropathogens isolated from hospital and community patients with urinary tract infections in two large hospitals in Kuwait. *Medical Principles and Practices*, 14: 401–407.

Francesco, M.A., Ravizzola, G., Peroni, L., Negrini, R. and Manca, N. (2007). Urinary tract infections in Brescia, Italy: etiology of uropathogens and antimicrobial resistance of common uropathogens. *Medical Science Monitor*, 13: 136–144.

Goldman, D.A. and Huskins, W.C. (1997). Control of nosocomial antimicrobial-resitant bacteria:A strategy priority for hospitals worldwide. *Clinical Infectious Diseases*, 24: 139–145.

Gupta, K. (2001). Increasing antimicrobial resistance and the management of uncomplicated community acquired urinary tract infections. *International Journal on Antimicrobial Agents*, 135: 41–50.

Hvidberg, H., Struve, C., Krogfelt, K.A., Christensen, N., Rasmussen, S.N. and Frimodt-M ler, N. (2000). Development of a long-term ascending urinary tract infection mouse model for antibiotic treatment studies. *Antimicrobial Agents on Chemotheraphy*, 44: 156–163.

Jones, R.N., Inabo, H.I. and Obanibi, H.B.I. (2006). Antimicrobial susceptibility of some urinary tract clinical isolates to commonly used antibiotics. *African Journal of Biotechnology*, 5: 487–489.

Kolawale, A.S., Kolawale, O.M., Kandaki-Olukemi, Y.T., Babatunde, S.K., Durowade, K.A. and Kplawale, C.F. (2009). Prevalence of urinary tract infections among patients attending Dalhatu Araf Specialist Hospital, Lafia, Nasarawa State, Nigeria. *International Journal on Medical Sciences*, 1: 163–167.

Kunin, C.M. (1994). Urinary tract infections in females. *Clinical Infectious Diseases*, 18: 1–12.

Maniatis, T., Fritsch, E.F. and Sambrook, J. (1989). *Molecular Cloning: A Laboratory Manual.* Cold Spring Harbor, NY: Cold Spring Laboratory.

Medeiros, A.A. (1997). Evolution and dissemination of β-lactamases accelerated by generations of β-lactam antibiotics. *Clinical Infectious Diseases*, 24: 19–45.

NCCLS (2000). Performance Standards for antimicrobial disc susceptibility tests. 7th ed. Approved Standards M2–A6. National Committee for Clinical Laboratory Standards. Wayne Pa.

Stratton, C.W. (1996). Mechanisms of action for antimicrobial agents: general principles and mechanisms for selected classes of antibiotics. In *Antibiotics in Laboratory Medicine*, ed. V. Lorian. Baltimore: Williams and Wilkins.

Synthesis of Antimicrobial Silver Nanoparticles from Thulsi Plant (*Ocimum sanctum* L.)

M. Aynul Rifaya* and R.M. Meyyappan

Department of Chemical Engineering,
Annamalai University Annamalainagar, Tamil Nadu
E-mail: aynul_bioinfo@yahoo.co.in

INTRODUCTION

Synthesis of nanoparticles using biological entities has great interest due to their unusual optical (Klaus Joerger *et al.*, 2001), chemical (Kumar *et al.*, 2009), photo electrochemical, and electronic properties. A wide variety of physical, chemical and biological processes results in the synthesis of nanoparticles, some of these are novel and others are quite common. Nature has devised various processes for the synthesis of nano and micro length scaled inorganic materials which have contributed to the development of relatively new and largely unexplored area of research based on the biosynthesis of nanomaterials (Silver *et al.,* 2006). The synthesis and assembly of nanoparticles would benefit from the development of clean, nontoxic and environmentally acceptable "green chemistry" procedures, probably involving organisms ranging from bacteria to fungi and even plants (Nedelcu and Narayan 2008). Green synthesis provides advancement over chemical and physical method as it is cost effective, environment friendly, easily scaled up for large-scale synthesis and further there is no need to use high pressure, energy, temperature and toxic chemicals. Hence, the present study was made to synthesis the antimicrobial silver nanoparticles by using leaf extract of the Thulsi plant, *Ocimum sanctum*.

*Corresponding author.

MATERIALS AND METHODS

Chemicals

All analytical reagents and media components were purchased from Hi-Media (India).

Collection of Plants

The medicinal herbs *Ocimum sanctum* was collected from University campus. Annamalai Nagar Chidambaram.

Synthesis of Silver Nanoparticles

Leaf extract was prepared by grinding 20 g of fresh leaves, and centrifuged at 3000 rpm to get the leaf extract. Five milliliters of leaf extracts was separately added to 45 ml of 10^{-3} M aqueous $AgNO_3$ solution. The reduction of silver ions was monitored by measuring the absorbance of the reaction mixture in a range of wavelength from 200 to 800 nm using UV–vis spectrophotometer (Shimadzu, Japan) to find the absorbance peak different intervals (0 min, 30 min, 60 min, 2 h, 4 h, 8 h, 16 h, and 24 h).

Characterization of Silver Nanoparticles

For Fourier transform infrared (FTIR) spectroscopy measurement, the following method was adopted. The silver nanoparticles were synthesized after 24 h of incubation with the leaf extracts, centrifuged at 10,000 rpm for 15 min, and their pellets were were then dried and the powder was subjected to FTIR spectroscopy measurement (Perkin Elmer spectrophotometer) in the reflectance mode at a resolution of 4 cm^{-1} in KBr pellets.

After 24 h of incubation leaf extracts were analyzed under a Scanning Electron Microscopic (JOEL) at a voltage of 120 kV.

Antimicrobial Assay

The antimicrobial assay was performed by disc-diffusion technique. In this technique, 50 µl of silver nanoparticle prepared from leaf extract, applied to sterile paper discs of 5 mm diameter. The discs were then placed on Nutrient Agar inoculated with clinical strains of bacteria (*E. coli* and *B. subtlis*). The plates were incubated at 37°C for overnight. The zone of inhibition was measured in millimeter after the 24 h of incubation and recorded.

RESULT AND DISCUSSION

Synthesis of Silver Nanoparticles by Using Silver Nitrate

The color change was observed by visual examination the leaf extracts after incubated with $AgNO_3$ solution (Figure 1). Absorption spectrum of leaf extracts at different wavelengths ranging from 200 to 800 nm revealed a peak at 420 nm (Figure 1). The color intensity at 420 nm increased with duration of incubation (Table 1). The color of the extract changed to intense brown after 24 h of incubation and there was no significant change afterwards (Figure 1).

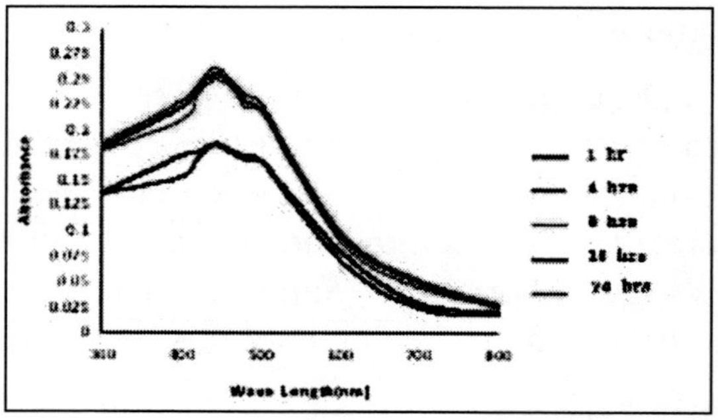

Fig. 1: UV-Visible Spectra of Silver Nanoparticles Synthesized through Leaf Extract of *Ocimum sanctum* at Different Time Interval

Table 1: Effect of Leaf Extract of Thulsi on Colour Changes in Silver Nitrate Solution at Different Time Interval

S. No.	*Time Interval*	*Colour Change*
1.	0 min	Dark green
2.	10 min	Pale green
3.	30 min	Reddish green
4.	1hr	Red
5.	2 hrs	Red
6.	4 hrs	Reddish brown
7.	8 hrs	Reddish brown
8.	16 hrs	Brown Threads
9.	24 hrs	Brown Threads

Characterization of Silver Nanoparticles

The shape and size of silver nanoparticles were analyzed after 24 h of incubation using SEM is showed in Figure 3. In general, the nanoparticles were in spherical

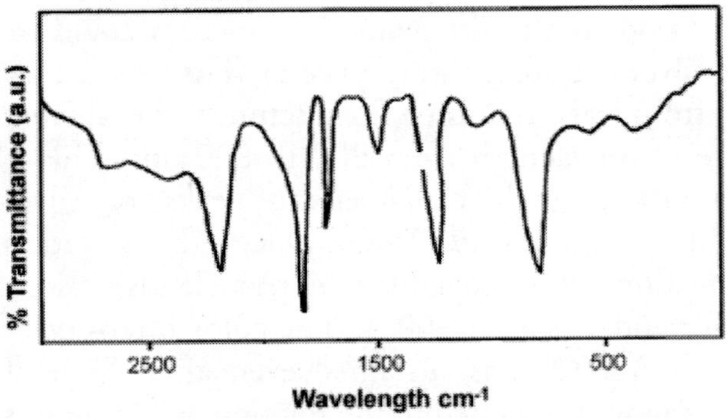

Fig. 2: FTIR Spectra of Silver Nanoparticles Synthesized through Leaf Extract of Ocimum Sanctum

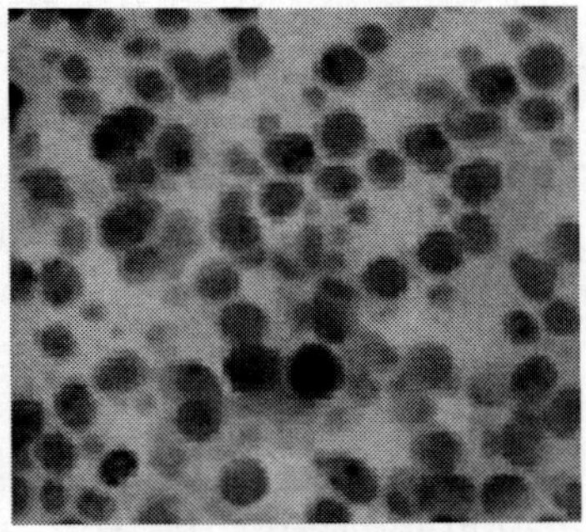

Fig. 3: SEM Image of Silver Nanoparticles Synthesized from Leaf Extract of *Ocimum santum*

shape with varying size ranged from 7 to 25 nm. Most of the nanoparticles were combined with only a few of them were scattered, as observed under SEM.

FTIR spectra of silver nanoparticles exhibited prominent peaks at 2752, 2318, 2015, 1713, 1495, and 812 cm^{-1} for leaf extract (Figure 2). The peaks corresponding to protein and silver nanoparticles were found commonly present in the nanoparticles synthesized by leaf extract.

Antimicrobial Activities of Callus and Leaf Extracts

Antimicrobial activities of silver nanoparticles synthesized by leaf extracts from O sanctum are analyzed. The antibacterial activity of silver nanoparticles showed, the highest inhibition zone of 12 mm diameter was formed against *E.coli* by the nanoparticles synthesized by leaf extract, and the lowest of 7 mm was produced against *B. subtilis* by nanoparticles synthesized by leaf extract.

Nature has devised various processes for the synthesis of nano and micro length scaled inorganic materials which have contributed to the development of relatively new and largely unexplored area of research based on the biosynthesis of nanomaterials (Silver *et al.,* 2006). The synthesis and assembly of nanoparticles would benefit from the development of clean, nontoxic and environmentally acceptable "green chemistry" procedures, probably involving organisms ranging from bacteria to fungi and even plants (Nedelcu and Narayan, 2008). Hence, both unicellular and multicellular organisms are known to produce inorganic materials either intra-or extracellularly (Minaeian, 1996).

Nanoparticles are known to be synthesized by leaf extract (Shankar *et al.*); fruit extract (Huang *et al.*, 2007) and seed extract (Bar *et al.,* 2009) of various plants. The present work proved that leaf extracts from the thulsi plant, O. sanctum could be able to reduce silver ions and produce silver nanoparticles (Figure 1). The production of the silver nanoparticles synthesized from leaf extracts of O. sanctum was evaluated through spectrophotometer in a range of wavelength from 200 to 800 nm. This is similar to the surface plasmon vibrations with characteristic peaks of silver nanoparticles prepared by chemical reduction (Petit *et al.,* 1993). Due to the excitation of plasma resonances or interband transitions, some metallic nanoparticle dispersions exhibit the unique bands/peaks (Kuber and D'souza, 2006). The color intensity of the callus and leaf extracts added with $AgNO_3$ was sustained even after 24 h of incubation indicating that the particles were well dispersed in the solution and there was not much aggregation. Monodispersity is an important characteristic of the nanoparticles. The silver nanoparticles with very good monodispersity have earlier been reported (Huang *et al.*, 2007).

Silver is well known as one of the most universal antimicrobial substances. Silver ion and silver-based compounds are highly toxic to microorganisms, showing strong biocidal effect against microbial species. Nanosilver is an effective killing agent against a broad spectrum of gram negative and gram positive bacteria, including antibiotic-resistant strains Gram-negative bacteria include genera such as *Escherichia, Psudomonas, Salmonella,* and *Vibrio* gram positive bacteria include many well known genera such as *Bacillus, Clostridium, Enterococcus, Listeria, Staphylococcus* and *Strepto coccus.* Silver nanoparticles (diameter 5–32 nm, average diameter 22.5 nm) enhance the antibacterial activity of various antibiotics (Bar *et al.,* 2009).

Silver nanoparticles exerts their antibacterial effects by anchoring to and penetrating the bacterial cell wall and modulating cellular signaling by dephosphorylating putative key peptide substrates on trypsin residues is main mechanism by which silver nanoparticles exhibit antibacterial properties. In gram negative bacterial, silver nanoparticles act in three ways. (a) Silver nanoparticles attach to the cell membrane surface and distrupts its function. (b) They are able to penetrate inside bacteria so there they tend to bind to sulphur and phosphorous containing compounds like DNA and damage them. (c) They release silver ions with additional contribution to the bactericidal effects (Chandra, *et al.*, 2006).

Bacterial cell lysis could be one of reason for observed antibacterial property. Nanoparticles modulated phosphotyrosine profile of bacterial peptide that in turn affects signal transduction and inhibited growth of micro-organisms. Antibacterial effect is dose-dependent and is independent of acquisition of resistance by bacteria against antibiotics.Sondi and his co-workers demonstrated that *E. coli* cells treated with silver nanoparticles found to be accumulated in the bacterial membrane which results in the increase in permeability and resulted in dealth of cell. They suggested the silver nanoparticles are the most suitable bactericidal agent (Huan *et al.*, 2007). The approach of bio-synthesis nano silver from plant materials has many advantages like ease with the process can be scaled up and economic viability. Biosyntheis of nanoparticles has an emerging highlight of the intersection of nanotechnology and biotechnology has received increased attention due to growing need to develop environmentally begin technology in material synthesis. The synthesis process was quiet fast and nanoparticles were formed within hours of silver ion coming in contact with the cell filtrate. The process of the nanoparticles production is eco-friendly as it is free from any solvent or toxic chemicals, also easily amenable on large scale production.

REFERENCES

Anal, K., Jha and Prasad, K. (2010). Green Synthesis of Silver Nanoparticles Using Cycas Leaf, *International Journal of Green Nanotechnology: Physics and Chemistry*, 1: 2, 110 –117

Arangasamy, Leela and Munusamy, Vivekanandan (2008). Tapping the unexploited plant resources for the synthesis of silver nanoparticles, *African Journal of Biotechnology*, Vol. 7(17), 3162–3165.

Bar, H., Bhui, D.K., Sahoo G.P., Sarkar, P., De, S.P. and Misra, A. (2009). Colloid Surf. A: Physiochem. *Eng. Aspects* 339 134.

Chandran, S.P., Chaudhary, M., Pasricha, R., Ahmad, A. and Sastry, M. (2006). Synthesis of gold nanotriangles and silver nano particles using Aloe vera plant extract, *Biotechnology progress*, 22, 577–583.

Dabhade, G.T. (1998). Mosses of Khandala and Mahabaleshwar in the western Ghats (India) Dubey Manish, Seema Bhaduria, B.S. Kushwah, Green synthesis of nanosilver particles from extract of Eucalyptus hybrida (Safeda) leaf, *Digest Journal of Nanomaterials and Biostructures*, September 2009, Vol. 4, No. 3, 537–543.

Devendra, Jain., Hemant Kumar Daim., Sumita, Kachhwaha and Kotharia, S.L. Synthesis of plantmediated silver nanoparticles using papaya fruit extract and evaluation of their anti microbial activities. *Digest Journal of Nanomaterials and Biostructures*, 2009, 4, 4, 723.

Durán, Nelson, Marcato Priscyla, D., Oswaldo, L Alves, Gabriel, IH De'Souza and Esposito, Elisa. (2005). Mechanistic aspects of biosynthesis of silver nanoparticles by several Fusarium oxysporum strains, *Journal of Nanobiotechnology*, 3:8

Elumalai *et al.* (2010). Extracellular synthesis of silver nanoparticles using leaves of Euphorbia hirta and their antibacterial activities. *J. Pharm. Sci. and Res.* 2(9), 549–554.

Karala, Chalouupka; Youkeshkumar Malamand Alexander M. Seifalian. Nano silver as a new generation of nano product in biomedical applications, *Journal-Trends in Biotechnology*, November 2010, Vol. 28, No. 11.

Mahendra Rai, Alka Yadav, Aniketgade, "Silver nanoparticles as a new generation of antimicrobials", department of biotechnology, SGB Amravati University, Amravaati-444-602, Maharashtra, India.

Morones, J.R. *et al.* (2005). The bacterial effect of silver nano particles, *Nanotechnology* 16, 2346–2353.

Murali, Sastry, Absar, Ahmad, Shiv, Shankar S. and Akhilesh, Rai (2004). Rapid synthesis of Au, Ag, and bimetallic Au core–Ag shell nanoparticles using Neem (Azardiracta indica) leaf broth

Prabhu, N., Divya, T.R., Yamuna Gowri, K., Ayisha Siddiqua, S. and Joseph Puspha, D. (2010). Silver phytho nanoparticles and their antibacterial efficacy. *Digest Journal of Nanomaterials and Biostructures*, 5, 185–189.

Rai, M., Yadav, A. and Gade, A. (2009). Silver nanoparticles as a new generation of antimicrobials. *Biotechnology Advances*, 27, 1.

Sathyavathi, R., Balamurali Krishna, M., Venugopal Rao, S., Saritha, R. and Rao, D. Narayana (2010). Biosynthesis of Silver Nanoparticles Using Coriandrum Sativum Leaf Extract and Their Application in Nonlinear Optics, *Advanced Science Letters*, Vol. 3.

Shankar, S.S., Rai, A., Ahmad, A. and Sastry, M. (2004). Rapid synthesis of Au, Ag and bimetallic Au core-Ag Shell nanoparticles using Neem (Azadirachta indica) leaf broth. *J Colloid Interface Sci*; 275(2): 496–502.

Tripathi, A., Chandrasekaran, N., Raichur, A.M., Mukherjee, A. (2009). Antibacterial Applications of Silver Nanoparticles Synthesized by Aqueous Extract of Azadirachta Indica (Neem) Leaves, *Journal of Biomedical Nanotechnology*, Vol. 5, No. 1, pp. 93–98.

Verma, V.C., Kharwa, R.N and Gange, A.C. (2010). Biosynthesis of antimicrobial silver nanoparticles by the endophytic fungus Aspergillus clavatus.

Enriching the Human Kind – The Probiotics Way!

Rahul Kumar*, Hitanshu Das and Rekha Govindan
Department of Biotechnology, Aarupadai Veedu Institute of Technology, Vinayaka Missions University, Rajiv Gandhi Salai, Paiyanoor, Kanchipuram, Tamil Nadu
E-mail: neosam.rahul@gmail.com

INTRODUCTION

The intestinal microflora has always been associated with various pathological conditions, but certain species have been found to be beneficial to overall health. Extensive bifidobacterial gastrointestinal colonization of the infant gut proved more beneficial healthwise and has been coined the term Probiotics (David, Glen, 1999).

Probiotics can be considered the realization of Hippocrates's age old saying "Let food be thy medicine and medicine be thy food", in the current time frame. More precisely, probiotics are defined as "Live microorganisms which when administered in adequate amounts confer a health benefit on the host", as per the definition adopted by FAO/WHO. Alternatively, probiotics can be considered as live microbes which provide health enhancement, beyond basic nutrition. Definitions vary, but the most acceptable one would be "Live microbial formulations which provide the host with benefits by balancing the gut micro-floral balance (Fuller, 1989). Use of probiotics is not new, and dates back to the times of the Samburu and Masai Warriors, who consumed *Lactobacillus* fermented milk. Probiotics have now come up as novel alternatives to chemotherapy.

WHY PROBIOTICS?

The mankind these days is exposed to numerous health hazards, related either to occupation or lifestyle. Once uncommon, cardiovascular diseases, hypertensions etc

*Corresponding author.

are now highly prevalent in the population irrespective of the class. Chemotherapy relies on chemical compounds of biomolecular mimics to alleviate the pathological condition. However this has many side effects. Moreover, drug resistant varieties of many pathogenic species challenge the efficacy of chemotherapeutic drugs. Active researches focused on improved life styles and eating habits is being encouraged now. In this context, Probiotics are considered as alternative therapies or complementary therapies in many pathological conditions. They are natural inhabitants of the gut, reinforced in amounts to achieve health enhancement hence have negligible or nill side effects at all. Moreover, drug resistant varieties of many pathogenic species challenge the efficacy of chemotherapeutic drugs. Probiotics can help tackle this menace as live microbes can co-evolve to survive over the pathogenic microbes. Further, use of probiotics has been marked by an overall health improvement, be it gastrointestinal disorders or stress related behavioral disorders. Hence probiotics prove to be excellent alternative or complementary therapeutic agents.

Characteristics of Probiotic Agents

These are commensal microbes that have coevolved with humans and demonstrate e a high degree of interdependence with them. For their existence, these probiotic species need to effectively colonize the gut to exert their effects. They must be non-pathogenic and resistant to gastric acidity and also bile tolerant. They must remain viable for shelf storage period. Any probiotic agent must evade flushing off due to the gut peristalsis (Anderson, Gilliland, 1999). Most importantly, it should confer health benefits to the host (Mukesh Kumar *et al.*, 2012).

Common Probiotic Species

Lactobacilli being the natural residents of the gut are most preferable. Several *Bifidobacterium* species have also been found to be potent probiotic agents. Some commonly used probiotic species are: *L. acidophilus, L. casei, L. delbruckeii, L. plantarum, L. fermentum, L. reuteri, Entercoccus faecium, Staphylococcus intermedius, Bifidobacterium bifidum, B. longum* (Collins and Gibson, 1999, Holzer *et al.*, 2003).

Probiotic Health Benefits

Probiotics offer a near infinite list of health benefits and hence make an ideal choice as a therapeutic agent. Probiotic consumption has been associated with proper digestion and better nutrient bioavailability. Lactic acid bacteria in particular, increase the pH of the gut, which results in better calcium uptake. Probiotics are also effective against allergies, especially food allergies. Also, probiotics can alleviate lactose intolerance, by catabolizing lactose. Lactose intolerance occurs due to the lack of the enzyme lactase or β-galactosidase.

Probiotics have been found to be very effective anti hypercholesterolemic agents. Up to 2.9% reduction in serum cholesterol levels is observed with the regular intake of *Lactobacillus* inoculated yoghurt (Anderson and Gilliland, 1999). The reduction in cholesterol levels is explained either by direct cholesterol breakdown or by bile salt deconjugation. *L. acidophilus* was found to have the highest deconjugation and bile salt hydrolytic activity (Rolfe, 2005). Various *Lactobacilli* are reportedly effective in cases of hepatic encephalopathy.

Probiotics have been reported to be effective in cases of antibiotic and infective diarrhea (Allen *et al.*, 2003). Repeated infections by *Candida albicans* is prevented upon regular probiotic consumption (Mallen *et al.*, 1992). Their activity against infectious agents can be competitively inhibitive or Immunomodulatory or due to released toxins (Kailasapathy and Chin, 2000). Immunomodulatory activity of probiotics causes increase in IL-10 levels, which has an anti-inflammatory effect. Also, intake of lactic acid bacteria has been reported to increase lymphocyte and interferon synthesis (Aattouri *et al.,* 2001). Probiotic microflora has immuno-stimulatory activity also (Rolfe, 2000). Elevated antibody levels have been reported in case of mice fed yoghurt regularly (Wade *et al.,* 1984). Interestingly, intake of probiotics helps increase CD4 counts in HIV positive subjects. Also they are effective in cancer suppression. Consumption of *Bifidobacterium longum* reportedly suppressed colon cancer induced by a chemical mutagen (Rivenson and Reddy, 1993).

PROBIOTICS IN INDIAN MARKET

Probiotics are strengthening their presence in the Indian market. Several pharmaceutical companies have launched probiotic formulations. Also, food industries have come up with probiotic food supplements. Usual food stuffs like curd and ice creams are also now probiotic. This embarks future possibilities for the probiotic market.

Some probiotic formulations in the Indian Market (Barun and Bhattacharyya, 2009):

1. *Eubioz:* L. acidophilus, L. rhamnosus, B. bifidum, B. longum, S. thermophilus, S. boulardii.
2. *Bifilac:* L. sporogens, S. faecalis, C. butyricum, B. mesentricus.
3. *Equpiro:* L. acidophilus, S. faecalis, C. butyricum, B. mesentricus.
4. *Goodlac:* L. acidophilus, L. rhamnosus, B. longum, S. boulardii.
5. *Econove:* L. reuteri RC-14, L. rhamnosus GR-1.

6. *Goodlac:* L. acidophilus, L. rhamnosus, B. longum, S. boulardii.
7. *Lactisyn:* L. acidophilus, L. lactis, S. thermophilus, S. lactis.
8. *Prepro:* L. acidophilus, S. faecalis, C. butyricum, B. mesentricus.
10. *Actigut:* L. acidophilus, L. rhamnosus, B. longum, B. bifidum, S. boulardii, S. thermophilus.

CONCLUSION

Gaining on by positive research reports, probiotics are proving to have a better therapeutic role. They have proved as excellent complementary therapeutics. They are effective against numerous infectious agents and have proved efficient even in case of autoimmune disorders. They provide both the medical professionals and patients with better treatment options. Identification of more probiotic species with further research will provide for better probiotic formulations. Rigorous research would help understand underlying mechanisms of action and that in turn would bring about the development of more specific formulations. Gene manipulation techniques will allow engineering microbes for better probiotic potential. Increasing market demand only encourages further research. Major pharmaceutical companies have already come up with numerous formulations and industry competition would lead to the development of more efficient probiotics. Further research is needed to understand the molecular mechanisms under the veil. Awareness amongst the consumers can help them choose probiotics for a healthy living.

REFERENCES

Aattouri, N., Bouras, M., Tome, D., Marcos, A. and Lemonnier, D. (2001). "Oral ingestion of lactic acid bacteria by rats increases lymphocyte proliferation and interferon production", *British J.*, 87; 367–373.

Allen, S.J., Okoko, B., Marinez, E., Gregorio, G. and Dans, L.F. (2003). "Probiotics for treating infectious diarrhea", *Cochrane Database of Systematic Reviews* 2, CD003048.

Anderson, J.W. and Gilliland, S.E. (1999). "Effect of Fermented Milk (Yogurt) Containing Lactobacillus acidophilus L1 on Serum Cholesterol in Hypercholesterolemic Humans". *Journal of the American College of Nutrition*, 18; 43–50.

Barun, K. and Bhattacharyya (2009). "Emergence of Probiotics in Therapeutic Applications", *International Journal of Pharmaceutical Sciences and Nanotechnology*, 2: 383–389.

David Collins, M. and Glen R. Gibson (1999). "Probiotics, prebiotics and synbiotics: Approaches for modulating the microbial ecology of the gut", *Am. J. Clin.Nutr.*, 69; 1052S–1057S.

Fuller, R. (1989). "A Review: probiotics in man and animals", *J. Appl. Bacteriol.*, 66; 365–378.

Holzer, M., Mayrhuber, E., Danner, H. and Braun, H. (2003). "The role of Lactobacillus buchneriin forge preservation", *Trends Biotechnol.*, 21; 282–286.

Mallen, A., Jarstrand, C. and Pahlson, C. (1992). "Treatment of bacterial vaginosis with lactobacilli", *Sex Trans. Dis.*, 19; 146–148.

Mukesh Kumar, D.J., Rahul Kumar, V., Poovai, P.D. and Kalaichelvan, P.T. (2012). "Probiotics and the multitude of health benefits", *J. Res. Biol.*, 2; 102–113.

Rivenson, A. and Reddy, B.S. (1993). "Inhibitory effect of Bifidobacterium longum on colon, mammary and liver carcinogenesis induced by 2-amino, 3-methylimidazo [4,5-f] quinolone, a food mutagen", *Cancer Res.*, 1; 3914–3918.

Rolfe, R.D. (2000). "The role of probiotic cultures in the control of gastrointestinal health", *J. Nutr.*, 130: 396S–402S.

Wade, S., Corthier, G., Moreau, L. and Besnier, M.O. (1984). "L'ingestion de yaout vivant modifie-t-elle la response immunitaires?", *IDF Bull.* 179.

Evaluation of Protective Effect of *Tabernaemontana heyneana* Wall. Leaf Extract against Paracetamol Induced Hepatic Injury in Rats

T. Sathishkumar* and R. Baskar

Department of Biotechnology, Kumaraguru College of Technology, Coimbatore, Tamil Nadu

E-mail: sathishkumart29@gmail.com

INTRODUCTION

The development of new diagnostic and therapeutic methods is regularly improving the management and prognosis of most diseases but it is also associated with the occurrence of new iatrogenic diseases. Liver toxicity is permanently developing and remains the first cause of drug-induced death. Despite improvement in toxicological studies and in the safety analysis of clinical trials, the frequency of drug hepatotoxicity has not decreased in the last 10 years. The spectrum of liver damage caused by drugs is very broad and indeed, all cells present in the liver can be affected by drugs (Larrey, 2000).

Paracetamol (*N*-acetyl *p*-aminophenol, APAP) also known as acetaminophen is considered to be a safe analgesic and antipyretic drug (Rocha *et al.*, 2005). Paracetamol has taken in overdoses results in hepatotoxicity and nephrotoxicity in men and in experimental animals. Liver failure rather than kidney failure is believed to be the cause of death. Excess APAP is oxidatively metabolized in the liver via the mixed function oxidase P450 system to a toxic metabolite N-acetyl-benzoquinoneimine (NAPQI). NAPQI has an extremely short half-life and is rapidly conjugated with

*Corresponding author.

glutathione, a sulphydryl donor and also covalently binds to cellular proteins including mitochondrial proteins and the lipid bilayer of hepatocyte membranes. This covalent binding is thought to be responsible for the mitochondrial dysfunction observed after APAP overdose including inhibition of mitochondrial respiration, ATP depletion, release of intermembrane proteins, and mitochondrial oxidant stress and peroxynitrite formation. The result is hepatocellular death and centrilobular liver necrosis (McConnachie, 2007).

Tabernaemontana heyneana Wall. (Apocynaceae) known as *Kundalam Paalai* in Tamil, possesses antimicrobial, anthelmintic and antioxidant properties (Sathishkumar *et al.,* 2012). The stem bark decoction has been used for cleaning cuts and wounds. The ethnobotanical studies have revealed the curative effect of leaf and stem against skin diseases (Ignacimuthu *et al.,* 2006). Previous scientific documentation has revealed the presence of alkaloids like 15-β-stemmadenine, tabernoxidine, coronaridine and iboganine from *T. heyneana*. Recent study has shown the presence of polyphenolics like rutin and quercetin related compounds and a few unknown phenolic acids in the optimal ethanolic leaf extracts of *T. heyneana* Wall. With the literature background and because of unavailability of scientific reports/documentation, the present study was formulated, focused and systematically investigated to explore the hepatoprotective properties of the leaves of *Tabernamontana heyneana* Wall.

MATERIALS AND METHODS

Plant Material and Optimal Extraction Process

The leaves and flowers were collected from the medicinal garden of Kumaraguru College of Technology, Coimbatore, India. The species was identified and confirmed at Botanical Survey of India (BSI), Southern Circle, Coimbatore, India (BSI/SC/5/23/06–07/Tech. 478). The optimal process was adopted for the extraction of flavonoid in the leaves (85°C, 2hrs extraction duration, 85% ethanol concentration, a material ratio 1:05 and 4 times of extraction cycle) and used for further analysis. All the chemicals and solvents used for experimental analysis were of analytical grade.

Animals and Treatments

All experiments were carried out with Male Wistar-albino rats weighing 90–120 g. The rats were housed in stainless steel cages under a 12 h light/12hdark cycle at 25–28°C for 7 days and were fed with standard pellet (Lipton India Ltd., Calcutta) and water *ad libitum*. They were divided into six groups of 6 animals each. The research was conducted in accordance with the internationally accepted guidelines for Laboratory animal use and care (Animal house registration number: 158/1999/CPCSEA – Ethical committee approval Dated 07.08.2008).

Paracetamol Induced Hepatotoxicity

Paracetamol (Acetaminophen) was obtained from GlaxoSmithKline Pharmaceuticals Limited, Mumbai, India. The animals were divided into 6 groups. Group-I was control rats fed on normal diet. The animals of group II was treated with 750 mg/kg/day paracetamol (20% saline as vehicle) orally by means of a stomach tube for 7 days to induce hepatotoxicity.

The animals of the test group (Group III and IV) were given the pretreatment of optimal ethanolic leaf extracts of *Tabernaemontana heyneana* Wall. (125 and 250 mg/kg/day) in 20% saline once daily for 7 days in succession followed by single administration of paracetamol 750 mg/kg, 1 h after optimal ethanolic extract administration. The animals of group V was treated with 25mg/kg/day of rutin in 20% saline (stabilized by 0.2% gum) for 7 days in succession followed by single administration of paracetamol 750 mg/kg, 1 h after rutin administration. The animals in group VI was treated with optimal ethanolic leaf extracts (250 mg/kg) alone for 7 days. Survival of animals was monitored for 24 h. After 24 h, i.e. the last dose of paracetamol administration, the surviving animals was anaesthetized with ketamine (100 mg/kg, i.m.) and blood (3 ml) was collected using sterile disposable syringes. Serum was separated by centrifugation (3000 rev/min for 15 min) and used for biochemical analysis. Later on, the experimental animals were sacrificed and liver and kidney was used for histopathological studies.

Biochemical and Histopathological Studies

The biochemical parameters selected for serum analysis were SGOT, SGPT, SALP, LDH, total proteins and total bilirubin. The liver tissues was assessed for enzymatic antioxidants like superoxide dismutase, glutathione peroxidase, glutathione reductase, glutathione-S-transferase and catalase. All serum enzymes and compounds assays were done by using commercial kits (Span Diagnostics Ltd., India and Right Choice Diagnostics, Israel) according to manufacturer's instruction. Samples of liver and kidney tissues were obtained from animals after sacrification at the end of the specified time. Tissue blocks for light microscopy were fixed in 10% buffered neutral formalin solution. All samples were embedded in paraffin, cut in sections of 3–5 μm thickness and stained with hematoxylin and eosin (H&E stain) for histopathological examination.

Statistical Analysis

All determinations were carried out at least three separate experiments (Triplicates). The results were expressed as means ± SD and the mean values were plotted in all figures. The level of significance was expressed using Mann-Whitney U-Test (Control and APAP induced models) and Turkey's Multiple Comparison post hoc test (APAP induced with rutin, EETH-L + APAP and EETH-L). All the analysis was carried out using GraphPad Prism 5 software.

RESULTS AND DISCUSSION

The effect of EETH-L on serum glutamate oxaloacetate transaminase (SGOT), serum glutamate pyruvate transaminase (SGPT), serum alkaline phosphatase (SALP), lactate dehydrogenase (LDH), serum bilirubin (SB) and serum total protein (STP) levels in APAP induced liver damage in rats were summarized in Table 1. There was a significant ($p < 0.05$) increase in the SGPT, SGOT, SALP, LDH and SB levels and a decrease in STP level ($p < 0.05$) of APAP induced group as compared with normal group. It was observed that the effects were totally changed reversibly in a significant manner ($p < 0.01$) on administration of EETH-L (125 mg and 250 mg/kg) in the experimental models after inducing with APAP. Whereas, there is no significant difference at 1% level ($p > 0.01$) in the serum enzymes, SB and SPT between rutin and EETH-L + APAP induced models proving noteworthy hepatoprotective effect of EETH-L.

Table 1

	Treatment/Groups (n = 6)	*Biochemical Parameters (Mean ± SEM), n=6*					
		SGOT (U/l)	*SGPT (U/l)*	*SALP (IU/l)*	*LDH (U/l)*	*SB (mg/dl)*	*STP (g/dl)*
1.	Control	33.6±3.14	32.3±4.12	54.4±4.98	349.1±31.31	0.73±0.09	6.9±0.33
2.	APAP induced	99.5±6.73[a]	93.1±6.44[a]	162.8±11.92[a]	730.3±63.1[a]	2.03±0.19[a]	4.8±0.29[a]
3.	EETH-L (125 mg/kg) + APAP (Pre-treated)	40.8±2.16[b]	40.2±2.73[b]	74.2±8.16[b]	438.3±20.49[b]	1.13±0.11[b]	6.02±0.29[b]
4.	EETH-L (250 mg/kg) + APAP (Pre-treated)	31.8±2.64[b]	28.5±3.56[b]	45.3±5.43[b]	327.9±25.23[b]	0.82±0.09[b]	6.7±0.24[b]
5.	Rutin (25 mg/kg)	39.7±3.36 b	38.5±2.41[b]	66.1±4.24[b]	357.2±32.46[b]	0.92±0.10[b]	6.5±0.30[b]
6.	EETH-L (250 mg/kg)	42.8±2.05[b]	41.2±2.49[b]	65.7±6.24[b]	364.2±22.29[b]	0.68±0.11[b]	7.5±0.34[b]

Similarly, there was a significant ($p < 0.05$) decrease in the activities of SOD, GPx, GR, GST and CAT of APAP induced group as compared with normal group. It was noticed that the antioxidant enzymes activities were increased in a significant manner

($p < 0.01$) on administration of EETH-L (125 mg and 250 mg/kg) in the experimental models after inducing with APAP. Whereas, there is no significant difference at 1% level ($p > 0.01$) in the enzymes between rutin and EETH-L + APAP induced models proving remarkable hepatoprotective property of EETH-L (Figure 1(a) and (b)).

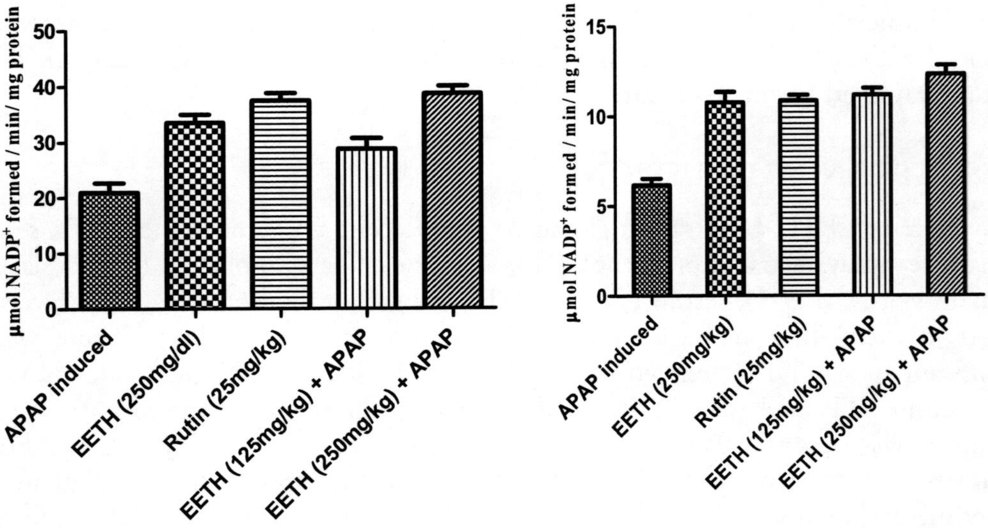

(a) (b)

Fig. 1: Effect of EETH-L on (a) GPx and (b) GR activities compared against APAP induced hepatotoxicity and other experimental groups. Values are mean ± SEM

Morphological observations showed an increased size and enlargement of the liver in APAP treated groups. These changes were reversed by treatment with EETH-L at the two different doses tested groups (125 mg/kg and 250 mg/kg). Histopathological profile of the normal animal showed normal hepatocytes with well preserved cytoplasm and there was no sign of inflammation. The APAP treated animals showed severe centrilobular necrosis, fatty infiltration and microvesicular steatosis. Treatment with EETH-L produced mild degenerative changes, absence of centrilobular necrosis, congested sinusoids and dilated central hepatic venules when compared with control. Treatment with rutin (25 mg/kg) and EETH-L alone showed normal hepatocytes and congested sinusoids. All these results indicate a hepatoprotective potential by the EETH-L.

Hepatotoxicity is a potential complication of APAP, which is widely, used in general medicine, and an assessment of its relative toxicity is important. A number of drugs or chemicals such as melatonin, vitamin E and N-acetyl-cysteine have been used to prevent APAP-induced hepatic injury (Sener *et al.,* 2003). In the present investigation, the serum level of hepatic enzymes like SGOT, SGPT, SALP, LDH and total bilirubin levels were increased and reflected the hepatocellular damage in the APAP-induced hepatotoxicity animal model. This is indicative of cellular leakage and loss of functional integrity of cell membrane in liver (Palani *et al.,* 2010). However the total protein level was decreased. There was a significant ($P < 0.01$) restoration of these enzyme levels on administration of the EETH-L in a dose dependent manner. The reversal of increased serum enzymes in APAP intoxicated hepatic damage by the EETH-L may be due to the prevention of the leakage of intracellular enzymes by re-establishing the membrane fluidity activity. The present results were well corroborated with previous studies conducted by Gupta *et al.* (2004) in *T. coronaria.*

Previous studies have demonstrated that oxidative stress is a major mechanism in the development of APAP-induced hepato and nephrotoxicity (Palani *et al.,* 2010). Mammalian cells contain endogenous antioxidant enzymes, including superoxide dismutase (SOD), glutathione peroxidase, glutathione reductase, glutathione-S-transferase and catalase. In the present study, administration of hepatotoxic doses of APAP to rats resulted in development of oxidative stress damage in hepatic tissues. Thus, oxidative stress is an early event related to radicals generated during the hepatic metabolism of APAP. Also the generation of reactive oxygen species has been proposed as a mechanism by which many chemicals can induce hepato and nephrotoxicity (Palani *et al.,* 2010). Previous study on the methanolic leaf extract of *T. coronaria* has proved its significance in scavenging superoxide and peroxide radicals through SOD and CAT activities (Gupta *et al.,* 2004) and the present investigation results were well agreed.

Effect of optimal extract of EETH-L on serum enzymes (SGPT, SGOT, SALP and LDH), bilirubin and protein in APAP intoxicated rats. APAP induced group compared with normal group [a]($p < 0.05$). Four different experimental groups compared with APAP induced group [b]($p < 0.01$).

ACKNOWLEDGEMENTS

The authors wish to thank the management of Kumaraguru College of Technology, Coimbatore, India and PSG Institute of Medical Sciences and Research, Coimbatore, India for their support to carry out the research work.

REFERENCES

Gupta, M., Mazumder, U.K., Kumar, R.S., Sivakumar, T. and Gomathi, P. (2004a). Antioxidant and protective effects of *Ervatamia coronaria* stapf., leaves against carbon tetrachloride-induced liver injury. *Euro. Bull. Drug Res.,* 12: 13–22.

Ignacimuthu, S., Duraipandiyan, V. and Ayyanar, M. (2006). Antimicrobial activity of some ethnomedicinal plants used by Paliyar tribe from Tamil Nadu, India. *BMC Complementary and Alternative Medicine*, 6: 35–41.

Larrey, D. (2000). Drug-induced liver diseases. *J. Hepatol*, 32: 77–88.

McConnachie, L.A., Mohar, I., Hudson, F.N., Ware, C.B., Ladiges, W.C., Fernandez, C., Chatterton-Kirchmeier, S., White, C.C., Pierce, R.H. and Kavanagh, T.J. (2007). Glutamate Cysteine Ligase Modifier Subunit Deficiency and Gender as Determinants of Acetaminophen-Induced Hepatotoxicity in Mice. *Toxicol. Sci.* 99: 628–636.

Palani, S., Raja, S., Karthi, S., Archana, S. and Senthil Kumar, B. (2010). *In vivo* analysis of nephro and hepato protective effects and antioxidant activity of Madhuca longifolia against acetaminophen-induced toxicity and oxidative stress. *J. Pharma. Res.*, 3: 9–16.

Rocha, J.B.T., Gabriel, D., Zeni, G. and Posser, T. (2005). Ebselen and diphenyl diselenide change biochemical hepatic responses to overdosage with paracetamol. *Environmen. Toxicol. Pharmacol*, 19: 255–261.

Sathishkumar, T. and Baskar. R. (2012). Evaluation of antioxidant properties in the leaves of *Tabernaemontana heyneana* Wall. *Indian J. Nat. Pro. Resour*, 3: 197–207.

Sener, G., Sehirli, A.Ö. and Ayanog˘lu-Dülger, G. (2003). Protective effects of melatonin, vitamin E and N-acetylcysteine against acetaminophen toxicity in mice. A comparative study. *J. Pineal Res.* 35: 61–68.

Screening of the Candidate Gene KIF21A in CFEOM Patients

C. Monica* and Chitra Kannabiran[1]

Department of Biotechnology, K.S. Rangasamy College of Technology, Tamil Nadu

[1]Kallam Anji Molecular Genetics Laboratory, L.V. Prasad Eye Institute, Hyderabad

E-mail: monicachristopher.bt@gmail.com

INTRODUCTION

Almost nothing need be said when you have eyes, says Tarjei Vesaas. The eye is the organ of vision, which produces an image of the external world and transforms it into nerve impulses and is thus an organ of immense importance. But there are several diseases of varying severity that affect the eye and pose a threat to our vision. Globally, the number of people of all ages living with sight loss is estimated to be 285 million, of which 39 million are blind. Every 5 seconds one person in the world goes blind. Some of the most common diseases that affect the eye are eye allergy, stye, age related macular degeneration, cataract, conjunctivitis, glaucoma, uveitis, keratoconus, eyelid cyst, corneal ulcer, strabismus, diabetic retinopathy, amblopia. On a global scale, uncorrected refractive errors, cataract, glaucoma is the cause of visual impairment of 43%, 33% and 2% people, respectively.

These eye diseases are caused due to several reasons like injury, complications due to other diseases, etc. Genetic reasons are one of these major causes. A genetic disorder is a condition that is passed on from the parents to their children through their genes. Genes contain coded information that determines the traits an individual express. Genetic disorders differ from other medical problems in that they tend to recur within families. Just as common traits such as facial features are passed from parent to child through the genes, an alteration in a gene that could lead to an eye disorder may be

*Corresponding author.

inherited. Often, the risk of other family members can be predicted. Some of the most common genetic eye diseases are congenital cataracts, retinal degenerations, congenital glaucoma, retinitis pigmentosa. Certain other rare genetic eye diseases include congenital fibrosis of extra ocular muscles (CFEOM), leber's congenital amaurosis.

The work was focused on genetic screening of the candidate gene KIF21A in patients affected by CFEOM. The CFEOM is a disorder that affects the extra ocular muscles surrounding the eye. In this condition, fibrosis or thickening or scarring of extra ocular muscles takes place. It is referred to as congenital as this disease is a condition that is present at birth due to hereditary influences. CFEOM is a very rare genetic ocular motility disorder. The most common form of CFEOM affects atleast 1 in 230000 people.

This disease is characterized by global non-progressive restrictive ophthalmoplegia with or without ptosis. The ophthalmoplegia is a paralysis or weakness of one or more muscles that control the eye movement leading to inability to move the eye in one or more directions. Ptosis is a condition in which patients have droopy eyelids. The clinical appearance of a patient with CFEOM is illustrated in Figure 1. The CFEOM phenotype was first described in the medical literature in 1840 (Baumgarten, M., 1840). It was first recognized to occur as a familial trait based on postmortem examination and it was shown that the extra ocular muscles were fibrotic and abnormally inserted (Heuk, G., 1879). This condition was described in detail by Brown in 1950 (Brown, H.W., 1950). Laughlin reported the condition in at least four generations in 1956 (Laughlin, R.C., 1956). After then, there have been lots of additional reports about this disorder in the literature.

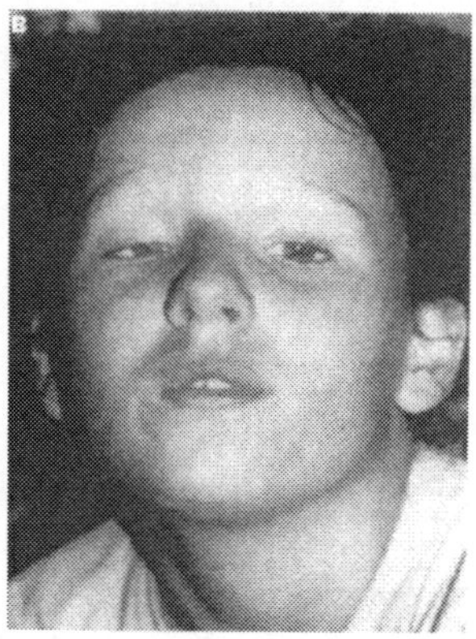

Fig. 1: Clinical Appearance of a Patient with CFEOM (patient exhibited "chin up" head posture with bilateral, asymmetrical ptosis and squint). Picture Taken from *J. Med. Genet.*, 35: 985–988, G C M Black *et al.*, 1998.

The disease manifests in three different types—CFEOM I, CFEOM II and CFEOM III. The most common form of CFEOM worldwide is the CFEOM1 which is often described as the "classic CFEOM" (Traboulsi, E.I. *et al.*, 2004). Engle identified three genetic CFEOM loci namely FEOM1, FEOM2, FEOM3. The FEOM1, FEOM2, FEOM3 disease genes are identified to be KIF21A, PHOX 2A, TUBB3 (Engle *et al.*, 2003). This disorder is characterized by bilateral ptotis and bilateral ophthalmoplegia. The affected individuals have an inability to supraduct above the horizontal midline. This disorder is an autosomal dominant fully penetrant trait. Screening of transcripts of the chromosome 12 in several affected families led to the discovery of heterozygous missense mutations in kinesin family member 21A (KIF21A) as the cause (Yamada, K. *et al.*, 2003). CFEOM2 is the rarest phenotype which is inherited recessively. This disorder is characterized by bilateral ptosis and restrictive ophthalmoplegia with the globes frozen in extreme abduction (Wang, S.M. *et al.*, 1998). CFEOM3 is also an autosomal dominant trait that is variably expressed and it has incomplete penetrance. This type of CFEOM has marked variability of phenotype in different disease severity.

The KIF-21A is a gene with 38 coding regions and is located at the chromosome 12q12. Previous studies show compelling evidences that certain mutations present in the KIF-21A are responsible for this ocular disorder. The work was aimed at screening for mutations in the disease gene KIF21A in patients affected by CFEOM. The exons 2–10, 20, 21 were initially chosen for the study based on previous literature.

MATERIALS AND METHODS

Clinical Investigation

The list of patients affected by CFEOM was obtained from the L.V. Prasad Eye Institute patient database and they were invited for the study. The Institute is approved by the Institutional Review Board to collect human samples for the study. All patients had undergone complete orthoptic and ophthalmic examination before the study. Informed consent was obtained from the participants and/or their guardians and 5 ml venous blood sampling for candidate gene testing was collected from the patient and their relatives. The methods adhered to the Declaration of Helsinki for research involving human subjects.

Molecular Genetic Studies

DNA Extraction, Primer Design and Optimization of PCR Conditions

The genomic DNA was extracted from the blood sample using the phenol chloroform extraction method and the spectrophotometer readings are taken again to check the quantity and quality of the DNA. The primers were designed for the selected exons 2–10, 20 and 21 using the primer 3 software. A total of seven sets of primers were designed for the exons 2_3, 4, 5_6, 7, 8, 9_10, 20_21 and they were purchased from Eurofins Genomics India Pvt. Ltd., Banglore. The Polymerase Chain Reaction (PCR) conditions such as annealing temperature, time, Magnesium Chloride ($MgCl_2$) concentration and primer concentration were optimized for these primers using control DNA samples in order to acquire adequate amplification of the desired product. The PCR conditions for all the seven primer sets were optimized.

Amplification of Desired Exons from Patient DNA Samples

The patient DNA samples were amplified by PCR using Veriti Thermal Cycler. The optimized conditions for each primer set were used to acquire desired amplification. The PCR procedure for PCR was briefly as follows:Initial denaturation at 95°C for 3 minutes, denaturation at 94°C for 30 seconds, annealing at the optimized temperature for 30 seconds, extension at 72°C for 30 or 45 seconds based on the PCR product size, final extension at 72°C for 7 minutes. The PCR products were checked for amplification by performing agarose gel electrophoresis and viewing the bands in the gel documentation system. Thirteen affected patient samples were amplified for all the seven primer sets.

Sequencing of Desired Exons

The sequencing was carried out using Applied Biosystems fluorescence-based cycle sequencing system which is an extension and refinement of the Sanger's dideoxy method. The sequencing PCR was carried out for 2 exons 8, 9 and 10 in the veriti thermal cycler (ABI). The sequencing PCR products were purified by the EDTA/Sodium acetate/Ethanol method. The purified sequencing PCR products were subjected to data collection in the ABI 3130 genetic analyzer. The sequences were analysed using chromas software.

Analysis of the Sequences

The obtained sequences were compared with the ENSEMBL sequences and the exons were screened for mutations. Chromas and Blast softwares were used for the analysis of sequences. For confirmation of any observed mutations 100 normal samples will be checked for variations.

RESULTS

The primers were used to amplify control samples at different experimental conditions and the conditions at which the amplification was effective were considered to be the

Table 1: The Primer Sequences for Different Exons and Their Product Size and Optimized Annealing Temperature

Exon	*Primer Sequences (5'–> 3')*	*Product Size (bp)*	T_m(°C)
2_3	F: tgggattaagaaagtcatctcatgt	821	60
	R: gcactatacttgaagccaacagt		
4	F: actacccagacaatttctgctaaa	274	60
	R: gcagattatcagcaaagctcact		
5_6	F: gccattacctcagactttggat	997	60
	R: ggttcagtatctgaaggatttcca		
7	F: atgtctctatattattctctg	408	50
	R: attagttcctcatcctattag		
8	F: agtcttcaagtacaggggctttt	746	64
	R: tgtgttcagaaagtgccagcct		
9_10	F: aaggctcctcaaccaacaagg	870	62
	R: gacaactgttgaacactaagccca		
20_21	F: gccttaccaacgccggcaac	333	66
	R: tgaaaagcaagcaggaagtggca		

optimized conditions. The list of the primers designed for the desired exons and the optimized PCR conditions are given in Table 1.

Bright bands were obtained on gel electrophoresis of most of the pcr products. Non specific bands and primer dimmers were present in few samples. In such cases the amplification step was repeated with modified operating conditions to obtain bright and clear bands. These ensure the quality of the PCR product and the amplification. The confirmation of amplification of patient samples by the gel electrophoresis for some of the patient samples are shown in Figure 2(a)–(d).

The sequencing was first standardized using control samples. Two primer sets containing three exons (8, 9, 10) were sequenced. It was inferred that all the patients involved in the study did not harbor any mutations in the exons 8, 9 and 10.

The KIF21A gene present at the FEOM1 locus spans ~150 kb of genomic DNA and has an open reading frame of 5,022 bp. It consists of 38 exons with alternative splicing of exon 12 and exons 29–31 (Yamada, K. *et al.*, 2003). This gene is predicted to encode a kinesin motor protein, KIF21A, which is responsible for the transport of membranous organelles, protein complexes, and mRNAs to specific destinations within the cell in a microtubule- and ATP-dependent manner. These functions are essential for normal morphogenesis and functioning of the cell (Miki, H. *et al.*, 2001).

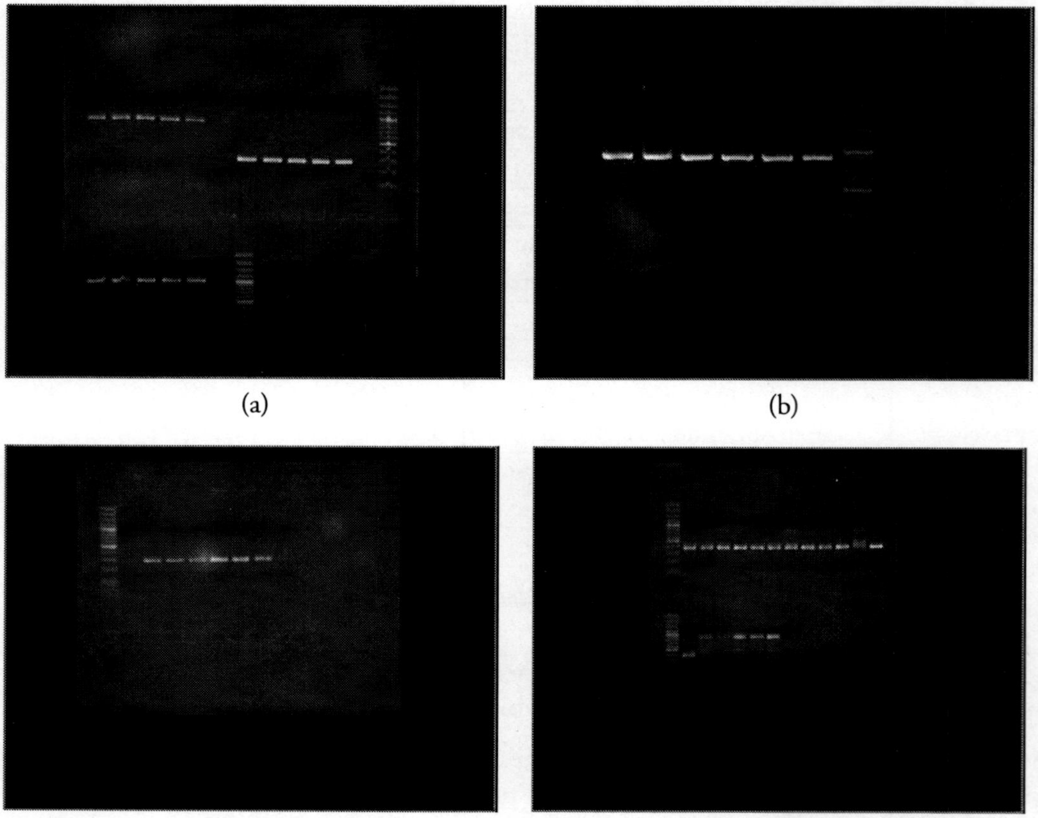

(a) (b) (c) (d)

Fig. 2: (a) PCR Products of Patient Samples for Exons 2_3, 4 and 5_6, (b) PCR Products of Patient Samples for Exons 20_21, (c) PCR Products of Patient Samples for Exon 7, (d) PCR Products of Patient Samples for Exons 9_10 Respectively

Several mutations have been reported in the KIF21A gene till date (Traboulsi, E.I. *et al.*, 2004; Yamada, K. *et al.*, 2003; Wang, S.M. *et al.*, 1998). Mutations are most frequently reported in the exons 8,20 and 21. The most common hotspot mutation that has been observed is 2860 C>T in the exon 20 (Traboulsi, E.I. *et al.*, 2004; Yamada, K. *et al.*, 2003; Wang, S.M. *et al.*, 1998). The second most common mutation that has been reported is 2861 G>A in the exon 20. Wai-Man Chan and Caroline Andrews reconfirmed the hotspot mutations 2860 C>T, 2861 G>A, and 1067 T>C in eight probands and reported three new mutations 2830 G>C, 2861 G>T, 3022 G>C in sporadic cases of Caucasian ethnicity (Wai-Man Chan *et al.*, 2007). The 2860 C>T mutation was reported in 32 probands and was established as a pathogenic missense mutation in the year 2003 (S.M. Wang *et al.*, 1998). The most common missense mutation 2860 C>T and the second most common missense mutation 2861 G>A were reported in two Chinese families consecutively (Xian Yang *et al.*, 2011). But in 2004, Yamada K reported absence of mutations in KIF21A mutations in 15 probands and a rare KIF21A single-nucleotide polymorphism, 3641 C>G in a sporadic individual (Koki Yamada *et al.*, 2004). Arif O. Khan *et al.,* reported lack of KIF21A mutations in CFEOM type 1 patients from consanguineous Saudi Arabian families (Arif O. Khan *et al.*, 2011). A missense mutation 1067 T>C has been reported in the exon 8 by Engle (Wai-Man Chan *et al.*, 2007). On the contrary, the study showed that in the selected patients harbored no mutations in the exons 8, 9, 10 of the KIF21A gene. If further study shows the presence of the common missense mutations in the exons 20 and 21, it can considered to reconfirm the pathogenicity of these hotspot mutations in Indian populations. In summary, the screening of thirteen CFEOM probands revealed absence of mutations in the exons 8, 9 and 10 of the candidate gene KIF21A. Thus, most CFEOM can be due to the most common hotspot heterozygous missense mutations. This information can be useful for genetic counseling of CFEOM patients. Further studies and ascertainments will be helpful to identify the pathogenic mutations and to validate published reports

ACKNOWLEDGEMENTS

I am thankful to the three science academies for giving this opportunity to broaden my exposure. I would also like to thank Dr. Chitra Kannabiran for accepting me as a summer research fellow under her guidance. I would also like to extend my gratitude to L.V. Prasad Eye Institute for providing all the necessary facilities. I also appreciate all the scientists, faculty, and staff of the institute for their encouragement and support. I am thankful to the head of our department Dr. P. Ponmurugan and the faculty for their support.

REFERENCES

Baumgarten, M. *et al.*, Erfahrungen uber den strabismus und die Muskeldurchschneidung am Auge in physiologischpathologischer und therapeutischer Beziehung. *Monatsschr Med Augenheilkd Chir* 1840, 3: 474–499.

Brown, H.W. Congenital muscle abnormalities. In: Allen JH,ed. Strabismus ophthalmic symposium. St Louis: CV Mosby, 1950: 229–33.

Heuk, G., Ueber angeborenen vererbten Beweglichkeits—Defect der Augen. *Klin Monatsbl Augenheilkd,* 1879, 17: 253–278.

Koki Yamada, Wai-Man Chan, Caroline Andrews, Elizabeth C. Engle *et al.* Identification of KIF21A Mutations as a Rare Cause of Congenital Fibrosis of the Extraocular Muscles Type 3 (CFEOM3). IOVS, July 2004, Vol. 45, No. 7.

Laughlin, R.C., Congenital fibrosis of the extra ocular muscles a report of six cases. Am J *Ophthalmol,* 1956; 41: 432–8.

Miki, H., Setou, M., Kaneshiro, K. and Hirokawa, N. All kinesin superfamily protein, KIF, genes in mouse and human. *Proc. Natl. Acad. Sci.* USA 2001; 98: 7004–11.

Traboulsi, E.I., Engle, E.C. Mutations in KIF21A are responsible for CFEOM1 worldwide. *Ophthalmic Genet,* 2004; 25: 237–9.

Wai-Man Chan, Caroline Andrews, Elizabeth C Engle *et al.* Three novel mutations in KIF21A highlight the importance of the third coiled-coil stalk domain in the etiology of CFEOM1. *BMC Genetics,* 8: 26: 1471–2156, 2007.

Wang, S.M., Zwaan, J., Mullaney, P.B., Jabak, M.H., Al-Awad, A., Beggs, A.H. and Engle, E.C. (1998). Congenital fibrosis of the extra ocular muscles type 2, an inherited exotropic strabismus fixus, maps to distal 11q13. *Am. J. Genet.,* 63: 517–525.

Xian, Yang; Koki, Yamada; Bradley, Katz *et al.*, KIF21A mutations in two Chinese families with congenital fibrosis of the extraocular muscles (CFEOM). *Molecular Vision 2011*; 17: 218–224

Yamada, K., Andrews, C., Engle, E.C. *et al.*, Heterozygous mutations of the kinesin KIF21A in congenital fibrosis of the extraocular muscles type 1 (CFEOM1). *Nat Genet 2003;* 35: 318–21.

Preliminary Screening of Fluorescent Pseudomonades from Groundnut Rhizosphere against Fungal Pathogens and Phenazine Gene Identification by Polymerase Chain Reaction (PCR)

R. Charulatha, H. Harikrishnan, P.T. Manoharan[1] and V. Shanmugaiah*

Department of Microbial Technology, School of Biological Sciences, Madurai Kamaraj University, Madurai, Tamil Nadu

[1]Department of Botany, Vivekananda College, Tiruvedakam (West), Madurai

E-mail: vshanmugaiah@yahoo.co.in

INTRODUCTION

Fluorescent pseudomonads are among the most effective rhizosphere bacteria because in addition to disease control, they exert beneficial effect on plant growth promotion (Dubeikovsky *et al.*, 1993; Raupach and Kloepper, 1998). The rhizosphere is the first line of defence for roots, against attack by pathogenic fungi. Therefore, there is an excellent opportunity to find rhizosphere component microbes that can act as potential bio pesticides. Fluorescent pseudomonads commonly isolated from the rhizosphere have been shown to be ideal biological control agents (Burr *et al.*, 1984; Gutterson *et al.*, 1990 and Weller, 1984) for the following reasons: (1) many of the secondary metabolites from pseudomonads inhibit other microorganisms, (2) a higher proportion of pseudomonads are inhibitory to a variety of microorganisms in comparison with other soil bacteria, (3) pseudomonads are selectively stimulated in the rhizosphere and when introduces via seed coating they can dominate the rhizosphere

*Corresponding author.

population, and (4) they extensively colonize the endorhizosphere (Leisinger *et al.*, 1979 and Weller, 1988).

Charcoal rot, caused by *Macrophomina phaseolina* is an economically important disease of groundnut. The soil borne fungus *M. phaseolina* is endemic to temperate and tropical regions of the world and can infect over 500 different hosts (Wyllie, 1998). Disease management strategies primarily depend on sanitary practices and well timed fungicide applications. However development of fungicide resistance within population of *M. phaseolina* became a problem, and alternative approaches that can be incorporated into integrated pest management of groundnut charcoal rot disease are needed. Biological control is coming up as an alternative strategy for disease management, which is also ecology-conscious and environment friendly.

Antibiotics are microbial toxins that can, at low concentrations, poison or kill other microorganisms. Most microbes produce and secrete one or more compounds with antibiotic activity. In some instances, antibiotics produced by microorganisms have been shown to be particularly effective at suppressing plant pathogens and the diseases they cause. To be effective, antibiotics must be produced in sufficient quantities near the pathogen to result in a biocontrol effect. *In situ* production of antibiotics by several different biocontrol agents has been measured (Thomashow *et al.* 2002); however, the effective quantities are difficult to estimate because of the small quantities produced relative to the other, less toxic, organic compounds present in the phytosphere.

MATERIALS AND METHODS

Isolation of Fluorescent Pseudomonades (FPs) from Rice Rhizosphere

10 g of soil samples collected from rice rhizosphere was taken in a conical flask. To which, 90 ml of sterile distilled water was added and the flasks were placed on a rotary shaker at 120 rpm for 30 min. The resulted suspension was serially diluted and spread plated on to King's B agar (KBA) (King *et al.*, 1954). The plates were incubated at room temperature (28 ± 2°C) for 48 h and observed under UV light. Distinct single colonies of fluorescent bacteria were picked, sub-cultured to purity and maintained in KBA slants and also in 30% glycerol vials.

Antifungal Activity in Dual Plate Assay

The selected isolate VSMKU-2013 was tested towards fungal pathogens for its antifungal activity against *Macrophomina phaseolina, Rhizoctonia solani, Fusarium udum, Alternaria alternata, Fusarium oxysporum* by dual plate assay (Shanmugaiah *et al.*, 2010).

Identification of Phenazine Gene by Polymerase Chain Reaction (PCR)

The identification of metabolite, phenazine gene from *Pseudomonas* sp.VSMKU 2013by heat lysis method and phenazine gene was determined by PCR amplification using the specific primer of the Phenazine forward and reverse primers Phz1a 5'–GGC GACATGGCTCGCTCAAC-3' and PHz2a 5'-CGCGTGGCGGCGGCGTATAT-3'. PCR amplification was performed using forward primer *Phz* and reverse primer *Phz2* (Medox, India), which was developed from the *Phz* gene sequence of *Pseudomonas fluorescens* Q2–87 (Raaijmakers *et al.*, 1997). The conditions used in PCR reaction are given below: PCR amplification was carried out in 15 µl of reaction was used an then the amplifications were performed with a Cyber lab (Smart PCR, USA) using the following PCR conditions: Initial denaturation for 2 min at 94°C followed by 25 PCR cycles of amplification, and each cycle had 1 min denaturation at 94°C 45 sec annealing at 56°C extension at 72°C for 1.75 min and a final extension at 75°C for 1min (Delaney *et al.*, 2001).

Gel Electrophoresis

Samples of the PCR products (6 µl) were separated on a 1.5% agarose gel in 1 X TAE buffer at constant voltage of 100 V for 2 h. The gel was stained with ethidium bromide for 20 min and destained in distilled water for 30 min. The gel was visualized under UV transilluminator and photographed.

Identification of Selected Strain

The selected isolate VSMKU-2013 was identified as *Pseudomonas* sp based on 16S rRNA, physiological and morphological characters exhibited in the Bergey's Manual of determinative bacteriology (Williams *et al.*, 1994).

Lytic Enzyme Production

The selected strain *Pseudomonas* sp VSMKU-2013 was streaked on the respective substrate (0.5%) amended agar medium and then the plates were incubated at room temperature. After 24 h and 48h, Protease, Amylase and Pectinase, Chitinase and Cellulase production was observed. By flooding Gram's iodine over the starch agar plate amylase activity was tested. Mercuric chloride was flooded over the gelatin and casein agar plates to check the zone of clearance for protease, similarly flooding 1% Congo red over the colloidal pectin, chitin and cellulose agar plates.

Screening of Phosphate Solubilization

To detect the phosphate solubilizing from selected isolate *Pseudomonas* sp VSMKU-2013 were streaked onto Pikovskaya's agar (Pikovskaya, 1948) medium, which contains (per liter): 0.5 g yeast extract, 10 g dextrose, 5 g $Ca_3(PO_4)_2$, 0.5 g $(NH_4)_2$ SO_4, 0.2 g KCl, 0.1 g $MgSO_4.7H_2O$, 0.0001 g $MnSO_4.H_2O$, 0.0001 g $FeSO_4.7H_2O$ and 15 g agar. After 3 days of incubation at 28°C, strains that induced clear zone around the colonies were considered as positive.

HCN and Siderophore Production

Production of hydrocyanic acid (HCN) was determined by method of Lorck (1948). Bacterial strains (24 h old) were separately streaked on nutrient sucrose agar with 4.4 g glycine with simultaneous addition of a filter paper soaked in 0.5% picric acid in 1% Na_2CO_3 in the upper lid of Petri plate. The plates were sealed with parafilm. The control plates did not receive inoculum. After incubation at 28±1°C changes in colour from yellow to reddish brown were examined. Siderophore production by different strains FPs *was* tested by chromo azural S (CAS) assay according to Schwyn and Neilands (1987). The strains were streaked on citramide agar and incubated for 48 h at 30°C. After incubation a thin layer of CAS reagent in 0.7% agar was spread on the bacterial growth and plates were again incubated for 24h at 30°C formation of yellow orange zone around the colonies indicates siderophore production.

IAA Production

It was done as per the protocols of Patten and Glick (1996). FPs was grown in Luria broth supplemented with L-tryptophan (1 μg/ml) for 72 h. At the end of the incubation, cultures were centrifuged at 10,000 g for 10 min and the supernatants collected. One ml of this culture filtrate was allowed to react with 2 ml of Salkowsky reagent (1 ml of 0.5 M $FeCl_3$ in 50 ml of 35% $HClO_4$) at 28±2°C for 30 min. At the end of the incubation, pink color developed which indicates the presence of IAA. Quantification of IAA was done by measuring the absorbance in a spectrophotometer at 530 nm.

Detection of Phenazine by TLC

Selected strain *Pseudomonas* sp VSMKU-2013 were grown with shaking in King's B broth for 48 h at 27°C, and phenazines were extracted with ethyl acetate as described previously (Mavrodi *et al.,* 1998) and slight modification. Commercial Phenazine compound (Sigma Aldrich, Mumbai) was used as a standard. The extracts were dried, suspended in methanol, and spotted on silica gel plates (Merck, Germany). Chromatography was carried out using benzene—acetic acid (95:5), and chromatograms were examined at 254 nm.

RESULTS

Isolation and Screening of Antagonistic Fluorescent Pseudomonads

A total of 386 FPs isolated from groundnut rhizosphere of southern district of Tamil Nadu. Out of 386, 92 fluorescent pseudomonades were showed antagonistic activity against charcoal rot groundnut pathogen *Macrophomina phaseolina.*

The zone of inhibition of mycelia growth of the above fungal pathogen ranged from 09mm to 29 mm. Among 92, based on the superior antagonistic performance, the strain VSMKU-2013 was selected for further studies. Further, the secondary screening of selected isolated VSMKU-2013 towards other fungal pathogens such as *Rhizoctonia solani, Fusarium oxysporum, Fusarium udum and Alternaria alternata* (Table 1).

Table 1: Antagonistic Activity of *Pseudomonas sp.*VSMKU-2013 against Fungal Pathogen

Fungal Pathogens	*Zone of Inhibition (mm)*
Macrophomina phaseolina	29
Rhizoctonia solani	18.8
Fusarium oxysporum	18
Fusarium udum	12.8
Alternaria alternata	17.5

Screening of Phenazine Gene by Polymerase Chain Reaction (PCR)

Phenazine gene from antagonistic bacteria VSMKU-2013 was identified by heatlysis method with using forward primer Phz1a and reverse Phz2a primers. *Pseudomonas* sp. VSMKU-2013 showed Phenazine positive with characteristic feature of 1Kb DNA fragments amplification compare with that of Phenazine positive bacteria *Pseudomonas aeruginosa* MML2212.

Identification of *Pseudomonas sp.*VSMKU-2013

The selected strain VSMKU-2013 was identified as *Pseudomonas* sp. based on the morphological, physiological tests and 16s rDNA sequence analysis (Table 2).

Production of Lytic Enzymes, Phosphate Solubilization, HCN, Siderophore and IAA

Production of lytic enzymes was investigated for the selected strain VSMKU-2013. Among five lytic enzymes, the isolate VSMKU-2013 showed the production of protease and amylase. Phosphate solubilisation, Hydrogen cyanide, Siderophore and Indole acetic acid production was observe in isolate VSMKU-2013 (Table 2).

Table 2: Physiological and Biochemical Traits of Lytic Enzymes and Secondary Metabolites Production by Rice Rhizosphere Isolate *Pseudomonas* sp VSMKU-2013

Biochemical lytic Enzyme and Metabolite Production	*Characters*
Gram staining	**Gram negative**
Motility	**Motile**
Fluorescent Character	+
Oxidase	+
Catalase	+
Secondary metabolites	*Characters*
Phosphate solubilization	+
HCN	+
Siderophore	+
Phenazine	+
Lytic enzyme and growth hormone	*Character*
Chitinase	–
Protease	+
Amylase	+
Cellulase	–
Pectinase	–
IAA	+

Phenazine Detection by TLC

The investigation of Phenazine compound from selected isolate VSMKU-2013 were observed with Rf value 0.5 was detected by bioanalytic method TLC. The current investigation deals with screening and selection of significant fluorescent pseudomonads from rice rhizospheric soil of groundnut with antagonistic activities against phytopathogenic fungi and prospective plant growth stimulating characteristics with purpose for further field application. Isolation of bacteria from innate of rhizosphere of the target crop is essential for successful identification of potential bio-control and bio fertilizing agents. The present result demonstrate that a total of 386 Fluorescent pseudomonads were isolated from rice rhizosphere. Of which 92 fluorescent pseudomonads were found to be effectively against *M.phaseolina.* Among 92 antagonistic isolates, a strain designated as VSMKU-2013 was selected because of its broad spectrum antagonistic activity against various phytopathogens.

The selected strain VSMKU-2013 was identified as *Pseudomonas* sp. based on the physiological tests, biochemical characteristics and 16S rRNA sequencing. Production of lytic enzymes such as Chitinase, Cellulase, Protease, Amylase, Pectinase, Phosphate solubilization, Hydrogen cyanide, Siderophore and IAA were recorded with 92 antagonistic fluorescent pseudomonads. Antagonistic FPs were also tested towards other fungal pathogens such as *R.solani, Fusarium oxysporum, Fusarium udum* and *Alternaria alternata* with zone of inhibition ranging from 0.7 cm to 1.9 cm, 0.7 cm to 2.9 cm, 0.9 cm to 2.9 cm and 0.2 cm to 3.0 cm. Phenazine derivatives aid in long-term survival and ecological competence of these strains in rhizosphere (Mazzola *et al.*, 1992).

The antimicrobial activity of phenazine depends on the rate of oxidative reductive, transformation of the compound coupled with the accumulation of toxic superoxide radicals in the target cells (Hassett *et al.*, 1992 and 1993). The ability of antagonistic bacteria to produce proteolytic enzyme, chitinase and cellulase was at high percentage and the composition of antagonistic mechanisms was specific for each isolates. Some of the tested isolates could exhibit more than plant trails, which may promote plant growth directly or indirectly or synergistically. Bio control agents such as *P. cepacia* (Fridlender *et al.*, 1993), *P. stutzeri* (Lim *et al.*, 1991), *P. fluorescens* (Velazhahan *et al.*, 1999; Meena *et al.*, 2001) and *P. aeruginosa* MML2212 (Shanmugaiah, 2007., Shanmugaiah *et al.*, 2010). Similarly, highest augment in plant growth and biomass of groundnut was observed with a rhizosphere isolate of *Bacillus firmus* and two phylloplane isolates of *Bacillus megaterium* and *Pseudomonas aeruginoisa* and *Pseudomonas aeruginoisa* MML2212 (Kishore, 2005, Shanmugaiah, 2007). The selected isolate hence proved to be effective in promoting the growth of a non-host, besides inhibiting *M.phaseolina* causing charcoal rot disease in groundnut.

ACKNOWLEDGEMENT

The authors greatly acknowledged the Head, Department of Microbial Technology, The Co-ordinator, NRCBS, The Chairperson, School of Biological Sciences, Madurai Kamaraj University, Madurai, Tamil Nadu, India.

REFERENCES

Burr, T.J. and A. Caesar. (1984). Beneficial plant bacteria. CRC *Crit. Rev. Plant Sci.* 2: 1–20.

Delaney, S.M., Mavrodi, D.V., Bonsal, R.F. and Thomashow, L.S. (2001). PhzO, a Gene for Biosysnthesis of 2-Hydroxylated Phenazine compounds in *Pseudomomnas aureofaciens* 30–84. *J. Bacteriol.* 183: 318–327.

Dubeikovsky, A.N., Mordukhova, E.A., Kochethovm, V.V., Polikarpova, F.Y. and Boronin, A.M. (1993). Growth promotion of black current soft wood cuttings by recombinant strain Pseudomonas fluorescens BSP53a synthesizing an increased amount of indole-3-acetic acid. *Soil Biol. Biochem.,* 25: 1277–1281.

Fridlender, M., Inbar, J. and Chet, I. (1993). Biological control of soil borne plant pathogens by a b-1,3 glucanase-producing Pseudomonas cepacia. *Soil Biology and Biochemistry*, 25: 1121–1221.

Gutterson, N. (1990). Microbial fungicides: Recent approaches to elucidating mechanisms. *Crit. Rev. Biotechnol.* 10: 69–91.

Hassett, D.J., Woodruff, W.A., Wozniak, D.J., Vasil, M.L., Cohen, M.S. and Ohman, D.E. (1993). Cloning of sodA and sodB genes encoding manganese and iron superoxide dismutase in Pseudomonas aeruginosa: demonstration of increased manganese superoxide dismutase activity in alginate-producing bacteria. *J. Bacteriol.* 175: 7658–7665.

Hassett, D.J ., Charniga, L., Bean, K., Ohman, D.E. and Cohen, M.S. (1992). Response of Pseudomonas aeruginosa to pyocyanin: mechanisms of resistance, antioxidant defenses, and demonstration of manganese –cofactored superoxide dismutase. *Infection and Immunity* 60: 328–336.

Jos, M. Raaijmakers; David, M. Weller. and Linda, S. Thomashow (1997). Frequency of antibiotic-producing pseudomonas spp.in natural environments. *Applied and Environmental Microbiology*, 63: 881–887.

King, E.O., Ward, M.K. and Raney, D.E. (1954). Two simple media for the demonstration of pyocyanin and fluorescin. *J. Lab Clin Med.,* 44: 301–307.

Kishore, G.K., Pande, S. and Podile, A.R. (2005). Phylloplane bacteria increase seedling emergence, growth and yield of field-grown groundnut (Arachis hypogaea L.). *Lett. App. Microbiol*, 40: 260–268.

Leisinger, T. and Margraff, R. (1979). Secondary metabolites of fluorescent pseudomonads. *Microbiol. Rev.,* 43: 422–442.

Lorck, H. (1948). Production of hydrocyanic acid by bacteria. *Physiol Plant,* 1: 142–146.

Lim Ho-Seong., Yong-Su Kim. and Sang-Dal Kim. (1991). Pseudomonas stutzeri YPL-1 Genetic Transformation and Antifungal Mechanism against Fusarium solani, an agent of Plant Root Rot. *Applied and Environmental Microbiology*, 510–516.

Mavrodi, D.V., Ksenzenko, V.N., Bonsall, R.F., Cook, R.J., Boronin, A.M. and Thomashow, L.S. (1998). A seven-gene locus for synthesis of phenazine- 1-carboxylic acid by Pseudomonas fluorescens 2–79, *J. Bacteriol.* 180: 2541–2548.

Mazzola, M., Cook, R.J., Thomashow, L.S., Weller, D.M. and Pierson, L.S. (1992). Contribution of phenazine antibiotic biosynthesis to the ecological competence of fluorescent pseudomonads in soil habitats. *Appl. Environ. Microbiol.*, 58: 2616–2624.

Meena, B., Radhajeyalakshmi, R., Marimuthu, T., Vidhyasekaran, P., Doraisamy, S. and Velazhahan, R. (2000). Induction of pathogenesis-related proteins, phenolics and phenylalanine ammonia lyase in groundnut by Pseudomonas fluorescens. *Journal of Plant Disease and Protection,* 107: 514–527.

Patten, C. and Glick, B.R. (1996). Bacterial biosynthesis of indole-3-acetic acid. *Can J. Microbiol,* 42: 207–220.

Pikovskaya, R.I. (1948). Mobilization of phosphorus in soil connection with the vital activity of some microbial species. *Microbiologia,* 17: 362–370.

Raupach, G.S. and Kloepper, J.W. (1998). Mixtures of plant growth promoting rhizobacteria enhance biological control of multiple cucumber pathogens. *Phytopathology,* 88: 1158– 1164.

Schwyn, B. and Neilands, J.B. (1987). *Anal Biochem,* 160: 47–56.

Shanmugaiah, V. (2007). Biocontrol potential of phenazine-1-carboxamide producing plant growth promoting rhizobacterium, Pseudomonas aeruginosa MML2212 against sheath blight disease of rice. PhD Thesis, University of Madras, Chennai, India.

Shanmugaiah, V., Mathivanan, N. and Varghese, B. (2010). Purification, crystal structure and antimicrobial activity of phenazine-1-carboxamide produced by a growth promoting biocontrol bacterium, Pseudomonas aeruginosa MML2212. *Journal of Applied Microbiology,* 108: 703–711.

Thomashow, L.S., Bonsall, R.F. and Weller, D.M. (2002). Antibiotic production by soil and rhizosphere microbes in situ. in: Manual of Environmental Microbiology (2nd ed.), ASM Press, Washington DC, pp. 638–647.

Velazhahan, R., Samiyappan, R. and Vidhyasekaran, P. (1999). Relationship between antagonistic activities of Pseudomonas fluorescens isolates against Rhizoctonia solani and their production of lytic enzymes. Zeitscrift fur Pflanzenkrankheiten und Pflanzenschutz, 106: 244–250.

Weller, D.M. (1984). Distribution of a take-all suppressive strain of Pseudomonas fluorescens on seminal roots of winter wheat. *Appl. Environ. Microbiol.,* 48: 897–899.

Weller, D.M. (1988). Biological control of soil- borne plant pathogens in the rhizosphere with bacteria. *Annu. Rev. Phytopathol,* 26: 379–407.

Williams, S.T. (eds) (1994). Bergey's Manual of Determinative Bacteriology, 9th edition Williams and Wilkins Co, Baltimore, USA.

Wyllie, T.D. (1998). Charcoal rot. In Compendium of soybean diseases, 3rd Edition. (Eds.: J.B. Sinclair and P.A. Backman), APS Press. St. Paul, MN.

SECTION–IV

IBS Award

Regulation of DNA Replication Machinery during Stress in Mammalian Cells

P. Suresh*, Ananya Kar[1] and Sandeep Saxena[1]

Department of Biotechnology, K.S. Rangasamy College of Technology, Tiruchengode, Tamil Nadu

[1]National Institute of Immunology, Aruna Asaf Ali Marg, New Delhi

E-mail: Subisuresh7@gmail.com

INTRODUCTION

Cell division is a very important process in all living organisms. During cell division, DNA replication and cell growth takes place. All these processes, i.e. cell division, DNA replication, and cell growth, take place in a coordinated way to ensure correct division and also formation of progeny cells containing intact genomes (Lodish *et al.*, 2000).

Cell cycle is divided into four major phases.

1. G1 phase
2. S phase
3. G2 phase
4. M phase

In cycling (replicating) somatic cells, cells synthesize RNAs and proteins during the G1 phase, preparing it for DNA synthesis. In proliferating cells, G1 is the period between "birth" of a cell following mitosis and the initiation of DNA synthesis, which marks the beginning of the S phase. At the end of the S phase, replicated chromosome consists of two daughter DNA molecules and associated chromosomal proteins. Each of the individual daughter DNA molecules and their associated

*Corresponding author.

Chromosomal proteins (not shown) is called a sister chromatid. The end of G2 is marked by the onset of mitosis, during which the mitotic spindle (red lines) forms and pulls apart sister chromatids, followed by division of the cytoplasm (cytokinesis) to yield two daughter cells. The G1, S, and G2 phases are collectively referred to as interphase. Most non proliferating cells in vertebrates leave the cell cycle in G1, entering the G0 state (Paul Nurse, 2000). After progressing through the G2 phase the cell has DNA content twice the diploid amount. The time fron G1-G2 is termed as Interphase, the chromatin is dispersed in the nucleus and is actively engaged in transcription.M is the phase in which the cell divides into two daughter cells, this phase is further divided into Prophase, Metaphase, Anaphase, Telophase and cytokinesis.. (Lodish *et al.,* 2000)

The maintenance of the eukaryotic genome requires precise and coordinated replication of entire genome each time cell division occurs. In eukaryotes, replication of DNA is initiated at hundreds to thousands of chromosomal elements called the origin of replication (Izumi *et al.,* 2000). Eukaryotic origins of replication direct the formation of a number of protein complexes leading to the assembly of two bidirectional DNA replication forks. These events are initiated by the formation of the pre-replicative complex (pre-RC) at origins of replication during G1. Pre-RC formation involves the ordered assembly of a number of replication factors including ORC, Cdc6p, Cdt1p, and Mcm2–7p (Wagaand Stillman, 1998). The regulation of pre-RC formation is a key element of the mechanisms coordinating DNA replication with the cell cycle. Pre Initiation is a phase in which S phase specific cyclinsCdk1 complex as well as Dfb4-Cdc7 kinase phosphorylate the initiation factors Sld2 and Sld3.Thseds factors form a tertiary complex with other initiation factors Dbp 11 in their phosphorylated state,and associate with the Origin of replication. Once formed, this complex awaits activation by at least two kinases that trigger the transition to DNA replication. As with the formation of the pre-RC, the transition to replication involves the ordered assembly of additional replication factors that facilitate unwinding of the DNA at the origin and culminate in the association of the multiple eukaryotic DNA polymerases with the unwound DNA (Zhou and Elledge, 2000).

DNA damage checkpoints were initially defined as non-essential regulatory pathways that control the ability of cells to arrest the cell cycle in response to DNA damage, allowing time for repair. However, recent evidence suggests that the historic definition of the checkpoint pathway is inadequate to explain the function of this pathway completely (Izumi *et al.,* 2000). In addition to controlling cell-cycle arrest, these pathways have been shown to control the activation of DNA repair pathways the composition of telomeric chromatin and the movement of DNA repair proteins to sites of DNA damage activation of transcriptional programmestelomere length and, in some cell types for reasons not fully understood, induction of cell death by apoptosis.Thus, it is now clear that the checkpoint comprises a subroutine integrated into the larger DNA damage response pathway that regulates a multifaceted response (Bartek and Lukas, 2001). In addition, several checkpoint genes are essential for cell and organism survival implying that these pathways are not only surveyors of occasional damage, but are firmly integrated components of cellular physiology (Lucas *et al.*, 2004).

MATERIALS AND METHODS

Cell Culture and Transfection

HeLa cells were grown at 37°C in DMEM supplemented with 10% fetal bovine serum. Cells were plated and when the confluency reached 30%, transfection of siRNA targeting endogenous genes was carried out using lipofectamine 2000. 40–80 nMsiRNA duplex were transfected on three consecutive days. Post third transfection drugs such as cyclohexamide, nocadazole, hydroxylurea treatment were given and the cells were harvested 12 hrs post drug treatment to evaluate the levels of proteins by immunoblotting.

Table 1: SDS-PAGE Gels Composition

	5%	*8%*	*10%*	*12.5%*	*15%*	*18%*
Sterile water	2.96	4.6	4	3.14	2.3	1.3
1.5 M TrispH (8.8)	1.25(0.5M Tris,pH6.8)	2.5	2.5	2.5	2.5	2.5
30% Acryl-bis mix	650 µl	2.7	3.3	4.17	5	6
10% SDS	50 µl	100 µl	100 µl	100 µl	100 µl	100 µl
10% APS	75 µl	100 µl	100 µl	75 µl	100 µl	100 µl
TEMED	8 µl	6 µl	8 µl	8 µl	8 µl	8 µl

Lysate Preparation

After 24 hour of last cyclohexane treatment the cells are harvested and collected in an eppendrof and centrifuged at 13200 rpm for 2 minutes at 4°C and the supernatant is removed. A dry spin is given at 13200 rpm for 1 minute and again the supernatant is removed. 1X Laemmli buffer is added to the pellet and the cells were disrupted using the disruptor for two minutes. The cells were then sonicated thrice for 7 seconds each using the ultrasonic processor. The sample is then denatured at 95°C for 3 minutes and then stored at –20°C.

SDS-PAGE

The casting apparatus was set using spacer plate and a thin plate. Separating gel was poured till ¾ of the total volume and water poured in the remaining space and left to polymerize Thenthe stacking gel was poured, following which comb was inserted and it was left for 10 minutes to get polymerized. The comb was lifted carefully, and the gel was ready for loading.In the electrophoresis apparatus prepared gels were tightly packed so that no leakage of buffer poured in the apparatus takes place. The apparatus was filled with 1X electrode buffer (gel running buffer). The apparatus was connected to the electrophoresis power supply and the input was maintained at 120 volt. The machine was run approximately for 3 hour. After completion of run the gel was transferred to the nitrocellulose membrane by using transferring apparatus.

Blotting and Developing Procedure

Once the transfer was over the blot was blocked with 5% milk in TBST for 1 hour. Then the primary antibody (usually 1:2000 in 1X TBST) was added and incubated for 6 hours or overnight at 4°C and then 1.5 hrs at room temp. Corresponding secondary antibody (usually 1:2000 in 1X TBST) was added and incubated for 1 hour. After incubation blot was placed on a developing sheet and 1 ml. ECL (1ml solution A and 5 µl solution B) was spread on it evenly. Finally the blot was developed on a photographic film in dark.

Immunoflourescence Assay

Immunofluorescence technique uses the specificity of antibodies to their antigen to target fluorescent dyes to specific biomolecule targets within a cell, and therefore allows visualization of the distribution of the targetmolecule through the sample. Confocal microscope is also widely used to analyze immunofluorescence samples. Secondary, or indirect, immunofluorescence uses two antibodies; the first (the primary antibody) recognizes the target molecule and binds to it, and the second (the secondary antibody), which carries the fluorophore, recognizes the primary antibody and binds to it.

RESULTS AND DISCUSSION

The candidate protein CIP was depleted using SiRNA strategy with GL2 as a negative control. HeLa cells were transfected with specific SiRNA duplexes on three consecutive days and the cell were harvested 24 hours post the last transfection.

Fig. 1: The Silencing of the Target Gene was Established by Immunoblotting

Fig. 2: The Depletion was Further Authenticated by Immunofluorescence

CHK1 phosphorylation at ser345 is mainly observed after UV insult or replication stress in cells which generate single strands. The apical kinase ATR phosphorylates CHK1 at various residues one of them being ser345 activating it as the effector kinase. Once active it further phosphorylates downstream effectors to bring about cell cycle block, which provides time for effective repair. Thus it can be inferred that the DNA damage machinery is activated on detecting the loss of CIP, which is a critical replication factor.

Fig. 3: It was Observed that Absence of CIP in Cells Caused Activation of Chk1 (ser345) that is a Marker for DNA Damage

By (Figure 4). The phosphorylation of chk1 is also establishesactivation of the ATR mediated pathway in response to CIP silencing.

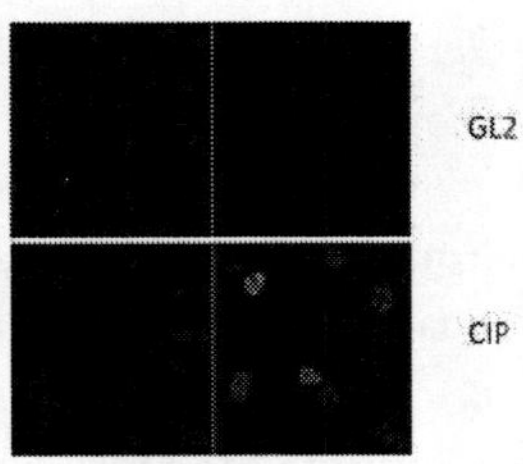

Fig. 4: Immunofluorescence Chk1 Foci was Detected in the CIP Depleted Samples as Compared to Gl2 Further Corroborating that Insufficient CIP in Mammalian Cells Invoke DNA Damage Pathway

As CIP is an important replication factor and its absence caused chk1 phosphorylation, its cell cycle profile was analyzed using flow cytometry. The propidium iodide cells revealed a G2/M block in CIP depleted sample (Figure 5). After any form of damage the cell cycle progression is delayed for the repair processes to effectively carry out their function. In absence of the cell cycle block the cells will continue with the faulty DNA accumulating genetic lesions that would ultimately cause genomic instability. Therefore, in the present study also the damage machinery employs a block to prevent genomic instability after perceiving the loss of CIP.

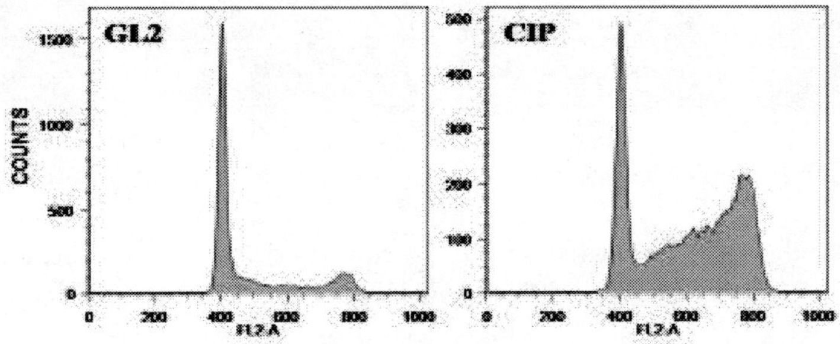

Fig. 5

So, in summation specific critical factors like CIP when absent in cells invoke ATR mediated response and employs a CHK1 mediated G2/M block. During the cell cycle phase chk1, CIP proteins are required for proper regulation, the proteins should be inhibited or degraded once their role in the replication is over. DNA Replication is well organized process which consists of well co-ordinate and regulated events that involves assembly and disassembly of particular proteins to the replication apparatus. Due to this reason replication takes place only once per cell cycle and prevent any re-replication.

REFERENCES

Bartek, J. and Lukas, J. (2001). *Curr. Opin. Cell Biol.*, 13: 738–747.

Izumi, M., Yanagi, K., Mizuno, T., Yokoi, M., Kawasaki, Y., Moon, K. Y., Hurwitz, J., Yatagai, F. and Hanaoka, F. (2000). *Nucleic Acids Res.*, 28: 4769–4777.

Lodish, H., Berk, A., Zipursky, S.L., Matsudaira, P., Baltimore, D. and Darnell, J. (2000). Molecular Cell Biology, 4th Ed., W.H. FreemanandCompany, NY, NewYork, ISBN 0-7167-31363.

Lucas, J., Lukas, C. and Bartek, J. (2004). *DNA Repair* 3: 997–1007.

Paul Nurse (2000). A Long Twentieth Century of the Cell Cycle and Beyond. *Cell* 100: 71–78.

Waga, S. and Stillman, B. (1998). *Annu. Rev. Biochem*, 67: 721–51

Zhou, B.B. and Elledge, S.J. (2000). *Nature*, 408: 433–39.

Molecular Diagnosis and Sequencing of Dengue Virus Based on *CprM* Gene

Arvind Rai, P. Anburajan* and G. Dharmaraj
Department of Biotechnology, Nandha Arts and Science College, Erode
E-mail: p.anburajan@gmail.com

INTRODUCTION

Dengue Virus (DENV) infection is now recognized as one of the most important mosquito-borne human infection of the current millennium. Global incidences of dengue has now increased enormously, and an estimated 100 million cases of dengue infections are now reported annually from more than 100 tropical and sub-tropical countries of the world (Gubler, 2006; WHO, 2002). Dengue is one of the most important re-emerging viral diseases, promulgated by increases in human population size and the expansion of global travel networks (Forattini, 2003, Gubler, 2002), and coupled with the global climatic changes that may alter the distribution of mosquito vector (Hales *et al.,* 2002, Jetten and Focks, 1997). The disease is caused by four antigenically distinct viruses designated as dengue virus type 1–4 (DENV 1–4). Dengue viruses belong to genus *Flavivirus* of family *Flaviviridae* (Lindenbach and Rice, 2003, Heinz *et al.,* 2000). Mature dengue virus consists of a single stranded ribonucleic acid genome (ssRNA) 10.7 kb in length, which has a positive polarity. Both the 5' and 3' ends contain noncoding (NC) regions. The genome encodes 10 proteins, 3 of which are structural (C, M, and E), and 7 of which are nonstructural proteins (NS1, NS2a, NS2b, NS3, NS4a, NS4b, and NS5). The M protein is synthesized as a precursor (prM) protein. The prM protein is processed to prM and M protein late in the virus maturation by a convertase enzyme (Furin). The genome is surrounded by an icosahedral nucleocapsid with a diameter of 30 nm. The nucleocapsid is covered by a lipid envelope of 10 nm thickness derived from host cell

*Corresponding author.

membranes and contains the envelope and membrane proteins (Cahour *et al.*, 1992, Biedrzycka *et al.*, 1987, Rice *et al.*, 1986). Although, all 4 dengue serotypes have been reported to circulate in the country (Dar *et al.*, 2006) but only DENV-2 and DENV-3 have been implicated in major DF/dengue hemorrhagic fever (DHF) outbreaks during the past decade (Singh *et al.*, 1999; Dash *et al.*, 2004; Dash *et al.*, 2005; Dash *et al.*, 2006). The resurgence of dengue infections with more virulence from newer areas in India is a cause of major public health concern (Parida *et al.*, 2002). For virus detection, virus isolation remains the "gold standard," although it is being rapidly replaced by the RT-PCR for rapid diagnosis (Shu and Huang *et al.*, 2004). Previous studies have indicated substantial genetic variation in the already existing serotype, to form distinct genotype, DENV-1 (3–5 genotype), DENV-2 (5–6 genotype), DENV-3 (4 genotype), DENV-4 (2 genotype); some having restricted geographical distribution, while others are more cosmopolitan, reflecting their dispersal across the tropical world (Rico-Hesse, 1990). Therefore, it is essential to carry-out molecular, phylogenetic and epidemiological studies that will enhance and update the knowledge about frequently circulating strains/variants of the virus in the area. This study will also reveal facts about the frequently circulating strains in the respective area, which shall provide important insights about the emergence and spread of the virus. The study may also prove to be helpful in tracing-back the route of transmission or importation of new strains. Knowledge about circulating/co-circulating strains in the area will be helpful in correlating with the usual/unusual clinical manifestations, severity and virulence. Information about newly emerging strains with altered antigenicity, genetic diversity, virulence and tissue tropism can be very useful for researchers involved in disease diagnostics and vaccine development. A molecular epidemiological study shall provide significant data about the prevalent strains in the area that would substantiate various improvisation strategies pertaining to diagnostics and disease control. For control and prevention of dengue virus infection, it is a prerequisite to specifically diagnose the infection early during the infection and outbreak. RT-PCR detection system provides near foolproof diagnosis compared to conventional and serologic methods. The PCR method which allows for multifold amplification of the viral RNA has been used to diagnose viral diseases rapidly (Deubel and Pierre, 1994; Deubel *et al.*, 1990; Garsen *et al.*, 1990; Henchel and Putnak, 1990; Ho-Terry *et al.*, 1990; Jansen *et al.*, 1990; Lifson *et al.*, 1990; Sumazaki *et al.*, 1989; Saiki *et al.*, 1988). Early diagnosis of dengue virus infection in patient sera allows the possibility of mounting a rapid response aimed at vector control in the affected areas for better management and disease containment.

MATERIALS AND METHOD

Sample Collection

The serum samples from suspected dengue patients were collected from department of Zoonosis, NCDC, Delhi.

Viral RNA Extraction

The RNA extraction was done by RNA extraction kit (Qiagen, Germany). The sample was first lysed under highly denaturing conditions to inactivate RNases and to ensure isolation of intact viral RNA. Buffering conditions were then adjusted to provide optimum binding of the RNA to the spin column membrane, and the sample was loaded onto the spin column. The RNA binds to the membrane, and contaminants are efficiently washed away in two steps using two different wash buffers. High-quality RNA was eluted in a special RNase-free buffer.

Viral RNA Extraction Protocol

Equilibrated the samples at room temperature (15–25°C). Equilibrated the Buffer AVE to room temperature for elution. Re dissolved precipitate in Buffer AVL/Carrier RNA by heating. All centrifugation steps were carried out at room temperature. 560 µ of prepared Buffer AVL containing Carrier RNA was pipette into a 1.5 ml microcentrifuge tube. Add 140 µl plasma, serum, urine, cell-culture supernatant, or cell-free body fluid was added to the Buffer AVL/Carrier RNA in the microcentrifuge tube. This was mixed by pulse-vortexing for 15 sec. Incubated at room temperature (15–25°C) for 10 min. The microcentrifuge tube was briefly centrifuged the 1.5 ml microcentrifuge tube to remove drops from the inside of the lid. 560 µl of ethanol (96–100%) was added to the sample, and mixed by pulse-vortexing for 15 sec. This was briefly centrifuged. Carefully 630 µl of the solution from above was applied to the spin column (in a 2-ml collection tube) without wetting the rim. Closed the cap, and centrifuged at 6000 x g (8000 rpm) for 1 min. The spin column was placed into a clean 2-ml collection tube, and discarded the tube containing the filtrate. The spin column was opened, and 500 µl of Buffer AW1 was added. Closed the cap, and centrifuged at 6000 x g (8000 rpm) for 1 min. The spin column was placed in a clean 2 ml collection tube, and discarded the tube containing the filtrate. The spin column was opened, and 500 µl of Buffer AW2 was added. Closed the cap and centrifuged at full speed (20,000 x g; 14,000 rpm) for 3 min. The spin column was placed in a clean 1.5-ml microcentrifuge tube. Discarded the old collection tube containing the filtrate. The spin column was opened and 60 µl of Buffer AVE was added and was equilibrated to room temperature. Closed the cap, and incubated at room temperature for 1 min. Centrifuge at 6000 x g (8000 rpm) for 1 min. The RNA was eluted in microcentrifuge tube.

Reverse Transcriptase PCR

RT-PCR (reverse transcription-polymerase chain reaction) technique was used for detection of dengue virus. Agpath ID™ one step RT PCR kit (Ambion) was used for this purpose. For detection of dengue virus CpreM gene (512 bp) was amplified. Two primers were used to carry out PCR OF Dengue virus genome

1. Forward Primer D1
   5'-TCA ATA TGC TGA AAC GCG CGA GAA ACC G-3'
2. Reverse Primer D2
   5'-TTG CAC CAA CAG TCA ATG TCT TCA GGT TC-3'

Loading of Samples in 3130 x 1 Genetic Analyzer

Each sample was resuspended in 12 µl of template suppression reagent (TSR), i.e. Hi-di formamide. The sample was vortexed to remove the drop from the lid.

Then, the sample was denatured at 96°C for 2 minutes and immediately chilled on ice for 15 minutes. The sample was placed on ice until ready to use. The sample was transferred to sequencing vails before loading.

BLAST (Basic Local Alignment Search Tool)

Basic Local Alignment Search Tool, or BLAST, is an algorithm for comparing primary biological sequence information, such as the amino acid sequences of different proteins or the nucleotides of DNA sequences. A BLAST search enables a researcher to compare a query sequence with a library or data base of sequences, and identify library sequences that resemble the query sequence above a certain threshold.

RESULTS AND DISCUSSION

Thirty acute phase serum samples were tested for infection of dengue virus by RT-PCR. Out of 20 samples, four samples were confirmed dengue positive by amplification of *CprM* gene region. The presence of *CprM* gene was confirmed by amplification of 511 bp product as visualized on 1.2% agarose gel.

Table 1: Result of Sequencing and Blast Analysis

Sample I.D.	*Serotype*	*Genotype*
IND/DEN/DEL/11/1	DENV-2	Genotype –IV
IND/DEN/DEL/11/9	DENV-1	Genotype-III
IND/DEN/DEL/11/14	DENV-1	Genotype-III
IND/DEN/DEL/11/15	DENV-1	Genotype-III

Blast analysis confirmed one sample as serotype 2 and genotype IV (sample I.D. *IND/DEN/DEL/11/1*). Remaining 3 samples were confirmed as serotype 1 and genotype III. The sequencing results are shown as follow:

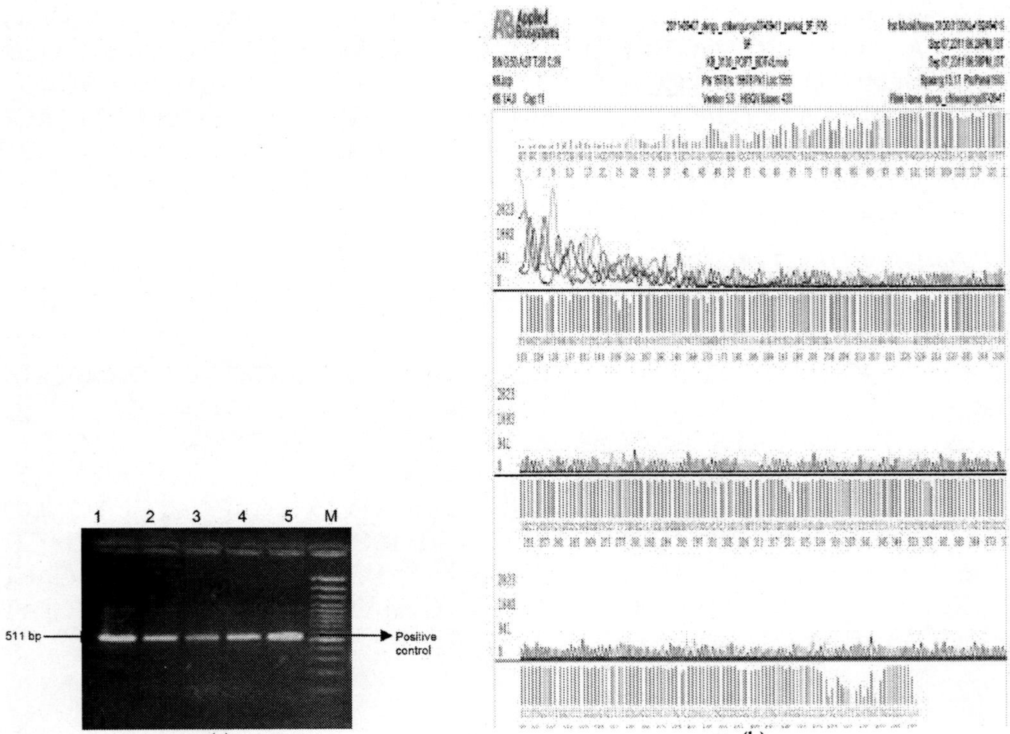

(a) (b)

Fig. 1: RT-PCR Result of *CprM gene region* (511 bp), lane M: 100 bp Ladder, Lane 1–4

IND/DEN/DEL/11/1

TGAAACGCGAGAGAAACCGCGTGTCGACTGTGCAACAGCTGACAAAGAGATT
CTCACTTGGAATGCTGCAGGGACGAGGACCGTTGAAACTGTTCATGGCCTTG
GTGGCATTCCTTCGTTTCCTAACAATCCCACCAACAGCAGGGATACTAAAAAG
ATGGGGAACGATCAAAAAGTCAAAAGCCATCAATGTCTTGAGAGGGTTCAGGA

AAGAGATTGGAAGGATGTTGAACATCTTGAATAGGAGACGCAGAACTGCAGG
CGTGATCATCATGCTAATTCCAACAGCGATGGCGTTCCATTTAACCACACGCA
ACGGAGAACCACACATGATCGTTAGCAGACAAGAGAAAGGGAAAAGTCTCTTG
TTCAAAACAGAGGATGGTGTGAACATGTGTACCCTCATGGCCATGGACCTTG
GTGAACT

(A) IND/DEN/DEL/11/1

>IND/DEN/DEL/11/9

ACTGGTTCACAGTTGGCGAAGAGATTCTCAAAAGGATTGCTCTCAGGCCAAG
GACCCATGAAAATGGTGATGGCTTTCATAGCATTCCTAAGATTTCTAGCCATA
CCCCCAACAGCAGGAATTTTGGCTAGATGGAGCTCATTCAAGAAGAATGGAGC
GATCAAAGTGTTACGGGGTTTCAAAAAAGAGATCTCAAGCATGCTGAACATAA
TGAACAGGAGGAAAAGATCCGTGACCATGCTCCTCATGCTGCTGCCCACAGC
CCTGGCGTTCCATTTGACCACACGAGGGGGAGAGCCACACATGATAGTTAGC
AAGCAGGAAAGAGGAAAGTCACTCCTGTTTAAAACCTCTGCAGGTGTCAACAT
GTGCACCCTCATTGCAATGGATTTGGGAGAGTTATGTGAGGATACAATGACC
TACAAATGCCCCCGGATCACTGA

(B) IND/DEN/DEL/11/9

>IND/DEN/DEL/11/14

GAGAAACCGCGTGTCAACTGGTTCACAGTTGGCGAAGAGATTCTCAAAAGGA
TTGCTCTCAGGCCAAGGACCCATGAAAATGGTGATGGCTTTCATAGCATTCC
TAAGATTTCTAGCCATACCCCCAACAGCAGGAATTTTGGCTAGATGGAGCTCA
TTCAAGAAGAATGGAGCGATCAAAGTGTTACGGGGTTTCAAAAAAGAGATCTC
AAGCATGCTGAACATAATGAACAGGAGGAAAAGATCCGTGACCATGCTCCTCA
TGCTGCTGCCCACAGCCCTGGCGTTCCATTTGACCACACGAGGGGGAGAGC
CACACATGATAGTTAGCAAGCAGGAAAGAGGAAAGTCACTCCTGTTTAAGACC
TCTGCAGGTGTCAACATGTGCACCCTCATTGCAATGGATTTGGGAGAGTTAT
GTGAGGATACAAT

(C) IND/DEN/DEL/11/14

>IND/DEN/DEL/11/15

TGCTCTCAGGCCAAGGACCCATGAAAATGGTGATGGCTTTCATAGCATTCCTA
AGATTTCTAGCCATACCCCCAACAGCAGGAATTTTGGCTAGATGGAGCTCATT
CAAGAAGAATGGAGCGATCAAAGTGTTACGGGGTTTCAAAAAAGAGATCTCAA
GCATGCTGAACATAATGAACAGGAGGAAAAGATCCGTGACCATGCTCCTCATG
CTGCTGCCCACAGCCCTGGCGTTCCATTTGACCACACGAGGGGGAGAGCCAC
ACATGATAGTTAGCAAGCAGGAAAGAGGAAAGTCACTCCTGTTTAAGACCTCT
GCAGGTGTCAACATGTGCACCCTCATTGCAATGGATTTGGGAGAGTTATGTG

AGGATACAATGACCTACAAATGCCCCCGGATCACTGAAGCGGAACCTGATGAC
ATTGACTGCTGGTG

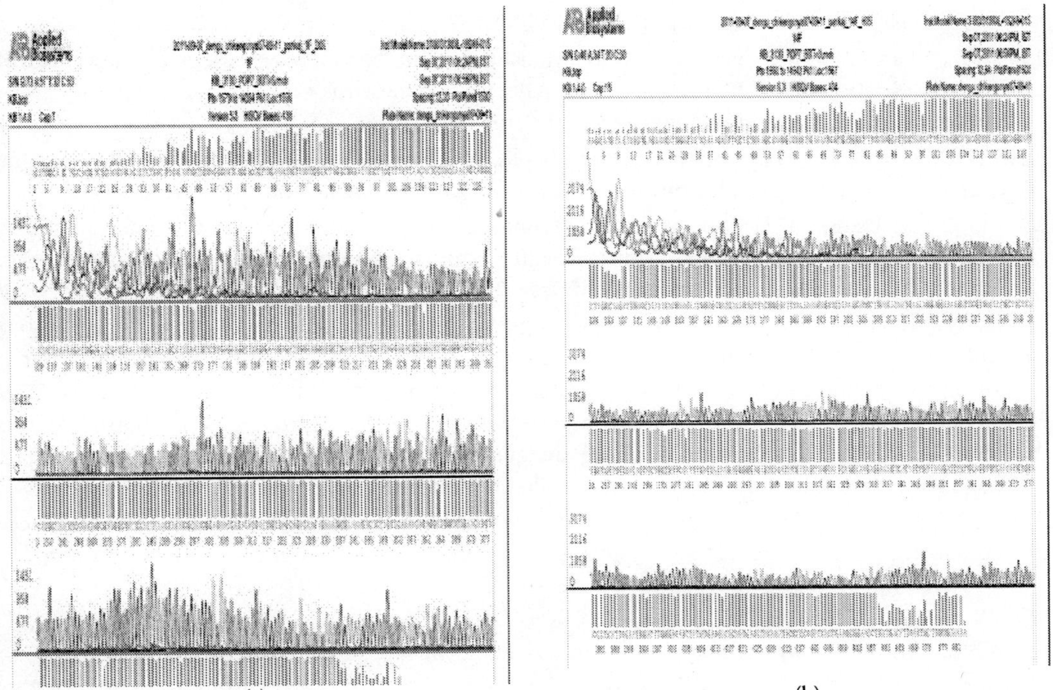

(a) (b)

Fig. 2

REFERENCES

Anzai, S., Fukuda, M., Otsuka, Y. and Eshita, Y. (2004). Nucleotide sequence and phylogenetic analyses of dengue type 2 virus isolated in the Dominican Republic. *Virus Genes.* 29(2): 219–227, *Arch. Virol.,* 1994; 134(1–2): 29–37.

Centers for Disease Control and Prevention (2010). Dengue Fever Symptoms. denguefeversymp toms.org. [Date visited 06/19/12 Available from: http://www.denguefeversymptoms.org

Central Ecology 8(1) Available: http://www.who.int/topics/dengue/en/[Date visited: 10/27/10]

Chambers, T.J., Hahnn, Galler, C. and Rice, C.M. (1990). Flavivirus genome organization expression and replication, *Annual Review of Microbiology,* Vol. (44) 649–688.

Chambers, T.J., Weir, R.C., Grakoui, A., McCourt, D.W., Bazan, J.F., Fletterick, R.J. and Rice, C.M. (1900c). Evidence that the N-terminal domain of nonstructural protein NS3 from yellow fever virus is a serine protease responsible for site-specific cleavages in the viral polyprotein. *Proc. Natl. Acad. Sci. USA.* 87(22): 8898–8902.

Chan, S.Y., Kautner, I.M. and Lam, S.K. (1994). The influence of antibody levels in dengue diagnosis by polymerase chain reaction. *J. Virol. Methods,* 49(3): 315–322.

Dar, L., Broor, S., Sengupta, S., Xess, I. and Seth, P. (1999). The first major outbreak of dengue hemorrhagic fever in Delhi, India. *Emerg. Infect. Dis.* 5(4): 589–90.

Dar, L., Gupta, E., Narang, P. and Broor, S. (2003). Co-circulation of dengue serotypes, Delhi, India. *Emerg. Infect. Dis.* 2006; 12(2): 352–353.

Dash, P.K., Parida, M.M., Saxena, P., Abhayankar, A., Singh, C.P., Tewari, K.N., Jana, A.M., Shekhar. K. and Rao, P.V. (2006). Remergence of dengue virus type 3 (genotypeIII) in India: Implication for increased incidence of DHF and DSS. *Virol. J.* 3: 55–64.

Dash, P.K., Saxena, P., Abhyankar, A., Bhargava, R. and Jana, A.M. (2005). Emergence of dengue virus type-3 in northern India. *Southeast Asian J. Trop. Med. Public Health,* 36(2): 370–377.

Deubel, V. and Pierre, V. (1994). Molecular techniques for rapid and more sensitive detection and diagnosis of flaviviruses. In: Rapid methods and automation in microbiology and immunology. Spencer RC, Wright EP, Newsom SWB, eds. Andover: *Intercept,* pp. 227–237.

Deubel, V., Kinney, R.M. and Trent, D.W. (1988). Nucleotide sequence and deduced amino acid sequence of the nonstructural proteins of dengue type 2 virus, Jamaica genotype: comparative analysis of the full-length genome. *Virology,* 165(1): 234–244.

Deubel, V., Laille, M., Hugnot, J.P., Chungue, E., Guesdon, J.L., Drouet, M.T., Bassot, S. and Chevrier, D. (1990). Identification of dengue sequences by genomic amplification: rapid diagnosis of dengue virus serotypes in peripheral blood. *J. Virol. Methods,* 30(1): 41–54.

Deubel, V., Nogueira, R.M., Drouet, M.T., Zeller, H., Reynes, J.M. and Ha DQ. (1993). Direct sequencing of genomic cDNA fragments amplified by the polymerase chain reaction for molecular epidemiology of dengue-2 viruses. *Arch. Virol.,* 129: 197–210.

Dietz, V., Gubler, D.J., Ortiz, S., Kuno, G., Casta-Velez, A., Sather, G.E., Gomez, I. and Vergne, E. (1996). In the 1986 dengue and dengue hemorrhagic fever epidemic in Puerto Rico: epidemiologic and clinical observations. *P.R. Health Sci J.* 15: 201–210 http://cmr.asm.org/cgi/external_ref?access_num=8994286&link_type=MED

Domingo, C., Palacios, G., Niedrig, M., Cabrerizo, M., Jabado, O., Reyes, N., Lipkin, W.I. and Tenorio, A. (2004). A new tool for the diagnosis and molecular surveillance of dengue infections in clinical samples. *Dengue Bull.,* 28: 87–95.

Gore, M.M. (2005). Need for constant monitoring of dengue infections. *Indian J. Med. Res.* 121(1): 9–12.

Grenfell, B.T., Pybus, O.G., Gog, J.R., Wood, J.L., Daly, J.M., Mumford, J.A. and Holmes, E.C. (2004). Unifying the epidemiological and evolutionary dynamics of pathogens. *Science,* 303(5656): 327–332.

Hill, A. Edward., Isolation of Dengue virus from Human Being in Trinidad. Virus disease in the West Indies. *The Caribbean MEDICAL Journal 1965;* Vol. 207, Nos. 1–4, pp. 83–85.

Palacios, G., Jabado, O., Reyes, N., Niedrig, M. and Gasco´n, J. (2006). Use of a short fragment of the C-terminal E gene for detection and characterization of two new lineages of dengue virus 1 in India. *J. Clin. Microbiol.,* 44: 1519–1529.

World Health Organization, 1966, Mosquito-borne hemorrhagic fevers of Southeast Asia and western Pacific, *Bull WHO,* 35: 17–30.

World Health Organization. Dengue hemorrhagic fever: diagnosis, treatment and control. Geneva, Switerland.

Molecular Characterization of Bestrophin Protein Targeting Vitelliform Macular Dystrophy

T. Raghuram*, V. Ranjith Kumar and I. Muthiah
Department of Biotechnology, K.S. Rangasamy College of Technology, Tiruchengode, Tamil Nadu
E-mail: muthiah@ksrct.ac.in

INTRODUCTION

Vitelliform macular dystrophy is an autosomal dominant disorder mostly present in juveniles. This disease is commonly known as 'Best disease'. The first pedigree was described by Dr. Franz Best, a German ophthalmologist in 1905. The onset of the disease is macular degeneration; macular region is act as a natural sun light blocker in eyes (i.e. blocking the entry of UV light into the retinal layer). It is characterized by the accumulation of lipofuscin-like material within and beneath the retinal pigment epithelium, which results in an egg-yolk-like appearance of the macula, this finally leads to the loss of central vision. The appearance of the disease is a yellow or orange yolk like lesion in the macula. It is a bilateral macular dystrophy with autosomal dominant inheritance characterized by the sub retinal accumulation of egg-yellowish material in the macular area. The disease is slowly progressive and eventually results in atrophy of the Retinal Pigment Epithelium (RPE) and photoreceptors, severely impairing central vision (Stone *et al.,* 1992). RPE is the pigmented layer of retina or retinal pigment epithelium (RPE) is the pigmented cell layer just outside the neurosensory retina that nourishes retinal visual cells, and is firmly attached to the underlying choroid and overlying retinal visual cells. Over time, the abnormal accumulation of this substance can damage cells that are critical for clear central vision. As a result, people with this disorder often lose their central vision and may

*Corresponding author.

experience blurry or distorted vision. Vitelliform macular dystrophy does not affect side (peripheral) vision or the ability to see at night. The diagnosis can be made in many affected individuals on the basis of the classic appearance of the retinal lesions. In patients without classic lesions, the diagnosis can be made by electro-oculography (EOG) (Gallemore *et al.,* 1962). There are two forms of Vitelliform macular dystrophy, one is early onset form (Best disease) and another one is adult onset form. The responsible mutation in Best disease is VMD2, or "bestrophin" gene, located on chromosome 11, which encodes a transmembrane chloride channel localized in basolateral membrane of the RPE. The protein bestrophin is encoded by gene *BEST I (anion channels).* Bestrophin protein is predicted to have a 585 amino acid and weight of 68 kDa. Bestrophins are transmembrane (TM) proteins that share a homology region containing a high content of aromatic residues, including an invariant arg-phe-pro (RFP) motif. Bestrophins are believed to function as chloride channels. moreover the challenges of the project is the exact mechanism of the dysfunction of the bestrophin protein have not yet been studied with certainty, It remains unclear how this defect is related to the build-up of lipofuscin and a progressive loss of central vision in people with Vitelliform Macular Dystrophy.

The main objective is to demonstrate the mode of function of bestrophin protein and to illustrate the model of interaction and the structural subunits responsible for the Vitelliform Macular Dystrophy, and also to identify and characterize the novel biological compounds for fighting against the best disease. The molecular dynamics of protein are also taken. To demonstrate the mode of interaction of biological compound with the bestrophin using bioinformatics tools.

MATERIALS AND METHODS

Sequence Allignment and Modelling

The sequence information of the Bestrophin protein was retrieved from the SWISS-PROT database. The accession id of the sequence is – O76090. The structure, molecular weight and formula are obtained from 'PubChem' which is a database of small chemical molecules.

MODELLER9v7 is used for homology or comparative modeling of protein three-dimensional structures, and ab-*initio* modeling is done by using I-Tasser server.

Structure Validaton

The model obtained was analyzed by using 'Ramachandran plot' The Ramachandran plot displays psi and phi backbone conformational angles for each residue in a protein.

The distance between two succession alpha carbon atoms in the backbone chain of a protein is approximately constant, as are the angles between the two bonds of such atoms. if the determination of protein structure is reliable, most pairs will be in the favored regions of the plot and only a few will be in disallowed regions.

Energy Minimization – GROMACS

The receptor (Bestrophin protein) is solvated in water and the same is then energy minimized. The native conformation of a protein is assumed to be its stable conformation with the least energy. Hence, it is usually referred to as the global minimum energy conformation. Here, we desire to model the protein in its natural state and thus, warrant energy minimization. GROMACS is a parallel molecular dynamics code designed for high-performance simulation of large bio molecular systems. GROMACS is an engine to perform molecular dynamics simulations and energy minimization.

Lead Optimization Study

The objective of this drug discovery phase is to optimize lead compounds i.e. new analogs with improved potency, reduced off-target activities, and physiochemical/ metabolic properties suggestive of reasonable in vivo pharmacokinetics. Lipinski's Rule of Five is a rule of thumb important for drug development where a pharmacologically active lead structure is optimized step-wise for increased activity and selectivity. The Lipinski's 5 rules are Molecular mass less than 500 Dalton, High lipophilicity (expressed as LogP less than 5), Less than 5 hydrogen bond donors, Less than 10 hydrogen bond acceptors, and Molar refractivity should be between 40–130.

Molecular Docking

PyRx was used for docking. PyRx is a suite of automated docking tool. It is designed to predict how small molecules such as substrate or drug bind candidates bind to the receptor of known 3D structure.

RESULTS AND DISCUSSIONS

Modelling: Homology Modelling of Bestrophin Protein

The three dimensional structure of bestrophin protein was generated by homology modeling method using three different templates. The generated theoretical structure was validated using Ramachandran plot.

Model 1: Template 1CIY with 28% Identity

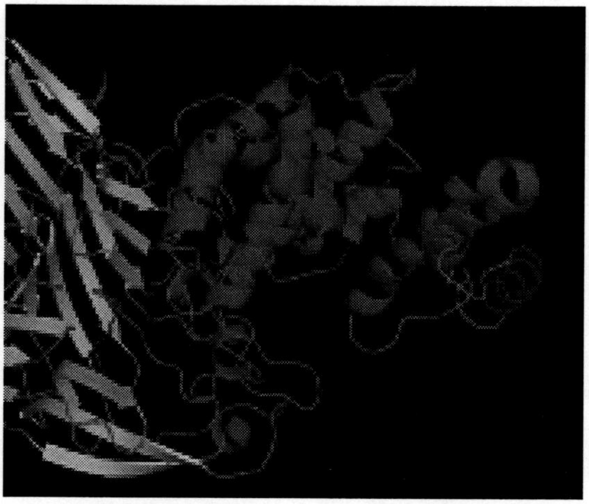

Fig. 1: Homology Modelled Protein (Model)

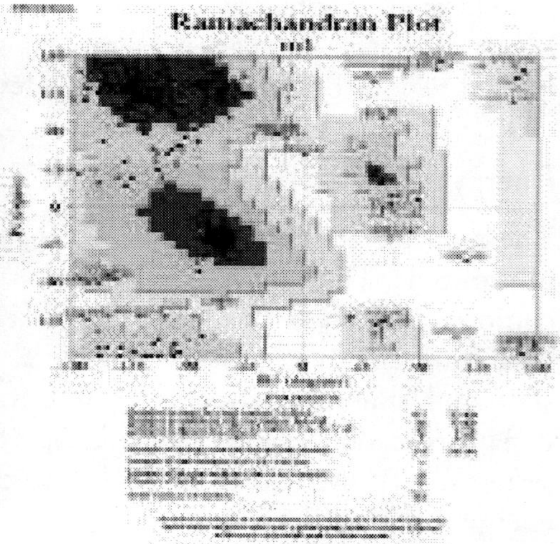

Fig. 2: Ramachandran Plot for Bestrophin Bestrophin Protein (Model 1)

Model 2: Template 2WCO with 26% Identity

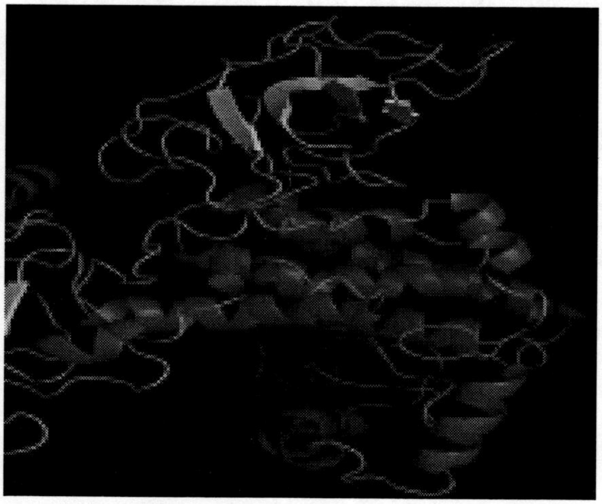

Fig. 3: Homology Modeled Bestrophin Protein (Model 2)

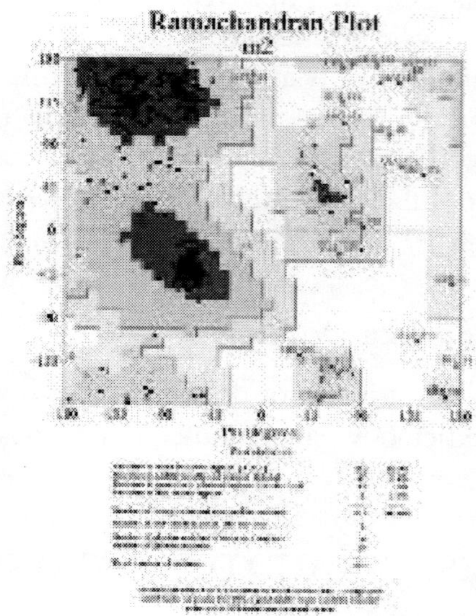

Fig. 4: Ramachandran Plot for Bestrophin Protein (Model 2)

Model 3: Template 3AO0 with 28% Identity

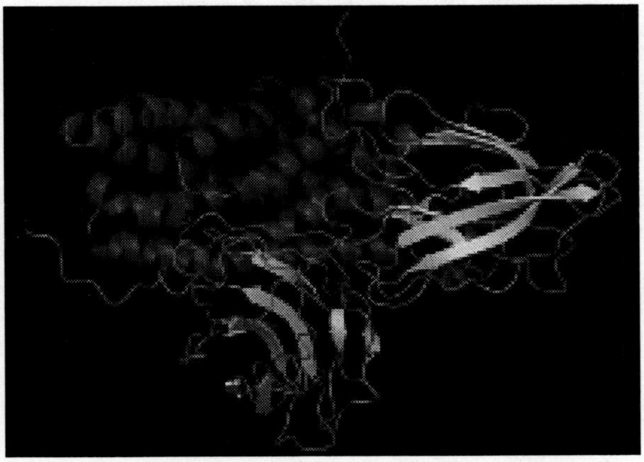

Fig. 5: Homology Modeled Bestrophin Protein (Model 3)

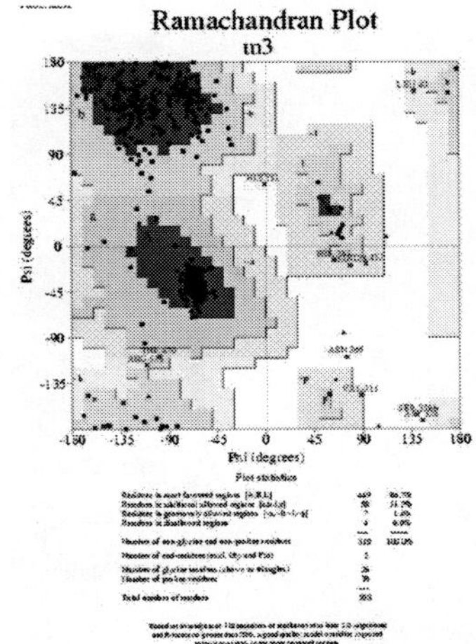

Fig. 6: Ramachandran Plot for Bestrophin Protein (Model 3)

Ab initio Modeling of Bestrophin Protein

Since the% identity of the template obtained for homology modeling is very low the three dimensional structure of the bestrophin protein was determined by *Ab initio* method using I-Tasser server.

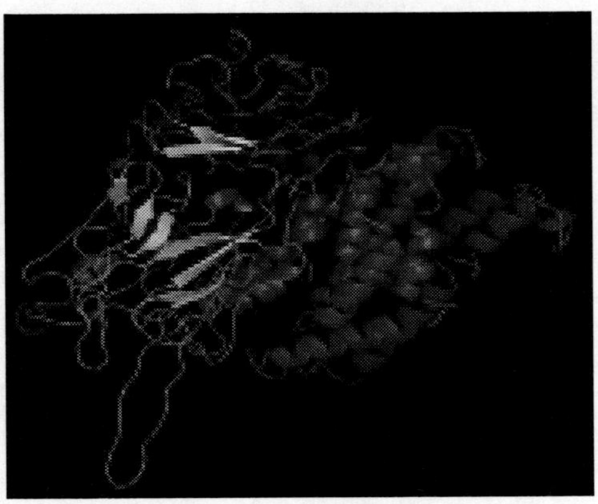

Fig. 7: *Ab Initio* Modeled Bestrophin Protein

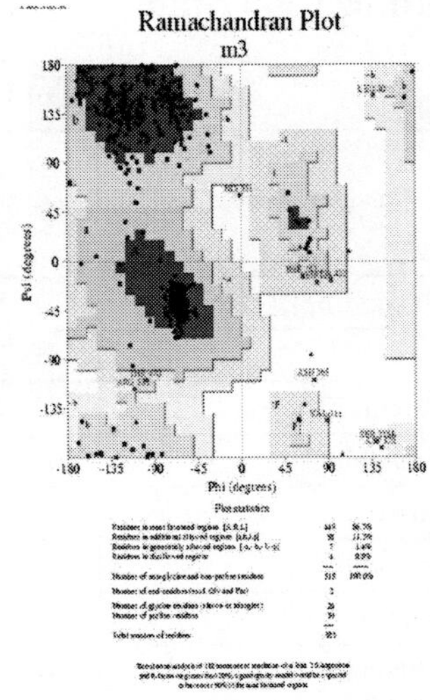

Fig. 8: Ramachandran Plot for *Ab Initio* Modeled Bestrophin Protein

Molecular Dynamics of the Bestrophin Protein

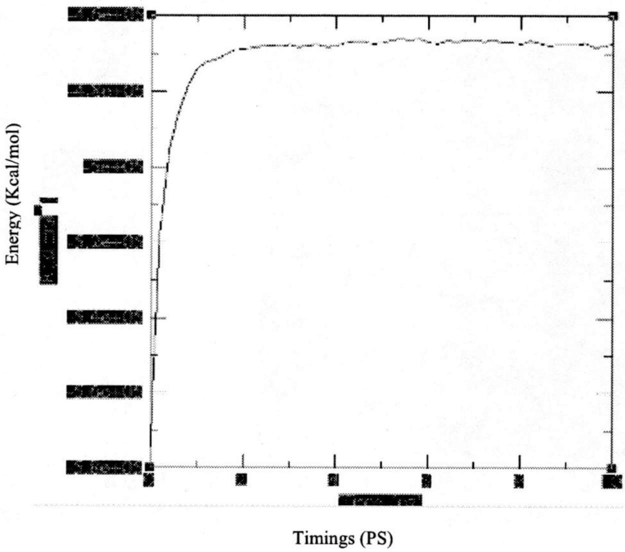

Fig. 9: Energy Graph of Bestrophin Protein

Total Energy of the bestrophin protein after energy minimization: –4.8107e+06.

Lead Optimization Study

Table 1: List of Compounds Satisfying Lipinski Rule

S. No.	*Drug Name*	*Accession No*	*Lipinski Rule of Five*				
			Log P	*H Donor*	*H Acce*	*Mol Wt*	*Mol Ref*
1.	Astaxanthin	CID 5281224	4.029	0	2	596.83	86.146
2.	Carnitnie	CID 288	–2.642	4	7	161.19	54.247
3.	Curcumin	CID 969516	3.854	1	8	368.37	119.387
4.	Dronabinol	CID 16078	0.672	2	7	314.46	112.435
5.	Lutein	CID 6433159	0.091	0	6	568.87	62.808
6.	Nipradilol	CID 72006	–2.307	3	7	326.34	83.247
7.	Quercetin	CID 5280343	3.438	0	0	302.23	56.421
8.	Tetrahydrocannabinol (THC)	CID 2977	2.459	1	6	314.46	106.694
9.	Zeaxanthin	CID 5280899	3.065	2	4	568.87	79.337

Energy Minimization of Lead Compounds

The physical, chemical and biological properties of a molecule depend upon the conformations adopted by molecules. Hence the compounds which had satisfied the Lipinski rule of 5 were energy minimized and conformed to stable conformation with the least energy. The energies of the compounds at its stable conformation are as follows:

Table 2: Energy Minimization of Natural Compounds

S. No.	*Drug Name*	*Accesion Number*	*Energy (Kcal/mol)*
1.	Astaxanthin	CID 5281224	–30608.10
2.	Carnitnie	CID 288	–38890.20
3.	Curcumin	CID 969516	–77437.02
4.	Dronabinol	CID 16078	–47033.77
5.	Lutein	CID 6433159	–10269.19
6.	Nipradilol	CID 72006	–67938.56
7.	Quercetin	CID 5280343	–83936.28
8.	Tetrahydrocannabinol (THC)	CID 2977	–48814.11
9.	Zeaxanthin	CID 5280899	–10423.05

Molecular Docking

Mapping the binding patterns of bestrophin protein (Receptor) with the various natural compounds which has activity against best disease using molecular docking.

Fig. 10. Docking of Astaxanthin – Bestrophin

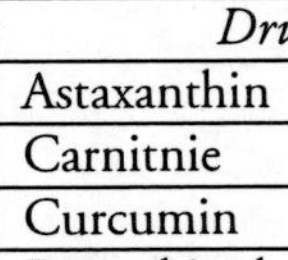

Fig. 11. Docking of Carnitnie – Bestrophin

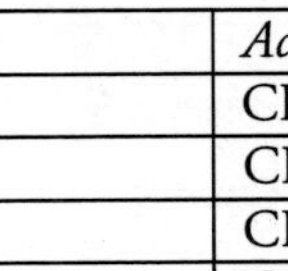

Fig. 12. Docking of Curcumin-Bestrophin

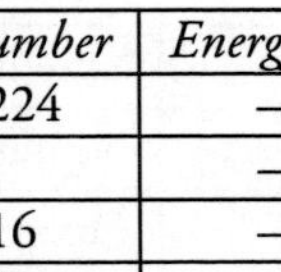

Fig. 13. Docking of Dronabinol – Bestrophin

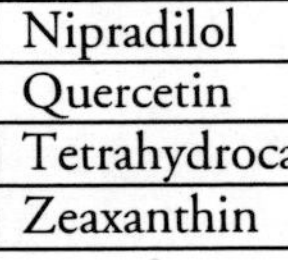

Fig . 14. Docking of Lutein – Bestrophin

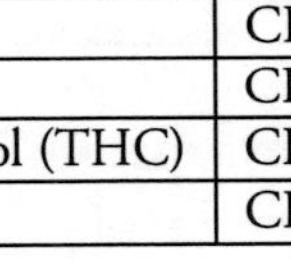

Fig . 15. Docking of Nipradilol - Bestrophin

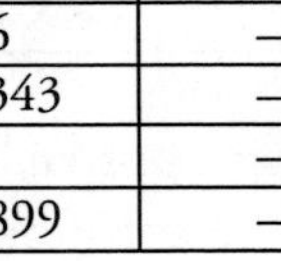

Fig. 16. Docking of Quercetin – Bestrophin

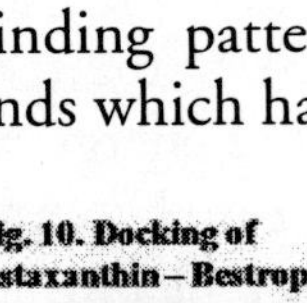

Fig. 17.Docking of THC – Bestrophin

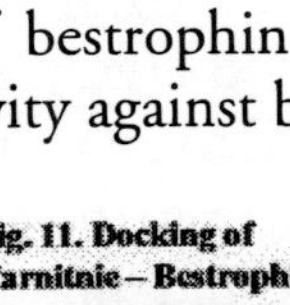

Fig . 18. Docking of Zeaxanthin – Bestrophin

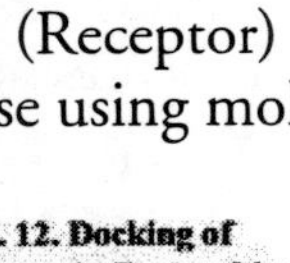

Table 3: Binding Energy of Lead Compounds with the Bestrophin Protein

S. No.	*Natural Compound*	*Binding Energy (Kcal/mol)*
1.	Astaxanthin	–7.34
2.	Carnitnie	–4.12
3.	Curcumin	–4.06
4.	Dronabinol	–3.78
5.	Lutein	–6.26
6.	Nipradilol	–5.65
7.	Quercetin	–5.12
8.	Tetrahydrocannabinol (THC)	–6.98
9.	Zeaxanthin	–7.76

One of the promising approaches in best disease drug discovery is the development of new inhibitor for bestrophin protein involved in the best disease. A homology model of bestrophin protein from human was built using homology modeling and *ab initio* which was validated through Ramachandran plot and RMSD value, to arrive at a reliable model for structure based drug design. Validations programs showed that the homology model scores are similar to crystal structure of the template. The 3D structures of bestrophin protein were solvated at 310 K and energy minimized using Gromacs. Docking the bestrophin protein with various natural compounds provided an insight into the nature of binding and interaction of ligands to the receptor. Docking studies also revealed that all the small molecule compounds are binding to the same receptor. Some differences in the residues lining extracellular region of the receptor were observed. The overall similarity of the binding pockets of these small molecule compounds and the bestrophin protein offers a promising opportunity to elucidate the mode of alteration of bestrophin protein. The results reported here serve to better define the parts of the bestrophin protein molecule that are essential for fighting against best disease. QSAR studies will be carried out to design an effective chemical compound from the active principles.

REFERENCES

Arden, G.B., Barrada, A. and Kelsey, J.H. (1962). New clinical test of retinal function based upon the standing potential of the eye. *Br. J. Ophthalmol.,* 46: 449–467.

Gallemore, R.P., Hughes, B.A. and Miller, S.S. (1998). Light-induced responses of the retinal pigment epithelium". In *The Retinal Pigment Epithelium.* M.F. Marmor and T.J. Wolfensberger, editors. Oxford University Press, Oxford. 175–198.

Multivariate Data Analysis for Biotechnology and Bio-processing

R. Bhanu Prakash*

CAMO Software India Pvt. Ltd., Domlur, Bangalore

E-mail: bhanup@camo.com

INTRODUCTION

The modern biopharmaceutical/biotechnology manufacturing facility contains many sophisticated control, data logging and data archiving systems. Massive amounts of data are collected from sources such as raw materials analysis, process outputs and final quality assessments, which are stored in data warehouses.

The sheer volume of data contained in these warehouses makes it a near impossible task to extract the information using simple charting and univariate methods of analysis. Such complex data requires methods of analysis that can cope with multiple variables simultaneously that not only reveal influential variables, but also reveal the relationship such variables have with each other. This is where Multivariate Analysis (MVA) is finding a much greater role in the analysis of complex bioprocess data.

With much more effort being put into the discovery and development of biotherapies and personalised medicines, biopharmaceutical and biotechnology companies are looking for ways to accelerate drug discovery, and through initiatives such as Quality by Design (QbD) and Data Driven Knowledge Discovery, reduce the regulatory approval time and be first to market. This means that data collected throughout the entire product lifecycle must be analysed and interpreted in order to gain extensive product and process understanding. This, in turn, leads to improved quality, greater confidence in the market for a company's products and ultimately market capitalisation.

*Corresponding author.

It is estimated that it takes approximately 12 years to bring a new drug or therapy to market. This usually involves three phases: 1. Discovery, 2. Clinical Trials, and 3. Registration.

Coupled with these phases is the development of a suitable manufacturing process that can consistently produce the highest quality product. This includes the development of a formulation that is robust under processing conditions, scale up considerations and technology transfer from facility to facility or even between different types of manufacturing equipment. Each of these phases can be improved and accelerated through the use of MVA and Design of Experiments (DoE).

Even before data is analysed, one of the biggest challenges facing the industry is getting this data into a format that is amenable to MVA. Many data collection and agglomeration systems are commercially available for compiling various forms of data and these can be seamlessly integrated into MVA packages such as The Unscrambler® X so that the vast array of graphical and analytical approaches can be applied to reveal the information it contains.

Multivariate Analysis in the Complete Product Lifecycle

Unlike small molecule drug product development, biotherapies are fundamentally more complex in terms of structure and application and suffer greatly from natural biological variability. For example, isolating and selecting cell cultures or bacterial strains to further develop into future products is aided greatly by the tools of MVA, including the monitoring of the processes (e.g. fermentation reactions) used to produce them. From there, the tools of DoE can be used to devise formulations that stabilise the active component(s) during manufacture and are also useful in product scale up studies.

Once the candidate therapy (cell cultures, antibody, virus strain, etc.) has been formulated into a stable matrix, MVA can be used to assist in the interpretation of clinical trial data and can even lead to accelerating the lengthy process through a much more comprehensive and holistic approach to data analysis, especially when combined with the principles of adaptive designs and the Critical Path Initiative endorsed by the US Food and Drug Administration (USFDA).

When the candidate therapy has been approved for market release, the tools of MVA are useful for assessing the success of technology transfer from R&D to production, or from one manufacturing facility to another. In the production environment, MVA is useful for assessing incoming or internally produced raw material quality and characteristics.

Combined with rapid spectroscopic or other characterisation methods, control strategies for the real time monitoring and adjustment of processes within the so-called 'design space' can be devised so that proactive quality control can be realised. DoE and MVA are then used in developing robust analytical methods for stability studies and other post production analyses.

Data collected over time from a manufacturing facility can be modelled to assess consistency from batch to batch and facilitate continuous improvement (CI) and preventive maintenance and corrective action (CAPA) programs. The entire process is summarised in Figure 1.

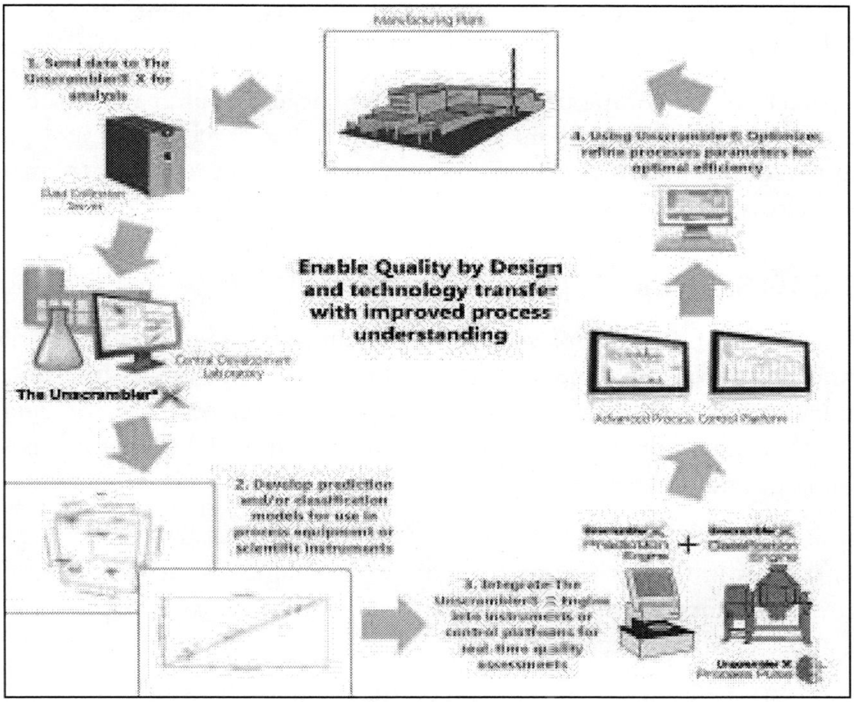

Fig. 1

Overview of How MVA and DoE Software can be Used Across Different Operational Areas

Multivariate analysis software can be seamlessly integrated into different areas of a biotech company to provide better understanding, process and quality controlThe scalability and flexibility of today's software systems allows powerful analytical tools to be integrated into existing systems such as MES/ERP systems, process equipment and scientific instruments e.g. spectrometers. This enables faster, more informed decision making from the laboratory to the shopfloor and quality department.

CONCLUSION

Multivariate data analysis and Design of Experiments are powerful tools ideally suited for understanding the complex behaviour and relationships in biological systems.

These methods can be used across the full biotech product lifecycle, from discovery and development, to scale up, production and quality control. Today's leading MVA and DoE solutions can be seamlessly integrated with other systems including process equipment, laboratory and spectroscopy instruments, enabling faster and more informed decision making.

Leading biotechnology companies which implement and exploit the power of MVA and DoE can realize substantial benefits including lower development and production costs, improved product quality, technology transfer, faster time to market and ultimately increased business value.

About CAMO Software

Founded in 1984, CAMO Software is a recognized leader in Multivariate Data Analysis and Design of Experiments software. Today, we have offices in Norway, USA, Japan, India and Australia. Multivariate analysis is a powerful set of data mining techniques that help identify patterns and understand the relationships between variables in large and complex data sets.

Our software is used by many of the world's leading companies, universities and research institutes in the life sciences, food and beverage, agriculture, energy, oil and gas, mining and metals, industrial manufacturing, pulp and paper, automotive, aerospace and technology sectors.

The Unscrambler® X software range is used by engineers, scientists and data analysts because of its ease of use, world-leading analytical tools and data visualization. Our solutions are used by more than 25,000 people in 3,000 organizations to analyze data, monitor process or equipment performance and build better predictive models. This gives them valuable insights to make more informed decisions, improve market segmentation, research and development, manufacturing processes and product quality.

REFERENCE

Please visit www.camo.com to know more about our products and services.

Structure Based Design of Potent PPARγ Binding Agent for Anti-Diabetic Effect

P. Vinothini, G. Ranjeetha, R. Snegabala and I. Muthiah*

Department of Biotechnology, K.S. Rangasamy College of Technology, Tiruchengode, Tamil Nadu

E-mail: muthiah@ksrct.ac.in

INTRODUCTION

Diabetes mellitus often simply referred to as diabetes—is a chronic disease caused by inherited or acquired deficiency in production of insulin by the pancreas or by the ineffectiveness of the insulin produced. The management of the blood glucose level is a critical strategy in the control of diabetes complications. Such a deficiency results in the increased blood glucose levels, which in turn can damage the body systems such as blood vessels, kidneys, and nerves. Diabetes is currently increasing in epidemic proportions throughout the world (Kunishima *et al.,* 2000). The metabolic disease was characterized by the high levels of blood glucose level and they cause several health problems.

CLASSIFICATION

There are two major classification of diabetes. The early classifications are genetically based disorder and diabetes caused due to dietary in discretion. Based on the etiology diabetes has been classified into type 1 diabetes or insulin dependent diabetes mellitus (IDDM) and type 2 diabetes or non-insulin dependent diabetes mellitus (NIDDM) (Chandra *et al.,* 2007).

---

*Corresponding author.

Type 1 Diabetes

Type 1 Diabetes (IDDM) is also called as immune mediated diabetes, juvenile onset diabetes or idiopathic diabetes which is caused b the cellular mediated auto immune destruction of bets cells of the pancreas. In this condition insulin is required for the survival to prevent the development of ketoacidosis, coma and eventually death. The presence of markers likes anti-glutamic acid decarboxylase, or insulin antibodies characterize the type 1 diabetes. Type 1 diabetes occurs in young age with acute onset, but it may also occur in any age, sometimes with slow progression (Kyvik *et al.*, 2004).

Type 2 Diabetes

Type 2 diabetes (NIDDM) is caused due to relative insulin deficiency. It is most common form of diabetes and is characterized by disorders of insulin action, insulin secretion, either of which may be the predominant feature. Decreased insulin secretion and decreased insulin resistance are involved in pathogenesis of type 2 diabetes. This form of Diabetes frequently goes for many years because hyperglycemia develops gradually and in earlier stages (Scheen *et al.*, 2003).

Table 1: Major Classification of Diabetes

S. No.	*Classification*	*Reasons*
1.	Type 1 Diabetes	Destruction of beta cells
2.	Type 2 Diabetes	Decreased insulin secretion or insulin resistance

PPAR RECEPTOR

Peroxisome proliferator-activated receptors (PPARs) are nuclear hormone receptors that are activated by fatty acids and their derivatives. PPAR has three subtypes (PPAR alpha, beta/delta, and gamma) showing different expression patterns in vertebrates. PPAR alpha plays a role in the clearance of circulating or cellular lipids via the regulation of gene expression involved in lipid metabolism in liver and skeletal muscle. PPAR beta/delta is involved in lipid oxidation and cell proliferation. PPAR gamma promotes adipocyte differentiation to enhance blood glucose uptake (Talpur *et al.*, 2002).

PPAR-GAMMA MODULATORS

PPAR gamma is the main target of the drug class of thiazolidinediones, used in diabetes mellitus and other diseases that feature insulin resistance. The PPAR-gamma receptor agonists, as a relatively new and perhaps still not very widely used class of antidiabetic agent in the Caribbean and particularly the Trinidadian context, possess pharmacologic properties that certainly have be shown to have impact on many of the inflammatory, metabolic, biochemical and structural macro vascular aberrations that occur in the type 2 diabetic. Activation of PPAR (gamma) nuclear receptors regulates the transcription of insulin-responsive genes involved in the control of glucose production, transport, and utilization (Ritchie *et al.,* 2000).

Modern Drugs

Several modern drugs are used for the treatment of diabetes. Drugs such as Sulfonylureas, tolazamide, biguanides, phenformin and metformin are used for the treatment of diabetes. Use of such modern drugs will lead to certain side effects. They are:

- Gastrointestinal side effects
- Bloating
- Abdominal discomforts
- Skin rashes
- Thrombocytopenia
- Anemia

Artificial Sweeteners

An Artificial sweetener or Sugar substitute is a chemical. It is added to some foods to make them taste sweet. It is made to taste like sugar or corn syrup. Usually it has a much lower energy value than sugar. It can be found in many products which say *Light* or *Diet* in their name. People with diabetes have problems regulating their sugar levels in the blood. Artificial sweeteners are better for them than sugar (Staszewski *et al.*, 2002).

Problems

Studies done with animals have shown that artificial sweeteners cause weight gain. A sweet taste causes insulin to be produced. This will store some of the sugar in the blood in tissues, including fat. With artificial sweeteners, the amount of sugar in the blood does not increase. This means that there will be too little sugar in the blood. This is known as hypoglycemia. After a while, rats given sweeteners have steadily increased the amount of calories. This increased body weight, and adiposity (fatness). The natural response to eating sugary foods is to eat less at the next meal and to use some of the extra energy to warm the body after the meal (Lisa Devitt *et al.,* 2004).

Sweet Proteins

The prevalence of obesity and diabetes has increased dramatically in recent years in the United States, with similar patterns seen in several other countries including India as well. Artificial sweeteners like Saccharin, Aspartame, Cyclamate and Acesulfame are used world-wide as low calorie sweeteners by patients affected by diseases linked to the consumption of sugar, e.g. diabetes, hyperlipemia, caries, obesity etc. but they have side effects such as psychological problems, mental disorders, bladder cancer, heart failure and brain tumors. Sweet proteins have the potential to replace these artificial sweeteners, by acting as natural, good, low calorie sweeteners, as we know that proteins do not trigger a demand for insulin in these patients whereas sucrose does. (Faus *et al.,* 2000) The sweet proteins are:

- Brazzein – Pentadiplandra brazzeana
- Monellin – Dioscreophyllum cumminsii
- Miraculin – Synsepalum dulcificum
- Curculin – Curculigo latifolia
- Thaumatin – Thaumatococcus danielli
- Mabinlin – Capparis masaikai

India leads the world in the number of patients with diabetes, with an estimated 19.4 million individuals affected by the disease. The number is expected to increase to 57.2 million by 2025.1 Before the introduction of the therapeutic use of insulin, diet was the main form of treatment for the disease and included the use of traditional medicines mainly derived from plants.2 Ancient Indian physicians such as Sushruta and Charaka knew about the importance of diet and exercise in the control of diabetes. However, not much is known about the origin or prevalence of alternative forms of therapy for this disease during ancient times (Kyvik *et al.*, 2004).

CHALLENGES AND MERITS OF THE PROJECT

- Sweet proteins are good, low caloric sweetener which is derived from natural compounds such as fruits.
- All the six sweet proteins reported in the literature are drastically differing in their primary and tertiary structures from each other.
- The key residues of sweet proteins responsible for their biological activities have not yet been identified.

INITIAL DATA SETS

Fasta Sequence of PPAR Gamma (1i7i) ID

>1I7I

ESADLRALAKHLYDSYIKSFPLTKAKARAILTGKTTDKSPFVIYDMNSLMMGEDKIK
EVAIRIFQGCQFRSVEAVQEITEYAKSIPGFVNLDLNDQVTLLKYGVHEIIYTMLASL
MNKDGVLISEGQGFMTREFLKSLRKPFGDFMEPKFEFAVKFNALELDDSDLAIFIAV
IILSGDRPGLLNVKPIEDIQDNLLQALELQLKLNHPESSQLFAKLLQKMTDLRQIVT
EHVQLLQVIKKTETDMSLHPLLQEIYKDLESADLRALAKHLYDSYIKSFPLTKAKAR
AILTGKTTDKSPFVIYDMNSLMMGEDKIKFKHIKEVAIRIFQGCQFRSVEAVQEITE
YAKSIPGFVNLDLNDQVTLLKYGVHEIIYTMLASLMNKDGVLISEGQGFMTREFLK
SLRKPFGDFMEPKFEFAVKFNALELDDSDLAIFIAVIILSGDRPGLLNVKPIEDIQDNL
LQALELQLKLNHPESSQLFAKLLQKMTDLRQIVTEHVQLLQVIKKTETSLHPLLQEIY

Structure of PPAR gamma (1i7i) ID

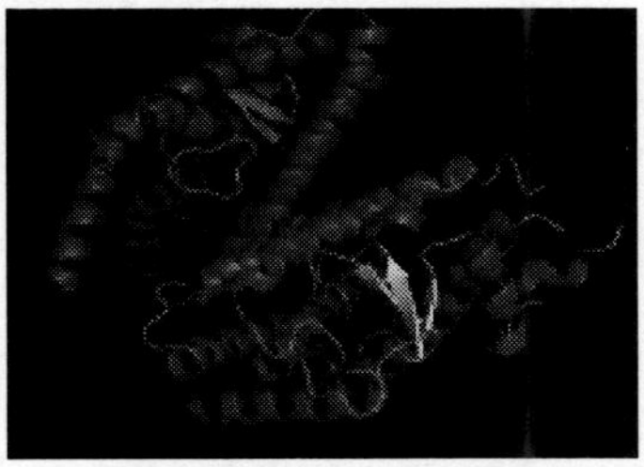

Fig. 1: Experimentally Determined Structure of 1I7I

Table 2: Sweet Proteins ID

Sweet Protein	*Seq. Length*	*PDB ID*	*Structure*
Brazzein	54	1BRZ	
Monellin	94	2O9U	
Thaumatin	207	1RQW	

RESULTS AND DISCUSSION

Molecular Dynamics Studies on the Structures of Sweet Proteins and the PPAR Gamma Receptor at Physiological Temperature in the Presence of Water

- The sweet proteins and PPAR receptor has been provided with condition inside the biological system of the organism to analyze the interactions.
- The 3D structures of sweet proteins and the 3D structure of PPAR gamma receptor were solvated and simulated at the physiological temperature 310 K (Gromacs).
- Simulation of sweet proteins and the receptor were carried out and the energy was converged at 10ps.

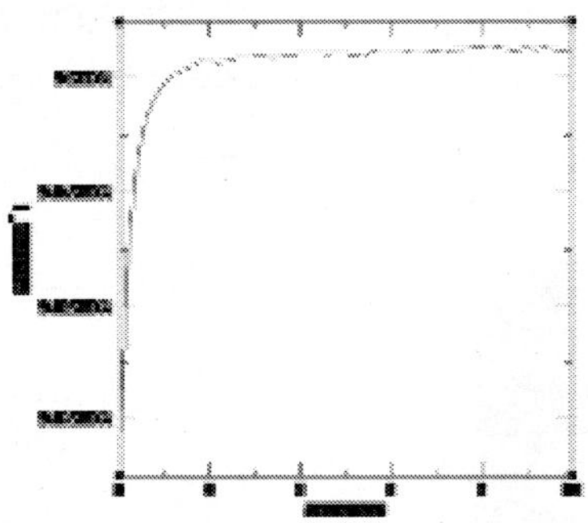

Fig. 2: Energy Graph of Brazzein

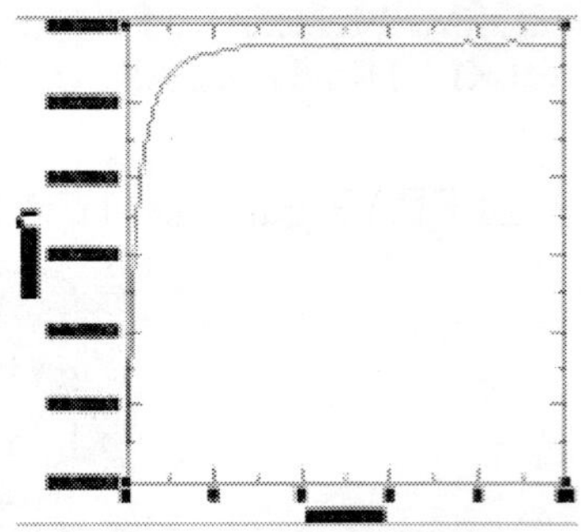

Fig. 3: Energy Graph of Monellin

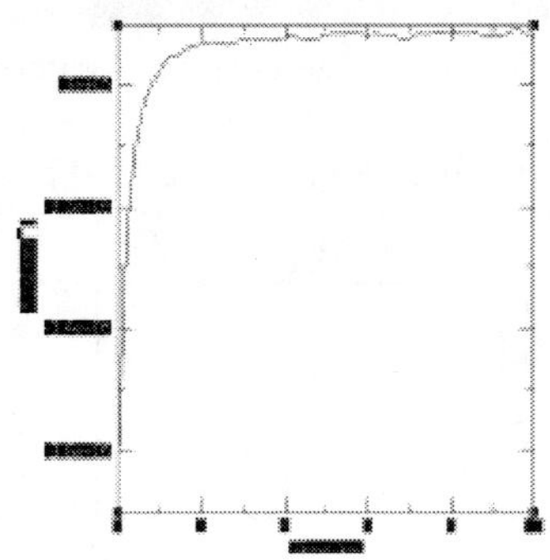

Fig. 4: Energy Graph of Thaumatin

Mapping the Binding Patterns of Sweet Proteins with the PPAR Gamma Receptor using Molecular Docking

- Docking studies were carried-out for various sweet proteins with the solvated receptor structure using docking software's (Patchdock).

- Various sweet proteins were docked with the PPAR receptor without any bias in order to identify the active sites.
- The interactions between the PPAR receptor and the sweet proteins were analyzed with the help of Patchdock online server.

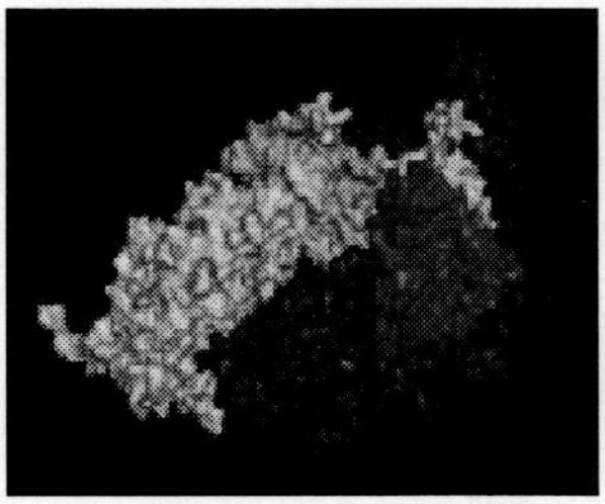

Fig. 5: Docking of PPAR – Thaumatin

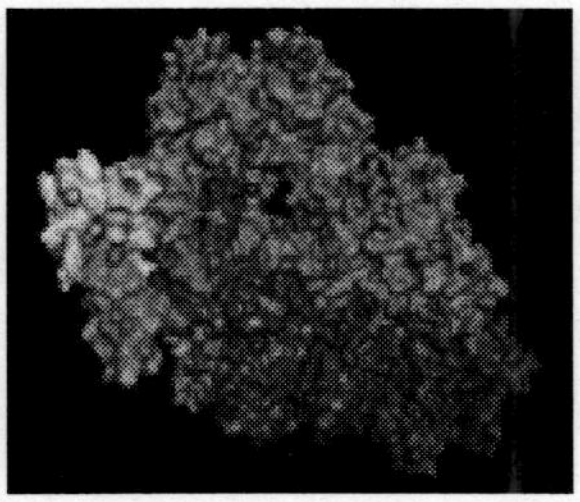

Fig. 6: Docking of PPAR – Monellin

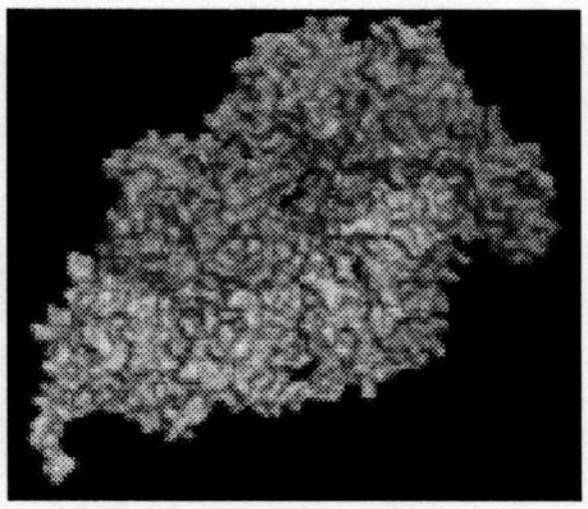

Fig. 7: Docking of PPAR – Brazzein

Table 3: Binding Energy of Sweet Proteins with the Receptor

Peptide	*Binding Energy [Kcal/mol]*
Thaumatin	–6.52
Monellin	–32.14
Brazzein	–34.84

Binding Regions of Dimer Receptor with Sweet Proteins

Dimer-Thaumatin

R 614, F 622, C 625, I 626, I 629, A 630, S 633, Q 635, I 636, A 639, S 644, R 645, P 647, R 648, S 651, V 654, R 655, C 862, T 863, S 866, G 869, L 871, W 872, L 874, A 875, M 878, S 1289, Q 1292.

Dimer-Monellin

Y 750, S 1377, W 1479, V 1484, Y 1557, A 1560, R 1561, L 1568, A 1569, Y 1570, F 1571, I 1572, I 1604, A 1607, F 1608, L 1610, P 1611, L 1616, L 1622, N 1623, T 1624.

Dimer-Brazzein

V 675, M 678, L 679, A 680, L 683, D 1394, S 1395, L 1397, V 1398, S 1401, G 1403, P 1404, A 1406, V 1411, L 1413, L 1415, Q 1439, N 1440, L 1443, A 1478.

Since these sweet proteins were linked with inducing or modulating the sweet taste, *de novo* peptides have been designed, by taking into considerations of physical-chemical properties of amino acids and overall fold of segment required for binding with the receptor based on the understandings of the mechanism of binding action of sweet proteins with PPAR receptor.

Sequence Alignment of Key Residues of Active Site of the Sweet Proteins

```
Thaumutin      ---YEYTG-KTEYR-RP-K 13
Monellin       ---YEFRE-IKGYEYQLY- 14
Brazzein       ----NYPV-SKQYEKRNL- 13
```

Fig. 8: Sequence Alignment of Active Residues of Sweet Proteins with PPAR Receptor

PEPTIDE DESIGN FOR SWEET PROTEIN

1. KDFHGSEHKGPSQV
2. HSYDWQDVHALNTW
3. RVWINAEVRPRCTA
4. REFKAFAKHPGVPW

The validity of the *de novo* peptides have been cross-verified using molecular dynamics and docking methods (patch binding energy values for receptor with the peptides were obtained from docking. From the analysis peptide 4 seems to be best as the binding energy is comparatively low.

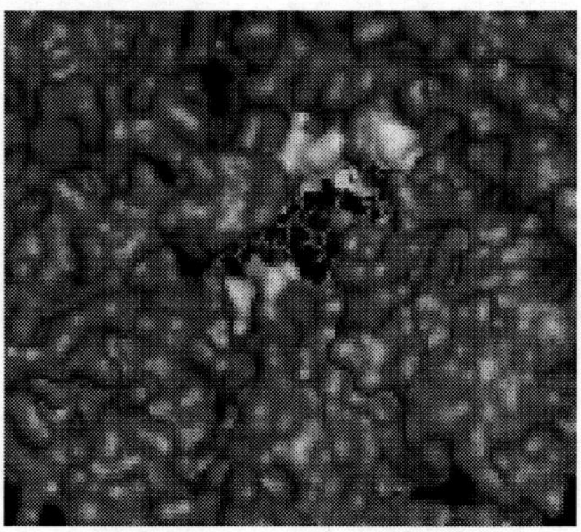

Fig. 9: Docking of Pep1

Fig. 10: Docking of Pep2

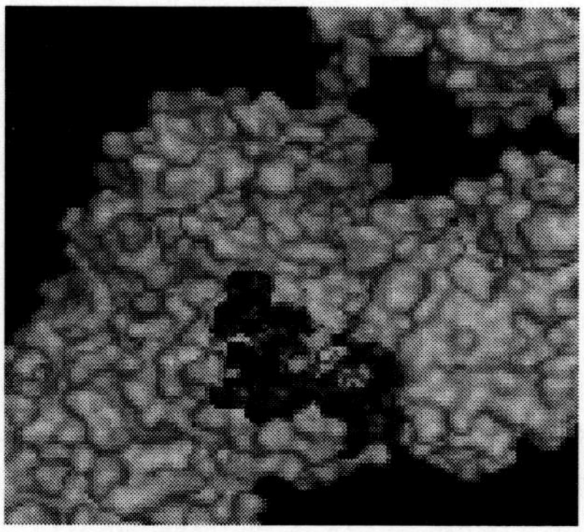

Fig. 11: Docking of Pep3

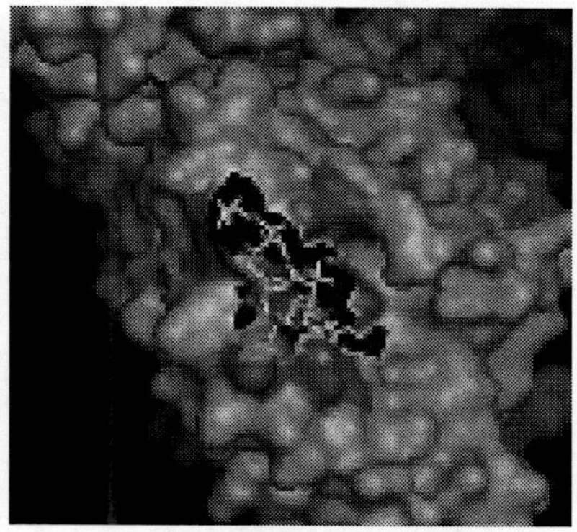

Fig. 12: Docking of Pep4

Table 4: Binding Energy of De-nova Peptides with the Receptor

Peptide	*Binding Energy [Kcal/mol]*
Pep1	–62.25
Pep2	–59.69
Pep3	–59.92
Pep4	–64.01

Peptide4 – REFKAFAKHPGVPW

In this study, an attempt was made to elucidate major aspects of sweet protein and PPAR receptor interactions. One of the promising approaches in anti-diabetic drug discovery is the development of new inhibitor for taste perception. The experimentally determined 3D structure of the PPAR gamma receptor (1I7I) was used in this study, to arrive at a reliable model for structure based drug design. The 3D structures of sweet proteins and the 3D structure of PPAR gamma receptor were solvated and simulated at the physiological temperature 310 K (Gromacs).Simulation of sweet proteins and the receptor were carried out and the energy was converged at 10ps. Docking the PPAR receptor with various sweet proteins (Thaumatin, Monellin, Brazzein, Curculin, and Mabinlin) provided an insight into the nature of binding and interaction of ligands to the receptor. Docking studies also revealed that sweet proteins are binding to the receptor. Some differences in the residues lining extracellular region of the receptor were observed. The overall similarity of the binding pockets of these sweet proteins and the PPAR receptor offers a promising opportunity to design and develop *de-nova* peptides (REFKAFAKHPGVPW) that could elucidate the sense of sweet taste with the PPAR receptor. As it has been found that sweet proteins are thousands of times sweeter than the artificial sweetener and are of low calorie value, these proteins can be used as natural low calorie sweeteners by people suffering from diseases linked to consumption of sugar e.g. obesity, diabetes and hyperlipemia.

REFERENCES

Chandra, A., Mahdhi, A.A., Ahmad, S. and Singh, R.K. (2007). Indian herbs result in hypoglycaemic responses in streptozotocin-induced diabetic rats. *Nutr Res.,* 27, 161–168.

Faus, I. (2000). Recent developments in the characterization and biotechnological production of sweet–tasting proteins, *Appl. Microbiol. Biotechnol,* 53, 145–51.

Kunishima, N., Shimada, S., Tsuji, Y., Sato, T., Yamamoto, M. and Kumasaka, T. (2000). Structural basis of glutamate recognition by a dimeric metabotropic glutamate receptor, *Nature,* 407, 971–977.

Kyvik, K.O., Nystrom, Gorus L., Songini, D., Oestman, M. and Castell, J.C. (2004). The epidemiology of type 1 diabetes mellitus is not the same in young adults as in children, *Diabetologia,* 47, 377–384.

Lisa Devitt, Denis Daneman, Jennifer Buccino (2004). Assessment of intakes of artificial Sweeteners in children with type 1 diabetes Mellitus, *Can.J. Diabetes,* 28, 89–97.

Ritchie, D.W. and Kemp, G.L. (2000). Protein Docking Using Spherical Polar Fourier Correlations', Proteins: *Struct. Funct. Genet.,* 39, 178–194.

Scheen, A.J. (2003). Pathophsiology of type 2 diabetes. *Acta Clin Belg.,* 58, 335–341.

Staszewski, L., Xu, H., Durick, K., Zoller, M. and Adler, E. (2002). Human receptors for sweet and umami taste. *Proc. Natl. Acad. Sci. USA,* 99, 4692–4696.

Talpur, N.A., Echard, B.W., Fan, A.Y., Jaffari, O., Bagchi, D. and Preuss, H.G. (2002). Antihypertensive and metabolic effects of whole Maitake mushroom powder and its fractions in two rat strains. *Mol Cell Biochem,* 237, 129–136.

Antioxidant Status of Oxidatively Stressed Goat Liver Slices Treated with Methanolic Extract of Leaves and Rhizomes of *Curcuma amada* Roxb. and Assessment of its Cytotoxicity

J. Sivaprabha*, B. Dharani, P.R. Padma and S. Sumathi

Department of Biochemistry, Biotechnology and Bioinformatics Avinashilingam Institute for Home Science and Higher Education for Women University, Coimbatore

E-mail: jsivaprabha@gmail.com

INTRODUCTION

Free radicals are involved in various pathophysiological disorders in humans such as cancer, artherosclerosis, Diabetes, arthritis and various other medical complications (Wadhkar, K.A. and Magdum, C.S., 2010). Natural antioxidants from plant sources have been used to combat various free radical mediated disorders (Nirmala, M.J. *et al.*, 2010 and Sukantha, T.A., 2012). Breast cancer is the major health problem among Women and there is need for new treatment strategies to combat this global health problem (Banshen, N.A., 2012) as cancer cells exhibits therapy resistance. Breast cancer occurs due to changes in the cell cycle and apoptotic machinery which apparently leads to accumulation of genetic errors and ultimately results in uncontrolled proliferation and loss of apoptosis (Lee, E.Y., 2010 and Hanahan, D., 2011). Phytotherapy which denotes the use of plant-derived products for the treatment of cancer has been gaining importance due to reduced risk of side effects which have been implicated in other forms of treatment (Gupta, S.C. *et al.*, 2010 and 2012). One such medicinal plant is *Curcuma amada* Roxb which has been used in traditional folk fore and has various medical applications. In the present study the

*Corresponding author.

antioxidant status of goat liver slices treated with methanolic extract of the leaves and rhizomes and anticancer activity of *Curcuma amada* has been analyzed.

MATERIALS AND METHODS

Preparation of Methanolic Extract

The components present in the leaves and rhizomes of *Curcuma amada* were extracted with methanol by cold extraction. Previous studies conducted by us showed that the methanolic extract of the leaves and rhizomes were rich in antioxidants. The methanolic extract prepared after evaporation of methanol was dissolved in DMSO (Dimethyl sulfoxide) [20 mg plant extract in 50 μl DMSO]. For each antioxidant assay 0.25 mg of rhizome extract and 0.15 mg of leaf extract was used and for cytotoxicity assays 0.2 mg of the extracts dissolved in 5 μl DMSO was used.

Preparation of Goat Liver Slices

The goat liver was collected fresh from a slaughter house, plunged into cold sterile PBS and maintained at 4°C till use. Thin slices of 1 mm thickness were treated with the oxidising agent H_2O_2 (200 μM) and/or rhizome and leaf extract and incubated at 37°C with mild shaking for one hour. Appropriate controls were also set up. The homogenate was used for the assay.

Maintenance of Cell Lines

MCF-7 and HBL-100 cell lines were purchased from National Centre for Cell Sciences (NCCS), Pune and maintained under standard in vitro growth conditions in our laboratory. The cells were grown in DMEM medium supplemented with 10% FBS, 1% Penicillin Streptomycin, 1% Non-essential amino acids and incubated at 37°C in 5% CO_2 and 95% humid atmosphere.

Treatment Groups for Liver Slices

1. Liver slices
2. Liver slices + H_2O_2
3. Liver slices + methanolic extract of *C. amada* leaves (CAL)
4. Liver slices + CAL+ H_2O_2
5. Liver slices + methanolic extract of *C. amada* rhizomes (CAR)
6. Liver slices + CAR+ H_2O_2

Treatment Groups for Cell Lines

1. Cells alone
2. Cells + tamoxifen (TAM)
3. Cells + methanolic extract of *C. amada* rhizomes (CAR)
4. Cells + CAR + TAM
5. Cells + methanolic extract of *C. amada* leaves (CAL)
6. Cells + CAL + TAM

The MCF-7 and HBL-100 cells were treated with 0.2 mg of rhizome and leaf extracts for 24 and 12h respectively and analyzed for cytotoxicity.

Antioxidant Analysis and Cytotoxicity Assays

The enzymic (SOD, CAT, POD, GST and GR) and non-enzymic antioxidants (Vitamin C, E, A and GSH) were analysed by standard procedures [9–17] and the cytotoxicty was analysed using MTT and SRB assays [18, 19].

Statistical Analysis

The parameters of the experiment are expressed as Mean ± S.D. Statistical evaluation of the data was done using one way ANOVA with the level of significance at $P < 0.001$ in sigma stat package version.

RESULT AND DISCUSSION

Enzymic Antioxidants

The activities of enzymic antioxidants namely SOD, CAT, POD, GR and GST were analyzed.

Table 1: Effect of *Curcuma amada* Roxb Rhizome and Leaf Extract on the Enzymic Antioxidant Activities in Goat Liver Slices Exposed to H_2O_2 *in vitro*

Treatment Groups	*Parameters*				
	SOD[1]	*CAT*[2]	*POD*[3]	*GR*[4]	*GST*[5]
Liver slices	11.49±1.2	360±28.28	32.28±0.2	7.25±0.16	0.44±0.03
Liver slices+ H_2O_2	8.45±0.56[a]	138.9±4.08[a]	23.18±0.72[a]	4.29±0.86[a]	0.26±0.002[a]
Liver slices+ CAR	16.22±0.46[a]	324.55±21.86	47.57±0.11[a]	9.98±0.12[a,d]	0.47±0.03
Liver slices+ CAR+ H_2O_2	13.52±0.09[a,b,c]	319.39±17.85[b]	47.64±2.38[a,b]	9.56±0.08[a,b]	0.38±0.007[b,c]
Liver slices+ CAL	13.91±0.34[a,d]	329.7±17.84	49.02±0.44[a]	8.96±0.32[a,d]	0.43±0.03
Liver slices+ CAL+ H_2O_2	11.7±0.19[b,c]	296.2±18.23[a,b]	46.15±2.38[a,b]	8.52±0.05[a,b]	0.41±0.08[b]
LSD ($p < 0.05$)	1.2	49.9	3.1	0.8	0.09

Values are Mean ± SD of triplicates.

1. 1 Unit is defined as the amount of enzyme that gives 50% inhibition of the extent of NBT reduction in 1 min
2. 1 Unit is defined as the amount of enzyme required to decrease the absorbance by 0.05 units at 240 nm
3. 1 Unit is defined as the Change in absorbance at 430 nm per minute.
4. 1 Unit is defined as the milli moles of NADPH oxidized/min.
5. 1 Unit is defined as the nano moles of CDNB conjugated/min.

Non-Enzymic Antioxidants

The levels of non-enzymic antioxidants namely Vitamins C, E, A and GSH were analyzed.

Table 2: Effect of *Curcuma amada* Roxb Rhizome and Leaf Extract on the Level of Non-Enzymic Antioxidants in Goat Liver Slices Exposed to H_2O_2 *in vitro*

Treatment Groups	*Parameters*			
	Vitamin-C (mg/g liver)	*Vitamin-E (μg/g liver)*	*Vitamin-A (μg/g liver)*	*Reduced Glutathione (mM/g liver)*
Liver slices	0.54±0.02	3.05±0.13	72.01±0.33	5.92±0.05
Liver slices+ H_2O_2	0.23±0.002[a]	1.07±0.19[a]	54.53±0.44[a]	3.95±0.05[a]
Liver slices+ CAR	0.71±0.02[a,d]	3.15±0.13	81.71±0.76[a]	6.35±0.04[a,d]
Liver slices+ CAR+ H_2O_2	0.64±0.012[a,b]	2.78±0.13[b]	76.16±0.98[a,b,c]	5.65±0.03[a,b,c]
Liver slices+ CAL	0.86±0.009[a]	3.19±0.33	78.4±0.66[a,d]	5.75±0.04[a]
Liver slices+ CAL+ H_2O_2	0.75±0.011[a,b]	2.64±0.2[b,c]	71.62±0.44[b,c]	5.52±0.02[a,b,c]
LSD ($p < 0.05$)	0.8	0.48	1.56	0.099

Values are Mean ± SD of triplicates.
(a) Statistically significant ($P < 0.05$) compared to untreated control
(b) Statistically significant ($P < 0.05$) compared to H_2O_2 alone treated group
(c) Statistically significant ($P < 0.05$) compared to the respective extract treated group
(d) Statistically significant ($P < 0.05$) compared to the rhizome extract treated group.

The results showed that the activities of enzymic antioxidants and the level of non-enzymic antioxidants which was decreased significantly on treatment with H_2O_2 was restored to the normal values on treatment with rhizome and leaf extracts. This confirms the antioxidant activity exhibited by the extracts and was also able to relieve the cells from oxidative stress caused by exposure to H_2O_2.

The result of MTT and SRB assays showed that on treatment with rhizome and leaf extract the cell viability of both MCF-7 and HBL-100 cells was reduced significantly and on co-administration of extracts with tamoxifen, a standard drug used in the treatment of ER positive breast cancer the viability was further decreased. The extracts also showed less cytotoxicity towards HBL-100 cells than MCF-7 cells.

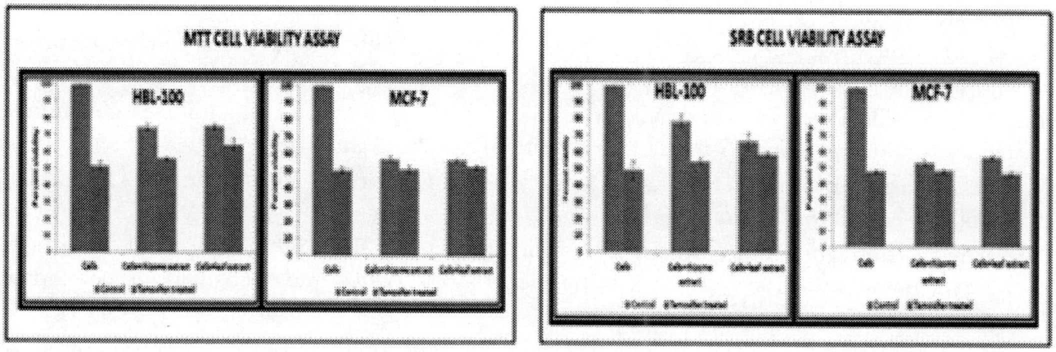

Fig. 1: Effect of *Curcuma amada* Roxb Rhizome and Leaf Extract on the Viability of MCF-7 and HBL-100 Cells as Determined by MTT and SRB Assays

The values of the negative (untreated) control group were fixed as 100 viability and the percent viabilities in the other groups were calculated relative to this.

REFERENCES

Banshen, N.A., Elkaldy, A.I., Abuzinadeh, O.A. and Mutwakil, M.H. (2012). Potential anticancer activity of medicinal herb, *Rhozya strieta* against human breast cancer, *Afr. J. Biotech*, 2, 8960–8972.

Bayfield, R.F. and Cole, E.R. (1980). Colorimetric estimation of vitamin A with trichloroacetic acid, *Meth. Enzymol*, 67, 189–195.

David, M. and Richard, J.S. (1983). In: Methods of enzymatic analysis, Bergmeyer, and Marianne Grab (eds.), Verlag Chemie Weinheim Deer Field, *Beach Floride*, p. 358

Gupta, S.C., Bazaid, S.A., Shohayeb, M.M., El-Sayed, M.M. and El-Wakil, E.A. (2012). Phytochemical studies and evaluation of atioxidant, anticancer and antimicrobial properties of Conocarpus erectus L. Growing in Taif, Soth Arabia, *European J. of Medicinal Plants*, 2, 93–112.

Gupta, S.C., Kim, J.H., Prasad, S. and Agarwal, B.B. (2010). Regulation of of survival, proliferation, invasion, angiogenesis and metastasis of tumour cells through modulation of inflammatory pathways by neutraceuticals, *Cancer Metastasis Rev*, 29, 405–434.

Habig, W.H., Pabst, M.J. and Jokoby, W.B. (1974). Glutathione transferase: A first enzymatic step in mercapturic acid III formation, *J. Biol. Chem.*, 249, 7130–7139.

Hanahan, D. and Weinberg, R.A. (2011). Hall marks of cancer—The next generation, *Cell*, 144, 646–674.

Lee, E.Y. and Muller, W.J. (2010). Oncogenes and tumor suppressor genes, Cold spring Harb, Perspect Biol, doi: 10.1101/cshperspect.a003236.

Luck, H. (1974). In: Methods in enzymatic analysis, II Edn., Bergmeyer, Academic Press, New York, pp. 805, Reddy, K.P., Subhani, S.M., Khan, P.A. and Kumar, K.B. (1995), Effect of light and benzyladenine on dark treated growing rice (*Oryza sativa*) leaveschanges in peroxidase activity, *Plant Cell Physiol,* 26, 987–994.

Misra, H.P. and Fridovich, A. (1972). The role of superoxide anion in the antioxidation of epinephrine and a simple assay for superoxide dismutase, *J. Biol. Chem*, 247, 3170–3171.

Moron, M.S., Depierre, J.N. and Mannervik, V.C. (1979) Levels of glutathione, glutathione reductase and glutathione S-transferase activities in rat lung and liver.

Nirmala, M.J., Samundeeswari, A. and Sarkar, P.D. (2011). Natural plant resources as anticancer therapy—A review, *Res. Plant biol,* 1,1–14.

Reddy, K.P., Subhani, S.M., Khan, P.A. and Kumar, K.B. (1995). Effect of light and benzyladenine on dark treated growing rice (*Oryza sativa*) leaves-changes in peroxidase activity, *Plant Cell Physiol,* 26, 987–994.

Roe, J.H. and Kuether, C.A. (1943). The determination of ascorbic acid in whole blood and urine through 2, 4-dinitrophenylhydrazine derivative dehydro ascorbic acid, *Biol. Chem.,* 147, 399–407.

Rosenberg, H.R. (1992). Chemistry and physiology of the vitamins, Interscience Publisher, New York, pp. 452–453.

Sukantha, T.A., Sripathi, S.K., Ravindran, N.T. and Balashanmugam, P. (2012). Antioxidant and antibacterial activities of *Trianthema decandra* Linn, *Int J. Pharm Pharm Sci,* 4, 410–413.

Wadhkar, K.A. and Magdum, C.S. (2010). Evaluation of total phenolic content, flavonoid content and antioxidant activity of stem bark of *Careya arbores* roxb, *Res J. Phrmacognosy and Phytochemistry*, 2, 49– 51.

Larvicidal Efficacy of *Solanum xanthocarpum* and *Cucurbita maxima* Plant Extracts Against Mosquito Species *Culex quinquefaeciatus*

R.P. Janani, R. Subha Harshni and K. Poongothai*

K.S. Rangasamy College of Technology, Tiruchengode

E-mail: poongothaikarunanithi@gmail.com

INTRODUCTION

Mosquitoes constitute a major public health problem as vectors of serious human diseases. WHO has declared the mosquito "public enemy number one" as they are responsible for the transmission of various dreadful mosquito diseases (WHO, 1996). Prevalence of mosquito borne disease is one of the world's health hazardous problems. Several mosquito species belonging to genera Anopheles, Culex and Adese are vectors for diseases, such as malaria, filariasis, dengue/dengue hemorrhagic fever (DHF), yellow fever and Japanese encephalitis and chikungunya contribute significantly to morbidity and mortality in many tropical and sub-tropical regions of the world (Bansal *et al.,* 2009). Many approaches have been developed to control mosquito menace. One such approach to prevent mosquito borne disease is by killing mosquito at larval stage by using chemical insecticides. Though these chemical insecticides are fast in action they are toxic. They cause cancer, birth defects and developmental problems in children (Mullai *et al.,* 2007). This has necessitated the need for a research and development of environmentally safe, biodegradable indigenous method for vector control.

The plant world comprises a rich storehouse of phytochemicals, which are widely used in the place of synthetic insecticides since continuous use of synthetic insecticides cause side effects to non-target organisms and insecticide resistance against mosquitoes

*Corresponding author.

(Kelm *et al.,* 1997). Phytoproducts on account of minimal hazardous effect on the environment and wide range of availability offer promises in future mosquito control programmes. They have revolutionized the fields of vector control asthey possess different bioactive components and can be used as general toxi cants against various larval stages of the mosquito (Sharma *et al.,* 2004; Mohan *et al.,* 2005).

In this study the mosquito larvicidal activity of two plants extracts namely *Solanum xanthocarpum* and *Cucurbita maxima* was investigated against *Culex quinquefaeciatus. Solanum xanthocarpum* belongs to the family *Solanaceae* (Kantakatri in Tamil). The fruits of this plant have wound healing activity (Neeraj Kumar *et al.,* 2010) and are eaten as an anthelminthic. Root is an expectorant, used for cough, asthma and chest pain. *Cucurbita maxima* belong to the family *Cucurbitaceae.* (Pusinikai in Tamil). This plant has been traditionally used as antidiabetic, antitumor, antihypertensive, anti-inflammatory, immunomodulatory and antibacterial agent (Jia *et al.,* 2003; Andrade-Cetto *et al.,* 2005). The present investigation assessed the larvicidal activity of these two plant leaf extracts against the larvae of *Culex quinquefaeciatus.*

MATERIALS AND METHODS

Plant Material

Fresh leaves of *Solanum xanthocarpum and Cucurbita maxima* were collected at the places near Vellore District, identified and authenticated by Tamil Nadu Agricultural University. The leaves were rinsed with distilled water, shade dried at room temperature and powdered using an electrical blender.

Preparation of Leaf Extract

The finely ground plant material (125 gm/solvent) was loaded in soxhlet apparatus and was extracted with four different solvents namely distilled water, methanol, ethanol and acetone individually (Vogel, 1978). The solvent from the extract was removed using a vacuum evaporator to collect the crude extract. The crude extracts were diluted in distilled water and made into 20 ppm, 50 ppm and 100 ppm concentrations. Triplets were made to these concentrations for repellency test.

Test Larvae

Larvae were collected from stagnant water and it was identified as *Culex quinquefaeciatus* using microscope with the resolution of 10X. The *larvae* was confirmed by the presence of Siphons and antenna which is absent in other larvae. It is then confirmed by Vector Control Research Centre at Pondicherry.

Larvicidal Bioassay

Triplets of each solvent with 20 ppm, 50 ppm, 100 ppm concentration were poured separately into sterile glass dishes. The totals of 5 larvae were added to each test solution along with a set of controls containing distilled water. After adding larvae, the glass dishes containing the larvae were kept in the laboratory at room temperature for different time exposures such as 24 hours, 48 hours and 72 hours. The dead larvae can be identified when they fail to move after being probed by a needle (Nandita *et al.,* 2007). By counting the number of dead larvae at different time exposures, the effects of all graded solutions were monitored.

Phytochemical Screening

Phytochemical tests were carried out on the powdered sample using standard procedures, to identify the constituents as described by (Edeoga *et al.,* 2005).

RESULTS AND DISCUSSIONS

Vector control is facing a serious threat due to the emergence of resistance in vector mosquitoes to conventional synthetic insecticides or development of newer insecticides. Mosquitoes in the larval stage are attractive targets for pesticides because mosquitoes breed in water and thus, it is easy to deal with them in this habitat. The use of conventional pesticides in the water sources, however, introduces many risks to people and/or the environment and due to the continuous increase in resistance of mosquitoes to familiar synthetic insecticides, better alternative means are sought (Arivoli *et al.,* 2012) Natural pesticides, especially those derived crude from plants, are more promising in this aspect (Amer and Mehlhorn, 2006). A considerable number of plant products/derivatives have shown to be effective against mosquitoes with a safe manner. The screening of plants for mosquitocidal property may eventually lead to their use in natural product-based mosquito abatement practices. Larvae from the three medically important mosquito genera *Aedes, Anopheles* and *Culex* are all susceptible to a greater or lesser extent to some phytochemicals (Shaalan *et al.,* 2005).

The comparison of larvicidal activity between aqueous, methanol, acetone and ethanol extract of *Solanum xanthocarpum* and *Cucurbita maxima* against *Culex* in 24 hours, 48 hours and 72 hours exposure was shown in Tables 1 and 2 respectively. The mosquito repellency of different extract was measured on the basis of the number of

Table 1: Mortality of Larvae of *C. quinquefaeciatus* from Different Concentrations of *Solanum xanthocarpum* Leaf Extracts

Solvent Extracts	*Concentration (PPM)*	*Exposures (Hours)*		
		24	*48*	*72*
		Mortality Rate		
Control	25	–	0.33±0.57	0.66±0.577
	50	0.66±0.57	1.00±0.23	1.33±0.577
	100	1.66±0.57	2.33±0.57	3.03±0.02
Methanol	25	2.00±1.73	3.00±1.02	4.00±0.03
	50	2.66±1.15	3.66±0.57	4.66±0.57
	100	4.66±0.57	5.00±0.12	5.40±0.02
Acetone	25	0.66±0.57	1.66±0.57	2.33±0.57
	50	1.66±0.57	2.66±0.57	3.66±0.57
	100	3.33±0.57	3.66±0.57	4.00±1.00
Ethanol	25	0.66±0.58	2.66±0.57	3.33±0.57
	50	1.67±0.57	3.13±0.23	3.66±0.57
	100	3.66±0.57	4.23±0.02	4.68±0.57
Aqueous	25	0.33±0.57	3.33±0.58	3.66±0.59
	50	1.33±0.57	1.66±0.57	2.33±0.57
	100	2.33±0.57	3.33±0.57	3.66±0.57

Mean ± SD of three replicates. – Nil mortality.

mosquitoes that fed within a specified time (hours) that was the accurate documentation of the duration of exposure.

The methanol leaf extract of *S. xanthocarpum* had larvicidal activity against the larvae of *Cx. quinquefasciatus* with a mortality rate of 4.664.66 ± 0.57, 5.005.00 ± 0.12 and 5.40 ± 0.02 at 24, 48 and 72 hr exposure time period. This is similar to the findings of Singh and Bansal (2003), who investigated the larvicidal activity of *S. xanthocarpum* crude extract against the mosquito larvae of malaria and dengue vectors. They found larvicidal activity against *Anopheles culicifacies, An. stephensi*and *Ae. aegypti.* They also described the different degrees of toxicity of the crude extracts from different parts of the plant.

The acetone leaf extract showed moderate mortality when compared to ethanol and aqueous leaf extracts. Our findings are also in accordance with Mohan *et al.* (2005), who studied the crude extracts of *S. xanthocarpum* against *An. stephensi* and *Cx. quinquefasciatus.* They found different solvents used during extraction resulted in different levels of larvicidal activity. Species usceptibility was described by Bansal and Singh (2005) who exposed the larvae of *An. culicifacies*, *An. stephensi*, *Ae. Aegypti* and *Cx. quinquefasciatus* to the methanolic fruit extract of *S. xanthocarpum* and found the larvae of *An. culicifacies* were 1.01, 2.29 and 3.04 times more susceptible than the larvae of *An. stephensi*, *Ae. aegypti* and *Cx. quinquefasciatus,* respectively the LC50 and LC90 by 48 hours of exposure were lower than at 24 hours of exposure. This implies the tested plants are more potent after 48 hours exposure (Lapcharoen *et al.*, 2005).

Table 2: Mortality of Larvae of *C. quinquefaeciatus* from Different Concentrations of *Cucurbita maxima* Leaf Extracts

Solvent Extracts	*Concentration (PPM)*	*Exposures (Hours)*		
		24	*48*	*72*
		Mortality Rate		
Control	25	0.33±058	1.00±1.02	1.33±0.58
	50	1.34±0.57	1.6667±0.59	2.00±0.03
	100	2.66±0.55	2.69±0.57	3.69±0.55
Methanol	25	4.33±0.57	4.42±0.62	4.66±0.52
	50	5.00±0.02	4.66±0.53	7.00±0.02
	100	5.02±0	6.00±0	5.02±0.03
Acetone	25	4.04±1.00	4.34±1.15	4.67±0.58
	50	4.34±0.54	4.67±0.57	4.68±0.52
	100	4.69±0.57	4.66±0.57	5.02±0.01
Ethanol	25	4.36±0.55	4.38±0.58	4.69±0.57
	50	5.06±0.02	4.66±0.57	5.02±0.02
	100	6.02±0.00	5.03±0.02	6.02±0.05
Aqueous	25	1.33±0.58	1.66±0.55	2.02±1.03
	50	1.33±0.59	1.66±0.58	2.02±0.03
	100	4.35±0.57	4.66±0.53	4.66±0.58

Mean ± SD of three replicates. – Nil mortality.

Both the plants showed maximum activity against the *Culex.* Low activity was found to be observed in aqueous extract. When comparing the both, *Cucurbita maxima* showed higher activity against *Culex.* This method of larvae destruction will be an eco-friendly approach. The exact components of these plants that are responsible for the larvicidal activity can be determined by phytochemical analysis.

Chemical composition of aqueous extract of the *S. xanthocarpum* and *C. maxima* were determined semiquantitively by phytochemical studies. As shown in Table 3, flavonoids, saponins, tannins and alkaloids were present in the aqueous extracts of *S. xanthocarpum* while saponins and alkaloids were absent in the aqueous extracts of *C. maxima.*

Table 3: Phytochemical Analysis of the Leaves of *S. xanthocarpum* and *C. maxima*

Phytochemical Constituents	*S. xanthocarpum*	*C.maxima*
Carbohydrate	+	+
Saponins	+	–
Tannins	+	+
Flavonoids	+	+
Anthraquinones	–	+
Alkaloids	+	–

+ = Positive, – Negative.

Both the plants showed maximum activity against the *Culex.* Low activity was found to be observed in aqueous extract. When comparing the both, *Cucurbita maxima* showed higher activity against *Culex.* This method of larvae destruction will be an eco-friendly approach. The exact components of these plants that are responsible for the larvicidal activity can be determined by phytochemical analysis.

The formulation of bio-products from the plants will help in preventing the mosquitoes in larval stage itself. This will provide the base for the reduction in the population of mosquito there by inhibiting its growth to emerge as an adult. Bio-insecticides will also promote the growth of plants by human at least to defend themselves from the mosquitoes and other insects.

REFERENCES

Amer, A. and Mehlhorn, H. (2006). Larvicidal effects of t various essential oils against *Aedes, Anopheles* and. *Culex* larvae (Diptera, Culicidae). *Parasitology Research,* 99: 466–472.

Andrade-Cetto A. and Heinrich, M. (2005). Mexican plants with hypoglycaemic effect used in the treatment of diabetes. *J. Enthnopharmacol,* 99: 325–348.

Arivoli, S., John Ravindran, K. and Samuel Tennyson (2012). Larvicidal Efficacy of Plant Extracts against the Malarial Vector*Anopheles stephensi* Liston (Diptera: Culicidae). *World Journal of Medical Sciences,* 7: 77–80.

Bansal, S.K., Karam, Singh, V. and Sherwani (2009). Evaluation of larvicidal efficacy of Solanum xanthocarpum storage against vector mosquitoes in north-western Rajasthan. *J. Env. Biol.* 30: 883–888.

Edeoga, H.O., Okwu, D.E. and Mbaebie, B.O. (2005). Phytochemical constituents of some Nigerian medicinal plants.Departments of Biological and Chemical Sciences, Michael Okpara University of Agriculture, Nigeria.

Hartzell, A. and Wilcoxon, F. (1941). A survey of plant products for insecticidal properties, Contrib. Boyce. *Thomp. Inst.* 12: 127–141.

Jacobson, M. and Crosby, D.G. (1971). Naturally Occurring Insecticides,Marcel Dekker Inc., New York, 585.

Jia, W., Gao, W. and Tang, L. (2003). Antidiabetic herbal drugs officially approved in China. *Phytother. Res.* 17: 1127–1134.

Kelm, M.A., Nair, M.G. and Schutzki, R.A. (1997). Mosquitocidal compounds from *Magnolia salicifolia. International Journal of Pharmacognosy.* 35: 84–90.

Lapcharoen, P., Apiwathnasorn, C., Komalamisra, N., Dekumyoy, P., Palakul, K. and Rongsriyam, Y. (2005). Three indigenous Thai medicinal plants for control of *Aedes aegypti* and *Culex quinquefasciatus. Southeast Asian J. Trop. Med. Public Health.* 36: 167–75.

Mohan, L., Sharma, P. and Srivastava, C.N. (2005). Evaluation of *Solanum xanthocarpum* extracts as mosquito larvicides. *Journal of Environmental Biology*, 26: 399–401.

Mullai, K. and Jebanesan, A. (2007). Larvicidal, ovicidal and repellent activities of the leaf extract of two cucurbitacious plants against filarial vector *Culex quinquefasciatus* (Say) (Diptera: Culicidae). *Tropical Biomedicine,* 24: 1–6.

Nandita Chowdhury, Indranil Bhattacharjee, Subrata Laskar and Goutam Chandra (2007). Efficacy of *Solanum Villosum Mill* (*Solanaceae: Solanales*) as a Biocontrol Agent against Fourth Instar Larvae of *Culex quinquefasciatus* Say. *Turk. J. Zool.*, 31: 365–370.

Neeraj Kumar, Dhan prakash and Pankaj Kumar (2010). Wound healing activity of *Solanum Xanthocarpum* Schard. and Wendl. fruits. *Ind. J. Natural Products and Resources,* 1: 470–475.

Shaalan, E.A.S., Canyonb, D., Younesc, M.W.F., Wahaba, H.A. and Mansoura, A.H. (2005). A review ofbotanical phytochemicals with mosquitocidal potential. *Environment International.* 31: 1149–1166.

Sharma, P., Mohan, L. and Srivastava, C.N. (2004). Larval susceptibility of *Ajuga remota* against anopheline and culicine mosquitoes.Southeast. *Asian Journal of Tropical Medicine and Public Health*, 35: 608–610.

Singh, K.V. and Bansal, S.K. (2003). Larvicidal properties of a perennial herb *Solanum xanthocarpum* against vectors of malaria and dengue/DHF. *Curr Sci.* 84: 749–751.

Vogel, A.I. (1978). Text book of practical organic chemistry. The English language book society and Longman, London, pp. 1368.

WHO (1996). Report of WHO informal consultation on the evaluation and testing insecticides. *CTD/WHO PES/IC/*96.1, 69.

Biofunctionalization of Nanoparticle Using ssDNA for Effective Drug Delivery

K. Sathishkumar*[1], J. Madhusudhanan, M. Thanigaivel, Robin Abraham and V.A. Veni

Department of Biotechnology, Shri Andal Alagar College of Engineering (SAACE), Mamandur

[1]Department of Chemical Engineering, Siva Subramania Nadar College of Engineering (SSNCE), Kalavakkam

E-mail: jmadhuj2008@gmail.com

INTRODUCTION

Nanotechnology is the understanding and control of matter generally in the 1–100 nm dimension range. The application of nanotechnology to medicine is known as nanomedicine. The use of engineered materials at this length scale to develop novel therapeutic.[1] Metal nanoparticles, especially nanoparticles, have attracted much attention because of their extraordinary electrical and optical properties. Such optical properties are termed surface plasmon induced by the collective oscillation of electron density. The efficient absorbance and scattering of light of the metal nanoparticles may be utilized in electronic and photonic devices. However, metal nanoparticles in solution are susceptible to aggregation themselves. The synthesis of stabilized nanoparticles, a lot of effort has been given to the preparation of well-dispersed, stable protected metal clusters. Using polymeric stabilizing species for metal nanoparticles has inspired the studies of various synthetic routes to link polymers to metal particles.[2] A major problem of the conventional cancer drugs relates to their toxicity which causes the death of healthy cells as well as cancerous cells. Moreover, their life time in the body is short. Up to 90% of intravenously delivered drugs may

*Corresponding author.

be gathered up by macrophages within the first 5 min of their introduction. Many drugs may be simply removed in the first blood circulation through the kidney and the liver and will break apart before reaching their targets. Inorder to overcome these problems, research has been aimed at modifying the drugs by methods such as conjugating the drug to different polymer-functionalized nanoparticle[3]. Nanoparticles in particular have been widely investigated due to their ability to be functionalized with many biocompatible materials and also their visible light extinction behavior. Among the cancer chemotherpeutics, chemotherapy is a widely used for anti-cancer treatment.[4] Although it kills the cancerous cells by inhibiting the synthesis of nucleic acids within cells, three main problems are encountered with drug, first, high toxicity and large volume of distribution, second short life time in the body and third low solubility. To overcome the non-specificity and high toxicity of drug, many researchers have proposed conjugation to hydrophilic polymers to reduce the toxicity level while sustaining the therapeutic efficacy.[5] The polymer which was completely water soluble, non-toxic and uncharged thus encountering less electrostatic interactions with nanoparticle. In this study a novel biodegradable drug conjugate was synthesized. The colloids of 12nm nanoparticles were first functionalized by polymer to give polymer-nanoparticle. Folic acid was selected to conjugate to the polymer because it has a high binding affinity for the folate receptor which is over expressed in the cancerous cells in the membrane of brain, kidney, breast, and lung.[5, 6] Finally folate-modified polymer-NP was conjugated with anti-cancer drug for targeted drug delivery applications and the drug release kinetics was studied.[7] Moreover, the targeting effect of drug conjugate was evaluated in vitro by measuring the cell cytotoxicity and cell viability for folate-receptor-positive cells.[4, 7]

MATERIALS AND METHODS

Chemical Synthesis of Nanoparticles

Auric chloride was added into distilled water, after stirring for one minute, then the NaBH4/Tri sodium citrate solution was added drop by drop until the mixture turned red with vigorous stirring, and kept stirring for five minutes, then stored at 4°C. The optical spectra if colloidal nanoparticle were acquired on UV-vis-spectrophotometer. TEM image was acquired. UV-vis of colloids measured immediately and two weeks later.

Preparation of ssDNA Modified Nanoparticle

Thiol modified poly-A-nucleotide was dissolved in double distilled water (100 µl), 120 µL of gold colloids (600 nM) was incubated with 134 µl of poly-A-nucleotide (10 µM) for 24 hrs. The reaction mixture was then added with NaCl solution to a final concentration of 0.005 M then incubated. Salt concentration was gradually increased to 0.1 M and to 2 M each incubated. Increase in the salt concentration causes increase in incorporation of more DNA on the Nanoparticle surface. Agarose gel electrophoresis was performed to confirm the binding of Poly-A-Nucleotide to the AuNP's. Agarose was prepared in TE buffer by heating the agarose containing solution until clear solution was obtained. Gel casting tray was prepared by giving diluted ethanol wash to the tray and its comb. Once the solution was heated, 20 μl of Ethidium bromide was added. Then the solution was poured into gel tray and left undisturbed to solidify. Comb was removed without disturbing the gel, and placed in agarose chamber followed by connecting the power card. Sample was loaded and electrophoresis was carried out at 50 V with tracking dye. After $3/4^{th}$ run, gel was viewed under UV transilluminator.

Conjugation Drug to Functionalized Nanoparticles

Drug was dissolved in buffer pH 7, and ssDNA modified nanoparticle was added and stirred at room temperature. The mixture was subsequently freeze dried.

Drug Loading Efficiency

The loading efficiency of drug onto the original and Polymer modified nanoparticle with folate were measured in 0 h and 24 h. Au nanoparticle (Np's) modified with ssDNA is mixed with drug with buffer for the given period of time and centrifuged to remove the nanoparticles. The supernatant were measured in a UV-Vis spectrometer for the concentration.

Drug Release Behaviour

The drug releasing kinetics of the nanoparticle—ssDNA-drug complex was then evaluated at different pH environment in buffer in 0, 3, 6, 12, 24, and 48 at 37°C. The drug releasing rate was calculated. Each absorbance was subtracted with blank.

Cytotoxicity

Cancer cell lines were obtained from NCCS, Pune. To study the in vitro cytotoxicity for Drug loaded nanoparticle-polymer-folate, Tryphan blue, Ethidium Bromide assay was performed on cell lines. Folate containing cell lines were maintained in RPMI1640 medium at 37°C in Carbon-di-oxide. Where as folate negative cell lines was maintained in Leibovitz L-15 medium, supplemented with FBS and penicillinsteptomycin and maintained at 37°C in carbon-di-oxide. Each cell line was seeded into 96-well plates. After 24 h the medium was changed and various concentration of drug loaded nanoparticle (12.5, 25, 50, 100% in culture medium were added). At intervals of 6 h the medium was removed and the cells were washed twice with buffer pH 7. And then Tryphan blue and dual staining was performed.

CHARACTERIZATION

Characterization of Nanoparticle

Polymer is a highly branched aliphatic polyamine whose amine groups exist in primary, secondary and tertiary forms. A most prominent feature of polymer is its extremely high cationic charge density which was used to protect nanoparticles in this work. Figure 1 describes the TEM image of nanoparticle. Since the ssDNA has been used to prevent the aggregation of nanoparticle, they also used to prevent the nanoparticle structure.

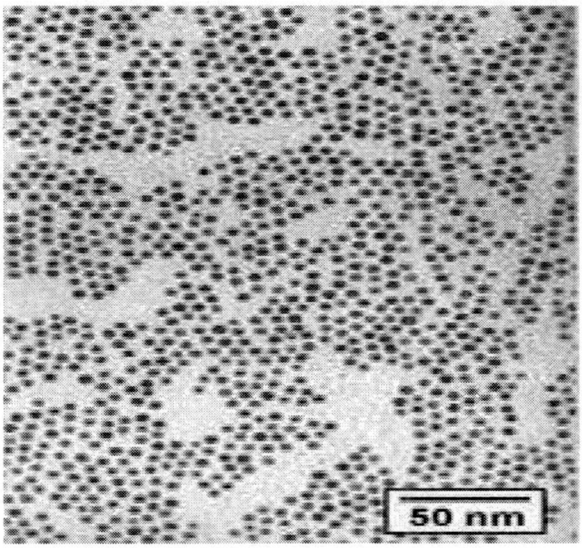

Fig. 1: Shows the TEM Image of a Nanoparticle

UV-vis spectra of as prepared colloids characterized as shown in the Figure 2. The Figure 2 showed UV-vis spectra analyses immediately after nanoparticle synthesis. The sharp peak indicates the colloids well-dispersed. From the maximum adsorption at 520 nm we can conclude the size of colloid is about 12 nm diameter.

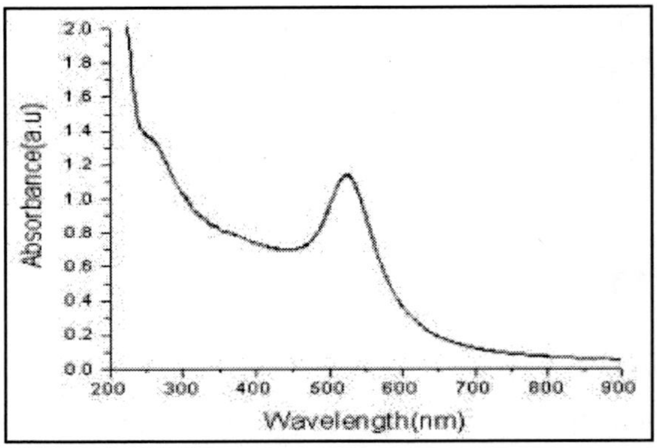

Fig. 2: Shows the UV-vis Spectra Analysis of Nanoparticle

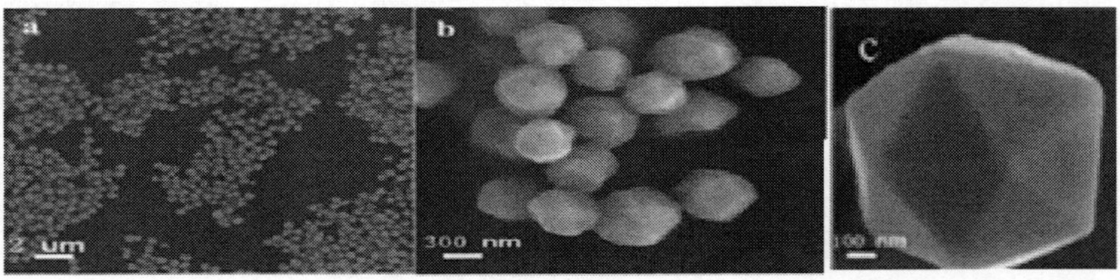

The SEM image of nanoparticle has been shown above,
a) SEM image of nanoparticle, b) magnified image of nanoparticle, c) single image of an nanoparticle.

Fig. 3: Shows the SEM Image of Nanoparticle

Characterization of Drug Loading Efficiency

The loading efficiency of drug onto the ssDNA modified nanoparticle was measured in 0 h and 24 h. The ssDNA modified nanoparticle were mixed with drug for the given period of time and centrifuged to remove the nanoparticles. The supernatant were measured in UV-vis spectrometer for the concentration of drug. The UV-vis value for drug in buffer @ 510 nm is noted.

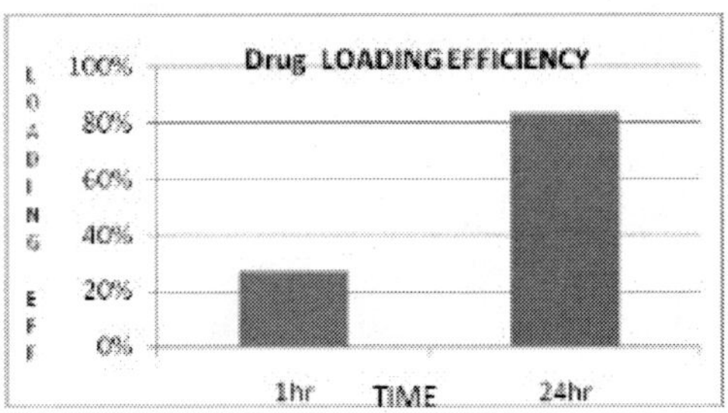

Fig. 4: Shows the Drug Loading Efficiency of ssDNA Modified Nanoparticle with Drug

The TEM image of functionalized nanoparticle with drug, says that there is an aggregation of nanoparticle with drug. This proves that there is an attachment of nanoparticle with drug. Therefore there is an increase in attachment of nanoparticle with drug using ssDNA modification Figure 5.

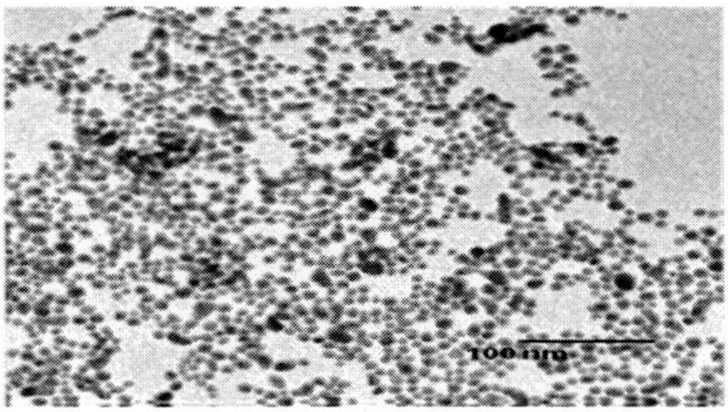

Fig. 5: TEM Image of Nanoparticle with Drug

Drug Release Behavior

In a recent report, we came to know that the use of porous iron oxide nanorods as a novel nano carrier for sustained or environmental salt concentration triggered drug release. The drug release mechanism was controlled through modulation of the inter-molecular space in the surface poly-electrolyte coating. In this study folate modified nanoparticle served as the affinity drug carrier and releasing control trigger at the same time. The polynucleotides serve as a natural biopolymer and decrease the potential toxicity, metabolic clearance and immunological issues when using synthetic biopolymers. In addition the spectrometer analysis revealed a sustained pH

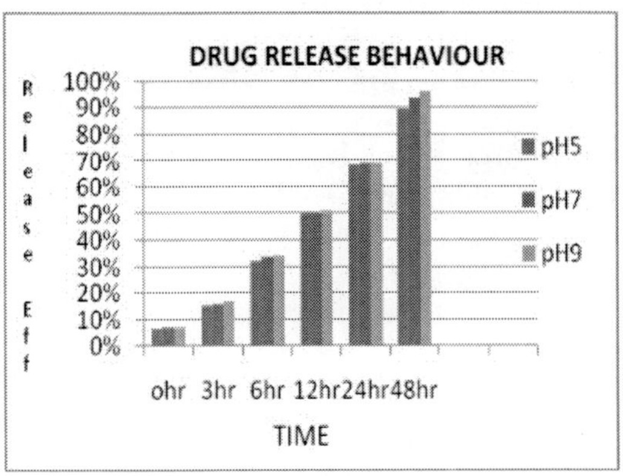

Fig. 6: Shows the Drug Release Behavior of ssDNA Modified Nanoparticle with Drug

dependent releasing manner of the nano carriers in buffer solution. The loaded drug has a significantly higher release rate in alkaline environment compare to neutral and aciding environment at 48 hrs. Thus the folate modified nanoparticle could serve as an intestine local delivery nano vehicle to pass through stomach and upper GI tract then release therapeutic agents in the lower GI tract, lung and oral.

Invitro Cytotoxicity

To evaluate the cell cytotoxicity of the free drug and the drug loaded nanoparticles, on cancer cells were used as invitro effect models. The Drug loaded nanoparticle on the viability of three model cell lines, respectively. Increasing the concentration of drug can decrease the viability of cancer cell lines. Since there is a comparative studies made between three cell lines and the effect or efficiency of drug to decrease the viability of cancer cell lines were analysed. The enhanced in vitro cytotoxicity of drug loaded nanoparticle for cancer cell lines could imply their successful effects to the cancerous cells. These results suggest that Drug loaded nanoparticle effectively decreased the in vitro cancer cell viability, which could imply the targeting effects of these nano-carriers in vivo.

CONCLUSION

Novel Nanoparticle-ssDNA conjugates with drug were prepared for targeted drug delivery. The Drug-loaded NPs showed significant invitro targeting effects for cell lines, which resulted in higher cytotoxicity than with free Drug. In addition, they caused higher cell viability at low concentration. And shows low cell viability at high concentration. Our results imply that the ssDNA functionalized nanoparticles could have high potential to be used for cancer therapy.

REFERENCE

Zhang, F.X., Gu1, J.M., Chan, A.Z., Wang, R.S., Langer and O.C. Farokhzad, Nanoparticles in Medicine: Therapeutic Applications and Developments.

Effect of Bioactive Compound from Endophytic Fungus in Anxiety Induced Zebrafish Animal Model

Jyothi S., Sivamani S., Benin Joseph, Basilia Gunalan, Varun Moganti and Bibhas Kar*

Cellular and Molecular Biology Lab, Department of Genetics, Madras Medical Mission, Chennai

E-mail: drbibhas_kar@yahoo.co.in

INTRODUCTION

Plant endophytic fungi are an important and novel resource of natural bioactive compounds with their potential applications in agriculture, medicine and food industry. In the past two decades, many valuable bioactive compounds with antimicrobial, insecticidal, cytotoxic and anticancer activities have been successfully discovered from the endophytic fungi (Strobel *et al.,* 2004). This chapter mainly describes the research progress on the endophytic fungi for producing bioactive compounds and used to treated anxiety induced zebrafish. The Anxiety is a psychological and physiological state characterized by cognitive, somatic, emotional, and behavioral components. These components combine to create an unpleasant feeling that is typically associated with uneasiness, apprehension, fear, or worry (Korte *et al.,* 2001). The general suitability of zebrafish as a model organism, as well as its use in the genetic and neuroanatomical analysis of larval behaviour has been comprehensively described elsewhere (Fero *et al.,* 2010). The present study aims at investigating the effect of bioactive compounds from endophytic fungus against anxiety induced zebrafish.

---

*Corresponding author.

MATERIALS AND METHODS

Isolation and Identification of Endophytes

Ten different fruit peels were collected in ten different sterile screw cap bottles containing water and a pinch of jaggery. It is allowed to grow for 24 hrs. Taking a sterile loop, the endophytes were inoculated in sterile petriplates containing agar medium and allowed to grow for 24 hrs. Meanwhile broth is filled in different screw cap bottles. The cultures which are allowed to grow in petriplates were transferred in broth. The cultures are taken on different slides and heat fixed. Giemsa stain is used to stain the slides. They are heat stressed at 45°C in the incubator and then allowed at room temperature to produce bioactive compounds.

Isolation of Bioactive Compounds

The bioactive compounds were poured in different screw cap bottles and butanol is added in all the screw bottles. After adding butanol they are allowed overnight to get a separation of two phases aqueous phase and butanol phase. The next day these compounds are poured in different petriplates and dried at 50°C and allowed to dry in hot air oven. After drying these compounds are reconstituted using Triton - X 100 and the compounds were scrapped up. These compounds are collected and stored in eppendorf tubes at -20°C.

Purification of Bioactive Compounds

Compounds which gave good results were taken and were run under thin layer chromatography. Silica gel was used on glass slide and a drop of bioactive compound was taken and was allowed to run on the plate. The plate was placed in BAW (Butanol: Acetic Acid: Water in the ratio of 4:1:5). The powdered form was taken in eppendorf tubes and distilled water was added. These tubes were centrifuged for 2000 rpm for 5mins and the supernatant was collected as it was the pure form of bioactive compounds.

Zebra Fish Assays

There were number of assays for studying anxiety in zebrafish. We have selected three different assays to choose the best one as a prototype. The assays selected were group preference test, light/dark preference test, and novel tank diving test. According to results produced by the above mentioned assays done in the lab, Group preference test has given good results.

RESULTS AND DISSCUSIONS

Novel Dive Tank Test

Ten fish have been taken from the tank. Anxiety was induced for these fish for a period of 48hrs under white light. These fish were subjected for dive tank test for 3 mins. Another ten control fish were taken and subjected to this test and the results are compared. The results were as below.

Table 1: Novel Dive Tank Tests for Control Fish

Group	*Lower Zone (sec.)*	*Middle Zone (sec.)*	*Higher Zone (sec.)*
AF1	180		
AF2	180		
AF3	170	10	
AF4	180		
AF5	175	5	
AF6	160	10	10
AF7	120	30	30
AF8	165	5	10
AF9	180		
AF10	180		
Total	1690	60	50
Avg	169	6	5

Table 2: Novel Dive Tank Tests for Anxiety Induced Fish

Group	*Lower Zone (sec.)*	*Middle Zone (sec.)*	*Higher Zone (sec.)*
F1	120	20	20
F2	90	45	45
F3	100	20	20
F4	135	15	15
F5	140	10	10
F6	100	35	35
F7	90	25	25
F8	125	20	20
F9	125	10	10
F10	70	20	20
Total	1095	220	220
Avg	109.5	22	22

Table 3: Comparison of Novel Tank Test in Control and Anxiety Induced Zebrafish

Experimental Group	*Lower Zone (sec.)*	*Middle Zone (sec.)*	*Higher Zone (sec.)*
Control	109.5	22	48.5
Anxiety induced fish	169	6	5

The comparative test result shows the control zebrafish spent more time on the higher zone (i.e. Top zone) as compared to the anxiety induced zebrafish.

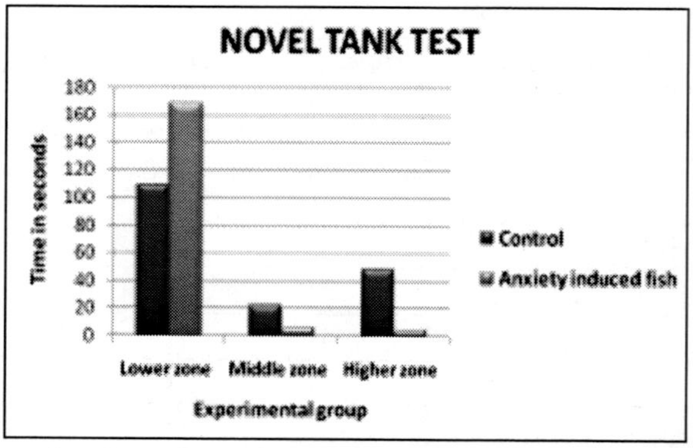

Graph 1: Graphical Representation of Comparative Novel Dive Tank Test

Light/Dark Preference Test

Ten fish were taken from the tank. Anxiety was induced for 48 hrs under white light. The fish were subjected for light/dark preference tank test for 3 mins. The tank setup was prepared with the help of black and white chart.

Table 4: Light/Dark Preference Test for Normal Fish

Group	*Light Preference (sec.)*	*Dark Preference (sec.)*
F1	50	130
F2	60	120
F3	60	120
F4	55	125
F5	90	90
F6	50	130
F7	40	140
F8	100	80
F9	90	90
F10	115	65

Table 5: Light/Dark Preference for Anxiety Induced Fish

Group	*Light Preference (sec.)*	*Dark Preference (sec.)*
AF1	20	160
AF2	25	155
AF3	30	150
AF4	70	110
AF5	40	140
AF6	45	135
AF7	40	140
AF8	35	145
AF9	50	130
AF10	10	170

Table 6: Comparison of Light/Dark Preference Test in Control and Anxiety Induced Zebrafish

Experimental Group	*Dark Surface (sec.)*	*Light Surface (sec.)*
Control	71	109
Anxiety induced fish	36.5	143.5

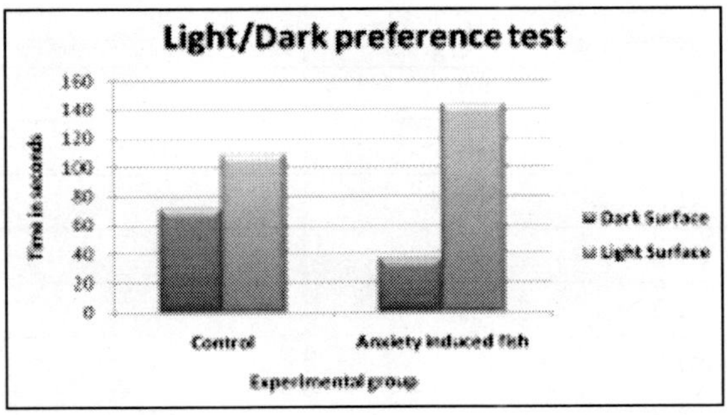

Graph 2: Graphical Representation of Comparative Light/Dark Preference Test in Control and Anxiety Induced Zebrafish

The Light/Dark preference test result shows, the anxious fish spent more time in dark surfaces and less time in light surfaces when compared to the control group zebrafish.

Group Preference Test

Ten fish were taken from the tank. Anxiety was induced for 48 hrs under white light. They were subjected for group preference test for 3 mins. Another ten control fish were taken and compared to the anxiety induced group. Results are tabulated below for both control fish and anxiety induced fish.

Table 7: Group Preference Test for Control Fish

Group	*Near to Group (sec.)*	*Away to Group (sec.)*
F1		180
F2	80	100
F3	40	140
F4	15	165
F5	120	60
F6	45	135
F7	105	75
F8	50	130
F9	10	170
F10	80	100

Table 8: Group Preference Test for Anxiety Induced Fish

Group	*Near to Group (sec.)*	*Away to Group (sec.)*
AF1	180	
AF2	140	40
AF3	145	35
AF4	140	40
AF5	145	35
AF6	105	75
AF7	90	90
AF8	60	120
AF9	180	
AF10	140	40

Table 9: Comparison of Group Preference Test in Control and Anxiety Induced Zebrafish

Experimental Group	*Near to Group (sec.)*	*Away to Group (sec.)*
Control	54.5	125.5
Anxiety induced fish	132.5	47.5

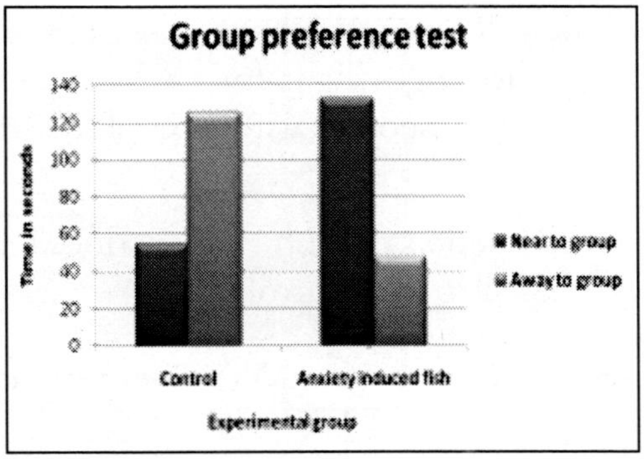

Graph 3: Graphical Representation of Comparative Group Preference Test in Control and Anxiety Induced Zebrafish

The comparative result of Group preference test displays the anxiety induced zebrafish spent more time near to the group when compared to control group zebrafish. Interestingly, the group performance test shows tremendous result, when compared to other examination. From the remarkable results we have decided the group preference test as a suitable method for screening of drug.

Drug Screening

Since we have isolated number of cultures (N = 24 pair) and induced heat stress at 45°C. The metabolic processes were arising after the heat stress, at the end of this reaction bioactive compound were produced and separated. The separated compounds were allowed to dry at 50°C for few minutes. The dried samples were reconstituted with 100 µL of Triton-X 100 and stored at –20°C until completion of the study.

Animals and Maintenance

Fish of uniform size of length (2.6 ± 0.2 cm) and weight (1.15 ± 0.1 g) were segregated from the stock and acclimatized for 10 days to lab conditions, temperature (26±2°C), pH (7.5–7.8) and almost normal photoperiod (12:12-h L/D). The fish were divided into two groups of each.

- *Group 1. Control (CON):* Control fish received control diet throughout experimental period.
- *Group 2. Anxiety induced (AI):* Anxiety was induced to fish by kept under white bright light for 48 hrs. The fish were received control diet throughout experimental period.

- *Group 3. Anxiety induced + Bioactive compound fed (AI + B):* Anxiety was induced to fish by kept under white bright light for 48 hrs. The anxieties induced fish received 100 µL of bioactive compound with control diet throughout experimental period.

At the end of the experimental period the fish were sacrificed and processed for group performance assay and compared to the control group.

Table 10: Time Comparison Spent by the Fish for Each Extract Present in Aqueous

Zone	*F1*		*F2*	
	Near (sec.)	*Away (sec.)*	*Near (sec.)*	*Away (sec.)*
AFC	140	40	135	45
AFa1	60	120	65	115
AFa2	40	140	45	135
AFa3	55	125	70	110
AFa4	70	110	60	120
AFa5	45	135	65	115
AFa6	90	90	115	65
AFa7	70	110	80	100
AFa8	105	75	90	90
AFa9	70	110	80	100
AFa10	100	80	55	125
AFa11	45	135	50	130
AFa12	60	120	85	95
Avg	**67.5**	**112.5**	**72.5**	**107.5**

Results as shown in the above tables and graphs are quiet interestingly, Out of the 24 different crude extracts we have obtained all extract of compounds have showed a genuine levels of decrease in anxiety among the given zebrafish. Aqueous phase compounds have shown a better decrement of anxiety when compared to the butanol phase compounds. Out of the 24 extracts we got we have selected three of the best compounds to identify the compound. The three best bioactive compounds were selected such as A 2, B 4, B 11.

Table 11: Time Comparison Spent by the Fish for Each Extract Present in Butanol Phase

Zone	*F1*		*F2*	
	Near (sec.)	*Away (sec.)*	*Near (sec.)*	*Away (sec.)*
AFC	140	40	135	45
AFb1	90	90	75	105
AFb2	100	80	120	60
AFb3	105	75	100	80
AFb4	50	130	35	145
AFb5	65	115	70	110
AFb6	60	120	70	110
AFb7	100	80	120	60
AFb8	95	85	75	105
AFb9	120	60	100	80
AFb10	55	125	65	115
AFb11	70	110	65	115
AFb12	85	95	80	100
Avg	**82.91**	**97.09**	**81.25**	**98.75**

AFC - Anxiety induced fish control group; AFa - Anxiety induced fish treated with aqueous phase bioactive compounds; AFb - Anxiety induced fish treated with butanol phase bioactive compounds

Table 12: Comparison between Aqueous and Butanol Phase Compounds

Type of Bioactive Compounds	*Average Time Near to Group (sec.)*	*Average Time Away from Group (sec.)*
AFC	137.5	42.5
AFa1-12	70	110
AFb1-12	82.08	97.91

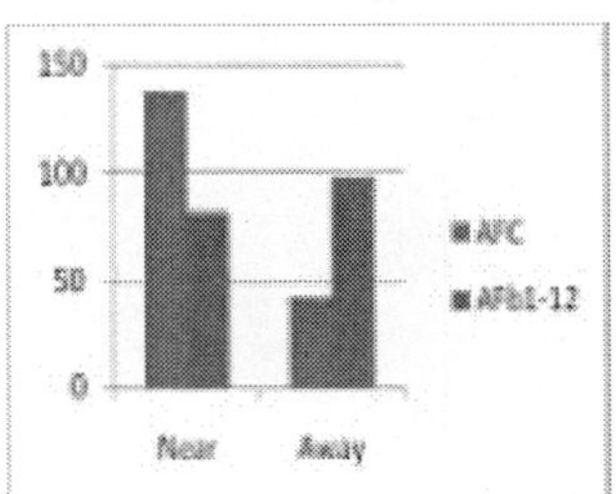

Graph 4: Comparison between Aqueous and Butanol Bioactive Compounds

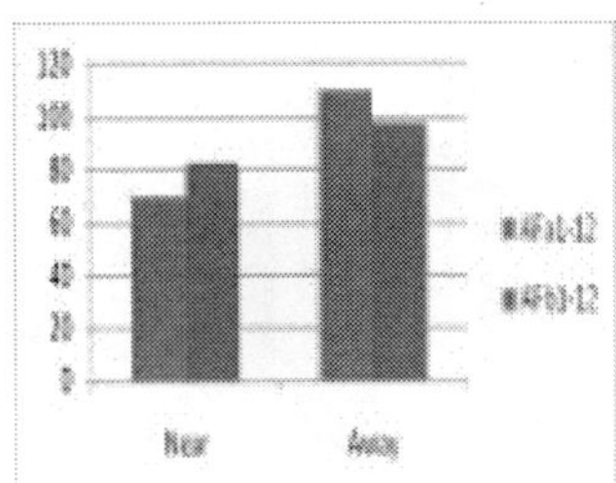

Graph 5: Comparison between Control and Aqueous Phase Bioactive Compounds

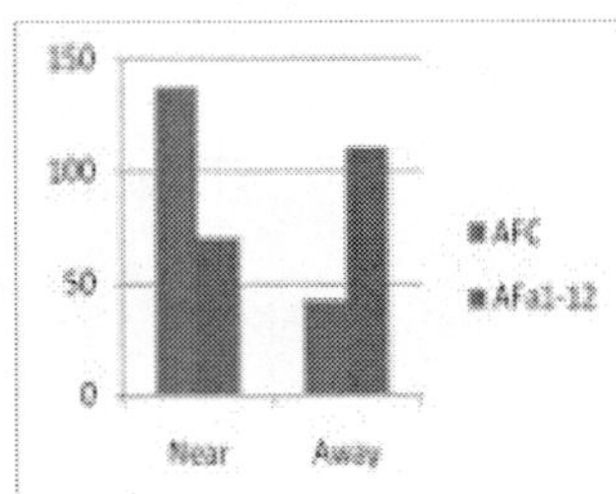

Graph 6: Comparison between Control and Butanol Phase Bioactive Compounds

Table 13: Control Group

Time Spent	*Near to Group (sec.)*	*Away from Group (sec.)*
AFc1	110	70
AFc2	125	55
Avg	62.5	117.5

Table 14: A2 Group

Time Spent	*Near to Group (sec.)*	*Away from Group (sec.)*
AFa2-1	65	115
AFa2-2	Expired	

Table 15: A2 Group

Time Spent	*Near to Group (sec.)*	*Away from Group (sec.)*
AFa11-1	25	155
AFa11-2	20	160
Avg	22.5	157.5

Table 16: B4 Group

Time Spent	*Near to Group (sec.)*	*Away from Group (sec.)*
AFb4-1	125	55
AFb4-2	90	90
Avg	107.5	72.5

Table 17: Comparison of Group Preference Test among Control to A11

Group of Fish	*Near to Group (sec.)*	*Away from Group (sec.)*
Avg of control	117.5	62.5
Avg of A11	22.5	157.5

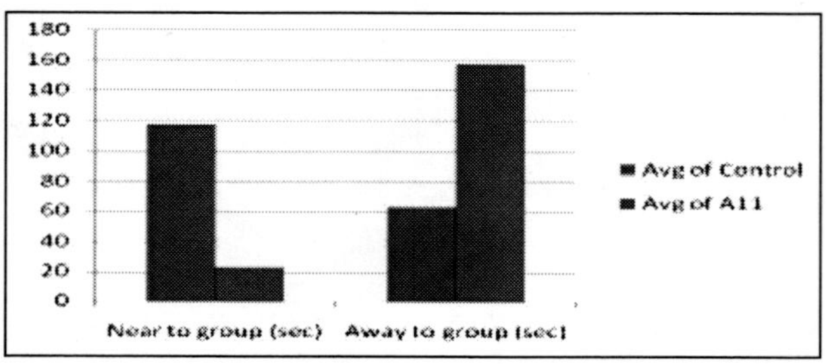

Graph 7: Graphical Representation of Comparative Group Preference Test among Control to A11

SUMMARY AND CONCLUSION

Fruit peels were collected and endophytes were allowed to grow in sterile conditions and the colonies were isolated and were placed in the nutrient medium. The colonies were heat stressed at 45°C to produce bioactive compounds. These compounds were isolated and butanol was added to separate polar and non-polar compounds in different phases, and they were dried at 50°C and reconstituted using Triton X-100. These compounds were taken and they were screened to the zebrafish at 100 μL. These zebrafish were induced with anxiety under bright light for 48hrs before giving the crude extracts of bioactive compounds. The fish were subjected to group preference tests to identify the improvement of anxiety levels in the zebrafish. There were promising significant results among the screened compounds.

Many novel and valuable bioactive compounds with antimicrobial, insecticidal, cytotoxic and anticancer activities have been successfully obtained from the endophytic fungi (Schneider *et al.,* 2008). Almost 10 years ago, Robert Gerlai noted the lack of behavioural studies using zebrafish, a species which showed great potential as behavioural genetic model (Gerlai, 2003). In the last decade, many different models of anxiety and fear in zebrafish appeared, and it is now coming clear that this species shows great potential as a pharmacological tool (Goldsmith, 2004). From the present study, it is quite apparent those endophytic fungi bioactive compounds are found to reduce anxiety in zebrafish. Apart from the huge number of research studies in the field of behavioural genetics and drug development, the field still needs more attention from scientists around the world.

ACKNOWLEDGEMENT

We would like to acknowledge the support of the Management Committee who provided us necessary facilities to conduct this research work.

REFERENCES

Abdessamad, D. (2009). "Bioactive secondary metabolites from the endophytic fungus *Chaetomium* sp. isolated from *Salvia officinalis* growing in Morocco", *Biotechnol. Agron. Soc. Environ.* Vol. 13(2), pp. 229–234.

Fero, K., Yokogawa, T., Burgess Ha. (2010). 'The Behavioural Repertoire of Larval Zebrafish. Zebrafish Models in Neurobehavioral Research', Cambridge University Presskalueff Av, Cachat Jm.

Gerlai, R. (2003). 'Zebra fish: an uncharted behavior genetic model', *Behav Genet,* Vol. 33, pp. 461–8.

Goldsmith, P. (2004). 'Zebrafish as a pharmacological tool: the how, why and when', *Curr Op Pharmacol,* Vol. 4, pp. 504–12.

Hundley, Nicholas J. (2005). "Structure elucidation of bioactive compounds isolated from endophytes of Alstonia scholaris and Acmena graveolens", Department of Chemistry and Biochemistry, Birmingham Young University 2005.

Korte, S.M. (2001). 'Corticosteroids in Relation to Fear, Anxiety and Psychopathology', *Neurosci Biobeha Rev.*, Vol. 25, pp. 117–42.

Mariana, R. (2011). "The Use of Endophytes to Obtain Bioactive Compounds and Their Application in Biotransformation Process", *Biotechnology Research International.* Article ID 576286.

Schneider, P., Misiek, M. and Hoffmeister, D. (2008). 'In vivo and in vitro production options for fungal secondary metabolites', *Molecular Pharmaceutics,* Vol. 5, pp. 234–242.

Strobel, G., Daisy, B., Castillo, U. and Harper, J. (2004). 'Natural products from endophytic microorganisms', *J. Nat. Prod.*, Vol. 67, pp. 257–268.

William Norton (2010). "Adult zebrafish as a model organism for behavioural genetics", *BMC Neuroscience,* Vol. 11, p. 90.

YIN Yan-yan (2007). "Bioactive compounds from Paecilomyces tenuipes regulating the function of the hypothalamo-hypophyseal system axis in chronic unpredictable stress rats", *Chin Med J.,* Vol. 120 (12), pp. 1088–1092.

An *in silico* Approach to Enhance the Bioremediation Ability of Versatile Peroxidases from *Pleurotus eryngii*

Nishant Saurab, C. Murugan and C. Nirmala*
Department of Bioinformatics, VMKV Engineering College, Salem
E-mail: nirmalabt@gmail.com

INTRODUCTION

Bioremediation is described as the process, whereby organic wastes are biologically degraded under controlled conditions to an innocuous state. PAHs are one of the major environmental pollutants. They are ubiquitous, persistent organic pollutants and are slow to degrade in the environment. They are highly hydrophobic in nature thus having low water solubility (Cerniglia, 1992). Many PAHs are found to be potent carcinogens. They are also cytotoxic and mutagenic that causes serious risk to human health.

The fungi are unique among microorganisms in secreting a variety of extracellular enzymes. Those enzymes are found to be capable of degrading a large number of different contaminants (Timothy, 2006). Fungi degrade lignin by secreting ligninases, an extracellular enzyme system. Among ligninases secreted by white rot fungi, Versatile Peroxidase (E.C 1.11.1.16) is a new ligninolytic (hybrid) enzyme, combining catalytic properties of manganese peroxidase (oxidation of Mn(II)), lignin peroxidase (Mn-independent oxidation of non-phenolic aromatic compounds) and plant peroxidase (oxidation of hydroquinones and substituted phenols). VPs oxidize not only Mn (II) but also Veratryl alcohol, phenolic, non-phenolic and high molecular weight compounds, including dyes in Mn-independent reactions (Wong, 2009).

*Corresponding author.

However, improvement of the catalytic properties is a current challenge for an efficient bioremediation of the PAHs. Owing to the complexity of PAHs, the enzymes that could fit as much as possible the bioavailability requirements in order to successfully remediate PAHs from the environment are designed.

It is known that the solubility and bioavailability of PAHs decrease with the increase of the PAHs molecular mass. By increasing the hydrophobicity of versatile peroxidase, the chance of the enzyme to meet the hydrophobic PAHs molecules can be increased. This would enhance the bioremediation efficiency of the enzyme against the PAHs. However, it is required to identify the residues and manipulate them without affecting the protein folding that could devastate the enzymatic function. In order to identify the suitable residues for mutagenesis experiments, the relationship between ET conservation, molecular surface accessibility and hydrophobicity pattern were analysed to design Versatile Peroxidase with higher hydrophobicity. The results would provide details on potential amino acid residues that could be manipulated without affecting its structure and function.

METHODOLOGY

Target Sequence and Homologous Sequence Search

The FASTA sequence of Versatile Peroxidase (E.C.1.11.1.16) from *Pleurotus eryngii* was retrieved from National Centre for Biotechnology Information (NCBI) sequence database with the accession number AAZ04666. The enzyme was used as the seed sequence for homologous sequence search that was performed using protein BLAST of SWISS PROT database (Altschul *et al.*, 1990). Protein sequences with more than 50% identity were selected (Bairoch *et al.*, 2004).

Homology Modeling of Target Versatile Peroxidases

The target sequence obtained from NCBI sequence database was used to build the model using SWISS-MODEL server. It is a fully automated protein structure homology-modeling server, accessible via the ExPASy web server, or from the program DeepView (Swiss Pdb-Viewer). The 3D structure of the custom enzyme was obtained.

Multiple Sequence Alignment and Evolutionary Trace (ET) Analysis

The protein sequences selected from Blast search were aligned together with ClustalW (Thompson *et al.*, 1994) using the Gonnet protein weight matrix (Gonnet *et al.*, 1992). A rooted phylogenetic tree was constructed based on the neighbor-joining algorithm from the multiple sequence alignment and visualized by PhyloDraw (Choi *et al.*, 2000).

The conserved residues obtained from multiple sequence alignment were mapped to the 3D structure of the Versatile Peroxidase protein using Evolutionary Trace Server. Evolutionary trace server divides the dendrogram into ten partitions. Based on the partitions the multiple sequences were clustered into three groups. Consensus residues from the multiple sequence alignment were classified as neutral, conserved and group-specific. The trace residues were then mapped onto the known 3D structures of Versatile Peroxidase obtained from SWISS MODEL server (Bertrand *et al.*, 2002) and the mapped structure was visualized by Rasmol (Sayle *et al.*, 1995).

Predicting Molecular Surface Accessibility

The molecular surface accessibility of the target enzyme was analyzed by WHATIF web server (Vriend, 1990). The accessible residues were mapped onto the modeled structure and visualized by Rasmol.

Hydrophobicity Analysis

The hydrophobicity of the residues in target enzyme was analysed using 3 different hydrophathy definition developed by Kyte and Doolittle (Kyte and Doolittle, 1982), (Rose *et al.*, 1985) and (Lins *et al.*, 2003). Residues are classified as hydrophilic and hydrophobic in Kyte-Doolittle definition. According to Rose *et al.* (1985), hydropathy classification, amino acid residues are divided into hydrophilic, hydrophobic and neutral as defined by Biological Magnetic Resonance Data Bank (Ulrich *et al.*, 2007). Lins *et al.* (2003), defined the residues as hydrophilic, hydrophobic, intermediate and special features. Hydrophobicity status are mapped onto the accessible residues of target enzyme and visualized by Rasmol.

RESULTS AND DISCUSSION

The target protein obtained from NCBI sequence database has 361 aminoacids with accession number AAZ04666. 19 homologous sequences with more than 50% identity to target protein were selected for multiple sequence alignment. A phylogenetic tree based on the neighbor-joining algorithm was generated and visualized by PhyloDraw. The multiple sequence alignment and the phylogram were used as input for ET analysis. ET server generates classes or groups of residues by dividing the phylogram with evolutionary time cutoff lines. A single ET partition cutoff was drawn to the phylogram that divided the tree into 3 groups of protein sequences (Figure 1).

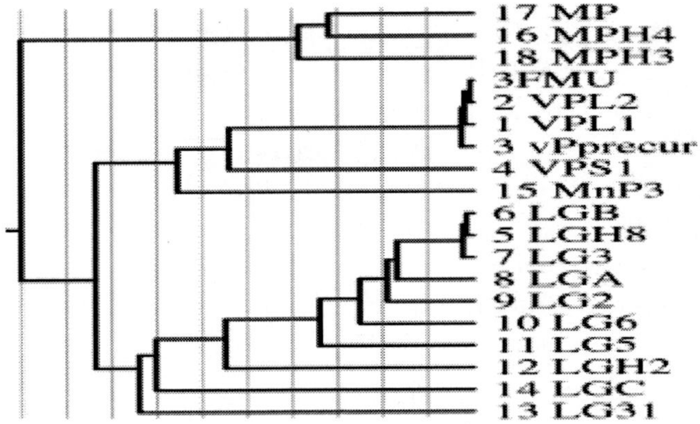

Fig. 1: Dendrogram Obtained from ETS

The members of Group 1 have 50 to 53% identity to AAZ04666. The members of Group 2 have more than 70% identity and the members of group 3 have 55 to 60% identity. Sequences from each group were separately aligned and compared to each other to develop a consensus sequence. The consensus sequence was aligned with the seed sequence to identify the position of conserved and group-specific residues on the seed sequence. 95 ET residues are identified at the cutoff partition with 75 conserved residues and 20 group-specific residues. ET analysis revealed that the selected versatile peroxidases are 20.7% evolutionarily conserved. Then the position of ET residues is mapped onto the known 3D structure of versatile peroxidase (3FMU). Most of the conserved residues are identified inside the 3D structure of the target enzyme.

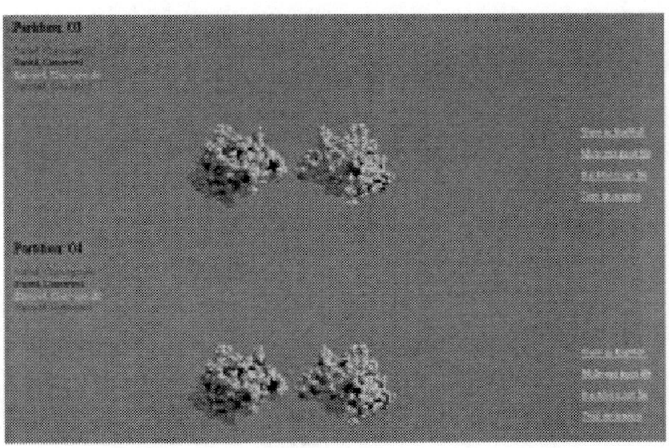

Fig. 2: Mapping of Conserved Residues on Modeled Enzyme

The 3D structure of target enzyme was run in the WHATIF web server to calculate the molecular surface accessibility. 200 residues were identified as surface accessible and mapped them onto modeled structure and visualized by Rasmol. Supporting the result, 97 out of 200 surface accessible residues are unconserved in ET analysis.

The hydrophobicity of the surface accessible residues was identified using different hydropathy scales. The Kyte-Doolittle scale is widely applied for delineating the hydrophobic character of a protein. The scale was derived from the physicochemical properties of amino acid side chains. On the other hand, the Rose scale was constructed by examining proteins with known 3D structures. Hydrophobic distribution of the molecular surface residues according to Kyte-Doolittle and Rose definition was mapped onto target enzyme.

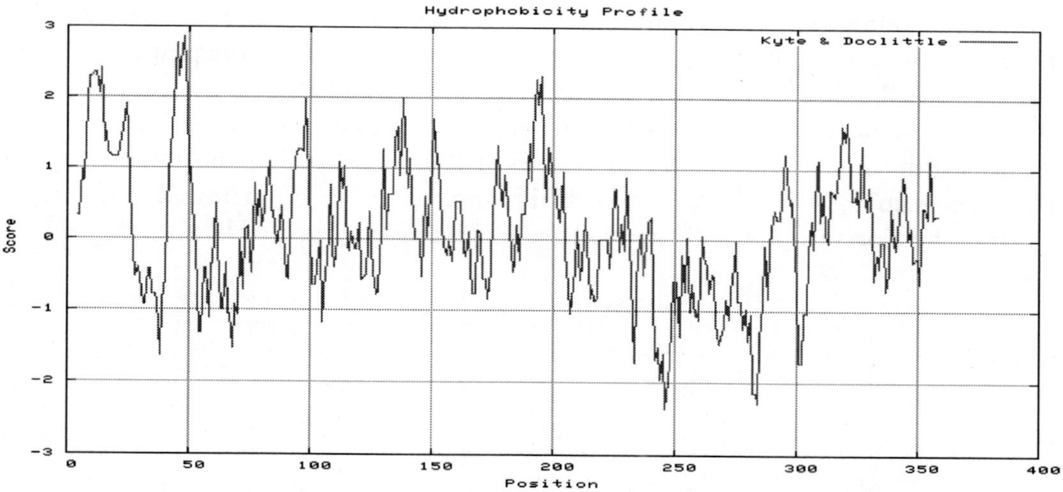

Fig. 3: Kyte-Doolittle Hydrophobicity Plot for Modelled Enzyme

There was not much difference found in the hydrophobic distribution between both hydrophobicity scales. Hydrophobic clustering on the surface of modeled enzyme was hardly identified. Assume that the hydrophilic distribution in modeled enzyme is scattered all around the molecule surface to ensure that the protein is soluble in water phase. Suppose those hydrophilic molecules that are not conserved in ET analysis are the possible residues to be manipulated in order to produce a higher hydrophobicity versatile peroxidase for bioremediation. The hydrophobicity analysis with different hydropathy definition as reported by Lins *et al.,* (2003) were tested and analyzed the relationship between amino acids hydrophobic and hydrophilic accessible surface in extended state and in folded protein. Based on the relationship, they classified 3 groups of amino acids. Ile, Leu, Val, Phe, Met, Ala and Gly are grouped as hydrophobic residues, while Asp, Asn, Glu, Gln and Arg are grouped as hydrophilic residues. The third group that holds an intermediate property was also identified, which consists of His, Tyr, Trp, Ser and Thr. However, Pro and Lys were reported to show a distinct property and were not classified into any group. There were no much changes found in the hydrophobic and hydrophilic distribution of Lins definition, compared to the hydrophobicity definition of Kyte-Doolittle and of Rose.

The residue of choice for the design of high hydrophobicity versatile peroxidase should be the one that was not conserved in ET analysis, located on the surface of the 3D structure and a hydrophilic residue. The hydrophobicity analysis showed that the residues Asp, Glu, Arg, Asn and Gln as candidate that can be chosen for mutagenesis experiments. However, the residues Asp, Glu and Arg can be avoided because modifying them would alter the pI of the protein. Considering ET, molecular surface accessible, hydrophobicity analysis and the factors mentioned above, Asn at the position 57, 104, 126, 244, 279, 280 and 286 and Gln 63, 249, 253, 259, 269, 281, 285, 300, 335 are concluded as possible candidates for site directed mutagenesis. Mutating the above residues with suitable hydrophobic molecules at favourable torsion angle, the chance of the enzyme to meet the hydrophobic PAHs molecules can be increased. This would enhance the bioremediation efficiency of the enzyme against the PAHs.

Table 1: Potential Residues for Mutagenesis Experiments

Residue No.	*Wild Type*	*Mutation*	*Predicted* $\Delta\Delta G$ *(kcal/mol)*
63	Gln	Met	0.02
249	Gln	Ala	0.24
253	Gln	Pro	0.55
259	Gln	Met	0.41
269	Gln	Ile	1.09
281	Gln	Ile	0.29
285	Gln	Ala	0.08
335	Gln	Leu	0.28
57	Asn	Val	0.11
104	Asn	Ala	0.81
126	Asn	Pro	1.64
244	Asn	Pro	0.55
280	Asn	Trp	2.21
286	Asn	Ile	1.1

A diverse group of enzymes from white rot fungi have the ability to oxidise PAHs, which have hydrophobic characteristics and low solubility. Versatile peroxidase is applicable for PAHs enzymatic bioremediation. By increasing the hydrophobicity of Versatile Peroxidase, the molecular mass of PAH can be increased thus solubilising them from the environment. The design of versatile peroxidase to optimize PAH enzymatic bioremediation is a challenge for today's biotechnology. Not all residues in a protein are carrying the same weight for a protein to function properly. Identifying which residue to be manipulated is a crucial step in designing a protein for specific purpose. The residues which were not conserved, present on the surface of the protein as well as which were hydrophilic in nature may satisfy our objective during their mutation into hydrophobic residues. The relationship between ET method, hydrophobicity studies and accessible surface analysis are studied to identify the suitable residues to be manipulated, in order to design versatile peroxidase with higher hydrophobicity. This analysis provides basic knowledge for designing high hydrophobicity versatile peroxidase which can be exploited for PAHs enzymatic bioremediation, as well as in industrial usage in future.

REFERENCES

Altschul, S.F., Gish, W., Miller, W., Myers, E.W. and Lipman, D.J. (1990). Basic local alignment search tool. *J. Mol. Biol.,* 215: 403–410.

Bairoch, A., Boeckmann, B., Ferro, S. and Gasteiger, E. (2004). Swiss-Prot: juggling between evolution and stability, *Brief Bioinform,* 5:39–55.

Bertrand, T., Jolivalt, C., Briozzo, P., Caminade, E., Joly, N., Madzak, C. and Mougin, C. (2002). Crystal structure of a four-copper laccase complexed with an arylamine: insights into substrate recognition and correlation with kinetics, *Biochemistry,* 41: 7325–7333.

Cerniglia, C.E. (1992). Biodegradation of polycyclic aromatic hydrocarbons. *Biodegradation* 3: 351–368.

Choi, J.H., Jung, H.Y., Kim, H.S. and Cho, H.G. (2000). PhyloDraw: A phylogenetic tree drawing system. *Bioinformatics,* 16: 1056–1058.

Gonnet, G.H., Cohen, M.A. and Benner, S.A. (1992). Exhaustive matching of the entire protein sequence database, *Science,* 256:1443–1445.

Kyte, J. and Doolittle, R.F. (1982). A simple method for displaying the hydropathic character of a protein, *J. Mol. Biol.,* 157:105–132.

Lichtarge, O., Bourne, H.R. and Cohen, F.E. (1996). An evolutionary trace method defines binding surfaces common to protein families, *J. Mol. Biol.,* 257: 342–358.

Lins, L., Thomas, A. and Brasseur, R. (2003). Analysis of accessible surface of residues in proteins, *Protein Sci.,* 12: 1406–1417.

Parthiban, V., Gromiha, M.M. and Schomburg, D. (2006). CUPSAT: prediction of protein stability upon point mutations, *Nucleic Acids Res.,* 34: W239–242.

Rose, G.D., Geselowitz, A.R., Lesser, G.J., Lee, R.H. and Zehfus, M.H. (1985). Hydrophobicity of amino acid residues in globular proteins. *Science,* 229: 834–838.

Thompson, J.D., Higgins, D.G. and Gibson, T.J. (1994). CLUSTAL W: improving the sensitivity of progressive multiple sequence alignment through sequence weighting, position-specific gap penalties and weight matrix choice. *Nucleic Acids Res.,* 22: 4673–4680.

Timothy, P. Ruggaber (2006). Enhancing Bioremediation with enzymatirc processes: A Review, Practice Periodical of hazardous,toxic, and radioactive waste management © ASCE, 73–85.

Vriend, G. (1990). WHAT IF: a molecular modeling and drug design program. *J. Mol. Graph,* 8: 52–56.

Wong, D.W. (2009). Structure and action mechanism of ligninolytic enzymes. *Appl Biochem Biotechnol,* 157: 174–209.

In-silico Study on ns-Synonymous SNPs of *APOE* Gene Responsible for Cardiovascular Disease

Sai Ramesh A.*, Padma Thiagarajan and Rao Sethumadhavan

School of Biosciences and Technology, VIT University, Vellore, Tamil Nadu

E-mail: sairamesh_24@yahoo.com

INTRODUCTION

World Health Organization reports that 36 million people die globally of non-communicable diseases and of these, 49% of mortality is due to cardiovascular diseases (CVD), Indian population is a major contributor to this figure (WHO 2011). Advances in the field of molecular genetics have been instrumental in identifying numerous polymorphisms in various genes, and their variations are strongly predisposed towards the development of CVD (Sai ramesh *et al.,* 2011). The variations in such genes resulting in deleterious effect can be analyzed to a certain extent by the process of computational biology (Ramnekey *et al.*, 2002). The genes Apo-lipoproteins E, selected for investigation, is a well-marked risk factor of CVD and been found to different between myocardial infarction survivors and the controls (A. M. Bennet *et al.,* 2007). The *APOE* polymorphism has been predicted to be responsible for 4%–8% of the variation in age- and sex adjusted total cholesterol (TC) and LDL cholesterol (LDL-C) levels in the general Caucasian population and may contribute to risk of CVD (J. Davignon *et al.,* 1988).

The study uses various computational tools such as SIFT, PolyPhen 2.0, PROVEAN 1.0 and PANTHER for identifying its deleterious effects. The RMSD value has been used to predict the deleterious influence on the gene, which in-turn may reflect on CVD predisposition.

---

*Corresponding author.

MATERIALS AND METHODS

SNP Retrieval and Function Prediction

Data for the gene *APOE* was retrieved from the NCBI-dbSNP (S.T. Sherry *et al.*, 2001) (http://www.ncbi.nlm.nih.gov/snp) by limiting the search only to non-coding synonymous (ns-SNPs) functional class. The ns-SNPs retrieved were investigated for their deleterious effect using standard computational tools listed below.

SIFT B Link

[http://sift.jcvi.org/www/SIFT_BLink_submit.html], an online server which helped in **S**orting the **I**n-tolerant **F**rom **T**olerant based on Homology Modeling (G.P. Jarvik *et al.*, 1998). The threshold value for predicting a substitution as deleterious was <0.05.

PolyPhen 2.0

The input for the web server PolyPhen 2.0 (**Poly**morphism **Phen**otyping Version 2.0) []http://genetics.bwh.harvard.edu/pph2/] was FASTA sequence of the selected genes. The server uses iterative greedy algorithm in predicting the damaging missense mutations. The cut-off score for any variation to be damaging was > 0.432.

PROVEAN 1.0

PROVEAN (**Pro**tein **V**ariation **E**ffect **An**alyzer) [http://provean.jcvi.org/about.php] assists in predicting the protein function effect caused due to their sequence variations. The server calculated and sorted them to be deleterious if the score was >2.282.

PANTHER

The PANTHER (**P**rotein **AN**alysis **TH**rough **E**volutionary **R**elationships) classified genes based on their functions and evolutionary relationships. In the case of prediction through evolutionary relationship, the tool estimated the likelihood of particular ns-SNPs which affects the protein function. Following subPSEC (substitution position-specific evolutionary conservation), score was calculated based on an alignment of evolutionarily related proteins.

Structure Modeling and RMSD Value Calculations

Only the ns-SNPs, predicted to be deleterious in all the above tools were used for further analysis. The 3D structures for *APOE* was obtained from SWISS-MODEL and validated using PROCHECK. The validated model was used to generate the mutant structures through Deepview/Swiss PdbViewer. Further, energy minimization through an online server No-mad Ref was carried out for both native and mutants models. Finally, the energy minimized structures of mutant was superimposed on native to calculate, Root Mean Square Deviation (RMSD) by Swiss PdbViewer. This is a simple method for detecting the effect of mutation that can be seen on a protein. Both the RMSD and Total Energy value computed defined how deviant a mutant was over the native and on stability of the protein structure.

RESULTS AND DISCUSSION

SNP Retrieval and Function Prediction

A total of 35 ns-SNPs were retrieved for the gene *APOE.* Of these, 16 ns-SNPs were predicted to be deleterious by both SIFT and PolyPhen 2.0. The deleterious and damaging ns-SNPs predicted by the SIFT and PolyPhen were further analyzed by PROVEAN 1.0 and PANTHER. The predictions finally identified 3 ns-SNPs of *APOE* to be deleterious in common by all tools and the same is shown as **bold** in Table 1. The study employed four tools for predicting the deleterious effect and this was because by combining more than one tool the accuracy of the prediction can be increased.

Structure Modeling and RMSD Value Calculations

The 3D structure of *APOE,* modeled using SWISS MODEL resulted in two models 2kc3A and 2l7b which had different residue ranges. The study employed both models for analysis, as deleterious nsSNPs were placed in those residue ranges. The validation for the 3D structure models performed using PROCHECK revealed maximum residues in most favored regions, there by validating the structures as good.

Based on the residue ranges, the models were selected for creating the mutants using SWISS PDBViewer, The RMSD value was calculated for the deleterious ns-SNPs of *APOE* which were predicted in common by all tools and is listed in Table 2. The analysis showed that, only one variant of *APOE* (**S215F**) had RMSD value >1.0Å. Many studies report that higher the RMSD value, greater are the deviations between the native and mutant.

It is now clear that from the screened variants, that **S215F** of *APOE* may be of high concern in affecting the structure and function of a protein, thereby making it a candidate gene for CVD. The molecular graphical representation of these variants were drawn using The PyMol Molecular Graphics System and are shown in Figure 1.

APOE - 17/35 (NP_000032.1)

S. No.	*rs ID*	*Allele*	*Position*	*Residue Change*	*SIFT Score*	*Polyphen 2 Score*	*Provean Score*	*Panther subPSEC*
1.	**rs11542029**	**C/T**	**50**	**R50C**	**0**	**1**	**–3.807**	**–3.74473**
2.	rs11542040	C/A	102	P102T	0	0.937	–6.417	–1.84238
3.	rs11083750	C/A	102	P102Q	0	0.984	–6.472	–2.55917
4.	rs11083750	C/G	102	P102R	0	0.995	–7.224	–2.5765
5.	rs28931577	G/A	117	A117T	0.02	0.897	–1.348	–1.55914
6.	rs11542041	C/A	132	R132S	0	1	–4.066	–2.57179
7.	rs41382345	A/T	139	E139V	0	0.993	–5.913	–1.70737
8.	rs11542034	A/G	150	E150G	0	0.991	–4.272	–1.51518
9.	rs28931578	G/A	152	R152Q	0	0.965	–1.289	–0.8305
10.	rs121918393	C/A	154	R154S	0.01	0.924	–3.751	–2.60905
11.	rs769455	C/T	163	R163C	0	0.995	–4.749	–2.47965
12.	rs121918394	A/C	164	K164Q	0.01	0.842	–2.246	–1.07687
13.	**rs7412**	**C/T**	**176**	**R176C**	**0.01**	**1**	**–2.693**	**–4.54079**
14.	rs11542032	G/A	189	E189K	0	0.999	–1.381	–0.55427
15.	**rs11542027**	**C/T**	**215**	**S215F**	**0**	**0.98**	**–2.351**	**–3.49266**
16.	rs121918395	C/T	246	R246C	0	0.927	–2.419	–1.24848
17.	rs121918398	G/A	292	R292H	0	0.981	–1.344	-1.2282

APOE - 3/17 (NP_000032.1)

S. No.	*rs ID*	*Residue Change*	*Swiss PDB*		
			Energy	*Backbone RMSD (Å)*	*Native (Å)*
1.	rs11542029	R50C	–14198	0.2	–14,353
2.	rs7412	R176C	–14123	0.6	**–23, 843**
3.	**rs11542027**	**S215F**	**–23,598**	**1.8**	

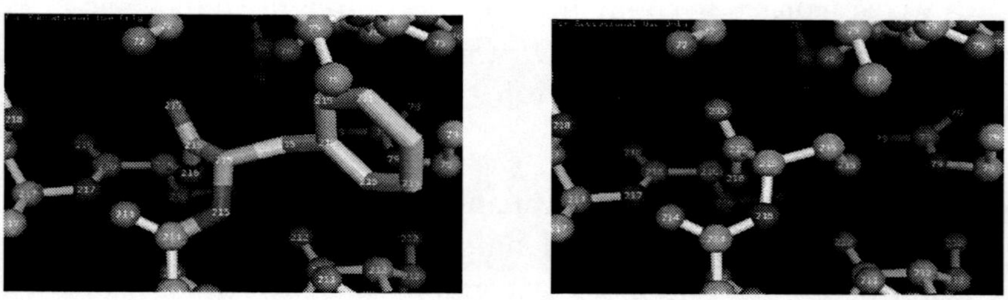

Fig. 1: Native and Mutant Structure of *APOE* (S215F)

CONCLUSION

The study examined *APOE* gene responsible for CVD through various computational evaluation tools. The gene had a total of 35 ns-SNPs, of which 3 ns-SNPs were predicted to be deleterious by all tools in common. Of the mutation analysis performed on the 3 deleterious ns-SNPs, only one ns-SNP (**S215F**) had RMSD value greater 1.0Å showing that the mutant structure is highly deviated from the native. It may be concluded that the ns-SNPs **rs11542027** may be the major target mutation for CVD caused by *APO E.*

REFERENCES

Adzhubei, I.A., Schmidt, S., Peshkin, L., Ramensky, V.E., Gerasimova, A. Bork, P., Kondrashov, A. S. and Sunyaev, S. R. (2010). A method and server for predicting damaging missense mutations, *Nat Methods*; 7 248–249.

Bennet, A.M., Angelantonio, E. Di, Ye, Z., Wensley, F., Dahlin, A., Ahlbom, A., Keavney, B., Collins, R., Wiman, B., de Faire, U., Danesh, J. . Association of apolipoprotein E genotypes with lipid levels and coronary risk, *JAMA* 298 (2007) 1300–1311.

Cumming, A.M., Robertson, F.W., Polymorphism at the apoprotein-E locus in relation to risk of coronary disease. *Clin Genet* 25, (1984), 310–313.

Davignon, J., Gregg, R.E. and Sing, C.F. (1988). Apolipoprotein E polymorphism and atherosclerosis, *Arteriosclerosis,* 8, 1–21.

DeLano, W.L. (2010). The PyMOL Molecular Graphics System, Version 1.3r1. DeLano Scientific LLC, South San Carlos, CA.

George, C., Doss, Priya and NagaSundaram, N. (2011). Investigating the Structural Impacts of I64T and P311S Mutations in APE1-DNA Complex: A Molecular Dynamics Approach, PLos ONE, 7 e31677.

Grillo, G., Turi, A., Licciulli, F., Mignone, F., Liuni, S., Banfi, S., Gennarino, V.A., Horner, D. S., Pavesi, G., Picardi, E. and Pesole, G. (2010). UTRdb and UTRsite (RELEASE 2010) a collection of sequences and regulatory motifs of the untranslated regions of eukaryotic mRNAs, *Nucleic Acids Res.* 38, D75–D80.

Guex, N. and Peitsch, M.C. (1997). SWISS-MODEL and the Swiss-PdbViewer: An environment for comparative protein modeling, *Electrophoresis.* 18, 2714–2723.

Han, J.H., Kerrison, N., Chothia, C. and Teichmann, S.A. (2006), Divergence of Interdomain Geometry in Two-Domain Proteins, *Structure* 14, 935–945.

Jarvik, G.P., Goode, E.L., Austin, M.A., Auwerx, J., Deeb, S., Schellenberg, G.D. and Reed, T. (1997). Evidence That the Apolipoprotein E–Genotype Effects on Lipid Levels Can Change with Age in Males: A Longitudinal Analysis, *Am. J. Hum. Genet.,* 61, 171–181.

Kiefer , F., Arnold, K., Künzli, M., Bordoli, L. and Schwede, T. (2009). The SWISS-MODEL Repository and associated resources, *Nucleic Acids Res.* 37, D387–392.

Kumar, P., Henikoff, S. and Ng, P.C. (2009). Predicting the effects of coding non-synonymous variants on protein function using the SIFT algorithm. *Nat Protoc.,* 4, 1073–1081.

Laskowski, R. A., MacArthur, M. W., Moss, D. S. and Thornton, J. M., (1993). PROCHECK—a program to check the stereochemical quality of protein structures, *J. App. Cryst.* 26, 283–291.

Laskowski, R.A., MacArthur, M.W. and Thornton, J.M. (2001). PROCHECK: validation of protein structure coordinates, in *International Tables of Crystallography, Volume F. Crystallography of Biological Macromolecules,* eds. M. G. Rossmann and E. Arnold, Dordrecht, Kluwer Academic Publishers, The Netherlands, 722–725.

Lindahl, E., Azuara, C. , Koehl, P. and Delarue, M. (2006). NOMAD-Ref: visualization, deformation and refinement of macromolecular structures based on all-atom normal mode analysis. *Nucleic Acids Res.* 34, W52–56.

Ng, P.C. and Henikoff, S. (2001). Predicting Deleterious Amino Acid Substitutions. *Genome Res.* 11, 863–874.

PROVEAN. Protein Variation Effect Analyzer. Available: http://provean.jcvi.org/index.php.

Ramavartheni, K. and Sethumadhavan, R. (2010), Identification of structurally and functionally significant deleterious nsSNPs of *GSS* gene: *in silico* analysis, Advances in Bioscience and Biotechnology. 1, 368–373.

Ramensky, V., Bork, P. and Sunyaev, S., Human non-synonymous SNPs: server and survey. *Nucleic Acids Res.* 30 (2002), 3894–3900.

Sai Ramesh, A., Vyos Dhamodaran, Padma Thiagarajan, Role of genetic variants in development of Cardiovascular Diseases. *International Journal of Pharmaceutical Science and Health Care,* 2 (2012) 45–53.

Sherry, S.T., Ward, M.H., Kholodov, M., Baker, J., Phan, L., Smigielski, E.M. and Sirotkin, K. (2001). dbSNP: The NCBI database of genetic variation. *Nucleic Acids Research,* 29, 308– 311.

Thomas, P.D. and Kejariwal, A. (2004). Coding single-nucleotide polymorphisms associated with complex vs. Mendelian disease: Evolutionary evidence for differences in molecular effects, *PNAS.* 101, 15398–15403.

Thomas, P.D., Campbell, M.J. and Kejariwal, A. (2003). PANTHER: A Library of Protein Families and Subfamilies Indexed by Function, *Genome Res.* 13, 2129–41.

Varfolomeev, S.D., Uporov, I.V. and Fedorov, E.V. (2002). Bioinformatics and molecular modeling in chemical enzymology: Active sites of hydrolases, *Biochemistry (Mosc),* 67: 1099– 1108.

World Health Organization. Non-Communicable Disease Country Profile 2011, World Health Organization report (2011).

Yuan, H.Y., Chiou, J.J., Tseng, W.H., Liu, C.H., Liu, C.K., Lin, Y.J., Wang, H.H., Yao, A., Chen, Y.T. and Hsu, C.N. (2006). FASTSNP: an always up-to-date and extendable service for SNP function analysis and prioritization. *Nucleic Acids Res.* 34, 635–641.

Preparation and Characterization of a Novel Bone Implant Material Using Nanohydroxyapatite and Beta Tri Calcium Phosphate Conjugated with Egg Yolk as a Pore Creating Agent

S. Srividya*, T.P. Sastry[1] and Kanchana Amarnath

Department of Biochemistry, Sathyabama University Dental College and Hospital, Chennai

[1]Bioproducts Lab, Central Leather research Institute, Chennai

E-mail: srividyabiomed@gmail.com

INTRODUCTION

Ceramics, polymers, metals and composite materials, with all advantages and drawbacks they have, are developed to be used for bone problems. There are a lot of concerns have been put in order for the researchers to consider the best way to respond to the demand on the biomaterials products since there are a lot of shortcomings derived from allograft and autograft as bone substitutes. Third-generation biomaterials have been developed with a focus on tissue regeneration to promote specific cellular responses at a molecular level with minimally invasive surgery. Bioactive materials compared to bone have very low mechanical strength whereas metals have superior mechanical properties but they are very corrosive. In spite of other desired properties such as wear resistance, biocompatibility and hardness, ceramic materials are brittle and possess low fracture toughness. Hence, the produced bone implants should meet several criteria before it can be applied to human body. These are the major challenges for the researchers in order to develop porous bone implant with desired porosity and acceptable mechanical strength. For artificial bone applications,

*Corresponding author.

bioceramics obtained from calcium phosphate family is frequently used due to its properties that mimic the natural human bone.

On the other hand, HA and βTCP are most favorable because of their good biocompatibility and osteointegrative properties (Amit *et al.,* 2001). However, βTCP is preferred for HA due to its high resorbing capacity than HA, as it shows least absorption and remains in the body for a long time as foreign particles (J.F. Keating, M.M. McQueen, 2001). In addition, an artificial bone implant for the replacement therapy, should exhibit porosity as to encourage cell growth within the three dimensional structure. In order to mimic the design of the mineral phase of living bone (B. Flautre, 2001) the pores must be interconnected to each other. Therefore a need to create and control the porosity of the artificial bone is desired for the recent research. Recently, many pore creating agents like polymethylmethacrylate micro beads (PMMA), sodium chloride (NaCl), carbon beads, starch, ice and naphthalene are used to achieve the desired porosity. In general, these pore creating agents will decompose from the prepared samples during sintering thus creating the porosities (I. Sopyan, 2007; X. Yao, 2006; K. Pal, S. Pal 2005). We, in this study used egg yolk which acts as an emulsifier in food industry promoted the idea that it can create porosity in addition to its excellent binding capacity when heated. The property of coagulation of egg when exposed to heat at the temperature between 60–70°C would contribute to the binding nature of the egg yolk to hold the HA and βTCP together. Furthermore, the film-forming property of egg yolk preserves the moisture between particles (Lynn, A.K., 1999). Henceforth, these unique properties of egg yolk, micro and macro-porosity is expected to be present in sintered βTCP and nano HA together with interconnecting porosity as to mimic human natural bones and promotes osteoinduction and osteoconduction.

This paper demonstrates the possibility of applying egg yolk as binder and pore creating agent [8] to develop porous βTCP co-precipitation method with desired porosity, to avoid the frequent migration of bioceramic particles from the implant site, reducing potential damage to soft tissue in the vicinity of the implant and to improve biodegradability, curing properties, mechanical strength and resorbability.

MATERIALS AND METHODS

Preparation of NanoHAP

To 0.5 M Calcium hydroxide [$Ca(OH)_2$] 0.3M of ortho phosphoric acid (H_3PO_4) was added drop by drop and the pH is brought to 12.5. This mixture is kept for stirring for 24 hours. The mixture is centrifuged, washed with distilled water and filtered. The obtained precipitate was dried in the oven at 100°C the dried precipitate was powdered to get a nanohydroxyapatite (nano HA).

Preparation of *β-TCP*

$(NH_4)_2HPO_4$ and $Ca(NO_3)_2.4H_2O$ were obtained from Sigma Aldrich and used in this synthesis. An aqueous solution of $(NH_4)_2HPO_4$ (325 ml) was added to an aqueous solution of $Ca(NO_3)_2.4H_2O$ (500 ml) under stirring. To this 5 ml of ammonia solution was added and stirred for 2 hrs. The mixture is filtered and dried in the oven at 60°C for 24 hrs. The flakes are then powdered and calcinated in the furnace at 850°C for 12 hrs followed by cooling to obtain single phase β-TCP.

Preparation of BCP

Biphasic calcium phosphate was prepared by mixing nano HA and β-TCP in the ratio of 60:40 and it is denoted as BCP.

Preparation of BCP-EY Bone Implant

4.0 g of BCP was made into a dough with 2.0 ml of egg yolk. This dough was extruded through a glass tube having 1.0 cm diameter and dried. These implants were dried initially at room temperature (30°C) for 2 h and later at 70°C for 5 h at which the binding nature of egg yolk was increased, later cooled and stored in a sterile container.

Preparation of Simulated Body Fluid (SBF)

Simulated body fluid (SBF) was used as a growth medium. SBF was prepared by dissolving reagent grade NaCl 11.994 g, $NaHCO_3$ 0·525 g, KCl 0·336 g, $K_2HPO_4.3H_2O$ 0·342 g, $MgCl.6H_2O$ 0·4575 g, $CaCl_2$ 0·417 g and Na_2SO_4 0·1065 g in deionized water [9, 10]. The solution was buffered at PH 7.4 with tris (hydroxylmethyl) amino methane $((CH_2OH)_3CNH_2)$ and 1M hydrochloric acid (HCl) at 36·5 ± 1°C (Duan *et al.,* 2005).

In vitro Bioactivity Test

The prepared bone implant BCP-EY was kept in SBF solution for 21 days at 37°C. Later the implant was removed from the solution, washed and dried initially at room temperature and later at 70°C.

Characterization of the Implant

UV Spectrophotometric Analysis

The UV analysis of the implant SBF supernatant was read at 210 nm at an interval of 3 days and it is plotted in the graph.

Fourier Transform Infrared Spectroscopy (FTIR)

The IR spectra of the samples prepared and used in the wave number range of 4000–400 cm^{-1} using Nicolet Impact 400 Fourier transform infrared spectrophotometer using KBr pellet containing 1–2 mg of the sample.

X-ray Diffraction (XRD) Analysis

X-ray diffraction (XRD) characterisation of prepared sample was conducted using an analytical X'Pert PRO alpha-1 with a RTMS X'Celerator detector. It used Ni-filtered Cu Kα radiation over the 2θ range of 20–80° at a scan rate of 2.4°/min and with a sampling interval of 0.002° at 40 mA and 45 kV.

Thermo Gravimetric Analysis (TGA)

The thermo-gravimetric analysis of the samples prepared was carried out using a Seiko SSC 5200 H in nitrogen atmosphere (80 ml/min) at a heating rate of 10°C/min. Primary weight loss of these materials as function of temperature was recorded using this study.

Scanning Electron Microscopy (SEM) Analysis

The surface morphology was analyzed with a Topcon, SM-300 scanning electron microscope. The copper disc was pasted with carbon tape and the sample was dispersed over the tape. The disc was coated with gold in ionization chamber before microscopic analysis.

Mechanical Strength

The implants were made cylindrical in shape with a length to diameter ratio of 2:1and subjected for testing mechanical strength using Instron 4501 model which was designed to minimize the end effect imposed by compressive loading and also with a 0.2 kN load cell or equivalent to check the tensile strength. The compressive strength was calculated from the break load and dimensions of the pellets.

RESULTS AND DISCUSSION

The implants for bone regeneration need to have a three-dimensional structure to allow bone ingrowth. The level of porosity, pore size distribution, and degree of pore interconnectivity of the scaffolds significantly influence the extent of bone ingrowth. Recently, several types of biomaterials, which have a variety of porous structures, have been reported to be efficient as scaffolds for bone regeneration. The Bcp/egg yolk used in this study also have a high porosity.

Figure 2a shows the UV Spectrophotometric analysis of the implant in which a gradual deposition of calcium and phosphorous attains a stationary phase from 14–21 days. This indicates that the implant once placed *in vivo* exhibits the diffusion of calcium and phosphorous which causes osteogenesis.

The FTIR spectra of nano HA, TCP and BCP-EY implant before and after SBF are shown in Figure 1a to 1d respectively. The band at 605 cm^{-1} can be assigned to the anti symmteric bending motion of phosphate groups in hydroxyapatite. The *ν*1 phosphate bands representing β-TCP can be seen at 942 cm^{-1}. The –OH stretching bands of hydroxyl groups are seen around 3500–3700 cm^{-1}. The FTIR spectrum of implant exhibits amide I and II absorption bands at 1653 cm^{-1} and 1557 cm^{-1},

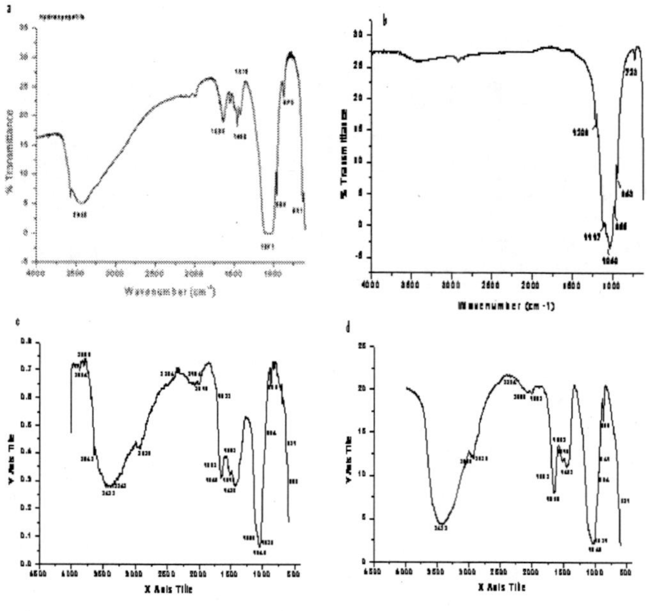

Fig. 1: FTIR Pattern for a) Nano HA b) βTCP c) BCP-EY Implant before SBF d) BCP-EY Implant after SBF

respectively. The hydroxyl groups of chitosan are merged with those of hydroxyapatite and are seen as broad band at 1020–1100 cm^{-1}. FTIR spectra of egg yolk showed absorption peaks at 3365 cm^{-1} (O–H), 2915 cm^{-1} (C–H), 1722 cm^{-1} (C=O), 1585 cm^{-1} (hydrogen bonded N–H) and 1028 (C–O).

X-ray diffraction (XRD) pattern of BCP-EY (Figure 2b) suggest the presence of nanocrystalline apatite along with the egg yolk before and after treating in SBF. The peaks at 32.07°, 33.199°, 34.69°, 47° and 49.75° indicate reflections from 211, 300, 202, 222 and 312 crystal planes, respectively thereby indicating the presence of β-TCP and HA. The peak at 26.10° indicates reflection from 002 indicates the presence of protein. The implants before and after SBF show the characteristic peaks of β-TCP suggesting that β-TCP phase did not change into other phases during preparation. This is important for achieving good porous mechanical and biological properties of produced scaffolds.

Figures 3a and b. depicts the SEM images of BCP-EY composites before and after SBF respectively. The SEM images showed that βTCP forms a composite with egg yolk thereby a porous network was formed in which nano HA particles were embedded.

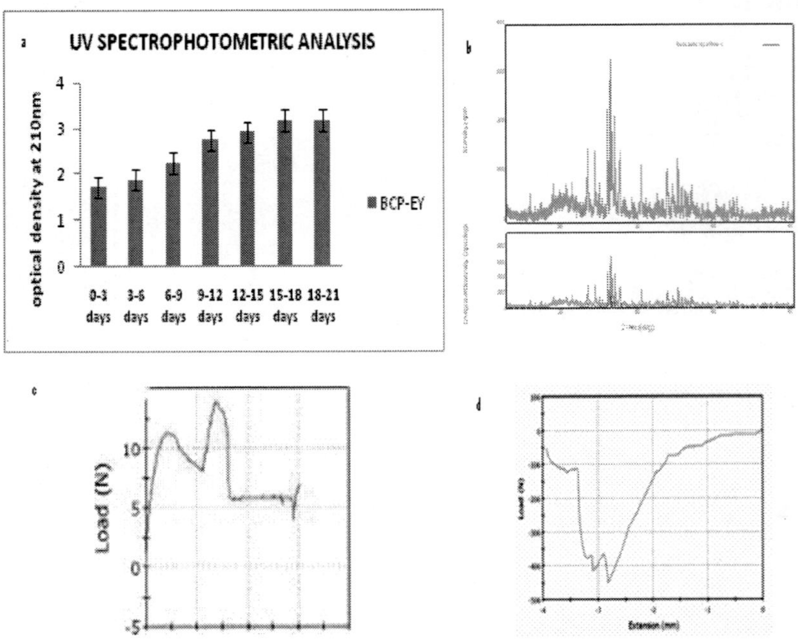

Fig. 2: Images of BCP-EY Implant a) UV Spectrophotometric Analysis b) XRD Pattern c) Tensile Strength d) Compression Strength

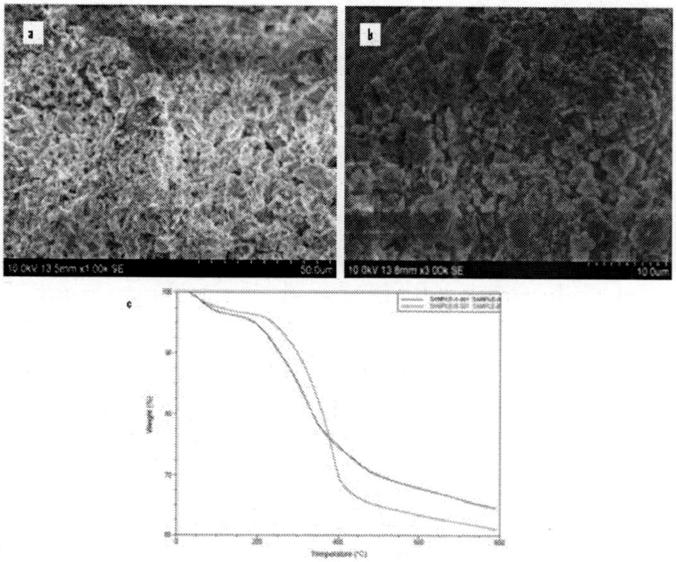

Fig. 3: (a) SEM Image of BCP-EY Implant before SBF (b) SEM Image of BCP-EY Implant after SBF (c) Thermogram before (black line) and after SBF (red line)

The SEM picture of sample after SBF clearly shows deposition of mineral phase (calcium phosphate) onto the crystals. The aggregation is seen more due to the interconnection of the crystals via calcium phosphate deposition. In addition, the matrix was seemed to continue to be porous and ceramic polymer interaction seems not to be disturbed.

Compression tests were performed to characterize the mechanical properties of the prepared scaffolds. The strength of the composite depends upon various factors like particle size of individual components, nature of the additives and percentage moisture present in sample. Figure 2d shows the compression Strength of BCP-EY implant before immersing in SBF. The implant shows the maximum compressive load of 449.55N which proves that the material can be used for non load bearing bone implant material which can be attributed to the presence of BCP and egg yolk.

Tensile properties shown in Figure 2c indicate how the material will react to forces being applied in tension (Sarkar K.T., 1997). Considering the structural or mechanical compatibility with tissues, metals or ceramics are chosen for hard tissue applications and polymers for the soft tissue applications. The implant was shown to have a tensile strength of 0.47 mPA which is suitable to be a non load bearing bone implant. Here the strength of the material is imparted by the egg proteins complexed with BCP. Similar studies were obtained by Fariza and his coworkers, who synthesized ceramic material using porous tri-calcium phosphate and egg yolk for artificial bone. They demonstrated the bioceramic material with desired porosity of 20–54.5% and acceptable compressive strength between 0.7–0.07 MPa and can attributed to the present study.

Thermo gravimetric Analysis (TGA) shown in Figure 3c measures the amount and rate of change in the weight of a material as a function of temperature or time in a controlled atmosphere. Thermogravimetry reveals loss of weight of a particular substance with increasing temperature [14, 15]. BCP-EY (Before and after SBF) were subjected to thermal analysis using TGA and their thermogram are shown in Figures 4.8.2. The first loss was observed between 226° C and 418°C by 2.6%. The second step weight loss (7%) was observed between 418°C and 780°C, around 61% of material was stable at 780°C. The thermal stability observed may be due to the crosslinking of Beta-TCP with egg yolk which imparts a strong binding capacity when the temperature increases [8]. The second weight loss was also may be due to the loss of water from egg yolk. In the case of implant processed after SBF also the two-step weight loss was observed first being (2.1%) from 85–185°C and second (6.4%) from 185–780°C and around 65% of the sample was stable at 780°C. The increase in the thermal stability reveals the ossification of BCP implant while the implant was immersed in SBF.

A nano hydroxyapatite synthesized by co-precipitation method was combined with βTCP in an appropriate ratio to form BCP. The obtained BCP along with egg yolk was casted into a bone implant material which was characterized by various physico chemical techniques. The results revealed that the bone implant shown to have a good mechanical strength for non load bearing purposes and the SEM analysis shows the porous surface of the implant imparted by the inclusion of egg yolk. Due to the presence of βTCP and egg yolk, the porosity as well as strength of the material increased which is more similar to the natural bone. This bone implant in future will be further subjected for toxicological and immunological properties in animal models.

REFERENCES

Amit, S., Ingle, A., Munim, K.R. Vaidya, S.N., Sharma, B.P. and Bhisey, A.N. (2001). Development of calcium phosphate based bioceramics, *Bulletin Materials Science,* 24: 653–657.

Fariza, A.R., Zuraida, A. and Sopyan, I. (2010). Egg Yolk as Pore Creating Agent for Porous Tri-Calcium Phosphate. *International Journal of Engineering and Science,* 1: 1.

Flautre, B., Descamps, M., Delecourt, C., Blary, M.C. and Hardouin, P. (2001). Porous HA Ceramic for bone replacement: Role of the pores and interconnections-Experimental study in the rabbit. *Journal of Materials Science: Materials in Medicine,* 12:P 679–68.

Harris, C.T. and Cooper, L.F. (2004). Comparison of bone graft matrices for human mesenchymal stem cell- directed osteogenesis. *J Biomed Mater Res A*, 68: 747–755.

Keating, J.F. and McQueen, M.M. (2001). Substitute for autologous bone graft in orthopedic trauma. *The Journal of Bone and Joint Surgery*, 82 3–8.

Krithiga, G., Antaryamijena, P. Selvamani and Sastry, T.P. (2011). In vitro study on biomineralization of biphasic calcium phosphate biocomposite cross linked with hydrolysable tannins of Terminalia. *Bull. Mater. Sci*, 34: 3: 589 –594.

Kwon, S.H., Jun, Y.K., Hong, S.H. and Kim, H.E. (2003). Synthesis and dissolution behavior of β-TCP and HA/β-TCP composite powders. *J. Eur Ceram Soc*, 23: 1039 –1045

Lynn, A.K. (1999). The egg and the equivalent-part two, Weeks Publishing Company.

Ohtsuki, C., Kokubo, T. and Yamamuro, T. (1992). Mechanism of HA formation of CaO–SiO_2–P_2O_5 glasses in simulated body fluid, *J Non-Cryst Solids*, 92: 143:84.

Pal, K. and Pal, S. (2006). Development of porous hydroxyapatite scaffolds. *Materials and Manufacturing Processes*, 21, 325–328.

Sarkar, K.T. (1997). Theory and practice of leather manufacture (Chennai: The Author), 309.

Scaglione, S., Braccini, A., Wendt, D., Jaquiery, C., Beltrame, F., Quarto, R. and Martin, I. (2006). Engineering of osteoinductive grafts by isolation and expansion of ovine bone marrow stromal cells directly on 3d ceramic scaffolds. *Biotechnol Bioeng*, 93: 181–187.

Sopyan, I., Mel, M., Ramesh, S. and Khalid, K.A. (2007). Porous hydroxyapatite for artificial bone applications. *Science and Technology of Advanced Materials*, 8, 116–123.

Tang, R., Hass, M., Wu, W., Gulde, S. and Nancollas, G.H. (2003). Constant composition dissolution of mixed phases. II: Selective dissolution of calcium phosphates. *J. Colloid Interface Sci.*, 260: 379–384.

Yao, X., Tan, S. and Jiang, D. (2005). Improving the properties of porous hydroxyapatite ceramics by fabricating methods. *Journal of Materials Science*.

Identification of Pathogens and Bacterial Endotoxins in Pharmaceutical Products

Kh. Nirupama Devi*, **P. Kesavakrishna and Suchitra Hijam**
Department of Biotechnology, Nandha Arts and Science College, Erode
E-mail: nirukhuman@gmail.com

INTRODUCTION

Drug is any substance that, when absorbed into the body of a living organism, alerts normal bodily function. In pharmacology a drug is "a chemical substance used in the treatment, cure, prevention, or diagnosis of disease or used to otherwise enhance physical or mental well-being." Humans have used drugs of one sort or another for thousands of years. The present study is to identify the pathogenic microorganisms present in Thalidomide drug and bacterial endotoxin in the ketorolac injection. Thalidomide is a sedative hypnotic and multiple myeloma medication. Thalidomide was said to have been developed by Germany German pharmaceutical company Grunenthal in Stolberg (Rhineland) near Aachen, although this claim has recently been challenged. Ketorolac is a type of no steroidal anti-inflammatory drug (NSAID) approved for the short-term relief of moderate to severe pain. This prescription medication is most often used to treat pain following a procedure, but it can also be used to relieve back pain, cancer pain, and pain caused by kidney stones.

A pathogen (from Greek pathos "suffering, passion", and gignomia (gen-) "I give birth to" an infectious agent, or more commonly germ, is a biological agent that causes disease to its host. After producing a drug, identification of specified pathogens is performed as per USP (1995) and in EP (1997) and in BP (1999). The specific pathogens include *Escherichia coli, Staphylococcus aureus,* and *Clostridium perfringens* and *Salmonella* species. These organisms must be removed from pharmaceutical producers of all kinds because

*Corresponding author.

they indicate particular infection hazard to the patient or because their presence is a criterion of the quality of raw materials. Endotoxins are high-molecular weight complexes of lipopolysaccharide (LPS) that constitute the major cell wall component in all Gram-negative bacterial families (McCuskev, 1996). These molecules have been intensively investigated by Rietschel (1996) because of the increasing appreciation of their potentially pathogenic role in wide variety of human disease states. These bacterial endotoxins were identified with the help of LAT (Limulus Ameobocyte test) test.

MATERIALS AND METHODS

Preparation of Sample

10 gms of sample (thalidomide capsules) was weighed and it was dissolved in 90 ml of SCDBPS (Soybean Casein Digested Broth Poly Sorbet), incubated for 2 hours.

Enumeration of Microorganisms

The microorganisms were isolated by pour plate method and were incubated at 37°C for 48 hours. And the colonies were counted by using colony counter and calculate the result for 1gm of sample.

IDENTIFICATION OF PATHOGENS

Required Media

Macconkey broth, Rappaport vassilidias medium, SCDB (Soybean Casein Digest Broth), Reinforced, MCA (Macconkey agar), XLDA (Xylose Lysine Deoxycholate Agar), MSA (Mannitol Salt Agar) and Columbia agar.

Sub Culturing on Selective Media

The 48 hrs incubated sample in enrichment medium was subculture on the selective medium that allows particularly the growth of specific microorganism. The selective medium for each microorganism is given below:

E.coli-Macconkey agar, Staphylococci-MSA, Salmonella-XLDA, Clostridium-Columbia agar.

BACTERIAL ENDOTOXIN TEST

Preparation of Sample or Dilution of Sample

The sample was prepared by diluting ketorolac injection as per MVD calculation MVD: is a calculation used to determine how much a parental product of raw material can be diluted and still detect the endotoxin.

$$MVD = \frac{EL \times Concentration}{\lambda}$$

EL = the maximum amount of endotoxin allowed in a product = 5.8

Concentration of the sample = 30 mg/ml

Lysate sensitivity = the maximum concentration which cause gelation = 12, therefore the MVD for ketorolac is 1392.

Procedure

In one assay tube 50 µl of diluted sample was added to 50 µl of LRW and to this 100 µl of lysate was added and named it as positive product control (PPC).

In second assay tube 50 µl of sample was added to 50 µl of LRW and to this 100 µl of lysate was added and named it as negative product control (NPC).

In third assay tube 100 µl of LRW was added to 100 µl of lysate and named it as negative control (NC). Maintainedduplicates for each assay tubes. These tubes were incubated at 37°C in heating block for 1hr, and observe the tubes for formation of gel clot.

Sterility Test

The sterility lab has been strictly sterilized before starting the sterility test and (20 ampoules) sample were passed into two sterile canisters threw the sterility machine and one was filled with SCDM and another one was filled with FTGM. The canister was closed and kept for incubation for 14 days at 37±2°C and the result were noted.

Result

In identification of pathogens the sample (Thalidomide drug) was tested for identification of *Escherichia coli, Staphylococcusaureus, Clostridium perfringens* and *Salmonella* species. All these microorganisms didn't show any growth on the selective media. That indicates the absence of pathogens in the sample (Figures 1–4).

Identification of Pathogenic Microorganisms

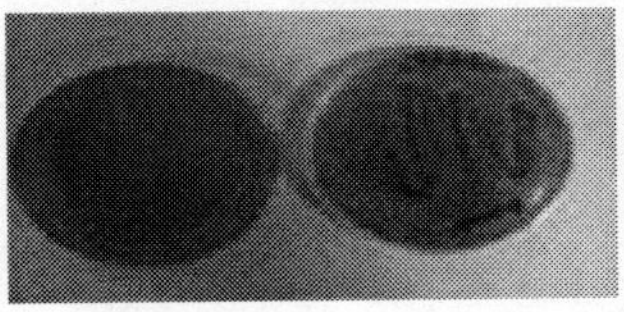

Fig. 1: *E.coli* on MCA

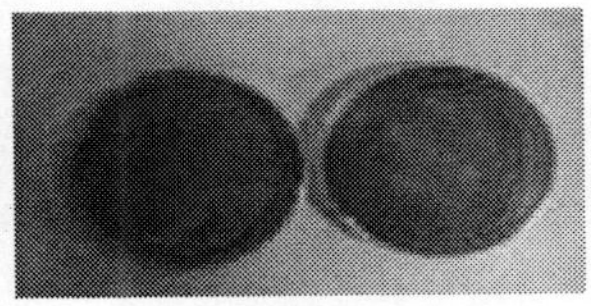

Fig. 2: *S.aureus* on MSA

Fig. 3: Salmonella on XLDA

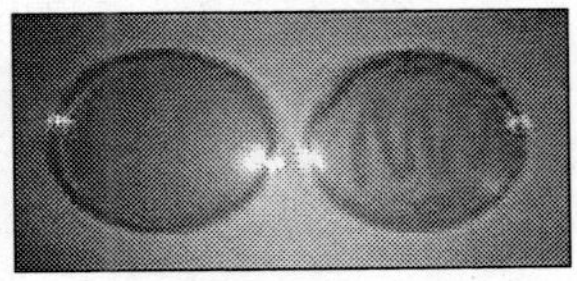

Fig. 4: Clostridium on CA

Endotoxin Limit Test

For endotoxin limit test, the sample which contains ketorolac, LRW and lysate are showed gelation, it indicates that the sample contain endotoxin within the limit (Figure 5).

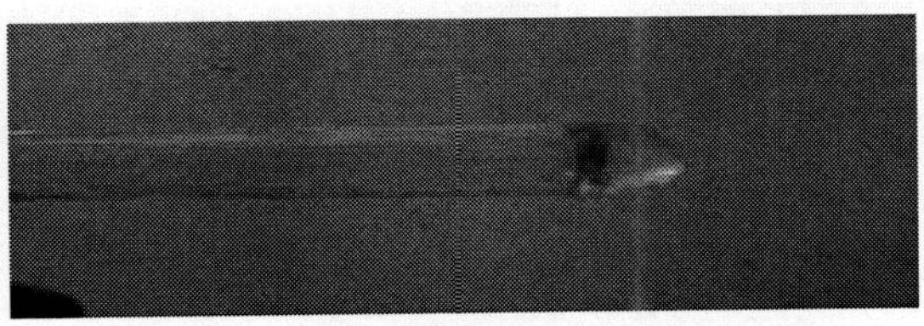

Fig. 5: Gelation

Sterility Test

After completion of 14 days incubation the medium was found sterile. And it didn't show any turbidity in SCDM and FTGM which indicate that the sample was in sterile condition (Figures 5–6).

Fig. 6: Sterility in SCDM

Fig. 7: Sterility in FTGM

Discussion

The formulated drug Thalidomide was analysed for Total microbial count,pathogens and the ketorolac injection was quantified for bacterial endotoxin and sterility. In Total microbial count the 10^{-1} dilution sample was subjected to pour plate technique to enumerate the number of microorganisms present in the drug (Thalidomide), the drug contain 300 Cfu/gm of sample.The identification of pathogens was performed by using enrichment and selective media. There were no colonies found and it indicates the absences of pathogens in the drug. In the quantification of bacterial endotoxin, the sample (ketorolac) was compared with the control standard endotoxin with the help of lysate enzyme, and it was quantified that the endotoxin in the sample was within the limit. The sterility of the sample was checked. The present study show that the selected drugs, i.e. Thalidomide contains limited number of microorganisms, and is free from pathogens. The ketorolac injection showed the endotoxin within the limit and the quality of the product was confirmed as sterile. So these two drugs can be used for treatment purpose.

REFERENCES

Amato, D., Loughran, M.S., Flynn, E. and Folkman, J. (1994). Thalidomide is an inhibitor of angiogenesis *proc Natl Acad Sci.*, 91: 4082–4085.

Arbuthnot (1970). *Staphylococcus* or haemolysin, *In t.c montie, JP,* pp. 189–242.

Borriello, S.P. (1998). Pathogenesis of *Clostridium difficile* infection. *J AntimicrobChemother,* 41:13–9.

Buckely, Rook A. and Krishan, P. (1990). The antiinflamatory actionof ketrolac in humans *Clinical investigation,* 63: 122–130.

Connor, D.G. and Kass, E.H. (1961). Effect of artificial fever in increasing susceptibility to bacterial endotoxin. *Nature (Lond.),* 12: 190, 453.

Darnton, N.C., Turner, L., Rojevsky, S. and Berg, H.C. (2007). On torque and tumbling in swimming *Escherichia coli. J. Bacteriol,* (5):1756–1764.

Eleutherakis-papaiakovou, V., Bamias, A. and Dimopolos, M.A. (2004). Department of clicinal therapeutics, university of Athens, school of medicine, Athens, grese. *Annals of Oncology,* 15: 1151–1160, 2004).

Fekety, R. (1997). Guidelines for the diagnosis and management of *Clostridium difficile*-associated diarrhea and colitis. *Am J Gastroenterol,* 92: 739–50.

Geitz, Meyer P., Dutoo (2002). Thalidomide prodcuces tramsfusion independence in long standing refractory anemias of patients with myeodysplastic syndrome Blood. *J. Cli Onco* 20: 1429–1450.

Kaper, J.B. and Hacker, J. (1999). Pathogenicity Islands and Other Mobile Genetic Elements of diarrheagenic *Escherichia coli* in Pathogenicity Islands and Other Mobile Virulence Elements *Society for Microbiology,* 3, pp. 33–5.

Murray *et al.* (2009). *Medical Microbiology* (6th ed.). Mosby Elsevier. ISBN 978-0-323-05470-6.

Ochman, P., Franks, K. and Tomoaia, A. (2000). Role of thalidomide in the treatment of metastatic prostate cancer. *IDF-bulletin,* 90: 03–15.

Poxton, I.R., McCoubrey, J.M. and Blair, G. (2001). Theathogenicity of *Clostridium difficile. Clinal Microbial Infection,* 7: 421–7.

Relationship of Body Temperature to the Lethal Action of Bacterial Endotoxin, *Journal of Clinical Investigation,* 43, 230–238.

Richardson, P., Hidershima, T., Anderson (2002). Thalidomide emerging role in cancer medicine, *Annu Rev.,* 3: 629–657.

Isolation, Identification and Characterization of Methicillin Resistant *Staphylococcus aureus* from Infected Wound Patients

Suchitra Hijam* and S. Arumugam[1]

Department of Biotechnology, Nandha Arts and Science College, Erode

[1]Department of Botany, Government Arts and Science College, Salem

E-mail: suchitrahijam@gmail.com

INTRODUCTION

The genus *Staphylococcus* consists of Gram-positive cocci, 0.5 to 1.5 mm in diameter. It occurs singly, in pairs, and in tetrads. The *Staphylococcus* characteristically divides in more than one plane to form irregular clusters. The cell wall contains peptidoglycan and teichoic acid. *Staphylococci* are facultative anaerobes. *Staphylococci* are among the most important bacteria that cause disease in humans. They are normal inhabitants of the upper respiratory tract, skin, intestine and vagina. *Staphylococci* can be divided into pathogenic and relatively non pathogenic strains based on the synthesis of the enzyme coagulase. Coagulase positive a yellow carotenoid pigment which has led to their being commonly called Golden staph and cause severe chronic infections. Coagulase negative *Staphylococci* such as Staphylococcus epidermis do not produce coagulase, are non pigmented and generally less invasive. Various toxins and enzyme are produced by *Staphylococci* such as coagulase, β-lactamase, Lipase, Hyaluronidase, Nuclease, Protein A and some of the toxins are Panton valentine leukocidin, exfoliative, toxic shock syndrome toxin-I, etc.

Methicillin is a narrow spectrum beta lactum antibiotic of the penicillin class. It was developed by Beecham in 1959, particularly beta-lactamase producing organisms such as

*Corresponding author.

Staphylococcus aureus, but is no longer clinically used.Methicillin is resistant to beta-lactamase (also known as penicillinase) enzyme secreted by many penicillin resistant bacteria. The presence of the ortho-dimethoxyphenyl group directly attached to the side chain carbonyl group of the penicillin nucleus facilitates the beta-lactamase resistance. The resistance to methicillin was due to a penicillin-binding protein coded for by a mobile genetic element termed the methicillin-resistant gene (mecA). In recent years, the gene has continued to evolve so that many MRSA strains are currently resistant to several different antibiotics. Data from around the world verify the escalating incidence of infections caused by Methicillin-resistant *Staphylococcus aureus* (MRSA). Since MRSA are spread primarily on the hands of health care workers, rates of infection is a function of infection control activities within institutions. Moreover, infections with MRSA are serious and often life-threatening. Thus, there are compelling medical and ethical reasons to invest in control measures. Currently available data suggest the efficacy of three measures: (a) identification of the entire patient reservoir (cases and carriers) for purposes of isolation; (b) strict hand washing between patients to prevent transmission; and (c) treatment of the carrier state in health care workers and patients during periods of high infection rates with safe and effective topical agents such as mupirocin (Richard *et al.,* 1991).

Statistical data suggest that as many as 19,000 people per year have died from MRSA in the U.S.; data supplied by the CDC in 2010 suggest this number has declined by about 28% from 2005 to 2008, in part, because of prevention practices at hospitals and home care.My present study lead to the isolation of *Staphylococcus aureus* from pus of wound infection collected from Governtment Hospital, Erode, Tamil Nadu,India and to identify and characterize MRSA.

MATERIAL AND METHODS

Collection of the Samples

The samples were collected from the pus of wound infected patients from Government Hospital, Erode, Tamil Nadu.

Isolation of Mrsa

Staphyloccocus aureus was isolated from the pus samples by serial dilution technique spreading on the mannitol salt agar. The single colony was streaked on the blood agar medium to check the hemolysis test.

Confirmation Test

Various tests were performed to confirm the presence of *Staphylococcus aureus:* (i) Gram staining test was performed to view the morphology of the bacteria and to distinguish gram positive and negative bacteria. (ii) Coagulase test-a loop of the culture was inoculated to the blood plasma to check the coagulate activity of the bacteria. (iii) Catalase test-a few drops of 3% hydrogen peroxide were added to the slant culture. (iv) Biochemical test such as indole test, methyl red test,voges-proskauar test and citrate utilization tests were performed. (v) Durham tube sugar fermentation tests—One loopful of *Staphylococcus aureus* culture was inoculated in all the sugar broths using sterile techniques, after keeping the Durham's tubes in broths and were incubated for 24 to 48 hours at 37°C in incubator.

Antibacterial Sensitivity Test

The effectiveness of an antimicrobial sensitive testing is based on the size of the zone of inhibition that surrounds a disk that has been impregnated with a specific concentration of the agent. The Kirby-Bauer method is a standardized system that takes all variables into consideration. It is sanctioned by the U.S. FDA and the subcommittee on Antimicrobial susceptibility testing of the National committee for clinical Laboratory standard.

Procedure

The Mannitol salt agar plates were prepared and labeled it. The surface of the medium was inoculated with the swab after expressing excess fluid from the swab by pressing and rotating the swab against the inside wall of the plate. The surface of agar was allowed for 3 to 5 minutes to dry before applying discs. The forceps was sterilized by flaming. The discs were kept on the plate. Light pressure was applied to each disc on the agar with the tip of a sterile forceps to secure it to medium. The plates were inverted and incubated for 16 hrs at 37°C.

RESULTS

Five samples were collected from pus of wound infected patients from Government Hospital, Erode. Out of five samples collected; only one showed the colonies.

The bacterial colonies were isolated from Pus culture by serial dilution method (Figure 1). A single colony of Methicillin-resistant Staphylococcus aureus was screened by streaking on Mannitol salt agar medium (Figure 2). The screened colonies were then confirmed with Grams staining test, β-hemolysis tests (Figure 3), Catalase test (Figure 4), Coagulase test (Figure 5), Mannitol fermentation test (Figure 6) and Biochemical test (IMVC) (Figure 7–10).

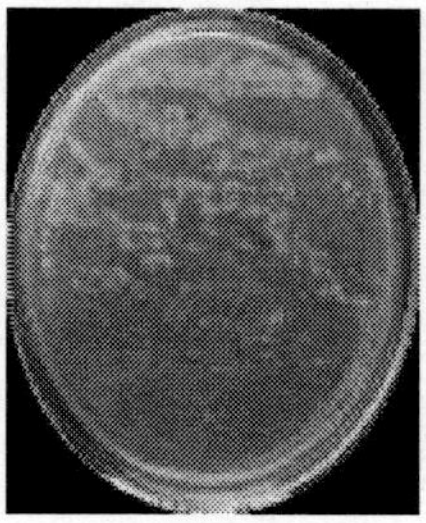

Fig. 1

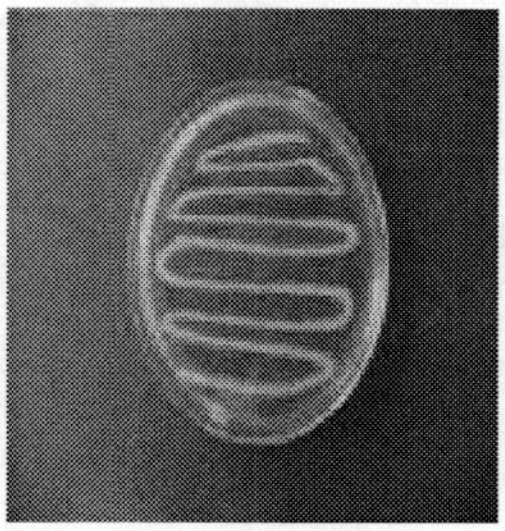

Fig. 2

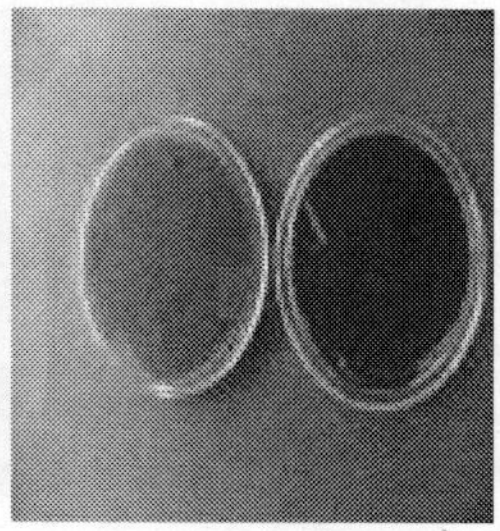

Fig. 3

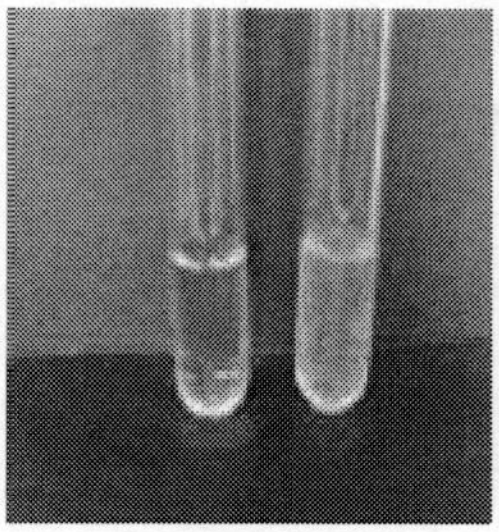

Fig. 4

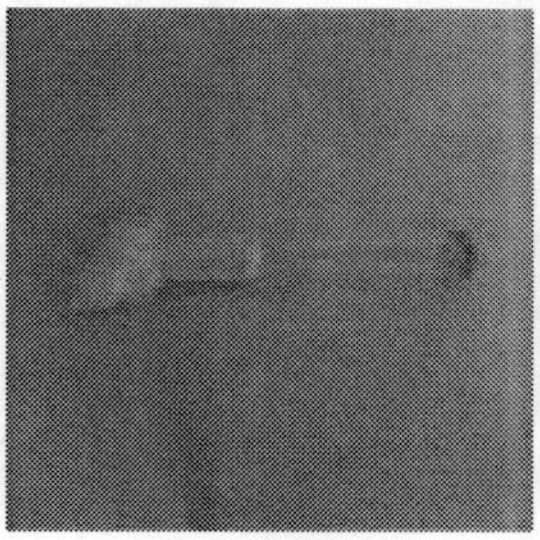

Fig. 5

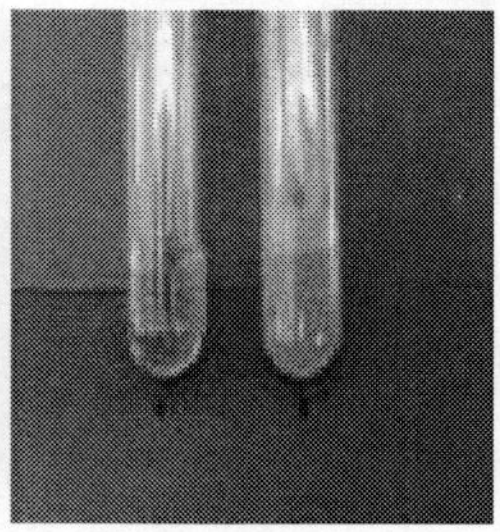

Fig. 6

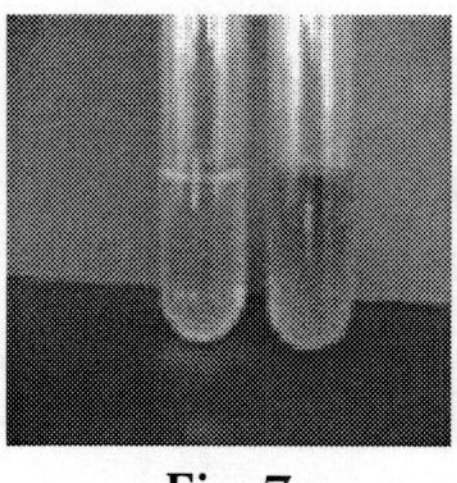

Fig. 7

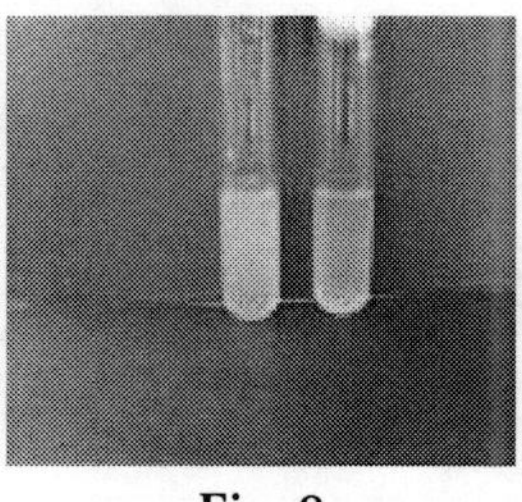

Fig. 8

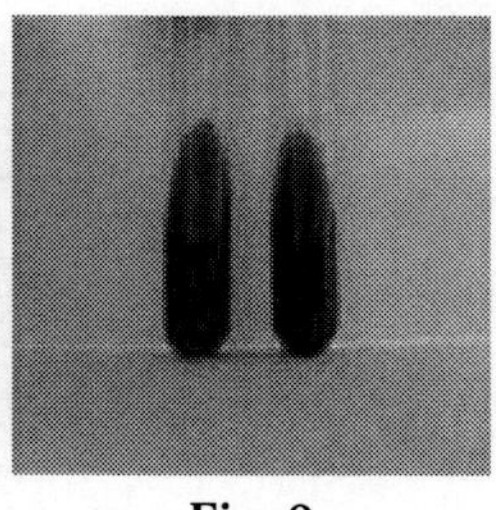

Fig. 9

The gram staining revealed that the isolated colony has gram positive bacteria in clusterform. The hemolysis tests and coagulase test showed positive results. The Mannitol fermentations test showed positive results but for only acid formation and not the gas. The colony when subjected to biochemical test, revealed positive response for Methyl Red and negative for Indole, VP and citrate utilization

The Antimicrobial sensitivity was carried out with seven antibiotics in accordance with the Kirby-Bauer method (Figures 11–17). Seven antibiotic, viz., penicillin, methicillin, norfloxacin tetracycline, co-trimozole, vancomycin and bacitracin were taken for this test. The results were observed and recorded in Table 1.

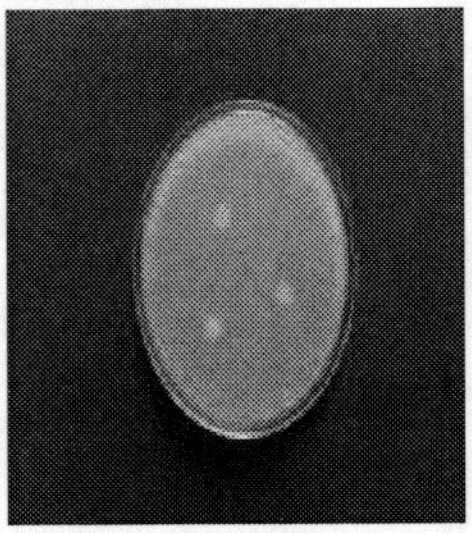

Fig. 11: Penicillin

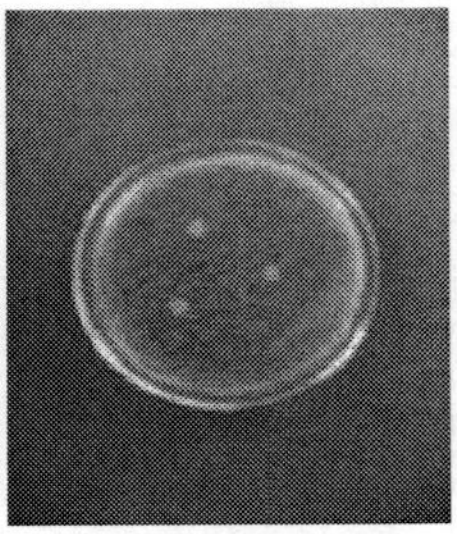

Fig. 12: Methicillin

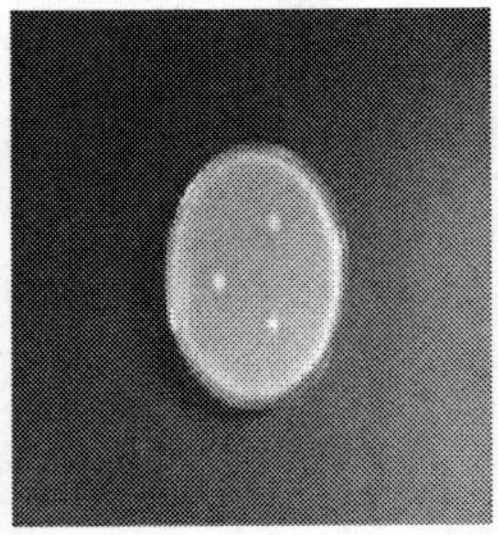

Fig. 13: Norfloxacin

Fig. 14: Tetracycline

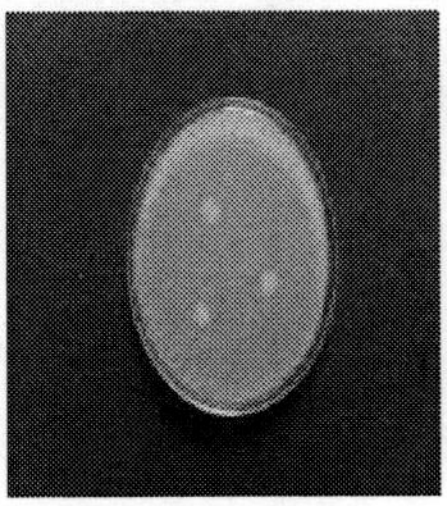

Fig. 15: Co-trimozole

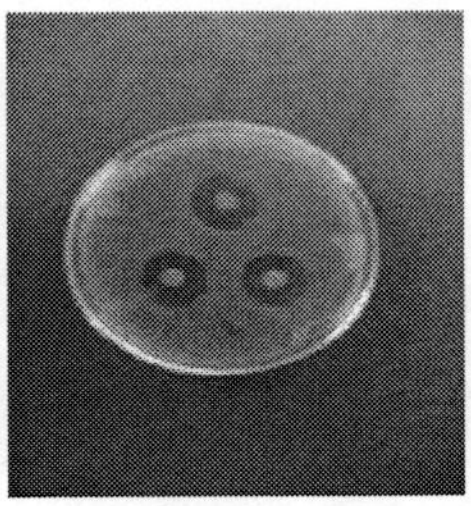

Fig. 16: Vancomycin

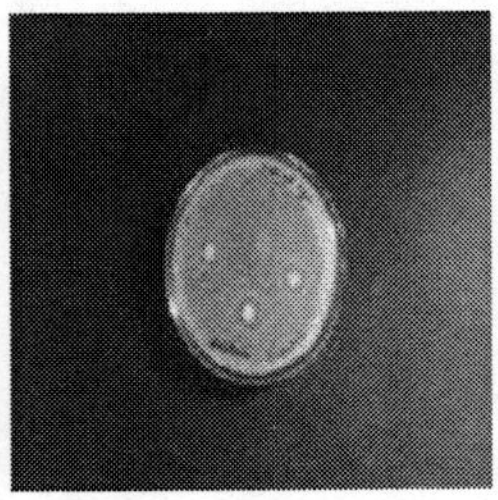

Fig. 17: Bacitracin

Table 1: Antibiotic Sensitivity Test for *Staphylococcus aureus*

S. No.	*Antibiotic*	*Zone of Inhibition*	*Result*
1.	Penicillin	No zone	Resistant
2.	Methicillin	No zone	Resistant
3.	Norfloxacin	No zone	Resistant
4.	Tetracycline	No zone	Resistant
5.	Co-trimozole	No zone	Resistant
6.	Vancomycin	22mm	Sensitive
7.	Bacitracin	6mm	Intermediate

DISCUSSION

Pus samples were collected from wound infection patients of Government Hospital. The collected samples were serially diluted and spread on Mannitol salt agar medium. Single colony of the bacteria was developed.The single colony was confirmed as

Staphylococcus aureus by Gram staining Catalase, Coagulase, Hemolysis, Mannitol fermentation and Biochemical test.The antimicrobial sensitivity test was done with 7 antibiotics. Methicillin, Penicillin, Co-Trimozole, Norfloxacin, Tetracycline showed resistant and Vancomycin showed sensitive and Bacitracin showed intermediate.So, the infection due to MRSA can be treated with Vancomycin.

REFERENCES

Ambler, R.P. (1980). The amino acid sequence of the *Staphylococcus aureus* Pencillinase. *Journal of Biochem* 151:197.

Arthur, P. Courvalion, Antimicrob Agents (1993), 37–1563.

Bootsma, M.C., Diekmann, O. and Bonten, M.J. (2006). "Controlling Methicillin resistant *Staphylococcus aureus*: quantifying the effects of interventions and rapid diagnostic testing". *Proc Natl Acad Sci USA* (14): 5620–50. PMID 16565219.

Chambers, H.F. (2001). The changing epidemiology of *Staphylococcus aureus, Energy infect DIS* 7: 178–182.

Didenberg *et al.* (1990). comparison of the sequence of class A, β-lactamase and of the secondary structure elements of penicillin recognizing proteins. *Antimicrob Agents Chemother,* 35: 294, 191.

Graham, P., Lin, S. and Larson, E. (2006). "A U.S. population-based survey of *Staphylococcus aureus* colonization". *Ann Intern Med.,* 144(5): 318–25, PMID16520472.

Hakenbeck, R., Grebe T., Zahner, D. and Stock, J.B. (1999). β-lactam resistance in *Streptococcus pneumonia.* Penicillin Binding Proteins and non-penicillin binding proteins. *Molecular biology* 33: 673–678.

Lyon, B.R., May, I.W. and Skurry, R.A. (1983). Analysis of plasmids in nosocomial strains of multiple antibiotic resistant *Staphylococcus aureus. Antimicrob Agents Chemother,* 23, 817–826.

Murry, B.E. (2000). Vancomycin resistance Enterococcal infections *N. Engl journal of Med:* 342: 710–721.

Schliefer, K.H. and Kloos, W.E. (1975). Isolation and characterization of *Staphylococci* from Human Skin. *Int Journal of Syst Bacteriol,* 25: 50.

Smith, T.L., Pearson, M.L., Wilcox *et al.* (1999). Emergence of vancomycin resistance in *Staphylococcus aureus, N. Engl Journal of Med.,* 340: 493–501.

Wang, P. and Novick, R.P. (1987). Nucleotide sequence and expression of the β-lactamase gene from *Staphylococcus aureus* plasmid p. 1258 in *Escherichia coli, Bacillus subtilis* and *Staphylococcus aureus. Journal of Bacteriol,* 169: 1763–1766.

Extra Cellular Biosynthesis of Silver Nano Particles and its Effect with Plant Yields Chemotherapeutics against Clinical Isolates of *MRSA* and *Candida albicans*

Y. Tamilselvi*, S. Prakash and P. Saravana Perumal

Department of Bio Technology, Udaya School of Engineering, Nagercoil, Tamil Nadu

E-mail: pughazselvi@gmail.com

INTRODUCTION

The microorganisms forming the biofilm, which is the major way of many infections shows resistance to antimicrobial drugs.

1. The biofilm serve as hide for disease producing microorganisms and it associated with antimicrobial resistance.
2. The methicillin resistance *Staphylococcus aureus* is one of the major resistant pathogens. Found on the mucous membranes and the human skin, it is extremely adaptable to antibiotic pressure. It was one of the earlier bacteria in which penicillin resistance. Methicillin was then the antibiotic of choice, but has since been replaced by oxacillin due to significant kidney toxicity.
3. *Candida albicans* is easily recognized as common medical yeast pathogen causes oral thrush and vaginitis, to systemic and potentially life-threatening diseases. *Candida* infections of the latter category are also referred to as candidemia and are usually confined to severely immunocompromised persons, such as cancer, transplant, and AIDS patients as well as non-trauma emergency surgery patients.

*Corresponding author.

4. The inhibition of biofilm is now considered as drug target and the pharmacological inhibition of biofilm development is now extensively studied for the treatment of various bacterial and fungal infections.
5. Now investigated the plant compounds, synthesised silver nanoparticles and chemotherapeutics are more effective against drug resistance microorganisms.
6. In this study the effect of sliver nanoparticle with aqueous extracts of Costus igneus, tulsi, ginger, garlic, coconut oil, with clotrimazole and miconazole for *C.albicans*, silver nanoparticle with tetracycline and chloramphenicol for *MRSA was* studied. Moreover the effect on biochemical composition of biofilm matrix mainly total carbohydrate and total protein was also studied.

METHODS

Source of Microorganism

The sea weeds were obtained from kanyakumari. The sea weeds were collected and used as source of microorganism. The isolated organism was maintained in Müller-Hinton medium.

Identification of Organism

The pure culture was isolated after the requisite period of incubation the isolated organism was identified based on the morphological, cultural and biochemical characteristics described by standard methods and This strain showed positive result for Nitrate reduction, Catalase test, Fermentation test for Lactose, Maltose, glucose, Sucrose and Mannitol and also positive for methyl red test VogeProskauer test and Growth on Mckonkey agar. They showed negative result with H_2S production, DNAse negative, many morphological and biochemical experiments were done for the identification of the test microbes (Table 1). The tests were done to identify the strain upto genus level. The observe characteristics were compared with Bergey's Manual of Determinative Bacteriology for proper identification of the organisms. The organism Maintained in Müller-Hinton medium as well as sub culturing from time to time regulate its viability in the microbiology laboratory. The isolated organism was allowed to perform 16s RNA and the species was identified. The 16S rRNA amplification of the strains obtained was given for analysis. The obtained sequence was undergone similarity search with the use of any bio informatics tools.

Preparation of Cell Free Microbial Extract

The culture medium was prepared, sterilized and inoculate with fresh culture of the strain. The cultured flasks were incubated at 37°C for 24 hours. After incubation time the cultures were centrifuged at 12000 rpm and their supernatants was used for further experiments. The slow-growing *S. pasteuri* strain was maintained in Müller-Hinton broth.

Biosynthesis of AgNPs

The supernatant was added separately to the reaction vessel containing silver nitrate ($AgNO_3$) at a concentration of 10^{-3} (1% v/v). The reaction between this supernatant and Ag^+ ions was carried out in bright conditions for 5 minutes. The bioreduction of the Ag^+ ions in the solution was carried out in the dark or bright condition. Change in color was observed to yellowish brown in the silver nitrate solution incubating with *S. pasteuri* and the nanoparticles was separated from the reaction mixture by centrifugation at 10,000 rpm for 10 minutes at 4° C. The synthesized silver nanoparticles optical density was measured by UV-Vis spectrophotometer, Atomic Force Microscope and FT-IR.

Inhibition of Biofilm Formation

Clinical samples of *C.albicans* were obtained from disseminated Candidiasis and from GH hospital Tamil nadu, India. The clinical isolate was identified based on the cultural and biochemical characteristics described by standard methods. Pure culture was maintained on Sabouraud maltose Yeast Extract Agar slant. *MRSA* was also obtained from clinical samples; pure culture was maintained on Trypticase Soy Agar.

Test Materials

Silver nanoparticle (AgNp) synthesizing with *Staphylococcus pasteuri*, aqueous extract of Costus igneus (Ci),ginger (G) (Zingiberofficinale), garlic (Gc) (Alliumsativum),tulsi (T) (Ocimumtenuiflorum), coconut oil (Cn) and antifungal drugs-miconazole (M) and clotrimazole (C) final concentration 100 mg/ml for *C.albicans*, antibacterial drugs- tetracycline (Tc), chloramphenicol (C) final concentration 100 mg/ml for MRSA. The biofilm inhibition assay was performed by the testing materials as single and combination. The testing materials as single and combination with silver nanoparticle for the biofilm inhibition assay was performed.

Biofilm Inhibition Assay

The Biofilm inhibition assay with the tested substance was done (Toole and Kolter, 1998). An overnight culture of *Candida albicans* in Sabouraud maltose Yeast Extract broth and methicillin resistant *Staphylococcus aureus* in Trypticase Soy broth resolve be diluted 1:100 ratio in respective fresh medium and was grow in an another hour. 100 µl of diluted respective strains was added into 96 well titre plate, 200 µl of respective test materials as single and combination (1:1 ratio) as described earlier was added and incubating at 37°C for 3 days. After the incubation the medium was removed and 100 µl of 1% w/v aqueous solution of crystal violet was added. Following staining at room temperature for 30 minutes the dye was removed and the wells was washed thoroughly, 95% ethanol was added and incubating for 15 minutes. The reaction mixture was read spectrophotometrically at 570 nm. Inhibition mediated reduction of biofilm formation was correlated to the color obtain without addition of test materials.

Effect of Test Materials on Biochemical Composition of Biofilm Matrix

The total proteins were calculated by Lowry's method and the total carbohydrates of the biofilm were calculated by Anthron's method (Mohammed *et al.,* 2008).

Results

The isolated organism was gram strained and confirmed as gram positive and grape like cluster cocci and characterization of the culture was done on bio chemical basis. The bio chemical experiment results were mentioned in Table 1. The observed characteristics were compared with Bergey's manual of determinative bacteriology for proper identification of the organism. The species of the organism was identified by 16srRNA.The identified organism was *S. pasteuri.*

The extra cellular biogenic synthesis of AgNPs by *Staphylococcus pasteuri* was carried out and synergistic antibacterial effect of AgNPs with chemotherapeutics plant products against clinical isolates of *MRSA, Candida albicans.* Figure 1 shows the optical photogram of color change of the supernatant when challenged with 1 mM $AgNo_3$ changed color from pale yellow to brown color. The control with out silver ions showed no change in color of the supernatant when incubated under same conditions. The characteristic brown color due to the excitation of Plasmon vibrations in the nanoparticles and provides a convenient signature of their formation. The formation of brown color suggests the presence of silver nanoparticles.

Table 1: Morphological and Biochemical Tests of the *S. pasteuri*

Test Name	*Result*
Pigmentation	V*
Colony morphology	Circular
Margin	Entire
Elevation	Convex
Surface	Smooth and shiny
Opacity	Opaque
Gram staining	Gram positive
IMViC	– + + –
TSI	Acid Butt, no gas
Catalase	Positive
Coagulase	Negative
DNase test	Negative
Sugar fermentation	Mannitol and lactose
Urease	Positive

+, Positive; –, Negative; V* – 11 to 89% of the strains are positive.

The formation and stability of the reduced AgNPs in the colloidal solution was monitored by using UV-Vis spectral Analysis. A UV-Vis spectrum is one of the important technique ascertain the formation of metal nano particles, provided surface plasmon resonance exits for the metal. The UV-Vis Spectra recorded from *S. pasteuri* reaction vessels was plotted and are shown in Graph 1. A Peak was noted around 410 nm.

The AFM micro graphs of nano particles obtained in the filtrate showed that silver nano particles are spherical shaped well distribute with out aggregation in solution with an average size of the about 50 nm Figure 2.

Fig. 1: Synthesis of Silver Nano Particles of *S. pasteuri* with $AgNO_3$

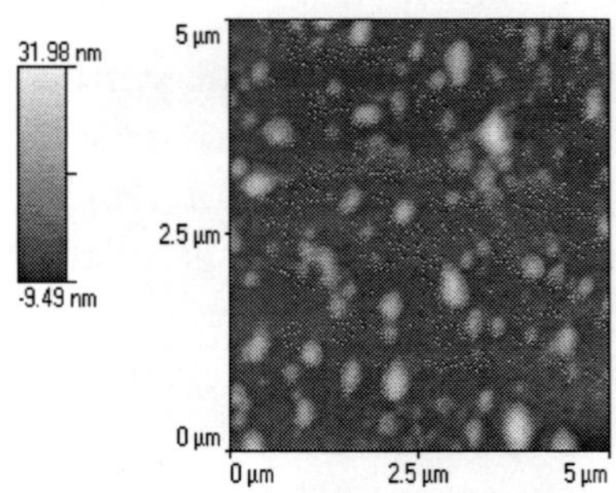

Fig. 2: AFM *Micrograph* of Biogenic Synthesized AgNPs

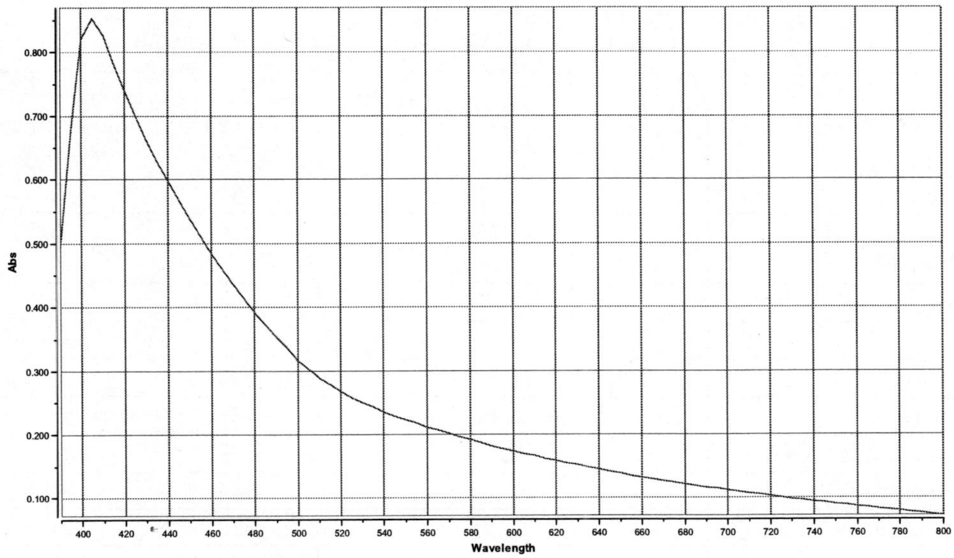

The Ag3d spectrum could be decomposed into a single spin-orbit pair.
The FTIR spectrum of extra cellular bio synthesized AgNps is shown in Graph

Graph 1: UV-Visible Spectroscopy

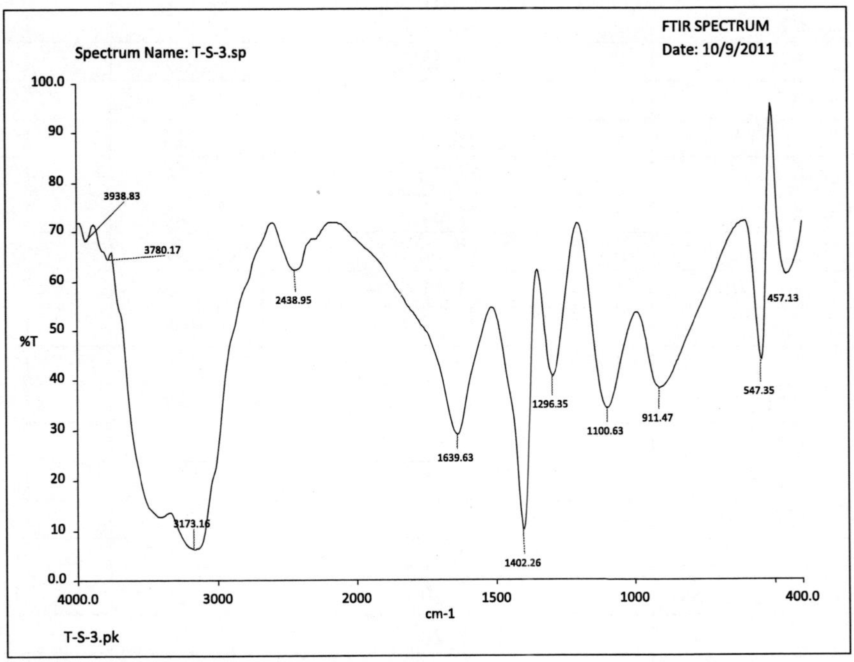

Graph 2: FTIR Spectrum Recorded by Making a KBr with Synthesized AgNPs

Table 2: Inhibition of Biofilm Formation by *C. albicans* and Changes in Total Protein and Carbohydrate Concentration of *C. albicans*

Treatment	*Biofilm OD 570 as a % of Control Value*	*Total Protein (µg/mg)*	*Total Carbohydrate (µg/mg)*
Control	1.91	731	339
Silver Nanoparticle (AgNp) T1	13.26	270	195
Costus igneus (Ci) T2	8.14	428	213
Tulsi (T) T3	5.1	650	320
Ginger (G) T4	11.2	693	184
Garlic (Gc) T5	14.24	583	157
Coconut oil (Cn) T6	10.2	637	201
clotrimazole (C) T7	13.33	584	148
Miconazole (M) T8	14.28	627	211
AgNp+ Ci T9	26.48	350	126
AgNp+ T T10	15.38	603	307
AgNp+ G T11	16.35	510	173
AgNp+ Gc T12	21.36	344	100
AgNp+ Cn T13	16.32	382	122
AgNp+ C T14	24.48	321	141
AgNp+ M T15	21.41	322	192
AgNp+ C+ Ci T16	29.55	114	106
AgNp+ C+ T T17	18.23	179	127
AgNp+ C+ G T18	31.61	126	114
AgNp+ C+ Gc T19	40.92	134	74
AgNp+ C+ Cn T20	21.41	218	113
AgNp+ M+ Ci T21	26.83	208	102
AgNp+ M+ T T22	22.85	296	171
AgNp+ M+ G T23	33.65	181	117
AgNp+ M+ Gc T24	36.83	106	92
AgNp+ M+ Cn T25	34.78	169	115
AgNp+ C+ M+ Ci T26	32.65	204	97
AgNp+ C+ M+ T T27	31.63	269	97
AgNp+ C+ M+ G T28	34.69	129	89
AgNp+ C+ M+ Gc T29	45.92	69	64
AgNp +C+ M+ Cn T30	37.76	98	102

Table 3: Inhibition of Biofilm Formation by *MRSA* and Changes in Total Protein and Carbohydrate Concentration of *MRSA*

Treatment	*Biofilm OD 570 as a % of Control Value*	*Total Protein (µg/mg)*	*Total Carbohydrate (µg/mg)*
Control	1.79	421	223
Silvernanoparticle (AgNp) T1	11.57	246	163
Costus igneus (Ci) *T2*	6.33	253	170
Tulsi (T) T3	4.25	302	176
Ginger (G) T4	5.43	245	169
Garlic (Gc) T5	7.45	257	158
Tetracycline (Tc) T6	11.85	240	143
Chloramphenicol (C) T7	13.76	283	131
AgNp+ Ci T8	16.82	232	112
AgNp+T T9	13.70	281	136
AgNp+G T10	15.92	211	129
AgNp+Gc T11	18.89	205	123
AgNp+Tc T12	17.26	213	119
AgNp+C T13	19.1	201	105
AgNp+Tc+ Ci T14	24.45	189	104
AgNp+Tc+T T15	15.34	215	114
AgNp+Tc+G T16	17.64	185	106
AgNp+Tc+Gc T17	33.12	155	99
AgNp+C+ Ci T18	28.59	170	109
AgNp+C+T T19	20.32	187	118
AgNp+C+G T20	28.65	150	113
AgNp+C+Gc T21	41.48	123	90
AgNp+Tc+C+ Ci T22	28.26	144	92
AgNp+Tc+C+T T23	21.27	133	95
AgNp+Tc+C+G T24	31.67	122	84
AgNp+Tc+C+Gc T25	45.21	112	54

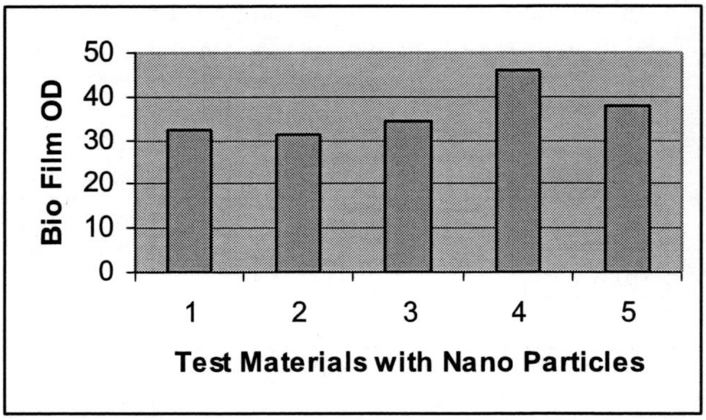

AgNp+C+M+C, AgNp+C+M+T, AgNp+C+M+G, AgNp+C+M+Gc, AgNp+C+M+Cn

Graph 4: Inhibition of Biofilm Formation by *c.albicans* with Four Test Materials Combination

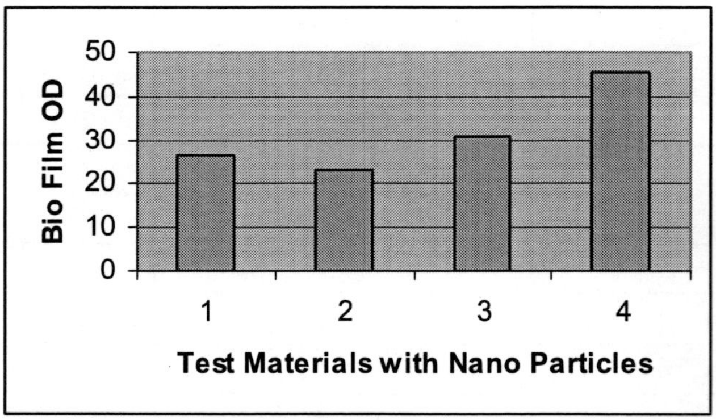

Graph 5: Inhibition of Biofilm Formation by *MRSA* with Four Test Materials Combination

The conical flask with *staphylococcus pasteuri* biomass were a pale yellow color before the addition of Ag ions and this change to a brown color clearly indicators the formation of silver nanoparticles in the reaction mixture (10) (Figure 1). Duran *et al.* (10) have shown the biosynthesis of AgNPs from *Fusarium oxysporum* strains by extracellular process.The conical flask with biomass is mixed with 10^{-3} M silver nitrate of final concentration and stored at dark conditions. Synthesized silver nanoparticles are characterized by UV-Vis spectroscope, a strong broad surface Plasmon peak located at 410 nm (Graph 1). The solution was extremely stable even for several weeks after reaction, it is known that silver cations area highly reactive and tend binds strongly to electron donor groups containing sulphur, oxygen or nitrogen. The method of synthesizing metallic nanoparticles by downstream process is simple and eco-friendly

(11). But Mineianetal, 2008 did not observed any extra cellular bio synthesis activity from *staphylococcus sp*, here it was observed that the silver was reduced by *Staphylococcuss pasteuri*. The elemental assay of silver nanoparticles was performed by FTIR; (Graph 2) shows the various chemical and molecular mechanisms of synthesised nanoparticles. The AFM micrograph at 30000 times magnification was shown in (Figure 2). In this micrograph observed rod nanoparticles in the size range of 50 nm.The absence of a higher binding energy Ag component clearly indicates that all the silver irons are fully reduced by *S.pasteuri*. It was observed from the FT-IR spectrum of AgNPs that the bands at 1642 correspond to a primary amine of NH band; similarly, 1409 and 1292 correspond to a secondary amine NH band and primary amine CN stretch vibrations of the proteins, respectively. The FTIR results indicate that the secondary structures of proteins were not affected as a consequence of reaction with Ag + ions are binding with AgNPs.In *C.albicans* the maximum effect on biofilm was recorded in silver nanoparticle, Clotrimazole, Miconazole and garlic (45.92%) followed by silver nanoparticle, Clotrimazole and garlic (40.92%). But 13.26%, 14.28%, 13.33%, 14.24% of inhibition was recorded in silver nanoparticle, Miconazole, Clotrimazole and garlic as single treatment. Moreover 37.76% of inhibition was recorded in silver nanoparticle, Clotrimazole, Miconazole and coconut oil combination(6). The inhibitory activity was increased to 31.61% in silver nanoparticle combined with Clotrimazole and ginger. 33.65% inhibition was in silver nanoparticle with ginger and Miconazole. 34.69% of inhibition was in silver nanoparticles with ginger, Clotrimazole and Miconazole. *Costus igneus* combined with silver nanoparticle and antifungal drugs shows distinct biofilmcidal effect. 29.55%, 26.83% and 32.65% of inhibition was recorded in silver nanoparticle with *Costus igneus* and *Clotri*mazole, silver nanoparticle with *Costus igneus* and Miconazole, and silver nanoparticles with *Costus igneus,* Clotrimazole and Miconazole respectively (Table 1) In *methicillin resistant staphylococcus aureus* distinct effect on biofilmcidal activity was observed in combination of silver nanoparticles with plant products and anti bacterial agents (12). Nanoparticle with tetracycline, chloramphenicol and garlic recorded maximum inhibition (45.21%) followed by 41.48% in silver nanoparticle with chloramphenicol and garlic (Table 2). The carbohydrate content was highly reduced in silver nanoparticle with Clotrimazole, Miconazole and garlic in *C.albicans* followed by silver nanoparticle with Clotrimazole, and garlic. The carbohydrate content of biofilm matrix of *S.aureus* was also reduced in silver nanoparticle with tetracycline, chloramphenicol and garlic, followed by silver nanoparticle with chloramphenicol and garlic. In the present study, both the tested strain biofilm formation and development was highly inhibited or reduced by the combined effect of Silver Nanoparticles with plant products and commercially available antimicrobial drugs.

REFERENCES

Adonizio, Kok-Fai Kong and KalaiMathee (2008). Inhibition of quorum sensing controlled virulence factor production in *Pseudomonas aeruginosa* by south Florida plant extracts. *Antimicrobial Agents and Chemotherapy*, Vol. 52, No. 1, 198–203.

Berkhout, De Schimmelgesl. Monilia, Oidium, Oospora en Torula, *Disset. Ultrecht: 44*, 1923.

Bozdogan, B.Ü.; Esel, D.; Whitener, C.; Browne, F.A. and Appelbaum, P.C. (2003). "Antibacterial susceptibility of a vancomycin-resistant Staphylococcus aureus strain isolated at the Hershey Medical Center". *Journal of Antimicrobial Chemotherapy,* 52(5): 864.doi:10.1093/jac/dkg457. PMID 14563898.edit

Duran, N., Marcato, P.D., Alves, O.L., De Souza, G.I.H. and Esposito, E., Mechanistic aspects of biosynthesis of silver nanoparticles by several Fusarium oxysporum strains, *J. Nanobiotechnol,* 3 (2005), pp. 1–7.

Karthick Raja Namasivayam Selvaraj, Elamathi Krishnamoorthi and Brijesh Kumar Singh (2011). Effect of biologically synthesized nanoparticles with plant products and chemotherapeutics against biofilm of clinical isolates of Staphylococcus aureus and Candida tropicalis. *IJPI's Journal of Biotechnology and Biotherapeutics.*

Kourkoumpetis, T., Manolakaki, D., Velmahos, G., *et al.* (2010). "Candida infection and colonization among non-trauma emergency surgery patients". Virulence 1 (5): 359–66.doi:10.4161/viru.1.5.12795. PMID 21178471.

Kuhn, D.M, George, T., Chandra, J., Mukherjee, P.K. and Ghannoum, M.A. (2002). Antifungal Susceptibility of Candida Biofilms; Unique Efficacy of Amphotericin B Lipid Formulations and Echinocandins. *Antimicrobial Agents and Chemotherapy*, Vol. 56, No. 6: 1773–1780.

Minaeian, S.A.R. Shahverdi, Nohi, A.S. and Shahverdi, H.R. (2008). Extracellular biosynthesis of silver nanoparticles by some bacteria, Vol. 17, No. 66.

Mohammed, A. Al-Fattani and Douglas, L. Julia (2006). Biofilm matrix of Candida albicans and Candida albicans: chemical composition and role in drug resistance. *Journal of Medical Microbiology*, 55: 999–1008.

Sastry, M.A., Ahmad, M.I. Khan and Kumar, R., Biosynthesis of metal nanoparticles using fungi and actinomycetes, *Curr Sci* 85 (2003), pp. 162–170.

Song, H.Y., Ko, K.K., Oh, I.H. and Lee, B.T. (2006). Fabrication of silver nanoparticles and their antimicrobial mechanisms, *Eur Cells Mater*, 11, p. 58.

Toole, G.A. and Kolter, R., Initiation of biofilm formation in pseudomonas fluorescenes WCS365 proceeds via multiple, convergent signaling pathways: a genetic analysis. M*ol, Microbiol* (1998), 28: 449–461.

Isolation and Virulence Determination of *Salmonella enterica* Serovar Typhi from Human

U. Shalini and Arunava Das*

Department of Biotechnology, Bannari Amman Institute of Technology, Sathyamangalam, Tamil Nadu

E-mail: drarunavadas@rediffmail.com; arunava20@gmail.com

INTRODUCTION

Salmonella enterica serovar Typhi, the most common causative agent of typhoid, remains an important public health problem. If untreated it may lead to fatal infection in children and adults (Hirose *et al.*, 2002). It causes 16 million cases of the disease and about 600,000 deaths, annually, all over the world. It has been found that all known *Salmonella* serovars has Invasion gene (*inv*) which enables the bacteria to invade host cells (Chiu and Ou, 1996). The O antigen gene (*tyv*) encodes CDP—tyvelose epimerase, which converts CDP—paratose to CDP—tyvelose which is present in both serovars Typhi and Paratyphi A. Vi capsular antigen gene posses in all virulent strains of *S. enterica* serovar Typhi causing typhoidial fever. Thus, the DNA sequence encoding the Vi antigen, pertaining to the *via*B region is useful in developing DNA based diagnostic tests for *S. enterica* serovar Typhi (Hashimoto *et al.*, 1995). In *S. enterica* serovar Typhi, the H antigen gene (*fli*C-d) *ie.*, phage-1 flagellin gene for d-antigen [H:d] encodes for flagellin which plays a key role for the Type III Secretion system (Hirose *et al.*, 2002). Antibiotics such as chloramphenicol has been a choice of drug for the treatment of typhoid fever for about 40 years, but recent studies have reported emergence of multi-drug resistant *S. enterica* serovar Typhi in India and all over the world (Senthilkumar and Prabakaran, 2005) .The emergences of the drug resistant *S.* Typhi strains possess major challenge in the treatment and prevention of typhoid fever which makes the use of alternative drugs for treatment necessary Therefore,

*Corresponding author.

it is vital to study the antibiotic susceptibility pattern periodically. In this study, *S. enterica* serovar Typhi were isolated from typhoidial human blood samples and their virulence was determined by single PCR and mPCR. The antibiotic susceptibility pattern of the *S.* Typhi strains was studied.

MATERIALS AND METHODS

Isolation and Identification of Bacteria

Blood specimens obtained aseptically before the antibiotic therapy from 16 patients (7 males and 9 females) with typhoidial fever were transported in an ice cold container and immediately processed for microbial investigation. A volume of three to five millilitres of venous blood with minimum blood-to-broth ratio of 1 to 10 was inoculated into 30 ml of brain heart infusion broth (Hi-Media, Mumbai). Blood culture broths were incubated at 37°C for 7 days. Once the visible growth was observed, it was streaked on sheep blood agar followed by streaking on xylose lysine deoxycholate (XLD) agar plates (Hi-Media, Mumbai) and incubated at 37°C for 24 h. Bacterial colonies were purified based on the size, shape, colour on XLD agar and patterns of haemolysis on blood agar and were subjected to Gram's staining. Bacterial isolates were identified by standard biochemical tests (Holt *et al.*, 1994). Ultra structure studies of the bacterial isolates were carried out by Scanning electron micrograph.

Detection of Virulence Genes by Polymerase Chain Reaction

The virulence of the isolates was determined by detection of the virulence genes such as invasion gene (*inv*A) by single PCR and phage-1 flagellin gene for d-antigen (*fli*C-d), tyvelose epimerase gene (*tyv*) and Vi antigen genes (*via*B) by multiplex PCR that was designed and performed according to the earlier research work (Ganeshkumar *et al.*, 2010). S. Typhi (MTCC 733) and *A.hydrophila* (MTCC 646), strains were used as positive and negative controls respectively. The specific forward and reverse primer pair sequences were obtained from the published literature. The *inv*A gene (Chiu and Ou, 1996); *tyv* gene, *fli*C-d genes and *via*B gene of 439 bp (Hirose *et al.,* 2002) were commercially synthesized (Bangalore Genei, Bangalore). The PCR amplicons (5μl) were electrophoresed in 1.5% agarose gel in TAE (pH 8) buffer, stained with ethidium bromide and observed in gel doc system (Universal Hood, BIORAD, Italy).

Antibiotic Susceptibility Test

Antibiotic susceptibility tests were performed by disc diffusion method (Bauer *et al.*, 1996) with little modification. Overnight cultures in peptone water were spread plated on nutrient agar (Hi-media, Mumbai) plates. The antibiotics discs (Hi-media, Mumbai) were purchased and used at the following concentrations: Gentamycin (10 g), Cefuroxime (30 g), Penicillin-G (2 Units/ml), Nalidixic acid (30 g), Clindamycin (10 g), Carbenicillin (100 g), Cephalothin (30 g), Kanamycin (30 g), Nitrofurantoin (100g), Tetracyclin (30 g), Ampillicin (10 g) and Chloramphenicol (30 g). The resistance breakpoints were those defined by the National Committee for Clinical Laboratory Standards (NCCLS, 1999) for Gram negative bacteria. *S. enterica* serovar Typhi (MTCC 733) and *A. hydrophila* (MTCC 646) were used as controls.

RESULTS AND DISCUSSION

In the present study, blood samples were collected from 16 patients of age group 12 to 50 years from Coimbatore, Erode and Salem districts, Tamil Nadu, India during the month of January and February and this end of dry season was considered to be the peak occurrence season of typhoidial fever. All the isolates were grown in BHI broth and were non haemolytic on sheep blood agar and showed pink colour colonies with black centre on XLD agar. Glucose, mannitol and sorbitol were fermented by all isolates and arabinose was fermented by few. The isolates also showed production of hydrogen sulphide and acid in the TSI agar and fermented glucose alone. Isolates were positive for oxidase test and methyl red test and were negative for indole production, urease production and citrate utilization. All the isolates were found to be Gram negative, flagellated and motile. Upon detailed bacteriological investigation, 16 isolates were tentatively identified as *S.* Typhi (Holt *et al.*, 1994) which was in complete agreement with previous studies (Wain *et al.*, 1998, Ganeshkumar *et al.*, 2010). The ultrastructure study of *S. enterica* serovar Typhi in SEM was observed to be in clusters of thick rods of variable length either singles or in pairs. In single step and mPCR assays, all the 16 isolates were found positive for the *inv*A, *tyv, fli*C-d and *via*B genes of desired fragments 244bp, 615bp, 750bp and 439bp respectively (Figure 1) which in full agreement with previous research work (Ganeshkumar *et al.*, 2010). Our study demonstrated that *inv*A gene can acted as a genus specific gene marker irrespective of clinical or healthy cases (Chiu *et al.*, 1996) while all the isolates of *S.* Typhi causing typhoid fever carried *tyv, fli*C-d and *via*B genes (Hirose *et al.*, 2002, Kumar *et al.*, 2006). It can be suggested from the present study that PCR is an effective and rapid diagnostic tool for the detection of *S. Typhi* from clinical samples.

The result of antibiotic susceptibility test revealed that isolates of *S.* Typhi were 100% resistant to ampicillin, 81.25% to cefuroxime and 68.75% resistant to penicillin-G and cephalothin respectively (Figure 2). The present result demonstrated that *S.* Typhi isolates are capacable of becoming resistant towards multiple drugs. In the present study, all the isolates were found to be moderately sensitive to nalidixic acid and nitrofurantoin. More recently, 76% of blood culture isolates of *S.* Typhi were reported to be resistant to nalidixic acid (Parry *et al.*, 1998). In India, although 29.47% and 28.42% of *S.* Typhi isolates has been reported to be resistant to ampicillin and chloramphenicol respectively (Nagshetty *et al.*, 2010), in the present study it was found 100% sensitive to chloramphenicol along with carbenicillin, clindamycin, gentamycin, kanamycin and tetracycline which is in agreement with earlier reports (Quintaes *et al.*, 2002). This also indicates the re-emergence of sensitivity of *S.* Typhi to chloramphenicol which has been earlier reported (Sood *et al.*, 1999).

The present study has confirmed the association of virulent strains of *S. Typhi* in the occurrence of the typhoidial fever in humans in Tamil Nadu. From this study this is also suggested that PCR technique can be used as a rapid diagnostic tool for the detection of *S. enterica* serovar Typhi in the diagnostic laboratories or clinics for the clinical diagnosis of typhoidial fever from patients. By using 12 antibiotics for

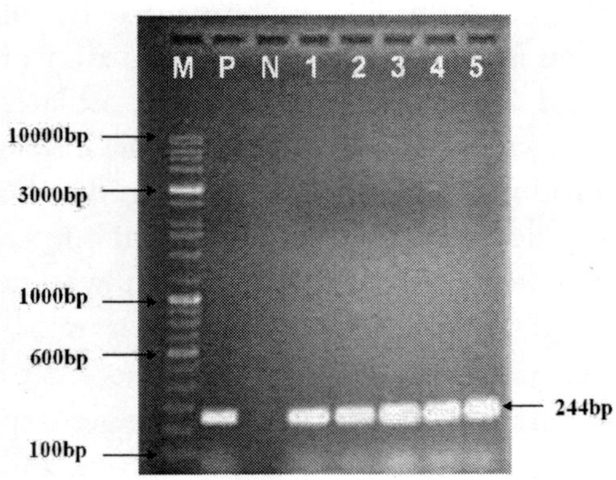

(A) Detection of *inv*A (244 bp) gene by PCR

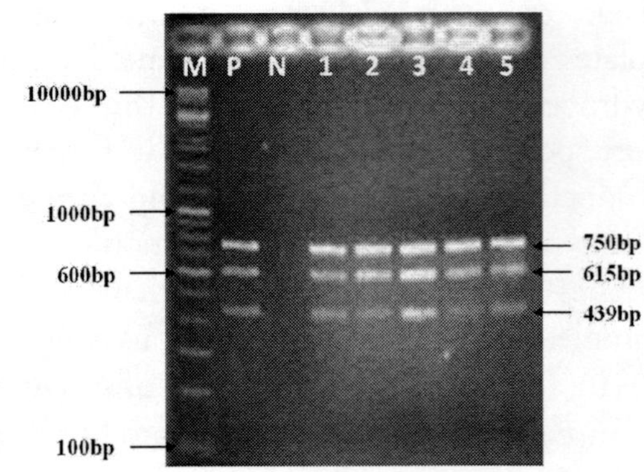

(B) Detection of *tyv* (615 bp), *fli*C-d (750 bp) and *via*B (439 bp) genes by mPCR

Lane P: Positive control (*S. enterica* serovar Typhi MTCC 733);

Lane N: Negative control (*A. hydrophila* MTCC 646);

Lane 1–5: Field isolates positive for virulence genes;
Lane M: High range DNA ruler.

Fig. 1: Detection of Virulence Genes from *S. enterica* Serovar Typhi by PCR

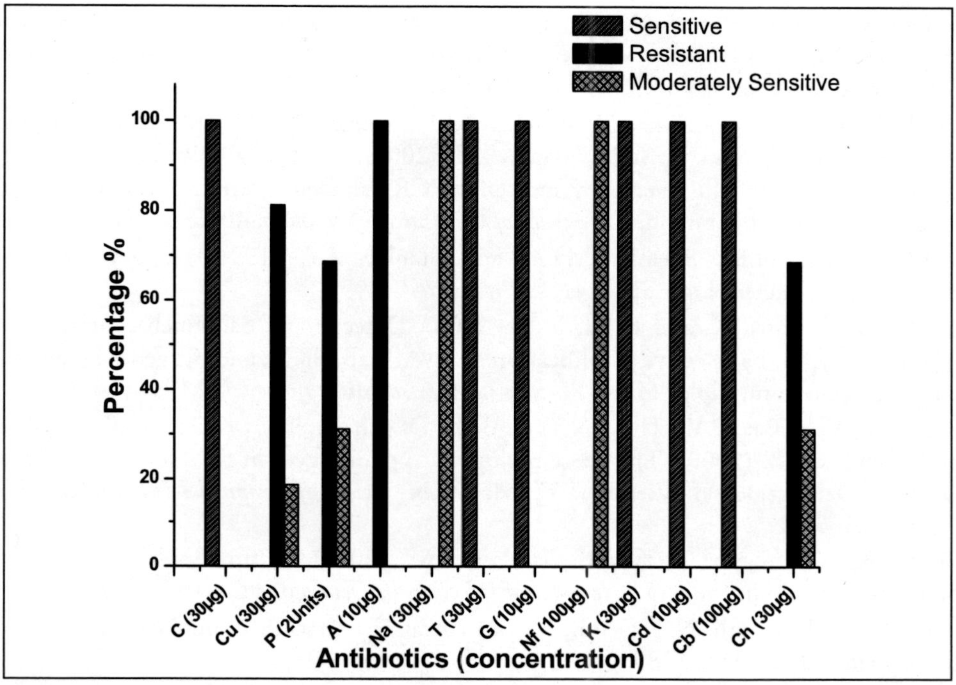

A: Ampillicin, C: Chloramphenicol, Cb: Carbenicillin, Cd: Clindamycin, Ch: Cephalothin, Cu: Cefuroxime, G: Gentamycin, K: Kanamycin, Na: Nalidixic acid, Nf: Nitrofurantoin, P: Penicillin G, T: Tetracycline.

Fig. 2: Antibiotic Susceptibility Test Results of *S. enterica* Serovar Typhi

susceptibility test, present findings helped to know the current status of typhoidial fever among the people in Southern part of India. Our study suggest continuous evaluation of sensitivity-resistance pattern of *S.* Typhi isolates is needed to make use of antibiotics such as chloramphenicol in the management of typhoidial fever in future.

REFERENCES

Agarwal, S.C. (1962). Chloramphenicol resistance of Salmonella species in India, 1956–61. *The International Journal of Public Health—World Health Organization* 17: 331–335.

Bauer, A.W., Kirby, W.M.M., Sheris, J.C. and Turck, M. (1996). Antibiotic suspectibility test by standardized single disk method. *American Journal of Clinical Pathology*, 36: 493–496.

Chiu, C.H. and Ou, J.T. (1996). Rapid identification of Salmonella serovars in faeces by specific detection of virulence genes, invA and spvC, by an enrichment broth culture-multiplex PCR combination assay. *Journal of Clinical Microbiology,* 34: 2619–2622.

Ganeshkumar, A. Shalini, U. Vaishnavi, K. Remya Ismail. Lakshmanaswamy, A. Vasanthi, N.S. and Das, A. (2010). Rapid detection of Salmonella enterica serovar Typhi from humans. *Journal of Pure and Applied Microbiology*, 4: 837–841.

Hashimoto, Y., Itho, Y., Fujinaga, Y., Khan, A.Q., Sultana, F., Miyake, M., Hirose, K., Yamamoto, H. and Ezaki, T. (1995). Development of nested PCR based on the ViaB sequence to detect Salmonella typhi. *Journal of Clinical Microbiology,* 33: 775–777.

Hirose, K., Itoh, K.I., Nakajima, H., Kurazono, T., Yamaguchi, M., Moriya, K., Ezaki, T., Kawamura, Y., Tamura, K. and Watanabe, H. (2002). Selective amplification of tyv (rfbE), prt (rfbS), viaB, and fliC genes by multiplex PCR for identification of Salmonella enteric Serovars Typhi and Paratyphi A. *Journal of Clinical Microbiology,* 40: 633–636.

Holt, J.C., Kreig, N.R., Sneath, P.H.A. and Stanley, J.Y. (1994). Bergey's Manual of determinative Bacteriology. Springer, USA, pp. 764–799.

Kumar, S., Balakrishna, K. and Batra, H.V. (2006). Detection of Salmonella enterica Serovar Typhi (S. Typhi) by selective amplification of invA, viaB, fliC-d and prt genes by polymerase chain reaction in multiplex format. *Letters in Applied Microbiology,* 42: 149–154.

Lin, F.C., Ho, V.A., Bay, P.V., Thuy, N.T.T., Bryla, D., Thanh, T.C., Khiem, H.B., Trach, D.D. and Robbins, J.B. (2000). The epidemiology of typhoid fever in the Dong Thap Province, Mekong Delta region of Vietnam. *The American Society of Tropical Medicine and Hygiene* 62(5): 644–648.

Nagshetty, K., Channappa, S.T. and Gaddad, S.M. (2010). Antimicrobial susceptibility of Salmonella Typhi in India. *The Journal of Infection in Developing Countries,* 4(2): 070–073.

Parry, C., Wain, J., Chinh, N.T., Vinh, H. and Farrah, J.J. (1998). Quinoloneresistant S. typhi in Vietnam. *Lancet* 351: 1289.

Quintaes, B.R., Leal, N.C., Reis, E.M.F., Fonseca, E.L. and Hofer, E. (2002). Conventional and molecular typing of Salmonella Typhi strains from Brazil. Journal of the Institute of Tropical Medicine of São Paulo 44: 315–319.

Senthilkumar, B. and Prabakaran, G. (2005). Multidrug resistant Salmonella Typhi in asymptomatic Typhoid carriers among food handlers in Namakkal District, Tamilnadu. *Indian Journal of Medical Microbiology,* 23: 92–94.

Sood, S., Kapil, A., Das, B., Jain, Y. and Kabra, S.K. (1999). Reemergence of chloramphenicol sensitive Salmonella typhi. Lancet 353: 1241–1242.

Wain, J., Diep, T.S., Ho, V.A., Walsh, A.M., Hoa, N.T.T., Parry, C.M. and White, N.J. (1998). Quantitation of Bacteria in Blood of Typhoid Fever Patients and Relationship between Counts and Clinical Features, Transmissibility and Antibiotic Resistance. *Journal of Clinical Microbiology,* 36: 1683–1687.

SECTION–V

Prof. Edathil Vijayan Award

Aborted Fetal Stem Cells as a Source of Neurons for the Transplantation in Parkinson's Disease

S. Saranya, S.M. Brinda Shree and R. Divya*
Department of Biotechnology, K.S. Rangasamy College of Technology,
Tiruchengode, Tamil Nadu
E-mail: divya.deto@gmail.com

INTRODUCTION

The discovery of these cells has spurred further research into the characteristics of neural stem cells from the fetus and the adult, mostly using rodents and primates as model species. The hope is that these cells may be able to replenish those that are functionally lost in human degenerative diseases such as Parkinson's Disease, Huntington's Disease, and amyotrophic Diseases of the nervous system including congenital disorders cancers and degenerative diseases affect millions of people of all ages congenital disorders occur when the brain or spinal cord does not form correctly during development cancers of the nervous system result from the uncontrolled spread of aberrant cells. Degenerative diseases occur when the nervous system loses functioning of nerve cells (Imitola *et al.,* 2004). Most of the advances in stem cell research have been directed at treating degenerative diseases. While many treatments aim to limit the damage of these diseases, in some cases scientists believe that damage can be reversed by replacing lost cells with new ones derived from cells that can mature into nerve cells, called neural stem cells. Research that uses stem cells to treat nervous system disorders remains an area of great promise and challenge to demonstrate that cell-replacement therapy can restore lost function. Lateral sclerosis (ALS, also known as Lou

*Corresponding author.

Gehrig's disease), as well as from brain and spinal cord injuries that result from stroke or trauma.

Scientists are applying these new stem cell discoveries in two ways in their experiments. First, they are using current knowledge of normal brain development to modulate stem cells that are harvested and grown in culture. Researchers can then transplant these cultured cells into the brain of an animal model and allow the brain's own signals to differentiate the stem cells into neurons or glia. Alternatively, the stem cells can be induced to differentiate into neurons and glia while in the culture dish, before being transplanted into the brain. Much progress has been made the last several years with human Embryonic Stem (ES) cells that can differentiate into all cell types in the body. While ES cells can be maintained in culture for relatively long periods of time without differentiating, they usually must be coaxed through many more steps of differentiation to produce the desired cell types. Recent studies, however, suggest that ES cells may differentiate into neurons in a more straightforward manner than may other celltypes. Second, scientists are identifying growth (trophic) factors that are normally produced and used by the developing and adult brain. They are using these factors to minimize damage to the brain and to activate the patient's own stem cells to repair damage that has occurred. Each of these strategies is being aggressively pursued to identify the most effective treatments for degenerative diseases much more research is necessary to develop stem cell therapies that will be useful for treating brain and spinal cord disease in the same way that hematopoietic stem cell therapies are routinely used for immune system replacement.

The majority of stem cell studies of neurological disease have used rats and mice, since these models are convenient to use and are well-characterized biologically. If preliminary studies with rodent stem cells are successful, scientists will attempt to transplant human stem cells into rodents. Studies may then be carried out in primates (e.g., monkeys) to offer insight into how humans might respond to neurological treatment. Human studies are rarely undertaken until these other experiments have shown promising results. While human transplant studies have been carried out for decades in the case of supply of transplantable cells. To isolate the embryonic nerve stem cells from the aborted foetus and to culture the nerve stem cells in the neural basal medium for the successive collection of passage cells in a periodic interval. when the subcultured nerve cells were implanted into the person affected by parkinson's disease the condition got reversed and the maximum recovery from the disorder was observed.

MATERIALS AND METHODS

Isolation of Neural Stemcells from Aborted Fetus

Neural Stem Cells (NSCs) reside a longer ventricular zone neuro epithelium during the development of the cortical plate in fetus. To isolate down syndrome NSCs from aborted human fetal cortices and grow them in culture by simple and reliable method. The methodology provides the specific aspects of harvesting the tissue dissection with limited anatomical land marks, cell sorting, plating and passaging of human NSCs.

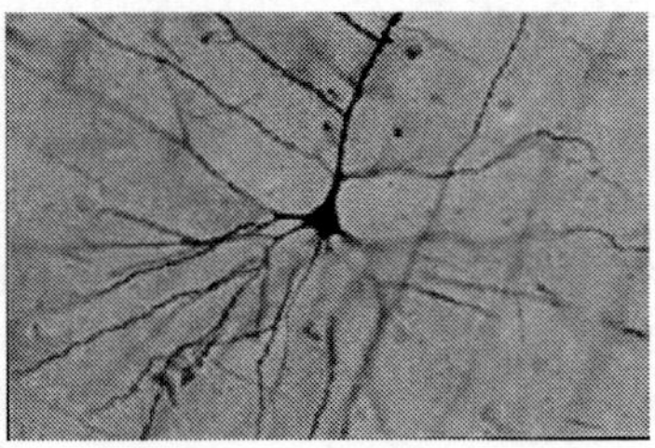

Fig. 1: Golgi Stained Passaged Neuron

Culturing of Neurospheres

The obtained neurons from the aborted fetus stem cells was grown in neurocult NSA media. The obtained passage cells were developed further without the growth factors of FGF and EGF differentiation of oligodendrocytes in specific NS-2-diffrentuation media to treat the neurological disorder.

Positron Emission Tomography

The intensive research aiming at curing parkinson's disease with stem cells triggered by the death of neurons in a brain region called the substantia nigra, Parkinson's disease begins with minor tremors that progress to limb and bodily rigidity and difficulty initiating movement. These neurons connect via long axons to another region called the striatum, composed of subregions called the caudate nucleus and the putamen (Keirstead *et al.,* 2005). These neurons that reach from the *Substantia nigra* to the striatum release the chemical transmitter dopamine onto their target neurons in the striatum. One of dopamine's major roles is to regulate the nerves that control body movement. As these cells die, less dopamine is produced, leading to the movement difficulties characteristic of Parkinson's disease. Currently, the causes of death of these neurons are not well understood (Arvidsson *et al.,* 2002).

Parkinson's disease patients with the drug levodopa (L-dopa) was treated, where the brain converts it into dopamine. Although the drug works well initially, levodopa eventually loses its effectiveness, and side-effects increase. Ultimately, many doctors and patients find themselves fighting a losing battle (Brustle *et al.,* 1999). For this reason, a huge effort is underway to develop new treatments, including growth factors that help the remaining dopamine neurons survive and transplantation procedures to replace those that have died we could measure an increase in dopamine neuron function in the striatum of these patients by using a brain-imaging method called positron emission tomography (PET).

The NIH has funded two large and well-controlled clinical trials in the past 15 years in which researchers transplanted tissue from aborted fetuses into the striatum of patients with Parkinson's disease. These studies included controls, where patients received "sham" surgery (no tissue was implanted), and neither the patients nor the scientists who evaluated their progress knew which patients received the implants. The patients' progress was followed for up to eight years. Unfortunately, both studies showed that the transplants offered little benefit to the patients as a group. While some patients showed improvement, others began to suffer from dyskinesias, jerky involuntary movements that are often side effects of long term L-dopa treatment, studies showed that the patient's immune system were attacking the graft.

(Before) (After)

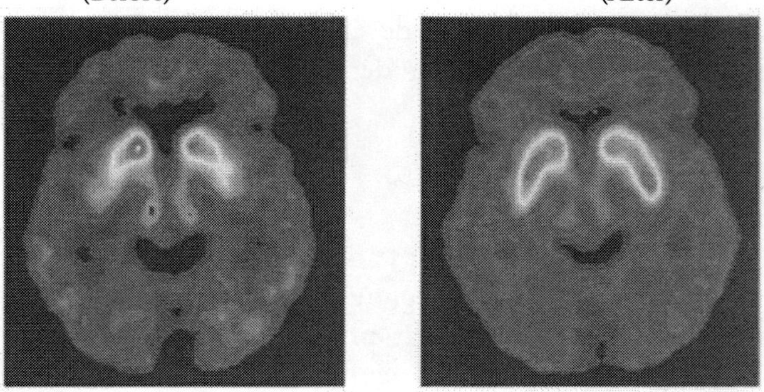

Fig. 2: Positron Emission Tomography (PET) Images from a Parkinson's Patient before and after Fetal Tissue Transplantation

RESULTS AND DISCUSSIONS

The major goal for Parkinson's investigators is to generate a source of cells that can be grown in large supply, maintained indefinitely in the laboratory, and differentiated efficiently into dopamine neurons that work when transplanted into the brain of a Parkinson's patient. Scientists have investigated the behavior of stem cells in culture and the mechanisms that govern dopamine neuron production during development in their attempts to identify optimal culture conditions that allow stem cells to turn into dopamine-producing neurons.

Preliminary studies have been carried out using immature stem cell-like precursors from the rodent ventral midbrain, the region that normally gives rise to these dopamine neurons. In one study these precursors were turned into functional dopamine neurons, which were then grafted into rats previously treated with 6-hydroxy-dopamine (6-OHDA) to kill the dopamine neurons in their substantia nigra and induce parkinson's-like symptoms. Even though the percentage of surviving dopamine neurons was low following transplantation, it was sufficient to relieve the Parkinson's-like symptoms.[11] Unfortunately, these fetal cells cannot be maintained in culture for very long before they lose the ability to differentiate into dopamine neurons.

Cells with features of neural stem cells have been derived from ES-cells, fetal brain tissue, brain tissue from neurosurgery, and brain tissue that was obtained after a person's death. There is controversy about whether other organ stem cell populations, such as hematopoietic stem cells, either contain or give rise to neural stem cells

Many researchers believe that the more primitive ES cells may be an excellent source of dopamine neurons because ES-cells can be grown indefinitely in a laboratory dish and can differentiate into any cell type, even after long periods in culture (Kerr *et al.*, 2003). Mouse ES cells injected directly into 6-OHDA-treated rat brains led to relief of Parkinson-like symptoms. Further investigation showed that these ES cells had differentiated into both dopamine and serotonin neurons.[12] This latter type of neuron is generated in an adjacent region of the brain and may complicate the response to transplantation. Since ES cells can generate all cell types in the body, unwanted cell types such as muscle or bone could theoretically also be introduced into the brain. As a result, a great deal of effort is being currently put into finding the right "recipe" for turning ES cells into dopamine neurons—and only this cell type—to treat Parkinson's disease. Researchers strive to learn more about normal brain development to help emulate the natural progression of ES cells toward dopamine neurons in the culture dish.

The recent availability of human ES cells has led to further studies to examine their potential for differentiation into dopamine neurons. Recently, dopamine neurons from human embryonic stem cells have been generated.[13] One research group used a special type of companion cell, along with specific growth factors, to promote the differentiation of the ES cells through several stages into dopamine neurons (Klein *et al.*,

2005). These neurons showed many of the characteristic properties of normal dopamine neurons.[13] Further-more, recent evidence of more direct neuronal differentiation methods from mouse ES cells fuels hope that scientists can refine and streamline the production of transplantable human dopamine neurons. One method with great therapeutic potential is nuclear transfer. This method fuses the genetic material from one individual donor with a recipient egg cell that has had its nucleus removed (Kelly *et al.,* 2005). The early embryo that develops from this fusion is a genetic match for the donor. This process is sometimes called "therapeutic cloning" and is regarded by some to be ethically questionable. However, mouse ES cells have been differentiated successfully in this way into dopamine neurons that corrected Parkinsonian symptoms when transplanted into 6-OHDA-treated rats.[14] Similar results have been obtained using parthenogenetic primate stem cells, which are cells that are genetic matches from a female donor with no contribution from a male donor.[15] These approaches may offer the possibility of treating patients with genetically-matched cells, thereby eliminating the possibility of graft rejection (Nakatomi *et al.,* 2002).

Hence the person affected by parkinson's disease was recovered by using neural stem cell transplantation than by other therapeutical methods.

REFERENCES

Arvidsson, A., Collin, T., Kirik, D., Kokaia, Z. and Lindvall, O. (2002). Neuronal replacement from endogenous precursors in the adult brain after stroke. *Nat Med.,* 8: 963–970.

Brustle, O, Jones, K.N. and Learish, R.D. (1999). Embryonic stem cell-derived glial precursors: a source of myelinating transplants. *Science,* 285: 754–756.

Imitola, J., Raddassi, K. and Park, K.I. (2004). Directed migration of neural stem cells to sites of CNS injury by the stromal cell-derived factor 1alpha/CXC chemokine receptor 4 pathway. *Proc Natl Acad Sci USA.,* 101: 18117–18122.

Keirstead, H.S., Nistor, G. and Bernal, G. (2005). Human embryonic stem cell-derived soligodendrocyte progenitor cell trans-plants remyelinate and restore locomotion after spinal cord injury. *J. Neurosci.,* 25: 4694–4705.

Kelly, S., Bliss, T.M. and Shah, A.K. (2004). Transplanted human fetal neural stem cells survive, migrate, and differentiate in ischemic rat cerebral cortex. *Proc Natl Acad Sci USA.*, 101: 11839–11844.

Kerr, D.A., Llado, J. and Shamblott, M.J. (2003). Human embryonic germ cell derivatives facilitate motor recovery of rats with diffuse motor neuron injury. *J. Neurosci.,* 23: 5131–5140.

Klein, S.M., Behrstock, S., McHugh, J. (2005). GDNF delivery using human neural progenitor cells in a rat model of ALS. *Hum Gene Ther.,* 16: 509–521.

Nakatomi, H., Kuriu, T. and Okabe, S. (2002). Regeneration of hippocampal pyramidal neurons after ischemic brain injury by recruitment of endogenous neural progenitors. Cell., 110: 429–441.

Long Term Effects of Neonatal Hypoxia on Cognition and Memory

T.R. Anju, M.S. Ajayan and Paulose C.S.*

Molecular Neurobiology and Cell Biology Unit, Centre for Neuroscience, Department of Biotechnology, Cochin University of Science and Technology, Cochin, Kerala

E-mail: cspaulose@cusat.ac.in; paulosecs@yahoo.co.in

INTRODUCTION

Hypoxic injury during neonatal development induces early changes in cerebral energy utilization which can lead to brain damage and subsequently to severe neuro-developmental impairments. Chronic hypoxia in advanced chronic obstructive pulmonary disease result in altered and reduced neuropsychological functioning, which, in turn, leads to memory impairment even when other mental faculties remain unaffected (Sandhu, 1986). Hypoxia or hypoglycemia has almost immediate effects on behaviour and brain function. Though dramatic, they are fully reversible if the hypoxia or hypoglycemia is not sustained. But longer or more severe energy deprivation leads to irreversible functional and indeed cellular damage (cell death)—which develop only after a delay of some days (Pulsinelli *et al.*, 1982). High incidence of minor neurological deficits, mainly regarding the fields of language and behaviour was reported in persistent pulmonary hypertension of the newborn (Berti *et al.*, 2010).

Hypoxia in newborn infants is becoming much easier to prevent, detect and treat. Nevertheless the successful management of potentially hypoxic fetuses and newborn infants remains the major challenge to all physicians concerned with perinatal care. What is at stake is not only that sick infants should survive, but equally or more importantly that the survivors should be normal children. Fetal and neonatal hypoxia

*Corresponding author.

is viewed as one of the major causes of the developmental disturbances of the brain. The long term effects of neonatal hypoxic shock on the cognitive and memory processing has been overlooked during the resuscitation processes. This happens as the primary aim during the physical stress management will be the management of hypoxic condition to initiate respiratory mechanism in newborn, during which the impact on central nervous system and its subsequent consequences are not taken into account. So it is of prime importance to understand the possible long term effects of neonatal hypoxic shock in order to devise a better resuscitation programme.

The present study tries to understand the impact of neonatal hypoxic insult on the behavioral response in a later stage of life. One month old rats exposed to a single dose of acute and severe hypoxia on postnatal day 4 was used to evaluate both motor and cognitive efficiency. The motor ability and coordination was assessed by Wire Maneuver Test and cognitive abilities were assessed by Morris water maze experiment. The histological evaluation of hypoxic brain soon after the stress was also examined using florescent nuclear dye TOPRO-3 to understand the extent of severity of hypoxia on CNS.

MATERIALS AND METHODS

Animals

Wistar neonatal rats of 4 days old (postnatal day 4, P4) weighing 6.0–7.5 g were used for hypoxic induction and maintained in room temperature and standard conditions for one month to study the long term effects of neonatal stress. The control and hypoxic neonatal rats were maintained with their mothers under optimal conditions 12 hour light and 12 hour dark periods and were fed standard food and water *ad libitum*. All animal care and procedures were in accordance with Institutional and National Institute of Health guidelines.

Induction of Acute Hypoxia in Neonatal Rats

Wistar neonatal rats of P4 used for the experiments were grouped as:

1. C – Control neonatal rats exposed to atmospheric air (20.9% oxygen) for 30 minutes
2. Hx – Hypoxia neonatal rats exposed to 2.6% oxygen for 30 minutes (Hx).

For each experiment, 4–6 rats were taken for each group (n = 4–6 rats).

For behavioral studies, one month old rats exposed to single dose of neonatal hypoxia on postnatal day 4 were used. For histological studies, the transcardially perfused, fixed brain from P4 rats exposed to the experimental conditions was used.

Body Weight and Food Intake Determination

Body weight was noted after the induction of experimental conditions (P 4 rats) and before performing the behavioral test (one month old rats) and food intake was noted during the period from P4 to one month.

Wire Maneuver Test to Evaluate Motor Coordination

For wire maneuver test, the rat was hung by both forelimbs on the wire 50cm high from the floor. The diameter of the wire was 1mm. The body of the rat was maintained parallel to the floor by hanging the tail by experimenter's hand, and the tail was then released. Its falling latency was measured as an index of forelimb strength. The animal was lifted by its tail and allowed to grasp the horizontal wire with its forepaws, and was then rotated partially downward and released. The ability of the animal to grasp to the wire also with its hind limbs was scored.

Morris Water Maze Experiment to Evaluate Spatial Learning and Cognitive Assessment

The water maze task is a common behavioral test to assess spatial learning, memory and cognition of experimental animals. The custom-constructed water maze pool measured 100 cm in diameter by 50 cm in depth and was filled with water to a depth of 35 cm. A 10-cm-diameter white platform was located 1.5 cm below the surface of the water. Nontoxic white paint was added to the water to visually obscure the location of the platform. The pool had been divided into four quadrants of homogeneous size and the platform was located in the center of one of the quadrants, halfway between the center and the wall of the pool. All swim latencies were recorded with a manual stopwatch, a technique routinely employed by others (Hort *et al.*, 1999). The water maze task consisted of 15 sessions conducted once daily over 15 successive days. Each session consisted of four trials separated by approximately 60 seconds. Rats were placed manually into the pool, facing the pool wall in the center of one of the quadrants that did not contain the platform. For any given rat, the location of the platform remained fixed across all trials and all sessions. The latency to find the platform was recorded as the time from release into the pool until the rat had reached the platform. A maximum of 60 second was allowed for each trial. Rats not reaching the platform within 60 second were guided to the platform and a score of 60 second was recorded for each of these experimenter-terminated trials. The rat was allowed to remain on the platform for the duration of the inter-trial interval. A 60-second probe test to determine the time spent in the platform quadrant after removing the platform from pool was conducted on the 13th, 14th and 15th day of the study (24 h after the last hidden platform session). Rats were released into the pool in the quadrant opposite to that previously associated with the escape platform. A manual time-sampling procedure (one measurement per second) was utilized to record the swimming bias of the rat in each of the four quadrants of the pool.

Histochemical Analysis of Cell Loss by TOPRO-3 Staining

10 μm sagittal sections of trancardially perfused, fixed brain of control and hypoxic neonatal rats (postnatal day 4) were stained with nuclear stain TOPRO-3. For nuclear staining, TOPRO-3 (1: 1000 dilution in PBS) was added and kept for 10 minutes at room temperature. The sections were observed and photographed using confocal imaging system (Leica TCS SP 5).

Statistical Analysis

The equality of all the groups was tested by the analysis of variance (ANOVA) technique for different values of p. Further the pair wise comparisons of all the experimental groups were studied using Students-Newman-Keuls test at different significance levels. The testing was performed using GraphPad Instat (Ver. 2.04a, San Diego, USA) computer program

RESULTS

Body Weight and Food Intake

Experimental rats on postnatal day 30 were used for behavioural study. The hypoxic group showed a significant decrease ($p < 0.01$) in the body weight compared to

Table 1: Body Weight of Control and Hypoxic Rats at P4 and One Month Old

Experimental Groups	*Body Weight (g)*	
	P 4	*One month old*
Control	8.56 ± 0.29	75.32 ± 0.35
Hypoxic	8.35 ± 0.54	65.53 ± 0.74 [a]

Values are Mean ± S.E.M of 4-6 separate experiments. Each group consist 4–6 rats.
[a]$p < 0.01$ when compared to Control.

control on postnatal day 30 (Table 1). There was no significant change in the net food intake in the control and hypoxic rats.

BEHAVIOURAL STUDIES

Wire Maneuver Test to Evaluate Motor Coordination

Evaluation of motor function by wire maneuver test showed that there is no significant difference in the time spent in the wire by hypoxic rats when compared with control (Figure 1). This indicates that the grasping capability and motor coordination in hypoxic rats did not vary significantly from the control rats.

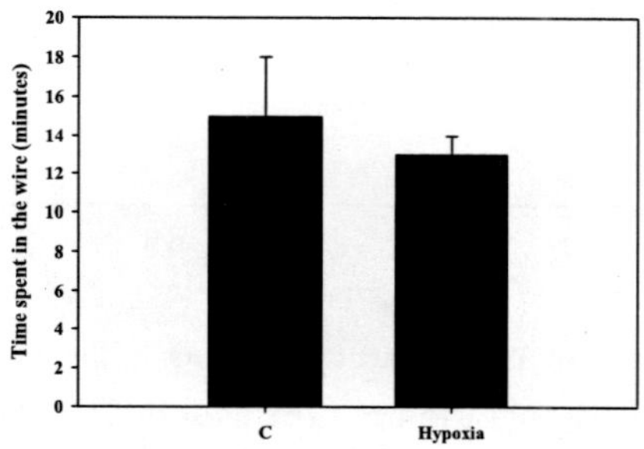

Values are Mean ± S.E.M of 4–6 separate experiments. Each group consist 4–6 rats.

Fig. 1: Response of One Month Old Rats Exposed to Neonatal Hypoxia in Wire Maneuver Test

Behavioural Response on Morris Water Maze

Morris water maze experiment showed that the escape latency was significantly increased ($p < 0.001$) in the hypoxic rats when compared to control. The increased escape latency indicates the impairment in spatial learning in neonatal rats exposed to hypoxic insult. The escape latency of hypoxic rats remained significantly high in trial 1, 2, 3 and 4 (Figure 2).

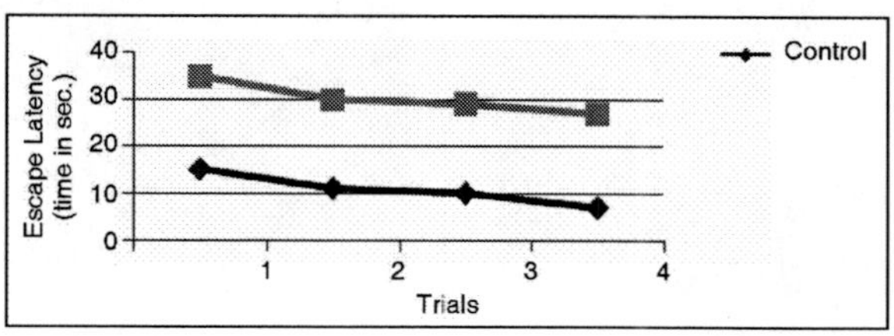

Values are Mean ± S.E.M of 4–6 separate experiments. Each group consist 4–6 rats.

Fig. 2: Escape Latency of One Month Old Rats Exposed to Neonatal Hypoxia in Water Maze Experiment

The time taken by hypoxic rats in recognizing the visual cues was also significantly high ($p < 0.001$) compared to control. In all the four trials the time taken by hypoxic rats was significantly high ($p < 0.001$) (Figure 3).

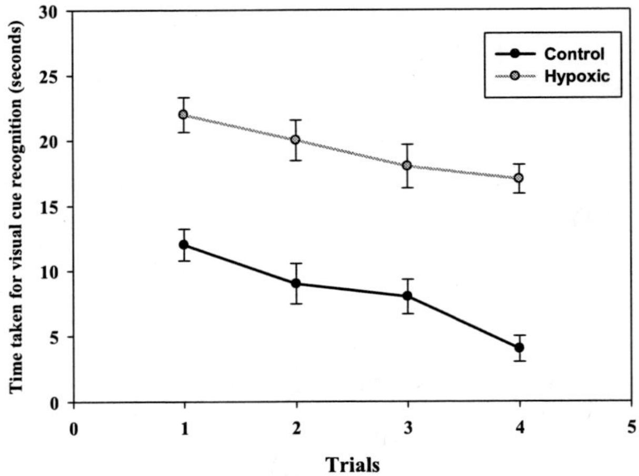

Values are Mean ± S.E.M of 4–6 separate experiments. Each group consist 4–6 rats.

Fig. 3: Visual Cue Recognition of One Month Old Rats Exposed to Neonatal Hypoxia in Water Maze Experiment

Histochemical Analysis by TOPRO-3 Staining

Histological analysis of brain sections of control and hypoxic rats at P4 with TOPRO-3 staining showed a significant decrease ($p < 0.001$) in the nuclear staining in the hypoxic neonatal rats. This indicated that hypoxic insult in neonatal rats caused a marked cell loss (Figure 4).

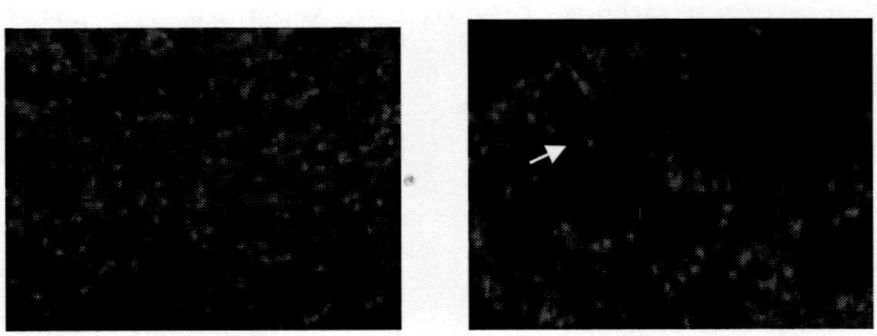

Control Hypoxic

The nucleus in the brain sections were stained red by fluorescent nuclear dye TOPRO-3.

Fig. 4: Histochemical Analysis of Brain Section of P4 Rats Using Nuclear Stain TOPRO-3

DISCUSSION

The brain regions associated with learning, memory and cognition is very sensitive to many types of cerebral insults and stress. The cell death in the brain regions in the early stages of development can drastically affect the memory and cognition in the later stages of life. Maintaining blood glucose level is important throughout the pregnancy, but it is particularly important during early developmental stage, when an embryo's organs are forming. Oxygen is needed by cells to break down glucose and produce energy. The oxygen requirement is more during embryonic development for all the tissues. Insufficient oxygen supply cause developmental abnormalities and birth defects. Thus hypoxia has a potential to cause damage to cells (Rulin *et al.*, 2005).

In the present study we observed a decline in the body weight of one month old rats exposed to neonatal hypoxic insult when compared to control rats, even though food intake was not affected. This point to the possibility of neuroendocrine disturbances attributed by severe and acute hypoxic insult during neonatal period. Earlier reports point out the endocrine regulation (Anju *et al.*, 2010) and the central nervous system disturbances in neonatal hypoxia (Joseph *et al.*, 2010).

Our studies on motor coordination using wire maneuver test showed that motor skills are not affected in a later stage of life due to neonatal hypoxic shock. Since brain stem is the major brain region involved in both respiratory control and motor coordination, the current observation can be related as the part of the adaptive response of the body to acclimatize the lack of oxygen. During the delay in pulmonary ventilation as in case of hypoxia, brain tries to overcome the situation by altering its signaling to promote pulmonary vasoconstriction (Anju and Paulose, 2011), preserving the ATP stores and routing it for the critical tissues thereby thriving the situation. Once the stress has been overcome, brain stem should be the prime part to resume its native function and override the complications. Thus, motor functions were not affected in a dramatic way during this neonatal stress condition to carry on to the later stages of life.

The Morris water maze has been used extensively to measure cognitive deficits in brain-damaged rats (Brandeis *et al.*, 1989). Performance in the Morris water maze depends on several elements, ranging from attention, learning and memory, to vision and motor coordination. Place navigation in the Morris water maze consists of two distinct components: declarative place representations as well as procedural learning (Morris *et al.*, 1999). The procedural aspects include learning to inhibit inborn nonadaptive behaviour, such as swimming along the wall (Whishaw and Mittleman, 1986; Paylor and Rudy, 1990), while selecting appropriate behavioural strategies, such as swimming across the pool or uniformly searching its surface. Other procedural components involved skills such as improved distance and angle judgment that are a necessary prerequisite for the cognitive demands of the task. The hippocampal formation is critical for computing place representations but is believed to be dispensable for procedural memories (O'Keefe and Nadel, 1978).

Morris water maze performance in one month old rats exposed to neonatal hypoxia showed the deficits in spatial memory and cognition in these rats. Hypoxic rats scored high escape latency with longer time for recognition of visual cues over trials.

The cognitive and memory deficits observed in one month old rats exposed to neonatal hypoxia might be a consequence of a series of alterations in the brain regions at the molecular level. The decrease in GABA and muscarinic receptor subtypes function along with the activation of apoptotic pathways under hypoxic stress leads to functional disturbances in neonatal brain, which manifest in a later stage of life as behavioural deficits. $GABA_A$ receptors are involved in anxiety and convulsion (Matsumoto, 1989) whereas $GABA_B$ receptors are related to depression and analgesia (Lloyd *et al.*, 1985; Sawynok, 1990). $GABA_A$ as well as $GABA_B$ receptors play an important role in learning and memory (Castellano and McGaugh, 1991). GABA receptor systems modulate learning and memory through influences on cholinergic systems (Yutaka Nakagawa *et al.*, 1995). Cholinergic system plays an important role during the early stages of memory formation (Miranda and Berm_udez-Rattoni, 1999). Long-term potentiation (LTP) has been proposed as a model of induction of the synaptic plasticity underlying memory formation (Escobar and Berm_udez-Rattoni, 2000). The neuronal plasticity needed for *in vivo* long-term potentiation is modulated by muscarinic acetylcholine receptors. Cholinergic system facilitate cortical plasticity by signaling stimulus relevance during memory formation. Improving the central cholinergic system and regulating the hippocampal monoamine neurotransmitters was reported to improve the learning and memory dysfunction (Wei *et al.*, 2010). Since both GABAergic and cholinergic stimulation is important in learning, memory processing and cognition, along with the neuronal death the reduction in GABAergic and muscarinic functions also contribute to the behavioural deficits observed in one month old rats exposed to neonatal hypoxia.

Hypoxic injury in term neonates, whether resulting from birth asphyxia, cardiac arrest or respiratory failure, is known to produce reduced brain growth and associated cognitive, motor and behavioral deficits later in life. Hypoxia during the first week of life can induce neuronal death in vulnerable brain regions usually associated with an impairment of cognitive function that can be detected later in life. Postnatal hypoxia resulting from lung immaturity and respiratory disturbances in infants is an important pathophysiological mechanism underlying the devastating neurological complications. This points the importance of a proper resuscitation program to overcome neonatal hypoxia for a better intellect in the later stages of life.

CONCLUSION

Our findings showed that the long term effect of neonatal hypoxic shock is affecting the cognitive abilities whereas motor abilities were not affected significantly. Neonatal hypoxic insult resulted in a marked cell loss in the brain which might have augmented the behavioural deficits observed in those rats in the later period of its development.

ACKNOWLEDGMENTS

This work was supported by the research grants from DBT, DST, ICMR, Govt. of India and KSCSTE, Govt. of Kerala to Dr. C.S. Paulose. Anju T.R. thanks Council of Scientific and Industrial Research for Junior and Senior Research Fellowship.

REFERENCES

Anju, T.R. and Paulose, C.S. (2011). Amelioration of hypoxia-induced striatal 5-HT(2A) receptor, 5-HT transporter and HIF1 alterations by glucose, oxygen and epinephrine in neonatal rats. *Neuroscience Letters* 502(3): 129–32.

Anju, T.R., Nandhu, M.S., Jes, P. and Paulose, C.S. (2010). Insulin and triiodothyronine regulation of neonatal hypoxia: Role of glucose, oxygen and epinephrine supplementation. *Fetal and Pediatric Pathology*. 30(5): 338–49.

Berti, A., Janes, A., Furlan, R. and Macagno, F. (2010). High prevalence of minor neurologic deficits in a long–term neurodevelopmental follow–up of children with severe persistent pulmonary hypertension of the newborn: a cohort study. *Ital J. Pediatr*, 36: 45.

Brandeis, B., Brandys, Y. and Yehuda, S. (1989). The use of the Morris Water Maze in the study of memory and learning. *Int. J. Neurosci.*, 48: 29–69.

Castellano, C. and McGaugh, J.L. (1991). Oxotremorine attenuates retrograde amnesia induced by post–training administration of the GABAergic agonists muscimol and baclofen. *Behav. Neural. Biol.*, 56: 25–31.

Escobar, M.L. and Berm_udez–Rattoni, F. (2000). Long–term potentiation in the insular cortex enhances conditioned taste aversion retention. *Brain Research*, 852: 208–212.

Hort, J., Brożek, G., Mares, P., Langmeier, M., Komárek, V. (1999). Cognitive functions after pilocarpine–induced status epilepticus: changes during silent period precede appearance of spontaneous recurrent seizures. *Epilepsia,* 40: 1177–1183.

Joseph, B., Nandhu, M.S. and Paulose, C.S. (2010). Dopamine D1 and D2 receptor functional down regulation in the cerebellum of hypoxic neonatal rats: Neuroprotective role of glucose and oxygen, epinephrine resuscitation. *Pharmacological Research*, 61: 136–141.

Lloyd, K.G., Thuret, F. and Pilc, A. (1985). Upregnlation of gamma–aminobutyric acid (GABA) B binding sites in rat frontal cortex: a common action of repeated administration of different classes of antidepressants and electroshock, *J. Pharmacol Exp Ther*, 235: 191–199.

Matsumoto, R.R. (1989). GABA receptors: are cellular differences reflected in function? *Brain Res Rev*, 14: 203–225.

Miranda, M.I., Bermudez–Rattoni, F. (1999). Reversible inactivation of the nucleus basalis magnocellularis induces disruption of cortical acetylcholine release and acquisition, but not retrieval, of aversive memories. *Proceedings of the National Academy of Sciences of the United States of America*, 96: 6478–6482.

Morris, T.A., Jafari, N., Rice, A.C., Vasconcelos, O. and DeLorenzo, R.J. (1999). Persistent increased DNA–binding and expression of serum response factor occur with epilepsy–associated long–term plasticity changes. *J. Neurosci*, 19: 8234–8243.

Paylor, R. and Rudy, J.W. (1990). Cholinergic receptor blockade can impair the rat's performance on both the place learning and cued versions of the Morris water task: the role of age and pool wall brightness. *Behav. Brain Res.*, 36: 79–90.

Pulsinelli, W.A., Brierley, J.B. and Plum, F. (1982). Temporal profile of neuronal damage in a model of transient forebrain ischemia. *Ann. Neurol.*, 11: 491–498.

Rulin, L., Martha, C., Sung-Kwon, J., Peter, J.S.S. and MaryLoeken, R. (2005). Hypoxic stress in diabetic pregnancy contributes to impaired embryo gene expression and defective development by inducing oxidative stress. *Am. J. Physiol Endocrinol Metab*, 289: 591–599.

Sandhu, H.S. (1986). Psychosocial issues in chronic obstructive pulmonary disease. *Clinics in Chest Medicine*, 7: 629–642.

Wei, J., Lu, D.X., Qi, R.B., Wang, H.D., Jiang, X.H. Chin J. Integr Med. (2010). Effect of Kangshuai Yizhi Formula I on learning and memory dysfunction induced by scopolamine in mice. *Epub.*, 16: 252–257.

Whishaw, I.Q. and Mittleman, G. (1986). Visits to starts, routes, and places by rats (Rattus norvegicus) in swimming pool navigation tasks. *J. Comp. Psychol.*, 100: 422–431.

SECTION–VI

KAPL Award

Isolation of Novel Protease Gene Isolated from Soil through Metagenomics

B. Kalpana*, S.N. Gautham Harinarayana, G. Saranya, S. Jalaja and S. Jamuna

Department of Biotechnology, K.S. Rangasamy College of Technology, Tiruchengode

E-mail: kalpujb@gmail.com

INTRODUCTION

Microbial ecology examines the diversity and activity of micro-organisms in Earth's biosphere. In the last 20 years, the application of genomics tools have revolutionized microbial ecological studies and drastically expanded our view on the previously underappreciated microbial world (Jo Handelsman, 2004). Despite being familiar and useful, soil is also one of the least understood habitats on earth (Venter J. *et al.,* 2004). The last 25 years of research have revealed that culturing is an excellent method to learn a lot about a tiny proportion of the microorganisms on earth. Many lines of evidence show that less than 0.1% of the microorganisms in soil are readily cultured using current techniques (Chung, E. *et al.,* 2008). And, most impressively, the other 99.9% of soil micro flora is emerging as a world of stunning, novel genetic diversity (Jo Handelsman Dec. 2004). Developing methods to culture the enormous diversity of soil micro flora will be slow and tedious and will require more knowledge of the physiology of the unknown microbes than is presently in hand. What is needed is a more direct, global and soil micro flora (Gray, J. *et al.,* 2005).

Several different laboratories have successfully isolated novel genes encoding different enzymes and secondary metabolites from microbial communities and their metagenome without cultivation of the microbes. The microbial niches studied were highly diverse and ranged from moderate environments, such as river soil to rather extreme

*Corresponding author.

environments, like the deep sea (Venter *et al.,* 2004, Ferrer M. *et al.,* 2009). However, a major difficulty associated with the methods employed is related to contamination of purified DNA with polyphenolic compounds which are co purified with the DNA (Jaouadi B *et al.,* 2008). These compounds are difficult to remove, and it is well known that polyphenols also interfere with enzymatic modifications of isolated DNA. As a result, construction of environmentally derived DNA libraries with large inserts is hindered due to the poor quality of the isolated DNA (Sharma *et al.,* 2007). These known difficulties associated with the construction of libraries directly derived from environmental DNA samples forced us to isolate DNA from the metagenome of a microbial community after pre-cultivation in the laboratory (Torsvik, V. *et al.,* 2002, Tringe S.G. *et al.,* 2005). Although it is expected that laboratory enrichment cultures have only limited biodiversity, this technique has proven to be highly efficient for rapid isolation of large DNA fragments and for cloning of operons and genes with great biotechnological value. Additionally, laboratory enrichments allow pre-selection of microbes that already carry the desired traits, resulting in a high frequency of gene detection and isolation.

MATERIALS AND METHODS

Sample Collection

Soil Samples were collected from the fields planted with ground nuts and rice. These samples were considered to be rich in proteins. The samples were inoculated on to M9 minimal media enriched with 1% casein and were incubated for a period of two days.

Metagenome Extraction

The enriched culture were pelleted, they were resuspended in Tris-EDTA (TE) buffer containing 20% (w/v) sucrose and lysed in DNA extraction buffer (100 mM Tris-HCl, 100 mM EDTA, 100 mM Na_2HPO_4, 1.5 M NaCl, 1% [w/v] sodium dodecyl sulfate [SDS]) for several hours. RNA was degraded with RNase A (10 mg/ml). The resulting DNA extracts were incubated with protease and sarcosyl (5%, w/v) in TE buffer overnight. Total genomic DNA was then repeatedly extracted with chloroform-phenol (1:1, v/v), washed once with chloroform, and dialyzed against 2 liters of TE buffer at 4°C overnight.

Miniprep of Plasmid DNA

LB medium supplemented with appropriate antibiotic was inoculated with a single colony of *E.coli* strain harboring plasmid and incubated at 37°C for overnight and the plasmid was extracted by Modified Birboim Dolly Method.

Ligation

Partial digest of the metagenome with *Sau3A1* and linearized plasmid DNA with *BamH1* was ligated using T4DNA Ligase.The ligation mixture was incubated at 16°C overnight and the reaction was stopped by heat inactivating the tube at 70°C.

Competent Cell Preparation and Transformation

A single bacterial colony of *E.coli* DH5α (MTCC 1652) was inoculated in LB broth and incubated for overnight at 37°C with 200rpm. The cells with O.D 0.4 were retrived and made competent with 100mM calcium chloride. The competent cells (100 µl) were taken and 10ng of ligated DNA were mixed well and incubated in ice for 30 minutes followed by heat shock at 42° for 2 min and incubated in ice for 10 min. The transformed competent cells were transformed to 800µl of LB medium and incubated it for 45min at 37°C. 100 µl of transformed competent cells were plated onto M9 casein agar containing 100 µg/ml ampicillin. The plates were incubated at 37°C for 12–16 hours.

Restriction Analysis of Recombinants

The positive clones were selected and subjected to plasmid DNA isolation using modified Binboim and Dolly method. Inserts were released from recombinant plasmids by restriction digestion single and double digestion with *XbaI* and *HindIII.*

RESULTS AND DISCUSSION

Microbiological Observation in Enriched Culture

The presence of different colony morphologies were observed on M9 Minimal media agar plates, supplemented with 1% casein. The presence of zone of clearance around the colonies in M9 casein medium confirms the organism produces the extra cellular proteases.

Extraction of Metagenome

To access the genomic information from the pool of microbes present in the enriched cultures the metagenome was extracted. The metagenome obtained was found to be pure and devoid of any inhibitors like humic acid. The quality and the quantity of the isolated metagenome were checked in 0.8% agarose gel electrophoresis (Figure 1).

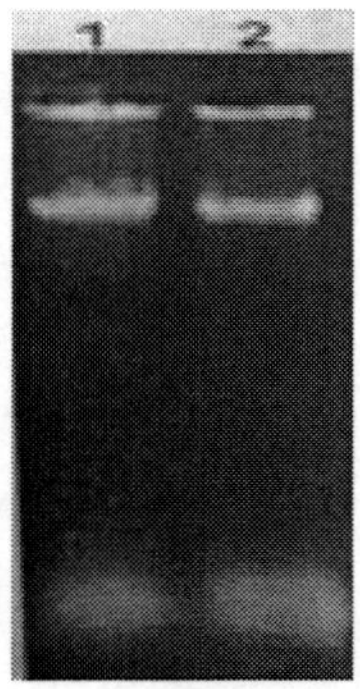

Fig. 1: Metagenome of Enriched Microbial Consortia

Partial Digestion

In order to generate random fragment size the metagenome was partially digested with *Sau3*A1. The size of the partials obtained was verified through 1% agarose gel (Figure 2). DNA of fragment size 6 Kb to 8 Kb were eluted from the gel and used as insert DNA.

Lane1: metagenome, Lane2: metagenome/*Sau3*AI

Fig. 2: Partial Digestion of Metagenome with Sau3AI

Vector Preparation

The plasmid DNA pBluescript SK+ that act as a vector to carry the partials of metagenome was digested with *Bam*H1 that has the compatibile end for *Sau3*A1. Linearization of the plasmid was confirmed in 0.8% agarose gel electrophoresis (Figure 3).

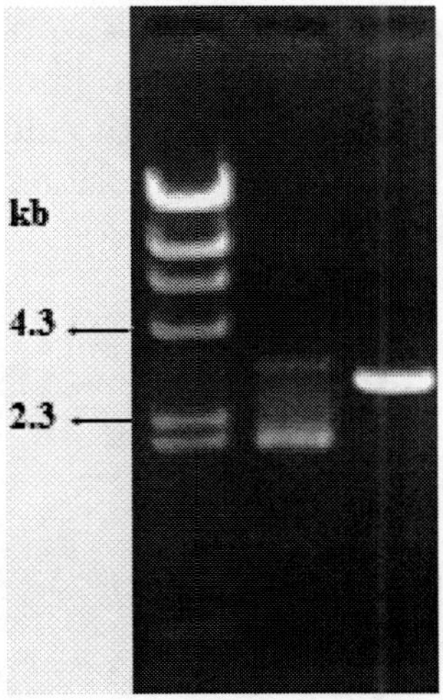

Lane1: λ *Hind*III Marker, Lane2: pBluescript SK+,
Lane3: pBluescript SK+/*Bam*H1) Cloning of metagenome

Fig. 3: Restriction Digestion of pBluescript SK^{+} with *Bam*HI

The passenger DNA and the vector DNA was ligated with T_4DNA ligase. Competent *E. coli DH5α* were transformed with the ligated product. The recombinants was screened in the M9 minimal medium supplemented with 1% casein and ampicillin. The cells that harbours the gene of interest can only able to utilize the carbon source present in the medium and grow. In the selection plate 15 colonies were appeared. These recombinant colonies were designated as CMP1-CMP15.

Screening of Transformants

The recombinants obtained can serve as starting material for the development of novel process and products that are industrially important. The recombinants that had grown on the M9 plate supplemented with casein, unfortunately showed no detectable protease activity. In spite of the above mentioned problem the transformants were subjected to plasmid isolation to confirm the presence of insert. Plasmid DNA of CMP1-CMP-15 showed mobility shift when compare pBluescript (SK+). These recombinant plasmids were designated as PMP1-PMP15. Further restriction analysis of the recombinants shows the polymorphism exist in the insert DNA.

Further analysis of the protease gene through the sequencing of recombinant and the ORF will unravel the mystery of the protease gene from the hidden diversity. The coupling of classical enrichment and modern metagenomic technologies in combination with simple activity based screening is the good way to isolate a large number of diverse genes. The data acquired in this work will pave the way for further progress in the construction of metagenomic library for proteases. However, complete analysis of the soil metagenome through culture independent method will bring much more promising results for the isolation of many novel industrially important genes from the uncultured silent majority.

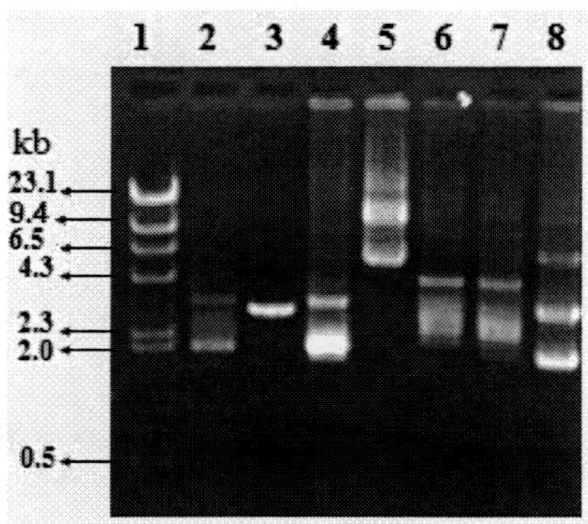

Lane1: λ HindIII, Lane2: pBluescript SK+, Lane3: pBluescript SK+/BamHI, Lane4: pMP1, Lane5: pMP2 Lane6: pMP3, Lane7: pMP4

Fig. 6: Isolation of Plasmid DNA from Recombinants

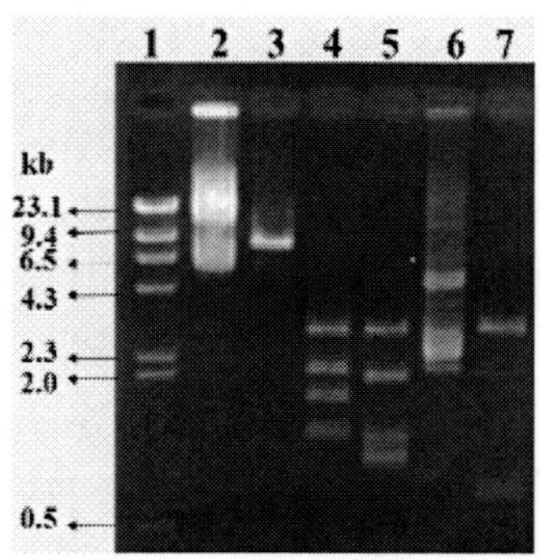

Lane1: λHindIII marker, Lane2: pMP2, Lane3: pMP2/*XbaI,* Lane4: pMP2/*XbaI/HindIII,* Lane5: pMP6/*XbaI/HindIII,* Lane6: pMP3, Lane7: pMP3/*XbaI/HindIII,* Lane8: pMP3/*XbaI/HindIII)*

Fig. 7: Restriction Analysis of Recombinant Plasmids

REFERENCES

Chung, E.J., Lim, H.K., Kim, J.C., Choi, G.J., Park, E.J., Lee M.H., Chung, Y.R. and Lee, S.W (2008). Forest soil metagenome gene cluster involved in antifungal activity expression in *E. coli*, *Application Environmental Microbiology*, Vol. 74, pp. 723–730.

Ferrer, M., Beloqui, A., Timmis, K.N. and Golyshin, P.N. (2009). Metagenomics for mining new genetic resources of microbial communities. *J Mol Microbiol Biotechnol*, Vol. 16, pp. 109–123.

Gray, J.P., Herwig, R.P., Uchiyama, T., Abe, T., Ikeruma, T. and Watanabe, K. (2005). Substrate-induced gene expression screening of environmental metagenome libraries for isolation of catabolic genes. *nat Biotechnol*, Vol. 23, pp. 88–93.

Jaouadi, B., Ellouz Chaabouni, S., Rhimi, M.S. and Bejar (2008). Biochemical and molecular characterization of a detergent-stable serine alkaline protease from Bacillus pumilus CBS with high catalytic efficiency" *Biochimie*, Vol. 90, pp. 1291–1305.

Jo Handelsman (2004). Metagenomics: Application of Genomics to Uncultured Microorganisms. *Microbiology and Molecularl Bology Reviews*, Vol. 68, No. 4.669–685

Riesenfeld, C.S., Schloss, P.D. and Handelsman, J. (2004). Metagenomics: genomic analysis of microbial communities. *Annu Rev Genet*, 38:525–552.

Sharma, P.K. and Capalash Kaur, J. (2007). An improved method for single step purification of metagenomic DNA, *Mol Biotechnol*, Vol. 11, pp. 587–591.

Torsvik, V. and Ovreas, L. (2002). Microbial diversity and function in soil: from genes to ecosystems. *Curr. Opin. Microbiol*, 5:240–245.

Tringe, S.G., von Mering, C., Kobayashi, A., Salamov, A.A., Chen, K., Chang, H.W., Podar M., Short, J.M., Mathur, E.J. and Detter, J.C. (2005). Comparative metagenomics of microbial communities. *Science*, 308: 554–557.

Venter, J.C., Remington, K., Heidelberg, J.F., Halpern A.L., Rusch, D., Eisen, J.A., Wu, D., Paulsen, I., Nelson, K.E., Nelson, W., Fouts, D.E., Levy, S., Knap, A.H., Lomas, M.W., Nealson, K. and Smith (2004). Environmental genome shotgun sequencing of the Sargasso Sea. Science, 304: 66–74.

Voget, S., Leggewie, C., Uesbeck, A., Raasch, C., Jaeger, K.E. and Streit,W.R. (2003). Prospecting for novel biocatalysts in a soil metagenome. *Appl. Environ. Microbiol*, 69: 6235–6242.

Leaching Behavior of Fly Ash for the Production of Enzymes Using Microbes

S. Mahesh*, M. Sithifathima[1] and G. Ayyappadasan[1]

School of Biotechnology, SASTRA University, Thanjavur, Tamil Nadu

[1]Department of Biotechnology, K.S. Rangasamy College of Technology, Tiruchengode

E-mail: biotechmagesh@gmail.com

INTRODUCTION

In India 70% of electricity generation is through coal/lignite based thermal power plants (*Bably Prasad 2004)*. Thermal power plant contains two different types of effluent. It may be bottom and fly ash. Characteristics and composition of bottom ash and fly ash differ depending on the sources of the feed coal or solid wastes and the method of combustion. In general, bottom ash and fly ash constitute up to 25 and 75% respectively, of the total ash which is enriched in most elements that are found in coal. Fly ash is a gray powder of very small particle size, with an average particle diameter below 3 μm (Gerald L. Fisher (1983). Class F fly ash and Class C fly ash. The chief difference between these classes is the amount of calcium, silica, alumina, and iron content in the ash. The chemical properties of the fly ash are largely influenced by the chemical content of the coal burned (i.e., anthracite, bituminous, and lignite) it is one of the numerous substances that cause air, water and soil pollution, disrupt ecological cycles and set off environmental hazard (Kanojia *et al.,* 2001). Ash may contain As, B, Be, Ca, Cd, Cr, Fe, Hg, Mg, Mo, Na, Ni, Pb, Ra, Se, Th, U, V, Zn, etc. either on the surfaces of ash particles or in the aluminosilicate matrix phase or in both, as a consequence of volatilization and condensation during combustion (Henry and Knapp 1980; Hansen *et al.*, 1984; Smith 1987).

*Corresponding author.

Thermal power generation through coal combustion produces minute particles of ash that causes serious environmental problems. Commonly known as fly ash, these ash particles consist of silica, alumina, oxides of iron, calcium, and magnesium and toxic heavy metals like lead, arsenic, cobalt, and copper.

The 80–odd utility thermal power stations in India use bituminous coal and produce large quantities of fly ash. According to one estimate, up to 150 million tones of fly ash will be produced in India. So fly ash pollute the environment it cause the severe health problem in humans as well as poultry.

Present study deals about the utilization of fly ash using micro organism is a useful method for clean the environment and other hand to produce the some economical important enzymes through degrade the fly ash.

MATERIALS AND METHODS

Collection of Sample

Fly ash Sample was collected from ENNORE and NLC power plant composting site in TAMIL NADU. Ennore power plant is based on the coal and its product. NLC power plant is based on the lignite material.

Isolation, Identification and Screening of Microorganism

Serial Dilution Method

0.85% NaCl is being preferred as a diluent, in place of sterile water, for the preparation of suspension of soil sample. Water being hypotonic results in increased osmotic pressure in the cells which can cause the rupture of sensitive cells. So, NaCl solution is the best possible option. 1g of FLY ASH (one month old) sample was transferred into the test tube which contains 9 ml of distilled water. Here the dilution was labeled as 10–1 and this has mixed properly. Using a sterile pipette, 1ml from 10–1 dilution was transferred to The next tube which contain 9 ml of distilled water [10–2]. Subsequent dilutions were made up to 10–6. Then the cultures were transferred into the plate for making pure culture.

Screening of Microorganism in the Fly Ash Degradation

MINERAL SALT AGAR media was prepared: Agar-15 gm, Sodium carbonat-2.0 gm, Dipotassium hydrogen Phosphate-1.2 gm, Magnesium sulphate-0.5 gm, Potassium chloride-0.5 gm, Potassium dihydrogen Phosphate-0.14 gm, Yeast extract-

0.02 gm, Ferrous sulphate-0.01 gm, Peptone-1 gm, Distilled water-1000 ml were prepared and streak the culture inoculate into the plate. Observe the clear zone which indicate the effect of fly ash degradation. The best culture suits for the production of enzymes for large scale level.

Identification of the Organism

Biochemical testing is a powerful tool for the identification of the genus and species of a particular bacterium. Since each different species of bacterium has different genetic code, they are able to synthesize different protein enzymes.

These enzymes catalyse all the various chemical reactions of which the organism is capable. Therefore the different species of bacteria carry out different and unique sets of biochemical reactions.

The type of biochemical reactions each organism undergoes acts as a thumb print for its identification.

Estimation of Protein from Degraded Fly Ash Sample by Lowry's Method

0.2–1.0 ml BSA (0.2 mg/ml) was pipetted to five different test tubes with a blank. Simultaneously test sample also taken in different test tubes. The volume was made up to 1 ml using distilled water. To this 5 ml for alkaline copper sulfate was added and incubated for 10 min. Then 0.5 ml of FC reagent was added and left for 30 min in dark. The absorbance was read at 660 nm. Then take the supernatant and discard the pellet. The supernatant was collected as source of extra cellular enzyme.

Peroxidase Assay

Switch on the spectrophotometer and kept 5–10 mints for initialization. Then add the reaction mixture as follows in the cuvette. Without the enzyme source maintain as blank. Instead of enzyme add distilled water. Take out the readings for ten minutes incubation.

Into a 1 cm quartz cell, pipette the following reaction mixture:

Buffer	2.80 ml
Guaiacol	0.05 ml
Substrate	0.05 ml

Equilibrate at 25°C A/minute and monitor

Enzyme at zero time	0.10 ml
Total reaction volume	3.00 ml

Record the rate of increase in absorbance at 436 nm using the linear portion of the curve after the initial lag phase.

Calculation

$$\text{Volume activity (U/ml)} = \frac{\Delta A_{436}/\text{min} \times 4 \times Vt \times \text{Dilution factor}}{\varepsilon n \times Vs}$$

RESULTS AND DISCUSSION

Thus Gram negative rods appeared as pink colour under the light microscope. Gram positive rods appeared as purple colour under the light microscope.

Table 1: Gram's Staining of Different Isolates

Strains	*Shape*	*Arrangement*	*Observation*
EC 10–2	Long rod	Single	Purple (gram positive)
EC 10–5	Rod	Single	Pink (gram negative)
NC 10–3	Long rod	single	Purple (gram positive)

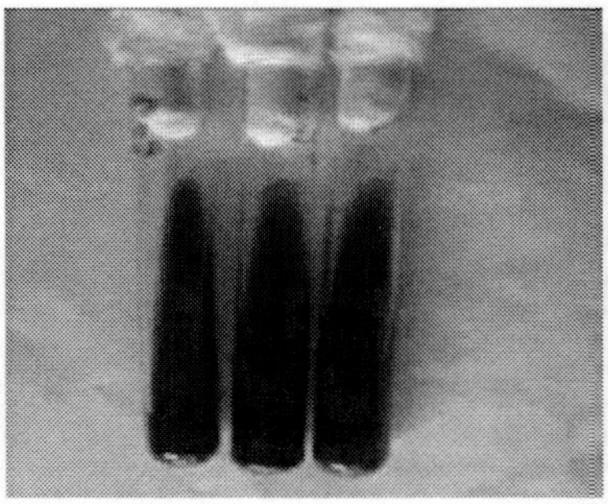
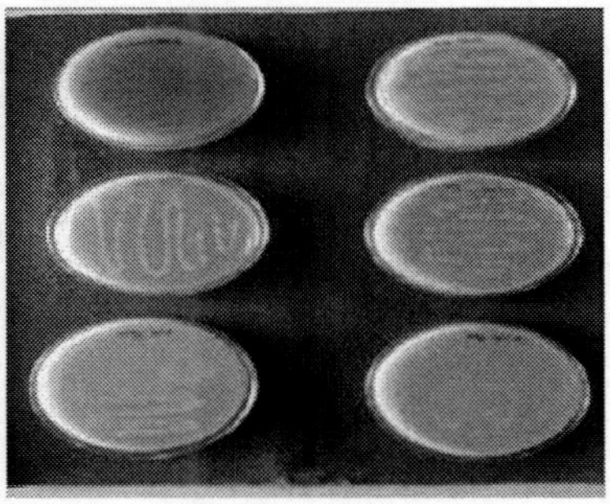

Fig. 1: Screening of Isolates for Fly Ash Degradation

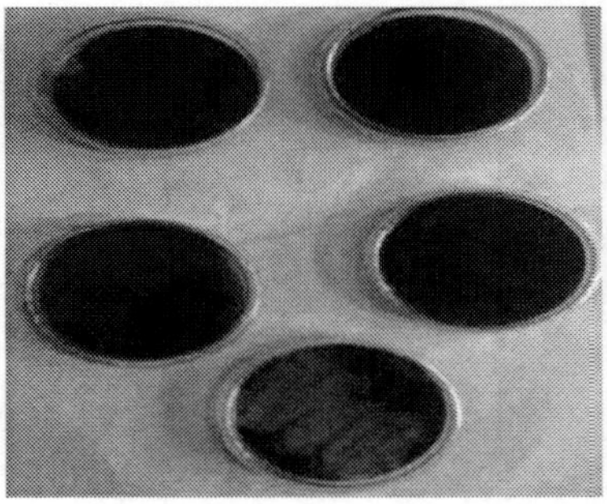
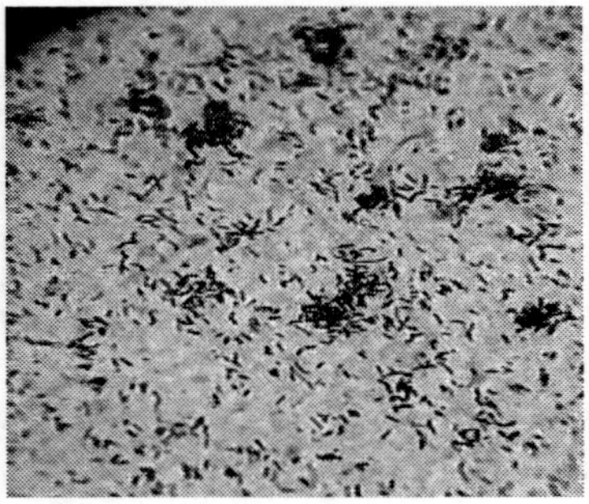

Fig. 2: Identification of Isolates

Table 2: Biochemical Results of the Isolates

Biochemiical Analysis	*Strain of Unknown Bacteria*		
	EC 10–2	*EC 10–5*	*NC10–3*
Indole test	Negative	Negative	Negative
Methyl red test	Positive	Positive	Positive
Vogues proskaur test	Positive	Negative	Negative
Citrate utilization test	Negative	Negative	Negative
Catalase test	Positive	Positive	Positive
Starch hydrolysis	Positive	Positive	Negative
Casein hydrolysis	Positive	Positive	Positive
Mannitol test	Negative	Negative	Negative
Sucrose test	Positive	Positive	Negative
Glucose test	Positive	Negative	Negative
	Bacillus Spp1	*Pseudomonas spp*	*Bacillus Spp2*

ESTIMATION OF PROTEIN BY LOWRY'S METHOD

Protein estimation is taken at every five days time intervals:

Table 3: 5th Day of Protein Estimation at Different Isolates

S. No.	*Content*	*Blank*	*S1*	*S2*	*S3*	*S4*	*S5*	*EC 10–2*	*EC 10–5*	*NC 10–3*
1.	Working standard	0	0.2	0.4	0.6	0.8	1.0	–	–	–
2.	Unknown sample	–	–	–	–	–	–	0.1	0.1	0.1
3.	Concentration ()	0	40	80	120	160	200	20	03	06
4.	Distilled water	1.0	0.8	0.6	0.4	0.2	0	0.9	0.9	0.9
5.	Lowry's buffer	5.0	5.0	5.0	5.0	5.0	5.0	5.0	5.0	5.0
6.	Folins reagent	0.5	0.5	0.5	0.5	0.5	0.5	0.55	0.5	0.5
	O.D at 660 nm	0	0.5	1.0	1.5	2.0	2.2	0.259	0.029	0.075

Table 4: 10th Day of Protein Estimation at Different Isolates

S. No.	*Content*	*Blank*	*S1*	*S2*	*S3*	*S4*	*S5*	*EC 10–2*	*EC 10–5*	*NC 10–3*
1.	Working standard	0	0.2	0.4	0.6	0.8	1.0	–	–	–
2.	Unknown sample	–	–	–	–	–	–	0.1	0.1	0.1
3.	Concentration	0	40	80	120	160	200	32	16	15
4.	Distilled water	1.0	0.8	0.6	0.4	0.2	0	0.9	0.9	0.9
5.	Lowry's buffer	5.0	5.0	5.0	5.0	5.0	5.0	5.0	5.0	5.0
6.	Folins reagent	0.5	0.5	0.5	0.5	0.5	0.5	0.55	0.5	0.5
	O.D at 660 nm	0	0.10	0.20	0.27	0.40	0.50	0.083	0.038	0.036

Enzyme Assay

In enzyme assay the enzyme reaction is taken by particular time interval respectively 1 min, 5 min and 10 min.

3rd Day

S. No.	*Content*	*Control*	*EC10–2*	*EC10–5*	*NC10–3*
1.	Phosphate buffer (ml)	3	3	3	3
2.	Guiacol solution (ml)	0.05	0.05	0.05	0.05
3.	Hydrogen peroxide (ml)	0.03	0.03	0.03	0.03
4.	Culture suspension (ml)	0.1	0.1	0.1	0.1
5.	O.D at 436 nm in 1 mins	0.000	0.001	0.001	0.009
6.	O.D at 436 nm in 5 mins	0.000	0.002	0.001	0.010
7.	O.D at 436 nm in 10 mins	0.000	0.002	0.002	0.010

6th Day

S. No	*Content*	*Control*	*EC10–2*	*EC10–5*	*NC10–3*
1.	Phosphate buffer (ml)	3	3	3	3
2.	Guiacol solution (ml)	0.05	0.05	0.05	0.05
3.	Hydrogen peroxide (ml)	0.03	0.03	0.03	0.03
4.	Culture suspension (ml)	0.1	0.1	0.1	0.1
5.	O.D at 436 nm in 1 mins	0.000	0.005	0.005	0.010
6.	O.D at 436 nm in 5 mins	0.000	0.005	0.007	0.011
7.	O.D at 436 nm in 10 mins	0.000	0.006	0.008	0.011

9th Day

S. No.	*Content*	*Control*	*EC10–2*	*EC10–5*	*NC10–3*
1.	Phosphate buffer (ml)	3	3	3	3
2.	Guiacol solution (ml)	0.05	0.05	0.05	0.05
3.	Hydrogen peroxide (ml)	0.03	0.03	0.03	0.03
4.	Culture suspension (ml)	0.1	0.1	0.1	0.1
5.	O.D at 436 nm in 1 mins	0.000	0.010	0.010	0.020
6.	O.D at 436 nm in 5 mins	0.000	0.010	0.011	0.021
7.	O.D at 436 nm in 10 mins	0.000	0.011	0.012	0.023

12th Day

S. No	*Content*	*Control*	*EC10–2*	*EC10–5*	*NC10–3*
1.	Phosphate buffer (ml)	3	3	3	3
2.	Guiacol solution (ml)	0.05	0.05	0.05	0.05
3.	Hydrogen peroxid e (ml)	0.03	0.03	0.03	0.03
4.	Culture suspension (ml)	0.1	0.1	0.1	0.1
5.	O.D at 436 nm in 1 mins	0.000	0.024	0.017	0.024
6.	O.D at 436 nm in 5 mins	0.000	0.024	0.018	0.025
7.	O.D at 436 nm in 10 mins	0.000	0.026	0.019	0.025

Isolated and screened the organism from the Ennore power plant fly ash composting site. It has identified as bacillus cereus and Staphylococcus aureus. Amylase, cellulase, Peroxidase, and dehydrogenease were produced during the fly ash degradation. Peroxidase and dehydrogenease were enzyme produced only stress condition. Peroxidase and dehydrogenease activity shows low activity. During stress condition, Protein production was very less compared to the other normal media. It has produced maximum 400 μg/ml of enzyme. Amylase and cellulase enzyme produced under normal conditions. Fly ash doesn't contain any carbohydrate source so without carbohydrates organisms can't able to produce amylase and cellulase. Organism may synthesis some carbohydrates during the degradation of fly ash or it might be some new enzyme. So we have to do protein sequencing to identify the new enzyme. Continuous reduction of pH indicates the mobility of metals ions during the growth and finally it reaches the acidic pH 3.52 and 4.32. *Bacillus* spp. having higher degradation efficiency of metal ions in fly ash compared to the *Pseudomonas species*.

REFERENCES

Bably, Prasad and Kajal, Kr. Mondal, Leaching Characteristics of Fluoride from Coal Ash, *Asian Journal of Water, Environment and Pollution,* 4, No. 2, pp. 17–21.

Christoph brombacher, Reinhard bachofen and helmut brand 1998, *Applied and Environmental Microbiology,* April. 1998, Vol. 64, 1237–1241.

Chun-Gang Yuan *et al.* (2008). Leaching characteristics of metals in fly ash from coal-fired power plant by sequential extraction procedure Microchim Acta.

Curutchet, G., Pogliani, C. and Donati, E. (1995). Indirect bioleaching of covellite by Thiobacillus thiooxidans with an oxidant agent. *Biotechnol. Lett.* 17:1251–1256

Hansen, L.D., Silberman, D., Fisher, D., Fisher, G.L. and Etaough, D.J., 1984. Chemical speciation of elements in stack-collected, respirable-size, coal fly ash. *Environ Sci Technol* 18: pp. 181–186.

Henery, W.M. and Knapp, K.T. (1980). Compound forms of fossil fuel fly ash emissions. *Environ Sci Technol* 4:450–456.

Hughes, M.N. and Poole, R.K. (1989). Metals and micro-organisms. Chapman and Hall, London, United Kingdom.

Kanojia, R.K., Kanawjia, S.K. and Srivastava, P.C. (2001). Utilisation of flyash in agriculture: a potential soil amendment for increasing crop yields, Indian Farming, pp. 29–32.

Lal, C. Ram *et al.* (2007). Leaching behavior of lignite fly ash with shake and column test, *Environ Geol* 51, pp. 1119–1132.

Smith, I.M. (1987). Trace elements from coal combustion: Emissions, chap 2, Source of trace elements, IEACR/01 IEA Coal Research, London.

TIFAC (1993). Technology Mission-fly ash disposal and utilization. Detailed project of the task force on fly ash as mine fill. New Delhi: Department of Science and Technology, GO1, p. 79.

Recovery and Reuse of Basic Dyes from Their Aqueous Solutions by Reverse Micellar Extraction

Prathap N.[**], Manimaran D.R.[*], Ramalingam P. and Sivamani S.

Department of Biotechnology, Kumaraguru College of Technology, Coimbatore

E-mail: prathi070@gmail.com; drmanichem@gmail.com

INTRODUCTION

The textile industry is one of the most chemically intensive industries on earth, the leading consumer of clean water after agriculture and the highest polluter of clean water. As many as 2,000 different chemicals are used in the textile industries, from dyes to transfer agents. The pollution load on the liquid effluent increases with subsequent steps of chemical processing, where the contribution of one single dyeing step amounts to about 8–20% of the total pollution load due to incomplete exhaustion of the dye (Anonymous, 2005).

The water consumption in chemical processing operation varies from 50–300 L for every kg of the processing goods, so that a mill processing 100,000 MT of cotton per day may require water varying from 2.5 to 4.0 million liters per day. About 40,000 to 80,000 tons of dyes are estimated to be discharged every year in textile processing units on a global basis. Dye production in India is estimated to be around 64,000 MT, which is about 6.6% of the world production. There are around 700 varieties of dyes and dye intermediates produced in India (Mathur and Bhatnagar, 2007).

Most of these dyes have not been characterized regarding their chemical nature, purity, possible toxicity or their impact on health and the environment. Yet, they are

*Corresponding author.

**Presenting author.

widely used by textile, leather, paint and even the food industry. The textile industry in India alone consumes up to 80% of the total dyestuffs produced (Mathur and Bhatnagar, 2007). Therefore, the problem is not only associated with the toxicity of the dyes released, also the colored effluent flowing through drains.

The various physical and chemical methods for the removal of dyes are Fenton's reagent, photochemical, electrochemical destruction, activated carbon, membrane filtration, ion exchange, irradiation, and electrokinetic coagulation (Robinson *et al.,* 2001). When compared with the methods available in the literature, the solvent extraction method has advantages in terms of efficiency and simplicity of the process design. The recovery of the solvent and reuse after its removal is related to the economic viability of the process.

Basic Dyes are cationic soluble salts of coloured bases. Basic dyes are applied to substrate with anionic character where electrostatic attractions are formed. Basic dyes are not used on cotton as the structures are neither planar nor large enough for sufficient substantivity or affinity (Shenai, 1997).

The reverse micellar extraction is an alternative method to conventional separation and purification techniques due to their potential for large scale use (Panditand Basu, 2004a). The removal of dye from water by reverse micelles is mainly controlled by the electrostatic interactions between the dyes and the surfactant head-groups. Thus, this process may be thought of as a phenomenon based purely on ion-exchange reactions. The backward extraction or recovery of the dye into the aqueous phase from the solvent phase by using a counterionic surfactant after the forward extraction is experimentally studied using a simple mixer and settler arrangement (Panditand Basu, 2004b).

The major objectives of this work are to remove basic dye magenta from their aqueous solutions by using isoamyl alcohol as solvent and Sodium Dodecyl Sulfate (SDS) as surfactant by reverse micellar extraction. Tostudy the effect of parameters such as dye concentration, surfactant dose, volume of solvent, pH, agitation speed, contact time and salt concentration on dye removal by employing design of experiments technique (Plackett–Burman and Box–Behnken) and todetermine the reusability of solvents for subsequent removal of dyes from their aqueous solutions.

MATERIALS AND METHODS

100 mL of aqueous solution with known dye concentration (feed) was added to a known volume of the solvent containing a known quantity of the cationic or anionic surfactants (extracting solvent). The feed and the extracting solvent were mixed thoroughly using a mechanical stirrer or a shaker at a fixed rpm for definite period of

time. The two-phase dispersion was formed and this mixture was transferred to a separating funnel. This resulted in the formation of two clear liquid phases—the solvent phase containing the dye encapsulated in the reverse micelles and the clear aqueous phase. The solvent and dyes are separated from the solvent phase in a flash evaporator and the solvent is recycled back for the extraction process and the dye is reused in the dyeing process.

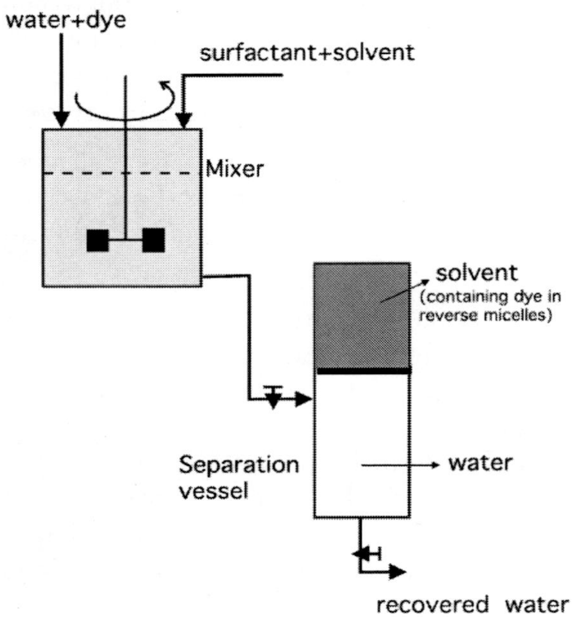

Fig. 1: Principle of Reverse Micellar Extraction

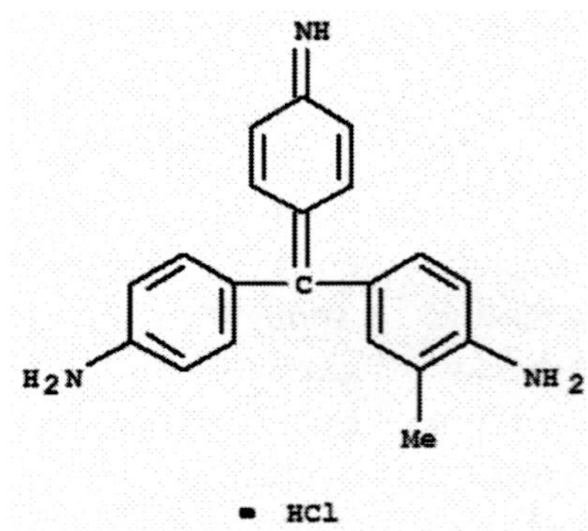

Fig. 2: Structure of Basic Magenta

Plackett-Burman Design

Plackett-Burman Design was used to identify which variable(s) has a significant effect on the percentage uptake of the dyes. Seven variables (dye concentration, surfactant dose, volume of solvent, pH, agitation speed, contact time and salt concentration) were used in this study and screened in 8 experimental designs.

RESULTS AND DISCUSSION

The medium components at different levels used in this design were given in Table 1. The design considerations and percentage uptake for the seven variables are mentioned in the Table 2.

Table 1: Medium Components at Different Levels Used in Plackett-Burman Experimental Design

Process Parameters with Designation	*High Level*	*Low Level*
A. Dye concentration (mg/100 ml of water)	20	10
B. Surfactant dose (mg)	20	10
C. Volume of solvent (ml)	30	50
D. pH of dye solution	9	4
E. Contact time (min)	15	5
F. Agitation speed (rpm)	300	200
G. Salt concentration (g/100 ml of dye solution)	5	1

Table 2: Plackett-Burman Experimental Design for Evaluation of 7 Variables

	A	*B*	*C*	*D*	*E*	*F*	*G*	*Dye Removal (%)*
1	H	H	L	H	L	H	L	72.79
2	L	H	H	L	H	L	H	12.27
3	H	L	H	H	L	H	L	1.82
4	L	H	L	H	H	L	H	88.49
5	H	L	H	L	H	H	L	24.93
6	L	H	L	H	L	H	H	70.67
7	H	L	H	L	H	L	H	21.22
8	L	L	L	L	L	L	L	86.80

Evaluation of Variables Affecting Percentage Uptake of Dye

The important variables that have a significant effect on the dye uptake were determined using PlackettBurman. The analysis of variance (ANOVA) table for basic magenta was presented in Tables 3. The F-value of 473.90 for basic magenta implies that the dye is significant. From Table 3, the F value for salt concentration was less than other variables which indicates the model terms are less significant for basic magenta uptake. This result agreed well for the uptake of dyes by solvent and surfactant depends on the contact time and pH of the solution. The volume of the solvent which showed the maximum F-value (Table 3) indicates the importance of the solvent in dye removal and the surfactant dose also helps in enhancing the removal.

Table 3: Analysis of Plackett-Burman for the Uptake of Basic Magenta Dye Solution

Process Parameters with Designation	*Difference*	*Effect*	*FMS*	*EMS*	*F-Value*
A. Dye concentration (mg/100 ml of water)	–137.46	–34.36	2361.85	4.98	473.90
B. Surfactant dose (mg)	109.44	27.36	1497.25	4.98	300.42
C. Volume of solvent (ml)	–258.50	–64.63	8353.06	4.98	1676.04
D. pH of dye solution	88.54	22.14	979.99	4.98	196.63
E. Contact time (min)	–85.15	–21.29	906.37	4.98	181.86
F. Agitation speed (rpm)	–38.57	–9.64	186.00	4.98	37.32
G. Salt concentration (g/100 ml of dye solution)	6.31	1.58	4.98	4.98	1.00

CONCLUSION

The percentage removal of dye is decreased with the increase in dye concentration in the feed. The increase in surfactant concentration resulted in higher dye removal, because more reverse micelles could be hosted in the organic phase. The increase in aqueous phase pH resulted in enhanced removal of basic magenta from water. The increase in KCl concentration resulted in decreased percentage removal of basic magenta. The effect of pH and salt concentration is explained based on charge transfer mechanism and electrostatic interactions and dye–surfactant complex formation. The usage of an experimental design permitted the rapid screening of a large experimental domain for optimization of conditions for the uptake basic magenta dye solution using reverse micelles.

REFERENCES

Mukherjee, A.K., Bhuvanesh, Gupta, S.M.S. and Chowdhary (1999). Separation of Dyes from Cotton Dyeing Effluent using Cationic Polyelectrolyte, American Dyestuff Reporter, 2 (Feb.), 25–28.

Nurpur, Mathur and Pradeep, Bhatnagar (2007). Mutagenicity Assessment of Textile Dyes from Sanganer, *Journal of Environmental Biology*, 28(1), 123–126.

Pandit, P. and Basu, S. (2004). Dye and Solvent Recovery in Solvent Extraction Using Reverse Micelles for the Removal of Ionic Dyes, *Ind. Eng. Chem. Res.*, 43, 7861–7864.

Pandit, P. and Basu, S. (2004). Removal of Ionic Dyes from Water by Solvent Extraction using Reverse Micelles, *Environ. Sci. Technol.*, 38, 2435–2442.

Shenai, V.A. (1997). Chemistry of dyes and principles of dyeing, Sevak publication, Mumbai, p. 34.

Tim Robinson., Geo McMullan., Roger Marchant, Poonam Nigam (2001). Remediation of dyes in textile effluent: a critical review on current treatment technologies with a proposed alternative, *Bioresource Technology*. 77, 247–255.

Polyhydroxy Alkonoates and Their Applications

Nityasree J., Jeyasri T., Nivedha Kumar, Narendrakumar G. and Antony V. Samrot*

Department of Biotechnology Sathyabama University, Jeppiaar Nagar, Sholinganallur, Chennai, Tamil Nadu

E-mail: antonysamrot@gmail.com

INTRODUCTION

Biopolymers are large macromolecules synthesized by microorganisms, plants and animals. Biopolymers produced by the microorganisms are divided into eight classes: (i) nucleic acids, (ii) polyamides (proteins and poly-amino acids), (iii) polysaccharides, (iv) organic polyoxoesters such as polyhydroxyalkanoic acids (PHA), poly (malic acid) and cutin, (v) polythioesters, (vi) inorganic polyesters with polyphosphate, (vii) polyisoprenoides such as natural rubber or Gutta Percha and (viii) polyphenols such as lignin or humic acid. PHAs, cyanophycin, glycogen, starch and polyphosphate are examples of biopolymers which are accumulated in the cytoplasm of cells (Chen *et al.,* 2007, Jamil *et al.,* 2007). Some of these biopolymers possess physical and chemical properties that are useful for food processing and pharmaceutical industries. They also have characteristics similar to those of the petrochemical plastics (Hikmet *et al.,* 2003). Polyhydroxyalkanoates (PHA) are polyesters accumulated by various microorganisms as carbon and energy storage material under unbalanced growth condition (Anderson and Dawes 1990, Chen *et al.,* 2001, Steinbuchel 1991, Samrot *et al.,* 2011) and these PHAs are normally consists of various 3-hydroxyalkanoates monomer. They have attracted attention because of their numerous industrial applications based on biodegradability, biocompatibility and piezoelectricity (Wu *et al.,* 2000, Zhao *et al.,* 2002, Deng *et al.,* 2002).

---

*Corresponding author.

PRODUCTION OF POLYMER

PHA are synthesized intracellularly and accumulated by several bacteria under unfavourable growth condition such as limitation of nitrogen, phosphorus, oxygen or magnesium in the presence of excess supply of carbon source (Du and Yu 2002a, Du and Yu 2002b,Lee, 1996, Samrot *et al.,* 2011). Limiting the nutrition is being followed as strategy to over produce these PHAs (Yu 2001, Du and Yu 2002a). Some bacteria such as *Alcaligenes.eutrophus, Alcaligenes.latus* and mutant strain of *Azotobacter vinelandii* are known to accumulate PHA during growth in the absence of nutrient limitation. Several other factors are also be needed for industrial production of these PHAs, those are product yield, complexity of the technology and hence the capital cost of the plant, and the ease or difficulty of product separation. This continues the selection of organism, substrate and optimized condition of PHA production. The yield values for PHB and cell biomass from a given substrate can be determined by experiment or by calculation based on the known metabolic pathways involved. Some workers have derived equation that predicts the PHA yield on several carbon sources (Yamane 1992, Yamane 1993, Yu and Wang 2001) which could be used for the preliminary calculation of PHA yields. Very few prominent PHAs such as PHB (poly-3-hydroxybutyrate) and PHV (poly-3-hydroxyvalerate) have been produced to a relatively high concentration using various microorganisms. More works are exploring cultivation strategies involving inexpensive, renewable carbon substrates in order to reduce production cost and obtain high productivity and purity.

BIOSYNTHETIC PATHWAYS OF PHA

PHAs are of three types they are short chain length (SCL) PHAs consist of 3–5 carbon atoms, medium chain length (MCL) PHAs consisted of 6–14 carbon atoms and long chain length (LCL) PHAs consist of more than 14 carbon atoms. There are three key enzymes involving in PHA biosynthesis in bacteria, they are ketoacylthiolase (encoded by *phbA*), aectoacetylcoA reductase (encoded by *phbB*) and PHA polymerase (*phbC*). Synthesis of PHA is believed to start from acetyl coA formation. Ketoacylthiolase converts two acetyl coA into acetoacetyl coA, acetoacetyl coA reductase converts acetoacetyl coA to 3-hydroxybutyryl coA and the polymerization is done by PHA polymerase. Medium chain length PHA (mcl–PHA) synthesis occurs through fatty acid oxidation. Here the intermediates in fatty acid oxidation are converted into other monomers by the enzymes such as enoyl-coA-hydratase (*phaJ*) and R-3-hydroxy acyl ACP-coA transferase (Tsuge, 2002). The type of the PHA is directly related to the structure of the carbon source used. Very few organisms can synthesize PHA of C3-C5 monomers. Conversion of succinate to propionyl coA through the action of methyl malonyl mutase and oxaloacetate transcarboxylase which leads to 3-HV units (*Poirier et al.,* 1995). The mcl-PHA synthesized from alkenes and alkanes utilizes de novo fatty acid biosynthesis and fatty acid oxidation (*De Smet et al.*, 1983). Studies have suggested that simultaneous accumulation and degradation of polyhydroxyalkanoates occurs, generating a potential futile cycle. An acyl-CoA synthetase associated with PHA granules has been identified that may convert PHA degradation products back into coenzyme A thioesters. Other granule-associated structural and regulatory proteins may also be involved. Possible regulatory mechanisms have been proposed and data suggested that in *Pseudomonas putida* GPo1 PHA metabolism is likely to be controlled by the ratios of acetyl-CoA to coenzyme A, and NADH to NAD+ which also regulate fatty acid oxidation (Ren *et al.,* 2009a, Ren *et al.,* 2009b).

FUNCTIONS OF PHA

It is well known that PHA is accumulated by wide range of microorganisms and serves as their carbon and energy source during starvation. PHA constitutes an ideal carbon-energy storage material due to its low solubility and high molecular weight which exerts negligible osmotic pressure to the bacterial cell. It is proposed that PHA can serve serves as a sink of reducing power in regard as a redox regulator within the cell. In *Azotobacteriaceae* family, the reductive step of PHA synthesis appeared to serve as an electron sink for the reducing power which accumulated when electron flow through the electron transfer chain was affected as a consequence of an oxygen limitation (Senior and Dawes, 1973). On the other hand, in the symbiotic nitrogen fixation process, PHA accumulation has been implicated as both an energy sources as well as a regulatory role controlling the availability of reducing power for the operation of nitrogenase.

APPLICATIONS OF PHAS

The potential application of PHA as replacement for petrochemical based polymers is gaining popularity in various fields involving packaging, medical and coating materials. These desirable properties in compounding and blending have broadened their performances as potential end-use applications.

PHA has been manufactured for non-woven materials, polymer films, sutures, and pharmaceutical products used in surgery, transplantology, tissue engineering, and pharmacology (Noda, 1998). PHA has a distinct advantage in the medical field over silicone, a traditionally used polymer, which has got more side effects (Zinn *et al.,* 2001). PHAs are found to be most useful in tissue engineering as they possess biocompatiblity and biodegradability. The biocompatibilities of two PHA polymers, namely, PHB and PHBV have been studied by a number of different research groups (Holmes and Bassett 1988). P(3HB-*co*-4HB) has been evaluated as scaffold in tissue-engineering (Williams *et al.,* 1999). PHA, the biodegradable polymers was found to function well in the pulmonary circulation as a pulmonary artery replacement (Shinoka *et al.,* 1998). The suitability of P(3HB) as nanoparticles in animals has been proven by toxicity tests according to ISO 10993 (Webb *et al.,* 2004). Apart from that, P(4HB) has also been found to be a suitable scaffold for preparing autologous cardiovascular tissue. These fast-absorbing P(4HB) scaffolds were seeded with autologous cells before implantation to augment the pulmonary artery in a juvenile sheep model (*Stock et al.,* 2000). It was agreed that patching of P(4HB) implant material is possible.

P(3HB) and P(3HB-*co*-15 mol% 3HV) copolymer sutures were implanted with a purpose to test animals intramuscularly. It was found to be biocompatible against suppurative inflammation, necrosis, and calcification of the fibrous capsule or malignant tumor formation (Potter; Steinbuchel 2006).

Chen *et al.* (2008) found PHB-PEG-PHB nanoparticles to show excellent biocompatibility and also found them to possess no cytotoxicity against animal cells, even when the concentrations were increased to 120 microg/mL. Lu *et al.* (2010) developed a sustained release system of PI3K inhibitor (TGX221) based on polyhydroxyalkanoate nanoparticles (NP) and used it to block proliferation of cancer cell lines. TGX221 was gradually released from PHA-based NP and growth of cancer cell lines was significantly slower in NP-TGX221-treated cells.

The delivery system of hEGF–PhaP–nanoparticles carrying RBITC was found to be endocytosed by the tumor cells in tumorous model mice. Thus, the ligand–PhaP–PHA specific drug delivery system was proven effective both *in vitro* and *in vivo* (Yao *et al.,* 2008). A hydrophobic drug thymoquinone was encapsulated in the mPEG-coupled and non-coupled PHA nanoparticles and its release kinetics was studied. The *in vitro* cytotoxicity evaluation of the nanoparticles on prenatal rat neuronal hippocampal and fibroblast cells revealed that biocompatibility of the amphiphilic nanoparticles was generally independent of the ratio of comonomer units in the PHA block. Shah *et al.* (2010) found that the amphiphilic nanoparticles contained the hydrophobic PHA segments buried in the core and could thus be used as safe carriers for the controlled release of a variety of hydrophobic drugs.

Yao *et al.* (2008) found that macrophages taking up the ligand–PhaP–PHA nanoparticles (a receptor-mediated drug specific delivery system), where the ligands were mannosylated human α1-acid glycoprotein (hAGP) and human epidermal growth factor (hEGF), The nanoparticle system was also clearly visible in the targeted cells and organs (liver or tumor) under fluorescence microscopy when rhodamine B isothiocyanate (RBITC) was used as a delivery model drug. He also found that the delivery system of hEGF–PhaP–nanoparticles carrying RBITC to be endocytosed by the tumor cells in tumorous model mice and proved ligand–PhaP–PHA specific drug delivery system was effective both *in vitro* and *in vivo*.

REFERENCES

De Smet, M.J., Eggink, G., Witholt, B., Kingma, J. and Wynberg, H. (1983). *J. Bacteriol,* 152: 870–878.

Deng, Y., Zhao, K., Zhang, X.F., Hu, P. and Chen, G.Q. (2002). *Bio-materials* 23: 4049–4056.

Du, G. and Yu, J. (2002a). *Process Biochem,* 38: 325–332.

Du G, Yu J. (2002b). *Environ. Sci. Technol,* 36: 5511–5516.

Hikmet, K., Belma, A. and Zehra, N.K. *et al.* (2003). *African J. Biotechnol* .2: 47–149.

Jamil, N., Ahmed, N. and Edwards, D.H. (2007). *J. Gen. App. Microbiol.*, 53: 105–109.

Noda, I. (1998). *United States Patent* 5: 849–854.

Poirier, Y., Nawrath, C. and Somerville, C. (1995). *Biotechnol.* 13: 142–150.

Potter, M. and Steinbüchel, A. (2005). *Biomacromolecules,* 6: 552–560.

Ren, Q., de, Roo, G., Ruth, K.,Witholt, B., Zinn, M., Thony-Meyer, L. (2009b), *Biomacromolecules* 10(4): 916–22.

Ren, Q. de, Roo, G., Witholt, B., Zinn, M., Thony-Meyer, L. (2009a). *Microb Cell Fact* 8: 60.

Samrot, A.V., Avinesh, Reddy, Sukeetha, S. and Senthilkumar (2011). *Applied Biochemistry and Biotechnology,* 163(1): 195–203.

Senior, P.J. and Dawes, E.A. (1973). *Biochem.* J., 134: 225–238.

Shah, M., Naseer, M.I., Choi, M.H., Kim, M.O. and Yoon, S.C. (2010). *International Journal of Pharmaceutics,* 400 (1–2): 165–175.

Shinoka, T., Shum-Tim, D., Ma, P.X., Tanel, R.E., Isogai, N., Langer, R., Vacanti, J.P. and Mayer, J.E. (1998). *The Journal of Thoracic Cardiovascular Surgery* 115: 536–728.

Steinbuchel, A. (1991). Stockton, New York, pp. 124–213.

Stock, U.A., Sakamoto, T., Hatsuoka, S., Martin, D.P., Nagasima, M., Moran, A.M., Moses, M.A., Khalil, P.N., Schoen, F.J., Vacanti, J.P. and Mayer, J.E. (2000). *The Journal of Thoracic Cardiovascular Surgery,* 120:1158–1168.

Tsuge, T. (2002). *Journal of Bioscience and Bioengineering,* 94(6): 579–584.

Webb, A.R., Yang, J. and Ameer, G.A. (2004). *Expert Opinion on Biological Therapy,* 4: 801–812.

Williams, S.F., Martin, D.P. and Horowitz, D.M. (1999). Peoples OP. *International Journal of Biological Macromolecules,* 25: 111–121.

Wu, Q., Sun, S.Q., Yu, P. and Chen, G.Q. (2000). *Acta Polym. Sin.*, 6: 751–756.

Yamane, T. (1992). *FEMS Microbiol. Rev.,* 103: 257–264.

Yamane, T. (1993). *Biotechnol. Bioeng,* 41: 165–170.

Yao, Y.C., Zhan, X.Y., Zhang, J., Zou, X.H., Wang, Z.H., Xiong, Y.C., Chen, J. and Chen, G.Q. (2008). *Biomaterials,* 29: 4823–4830.

Yu, J. (2001). *J. Biotechnol,* 86: 105–112.
Yu, J. and Wang, J. (2001). *Biotechnol. Bioeng,* 73: 458–464.
Zhao, K., Yang, X.S., Chen, J.C. and Chen, G.Q. (2002). *J. Mater. Sci. Mater. Med.,* 13: 849–854.
Zinn, M., Witholt, B. and Egli, T. (2001). *Advanced Drug Delivery Reviews,* 53: 5–21.
Anderson, A.J. and Dawes, E.A. (1990). *Microbial. Rev.,* 54: 450–472.
Chen, G.Q., Wu, Q. and Chen, J.C. (2001). *Tsinghua Sci. Tech.* 6:193–199.

Preparation of Biodegradable Starch-Clay Nanocomposites Film for Food Packaging

Thangaraja A.*, Chittu, Sorna, Umarani, Ganesh Babu S., Hemanth Kumar S., Jegatheesan[1] and Arunkumar C.[2]

Department of Biotechnology, St Peter's University, Tamil Nadu

[1]Department of Biotechnology, St Michael Engineering College, Kalayarkovil

[2]Department of Chemical Engineering, Annamalai University, Chidambaram, Tamil Nadu

E-mail: athangaraja@gmail.com

INTRODUCTION

Plastics are widely used packaging materials for food and non-food products due to their desirable material properties and low cost. Most synthetic polymers are produced from petrochemicals and are harmful to nature. Their synthesis produces hazardous waste and these materials are not easily degradable, causing environmental problems. However, the merits of plastic packaging have been overshadowed by its non-degradable nature, thereby leading to waste disposal problems. In order to solve the environmental impact generated by plastic waste the use of biodegradable polymers from renewable sources (Viviana *et al.,* 2004). Plants are potential sources for a wide varietyof polymers which are renewable and ecologically friendly. Theagro-based biopolymers are edible, biocompatible and biodegradable, which make them superior to synthetic polymers and particularly useful in disposable plastics, food, and medicine applications. Starches, the dominant carbohydrate reserve materials of higher plants, are abundant and relatively inexpensive biopolymers.

Biodegradable polymers such as Starch are completely and quickly biodegradable and easy to process because of its thermoplastic nature (Doane, 1994). Starch is obtained from many plants and its rather excessive production with regard to current needs

*Corresponding author.

and its low cost (Biqiong, Evans, 2005) which consists of two polysaccharides, the linear amylose and the highly branched amylopectin. The relative amounts of amylose and amylopectin depend upon plant sources and affect the material properties and gelatinization behavior of the starch. The pure native starch films are brittle so the additionof plasticizers such as glycerol, urea and formamide used to improve the barrier and mechanical properties of starch-based biodegradable packaging films which aid in the thermoplastic process and also increase flexibility of the final product by forming hydrogen bonds with starch that replace the strong interactions between its hydroxyl groups (Masclaux *et al.*).

In this context, the nanocomposites concept can be a promising option as it could lead to decrease the initial gas permeability of the plasticized film as well as to keep a low oxygen flux upon hydration. As reported in studies concerning various synthetic polymers, inorganic lamellar fillers like nanoclays can improve polymer gas and water barrier properties at rather low content due to their high aspect ratio (Averous *et al.*, Yilmaz *et al.,* Garcia *et al.*). Recently, starch/clay nanocomposites have received much attention because by adding minimal concentrations of clay, decreased water permeability and superior thermal and mechanical properties has been achieved (Chiu *et al.*). Starch is hydrophilic and forms nanocomposites with natural clays (Tronto *et al.*, Chang *et al.*).

Despite the existence of an extensive work describing gas barrier and structural properties for the most widely synthetic polymer based nanocomposites, there is a lack of proper understanding of the biodegradable properties in natural polymer based nanocomposites such as clay/starch based nanocomposites (Maurizio *et al*). Therefore, the scope of this work is to examine the combined effects of clay/starch nanocomposites on preparation of film and its biodegradability nature.

MATERIALS AND METHODS

All reagents (expect water) were purchased from the Himedia and used without modification. Deionized water was purified by milli–Q water system (Millipore Corporation) and filtered through various filter to remove any impurities. The materials used for the preparation of nanocomposite films are potato starch and purified clay mineral with cation-exchange capacity (43 meq/100 g).

Preparation of Starch-Clay Nanocomposites Film

On one side, native starch was dispersed in distilled water with 20 wt% of glycerol. Then the suspension was heated to 70°C for 30 min under continuous stirring to gelatinize the starch granules. On the other hand, the clay mineral was dispersed in distilled water by sonication during 10 min at room temperature. The clay dispersion was added to the aqueous dispersion of starch and the mixture was continued to sonicate for 30 min at room temperature. The solutions were then poured onto acetone to precipitate the resultant starch/clay nanocomposites, followed by the solvent evaporation to dryness at 65°C. Then, the films were obtained by casting pouring the hot suspension into rectangular moulds and evaporating it in an oven at 45°C until constant weight was reached. The starch/clay ratio was 90/20 relative to dry starch. The films were stored at room temperature in vacuum desiccators for two days in order to obtain reproducible results (Tronto *et al.*).

Characterization Methods

Portions of the conditioned films were dried at 110°C for 12 h. After weighting, they were conditioned at 25°C in adesiccators at 43% r.h., then removed at specific timeintervals and weighted. The differences between the initial and final masses correspond to the water loss, thus determining the amount of moisture retained by the films.

The FT-IR spectrum was recorded on Perkin Elmer FTIR 1650 spectrophotometer at ambient temperature using KBr pellet method. The pellet was scanned at 16 scans at wave number range of 400–4000 cm^{-1}.

XRD analyses of the composites were performed in a Bruker Discover diffractometer operating at 40 kV and 40 mA. The crystallographic study was performed on the synthesized Starch-Clay nanocomposites by X-ray diffractometer. The nature of the nanocomposites was calculated using Bragg's law and the crystallinitywas determined dividing the crystalline area by the total (crystalline + amorphous) area. Silicon powder was used as the internal standard.Thermo gravimetricanalysis were performed from 20–450°C at a heating rateof 20°C/min under N_2 flow using a Thermo gravimetric analyzer. All the specimens were dried at 40°C for 12 h prior to TGA tests (Braggs, W.L.).

A JEOL JSM-6100 scanning electron microscope (SEM) was used to investigate the morphology of the starch-clay nanocomposites. The specimens were frozen under liquid nitrogen and fractured and coated with gold on a ion sputter coater. SEM micrographs were obtained using 15 kV secondary electrons at a magnification of 20000 to 150000X to find out the shape and size of the particles.

Natural Soil Burial Test

The natural soil test of the Starch –clay nanocomposites was conducted for 30 days (Maurizio *et al).* The nanofilms were kept in glass container and kept the glass container under a controlled-environment chamber capable of maintaining a temperature of 30+2°C. After each natural 05, 10, 15, 20, 25 and 30 days duration, the buried specimens were dug out, removed from the soil and washed in distilled water. Finally, the specimens were dried in an air dried oven at 60+2°C for 24 h. The percentage weight loss was determined by Eq. Weight loss (%) = [(M_0 – M_d)/M0]100] where M_0 is the initial mass, and $M_d$is the degradation mass at each designated day (Tronto *et al*). The percentage weight losswas taken from the average of five samples (Sibel *et al.*).

Solubility Test

The solubilitynature of nanocomposites film was performed by kept the pre-weighted discs of 1 cm diameter nanocompositesfilm under various solventslike water, acetone, ethanol, HCl and sulphuric acid for 15 days. Five flasks were maintained for each treatment and left in a shaker. After 15 days of shaking, the discs were collected, washed thoroughly using distilled water, shade-dried and then weighed for final weight (Kazem *et al.*).

Microbial Degradation

To assess this, the pre-weighed discs of 1 cm diameter prepared from nanocomposites films was aseptically transferred to the conical flask containing 50 ml of culture broth medium, inoculated with different bacterial and fungal species separately. Nutrient broth medium was used for bacteria and Rose Bengal broth medium for fungi. Control was maintained with plastic discs in the microbe-free medium. Five flasks were maintained for each treatment and left in a shaker. After one month of shaking, the discs were collected, washed thoroughlyusing distilled water, shade-dried and then weighed for final weight. From the data collected, weight loss of the films, was calculated (Zhang *et al.*).

RESULTS AND DISCUSSION

Films under investigation were made by casting since film blowing appeared to be difficult. A rather stable film blowing process could be achieved in the laboratory, but the melt toughness was low and films became brittle and sticky straight after production. The latter is due to the highwater content.

A. Fourier Transform Infra-Red (FT-IR) Spectroscopy

FTIR results of the present study in Figure 1 shows the spectrum of Starch-Clay nanocomposites. The peak observed at 3567cm^{-1}which can assign to be –OH stretching vibration of water and hydroxyl groups. The peak at 945.45 cm^{-1} was attributed to C–O stretching (Jong, 2007). The bands in the region of 2933 cm^{-1} are due to C–H stretching while the bands in the region around 1465–1356 cm^{-1} are attributed to CH_2 bending in the plane (Bocchini *et al.*). The hybrid materials band at Si–O stretching was observed at 445 cm^{-1} and its intensity increases in the hybrid films with large amount of clay (Piyaporn *et al.*).

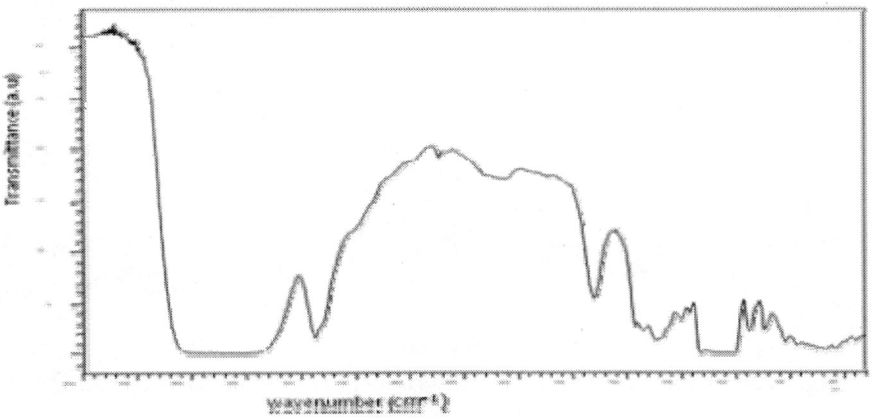

Fig. 1: FTIR Spectra of Starch-Clay Nanocomposites

B. X-Ray Diffraction Analysis (XRD)

X-ray di*ff*raction (XRD) is an effective method for the investigation of the intercalation existence of mineral clay. The X-ray diffraction pattern taken from Starch-Clay nanocomposites is shown in Figure 2. The X-ray Diffraction peaks of starch-clay nanocompositesdisplays at 2θ = 56.66; 64.88; 75.57 corresponding to d_{001} value of 1.63, 1.43, and 1.25 A° respectively (Figure 2). The mean values for starch-clay nanocomposites was calculated as Bragg's law (Chang *et al.*). Native starches generally exhibit two main crystalline types, namely, the A-type for cereal starches and the B-type for tuber and amylose-rich starches (Maurizio *et al.*).

The numbers of diffraction patterns were observed in the X-ray Diffraction spectrum indicating that the presence of mineral clay layer in the Starch. The characteristic difference between their XRDs is that B-type has a strong diffraction at 2θ = 5.5° (Tester and Karkalas, 2004). The diffraction studies confirm a high degree of crystallinity and uniformity in the particles. The intercalation of polymer chains usually increases the interlayer basal spacing of the clay mineral leading to a shift of the diffraction peak towards lower angle values (Lu *et al.*, 2009). We attribute the increase of the interlayer spacing to the intercalation of polymer into the clay (Bocchini *et al.*). In the case of MMT and CMMT the organic cation penetrate in to the interlayer space of clays that results in enlarging of the gallery space and shifts in peak (Colonna *et al.*).

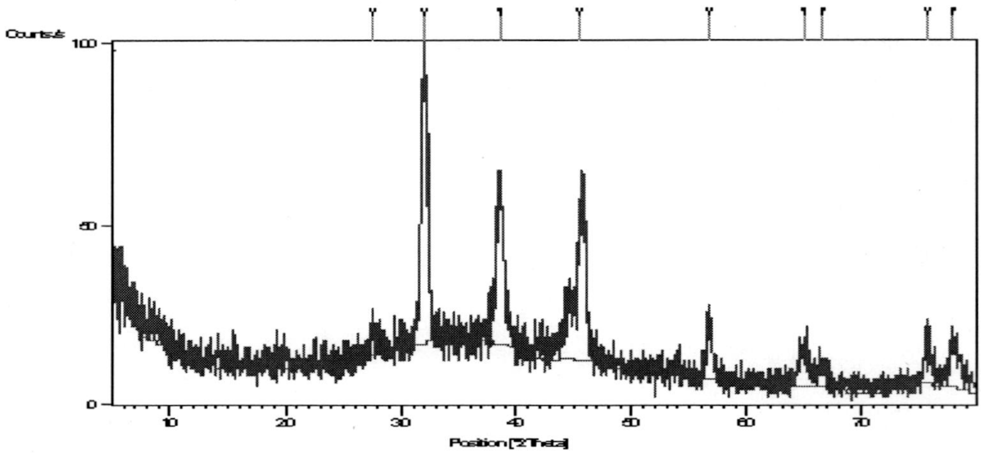

Fig. 2: X-ray Diffraction (XRD) Spectrum of Starch-Clay Nanocomposites

There are two basic structures reported for polymer–clay nanocomposites. In intercalated nanocomposites the polymer chains penetrate between the silicate layers keeping the clay stacks (tactoids) virtually unchanged except for an increase in the repeat distance (d-spacing). In exfoliated nanocomposites, the clay layers are delaminated and separated from one another by a distance greater than a few nanometers with no memory of the original clay tactoids (Karkalas *et al*). The present investigations revealed that the natural mineral clay-starch nanocomposites are shown in the form of intercalated types.

C. Scanning Electron Microscopy Analysis (SEM)

The surface morphology of Starch-Clay nanocomposites films under SEM as shown in Figure 3 was to investigate clay aggregation in the starch matrix and can be seen that nanocomposites have well defined monodispersed spherical shapes. The sizes of the Starch-Clay nanocomposites have a diameter of 20–40 nm. The variation in the granule size, shape and composition depends on origin of starch, because without

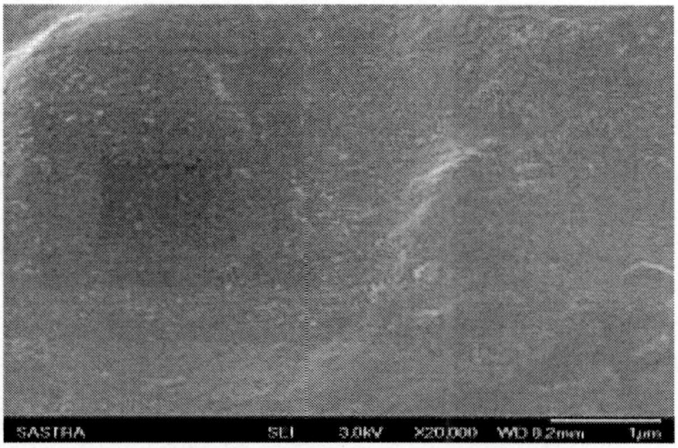

Fig. 3: SEM Photograph of Starch-Clay Nanocomposites Film

modification cannot be melt processing for film preparation (Darder *et al.*, Yi-Lin Chung *et al.*) The high degree of dispersion of clay in the polymeric matrix at low level of clay content [25–27].

Solubility Test

Figure 4 shows the solubility test in water and solvents such as acetone, ethanol, hydrochloric acid (HCl) andsulphuric acid (H_2SO_4). The Starch-Clay nanocomposites film was put in distilled water and left for 15 days, it remains as such which results the film is insoluble in water. When the nanocomposites film was kept in solvents ethanol (50%) and acetone (50%) the film is insoluble and no change its morphological nature for nearly 15 days but when placed in hydrochloric acid (50%) and Sulphuric acid (50%) the film immediately ruptured and dissolved in the acid.

Fig. 4: Solubility Test of Starch-Clay Nanocomposites Film

Biodegradability Test

Five microbial species were tested in the laboratory for their ability of degrading the nanocomposites films. The species tested were *E.coli*, Pseudomonas sps, *Staphylococcusaureus* species and two fungal species—*Aspergillusniger* and *Fusariumoxysporum*. These microbes were separately allowed to degrade the nanocomposites films under shaker cultures for a month. The results are shown in Table 1.

Table 1: Degradation of the Nanocomposites Films Incubated with Different Microbial Species in Shaker Cultures under Laboratory Conditions

S. No.	*Name of the Microorganism*	*Microbial Degradation (% weight loss/month)*
1.	*Escherichia coli*	22.42 ± 0.21
2.	*Pseudomonas sp.*	15.89 ± 0.10
3.	*Staphylococcusaureus*	10.54 ± 0.26
4.	*Fusariumoxysporum*	25.76 ± 1.23
5.	*Aspergillusniger*	18.24 ± 1.01

The bacteria caused the biodegradation ranging from 10.54 to 22.42% fornanocomposites film. Among the bacteria, *E.coli* was found most active in degrading 22.42% of nanocomposites polymers, in onemonth period. Among the species, *Fusariumoxysporum* was more active than *A. niger* in degrading 25.76% of nanocomposites within a month (Table 1). Starch can be readily metabolized by a range of microorganismsto fermentation products such as ethanol (Bai *et al.*), hydrogen (Yang *et al.*) and methane (Amon *et al.*). Several researches have been conducted on the biodegradation of starchnanocomposites (Mao *et al.*), reported starch glycerol samplesincubated in compost lost up to 70% of their dry weight within 22days (Russo *et al.*) investigated the degradability of thermoplastic starch and PVOHblends under anaerobic conditions to simulate the most commondisposal environment for household wastes.

Biodegradable starch-clay nanocomposites films were prepared by dispersing clay particles intothe starch matrix via melt extrusion processing. The film properties of the nanocomposites films were greatly influenced by clay content. The FT-IR, XRD and SEM results revealed that dispersion of clay in the starch matrix depended on the compatibility and the polar interactions among the starch and glycerol to form agglomerations. The Solubility test revealed that the film is insoluble in water which allows low humidity and the biodegradability nature of starch-clay nanocomposites films could be used as environmentally friendly food packaging materials for extending shelf life and reduce the menace caused by the non degradable plastics.

REFERENCES

Amon, T., Amson, B., Karyvoruchko, V., Machmuller, A., Hopfner-Sixt, K., Bodiroza, V. *et al.* (2007). Methane production through anaerobic digetion of variousenergy crops grown in sustainable crop rotations. *Bioresource Technology*, 98(17), 3204–3212.

Averous, L., Fauconnier, N., Moro, L. and Fringant, C. (2000). Blends of thermoplastic starch and polyesteramide: processing and properties, *J. Appl. Polym. Sci.*, 76: 1117–1128.

Bai, F.W., Anderson, W.A. and Moo-Young, M. (2008). Ethanol fermentation technologies from sugar and starch feedstocks. *Biotecnology Advances,* 26, 89–105.

Biqiong, C. and Evans, R.G. (2005). "Thermoplastic starch-clay nanocomposites and their characteristics", *Carbohydrates polymers*, Vol. 61, pp. 455–463.

Bocchini, S. Battegazzore, D. and Frache, A. (2010). "Poly (butylensuccinate co-adipate) thermoplastic starch nanocomposites blends", *Carbohydrate Polymers*, Vol. 82, pp. 802–808.

Braggs, W.L. (1913). "The diffraction of short electromagnectic waves by a crystal", *Math. Proc. Cambridge*, Vol. 17, pp. 43–57.

Buléon, A., Colonna, P., Planchot, V. and Ball, S. (1998). *Int J Biol Macromol,* 23: 85–112.

Chang, Y., Seema, L., Luisand, V. and Giannelli, P. (2010). "Preparation and properties of biodegradable starch-clay nanocomposites", *Carbohydrate Polymers*, Vol. 79, pp. 391–396.

Chiu, F.C. Lai, S.M. and Ti, D.T. (2009). "Characterization and comparison of metallocene-catalyzed polyethylene/thermoplastic starch blends and nanocomposites", *Polymer Testing*, Vol. 28, pp. 243–250.

Darder, M., Colilla, M. and Ruiz-Hitzky, E. (2003). Biopolymer–clay nanocomposites based on chitosan intercalated in montmorillonite. *Chemistry of Materials*, 15,

Doane, W.M. (1994), "Opportunities and challenges for new industrial uses of starch", *Cereal Foods Worlds*, Vol. 39(8), pp. 556–563.

Garcia, M.A., Martino, M.N. and Zaritzky, N.E. (1998). Plasticized starch-based coatings to improve strawberry (Fragaria × Ananassa) quality and stability, *J. Agric. Food Chem.*, 46: 3758– 3767.

Jong, W.R. (2011). "Effects of clay contents on mechanical and water vapour barrier properties of agar based nanocomposites films", *Carbohydrate Polymers,* Vol. 86, pp. 691–699.

Kathiresan, K. (2008). Polythene and Plastics-degrading microbes from the mangrove soil. *Rev. Biol. Trop.* 51(3): 629–634.

Kazem, M.A., Amir, H.N. and Sadeghi, F. (2010). "Optimization of mechanical properties of thermoplastic starch/clay nanocomposites", *Carbohydrate Polymers,* Vol. 79, pp. 547– 554.

Lu, D.R. Xiao, C.M. and. Xu, S.J. (2009). "Starch based completely biodegradable polymer materials", *Express Polymer Letters*, Vol. 3, pp. 366–375.

Mao, L.J., Imam, S., Gordon, S., Cinelli, P. and Chiellini, E. (2000). Extrudedcornstarch–glycerol–polyvinyl alcohol blends: Mechanical properties, morphologyand biodegradability. *Journal of Polymers and the Environment*, 8(4), 205–211.

Masclaux, C., Gouanve, F. and Espuche, F. (2010), "Experimental and modeling studies of transport in starch nanocomposites films affected by relative humidity," *J. Mem. Science*, Vol. 363, pp. 221–231.

Maurizio, A., Jan, D.V., Mara, F.V. and Paolo, V. (2005). "Biodegradable Starch-Clay nanocomposites flms for food packaging applications", *Food Chemistry*, Vol. 93, pp. 467–474.

Mondragón, M., Hernández, E.M., Rivera-Armenta J.L. and Rodríguez-González. F.J. (2009). Injection molded thermoplastic starch/natural rubber/clay nanocomposites: Morphology and mechanical properties. *Carbohydrate Polymers*, 77, 80–86.

Piyaporn, K., Duangdao, A., Duanghathai, P. and Kawee, S. (2006). "Preparation of cassava starch/montmorillonite composite film", *Carbohydrate Polymer*, Vol. 78, 134–143.

Russo, M.A., O'Sullivan, C., Rounsefell, B., Halley, P.J., Truss, R. and Clarke, W.P. (2009).

Sibel, T. and Osman, D. (2010). "Preparation and characterization of biodegradable methyl cellulose/montmorillonitenanocomposites films", *Appl. Clay science*, Vol. 48, pp. 414–424.

Tester, R.F., Karkalas, J., Qi, X.J Cereal (2004). *Sci.*, 39: 151–165.

The anaerobic degradability of thermoplastic starch: Polyvinyl alcohol blends:Potential biodegradable food packaging materials. *Bioresource Technology*, 100, 1705–1710.

Tronto, J. Perotti, G.F., Parra, D.F., Constantino, R.L. and Bizeto, M.A. (2007). "Starch and synthetic claynanocomposites", LNLS Activity Report.

Tronto, J., Perotti, G.F., Parra, D.F., Constantino, R.L. and Lugao, A.B. (2007). "Organic-Inorganic hybrid materisls: Biopolymer/Hectorite clay nanocomposites," *Polymer Processing Society*, Vol. 23.

Viviana, P.C., Liliana, B.M., Minh, T.T. and Analia, V. (2000). "Physical and mechanical properties of thermoplastic starch/montmorillonitenanocomposites films," *Carbohydrate Polymers*, Vol. 7, pp. 55–63.

Yang, H. and Shen, J. (2006). Effect of ferrous iron concentration on anaerobic biohydrogen production from soluble starch. *International Journal of Hydrogen Energy*, 31(15).

Zhang, Q.X., Yu, Z.Z., Xie, X.L., Naito, K. and Kagawa, Y. (2007). "Preparation and crystalline morphology of biodegradable starch/clay nanocomposites," *Polymer*, Vol. 48, pp. 7193–7200.

Studies on Pretreatment of Cotton Gin Waste for the Production of Reducing Sugars

Arun Pandian C. and Sivamani S.*
Chemical Engineering Laboratory, Department of Biotechnology,
Kumaraguru College of Technology, Coimbatore
E-mail: sivamani.s.bt@kct.ac.in

INTRODUCTION

India is the third largest cotton producing country in the world and has the large number of cotton mills. A huge quantity of waste is generated during the processing of the cotton in these mills. Due to the stringent environmental regulations, the disposal of cotton gin waste is one of the biggest problems faced by cotton based industries. The waste generated after the ginning of cotton fibers and recovery of cotton seeds is the cotton gin waste, lignocellulosic material which can be potentially utilized as feedstock for the production of fuel ethanol. Production of ethanol by fermentation will provide the cotton ginning industries with the waste management solution and an added bonus of value added product. Generally, bioconversion of lignocellulosics to ethanol consists of four major operations/processes: pretreatment, hydrolysis, fermentation and product recovery (Saha *et al.*, 2005).

Pretreatment is required to alter the biomass macroscopic and microscopic size and structure as well as the submicroscopic chemical composition and structure of biomass so that hydrolysis of carbohydrate fraction to monomeric sugars can be achieved more rapidly with greater yields. The pretreatment methods most commonly used are mechanical (grinding, crushing and chopping), thermal/hydrothermal (autoclaving, steam explosion and liquid hot water), chemical (acid, alkaline and ozone) and biological (microbial and enzymatic). Pretreatment affects the structure of biomass by

*Corresponding author.

solubilizing hemicellulose, reducing crystallinity and increase the available surface area and pore volume of the substrate. Pretreatment has been considered as one of the most expensive processing steps in biomass to fermentable sugar conversion (Harmsen *et al.*, 2010).

The hydrolysis methods most commonly used are acid (dilute and concentrated) and enzymatic. To improve the enzymatic hydrolytic efficiency, the lignin-hemicellulose network has to be loosened for the better amenability of cellulases to residual carbohydrate fraction for sugar recovery. Enzymatic hydrolysis offers major advantages over other chemical routes (e.g., acid hydrolysis) such as minimal byproduct formation, low energy requirements, mild operating conditions, and low chemical disposal costs (Ku *et al.,* 2008).

The main objectives of the present paper are (i) to characterize cotton gin waste for cellulose, hemicellulose and lignin; and (ii) to optimize pretreatment process and parameters involved in the pretreatment using Box-Behnken Design (BBD) in Minitab 15.

MATERIALS AND METHODS

Cotton gin waste is collected from nearby cotton industries in Coimbatore district. Muffle furnace and Hot air oven, procured from Inlab equipments (Madras Private Limited), were used for the purpose of incineration. Analytical weighing balance from Shimadzu Corporation (Model: AUY 120) was used for weighing the materials. Water bath, purchased from Technico Limited, was used for maintaining the temperature. UV-Visible spectrophotometer of Shimadzu (Model: UV-1800) was used for measuring the absorbance of solutions. Incubator shaker, purchased from Scigenics (India) Private Limited (Model: Orbitek), Chennai and autoclave were also used. Jaw crusher, purchased from Almech Enterprise, Coimbatore was used for milling the substrates. Centrifuge from Remi (Model: C-23BL) was used for removing the suspended solids from the substrate after gelatinizing the starch. A trial version of Minitab 15 was used for Box-Behnken Design experiments.

Pretreatment Studies

In steam explosion, high pressure saturated steam is injected into a batch reactor filled with 1 g of sample. During the steam injection, the temperature rises to 160°C. Subsequently, pressure is suddenly reduced and the biomass undergoes an explosive decompression with hemicellulose degradation and lignin matrix disruption as a result. Finally, dinitro salicylic acid (DNS) assay is performed (Miller, 1959). Liquid hot water processes are biomass pretreatments with water at high temperature and pressure. This process contacts hot compressed water with 1 g of sample for 15 min at temperature of 100°C. Autoclaving is carried out at 121°C and 15 psi for 15 min.

Experimental Design and Optimization

Box-Behnken design (BBD) was used in the optimization of pretreatment by autoclaving cotton gin waste for TRS production. Time (X_1, min), Slurry density (X_2, g/mL) and Temperature (X_3,°C) were chosen as the independent variables. Total reducing sugars (TRS) concentration (Y, mg/mL) was used as dependent output variables. 15 experiments were performed to optimize the parameters of autoclaving cotton gin waste. Among them, three replications were at center points (n_0=3). The coefficients of the polynomial model were calculated using the following equation:

$$Y = b_0 + \sum_{i=1}^{k} b_i X_i + \sum_{i=1}^{k} b_{ij} X_i^2 + \sum_{i<j}^{k} \sum_{j}^{k} b_{ij} X_i X_j \quad \ldots (1)$$

where Y was predicted response, and i, j were linear, quadratic coefficients, respectively. b and k were regression coefficient and the factor studied in the experiment, respectively.

The significance of each coefficient was determined using student's *t*-value. The results were analyzed by Minitab 15 (Minitab Inc., USA). Three-dimensional surface plots were obtained to study the interaction of one parameter with another. The optimum concentration based on the hump was identified in the three-dimensional plots.

Analytical Methods

Cellulose, total reducing sugars, hemicellulose, lignin, moisture and ash were determined by anthrone method (Updegroff, 1969), dinitrosalicylic (DNS) acid method (Miller, 1959), neutral detergent method (Goering; Vansoest, 1975), acid detergent method (Goering; Vansoest, 1975), AOAC method (1975).

RESULTS AND DISCUSSION

Characterization of Cotton Gin Waste

Characterization of cotton gin waste was done and found to be 26.1% of total reducing sugar (TRS), 22.4% of lignin, 26.8% of cellulose, 32.1% of hemicellulose, 10.2% of ash and 9.9% of moisture.

Optimization of Pretreatment Process

Figure 1 represents the effect of pretreatment methods including autoclaving, liquid hot water and steam explosion on TRS concentration. The cellulose and hemicellulose are cemented together by lignin. Lignin is responsible for integrity, structural rigidity, and prevention of swelling of lignocelluloses. Dissolved lignin due to pretreatment of lignocelluloses is also an inhibitor for cellulase, xylanase, and glucosidase. It is shown that, autoclaving method of pretreatment having higher TRS concentration compared with liquid hot water and steam explosion due to degradation of lignin.

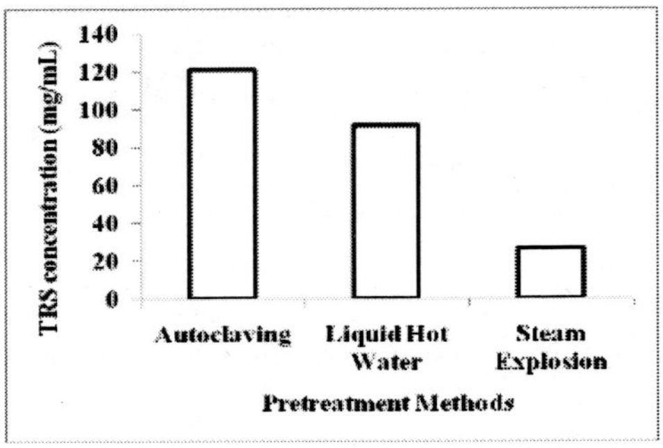

Fig. 1: Effect of Pretreatment Methods on TRS Concentration

Optimization of Pretreatment by Autoclaving for Cotton Gin Waste—Response Surface Analysis for the Optimization of Three Factors

Table 1 shows the actual ethanol yield in BBD. The regression coefficients and significance levels are given in Table 3. Although the model showed a satisfactory explanation (R^2 = 1), not all the effects of factors and their interactions on TRS concentration were significant (P < 0.05). Among them, the items that are underlined in Table 2 were found to be unimportant items.

In order to simplify the model as well as to enhance the effect of significant items, the ones which showed trivial effect were eliminated. Thus, the new model and corresponding information are shown in Table 2. It could be seen from the table, that all the items showed important effect on TRS concentration (P < 0.05). The second order polynomial equation giving the TRS (Y, g/L) as a function of time (X_1, min), slurry density (X_2 g/mL) and temperature (X_3,°C) was obtained:

$$Y = 172.149 + (2.23636 \times \text{Slurry Density}) - (2.64027 \times \text{Temp}) - (2.77191 \times \text{Time}) - (0.0248078 \times \text{Slurry Density} \times \text{Slurry Density}) + (0.0104104 \times \text{Temp} \times \text{Temp}) + (0.262567 \times \text{Time} \times \text{Time}) + (0.0398750 \times \text{Slurry Density} \times \text{Temp}) - (0.0555435 \times \text{Slurry Density} \times \text{Time}) - (0.0176500 \times \text{Temp} \times \text{Time}) \quad \ldots (2)$$

Table 1: Box-Behnken Design Matrix Employed for Three Independent Variables

S. No.	*Slurry Density (g/mL)*	*Temperature (°C)*	*Time (min)*	*TRS (mg/mL)*
1.	0.30	100	10	128.41
2.	0.50	100	5	231.67
3.	0.50	100	15	210.72
4.	0.04	100	5	19.54
5.	0.30	80	5	133.94
6.	0.50	80	10	193.41
7.	0.30	120	15	140.14
8.	0.30	80	15	129.94
9.	0.50	120	10	243.72
10.	0.30	100	10	127.41
11.	0.04	80	10	31.2
12.	0.04	120	10	8.14
13.	0.30	100	10	128.41
14.	0.04	100	15	24.14
15.	0.30	120	5	151.2

Table 2: Estimation Regression Coefficients for TRS

Term	*Standard Error*	*t*	*P*
Constant	0.2564	499.550	0.000
Slurry Density	0.1570	634.146	0.000
Temp	0.1570	43.558	0.000
Time	0.1570	–25.008	0.000
Slurry Density × Slurry Density	0.2311	–56.786	0.000
Temperature × Temperature	0.2311	18.019	0.000
Time × Time	0.2311	28.404	0.000
Slurry Density × Temperature	0.2220	82.611	0.000
Slurry Density × Time	0.2220	–28.768	0.000
Temperature × Time	0.2220	–7.949	0.001

R^2 = 100.00%; R^2 (pred) = 99.99%; R^2 (adj) = 100.00%.

Table 3: Analysis of Variance for TRS

Source	*Seq SS*	*Adj SS*	*Adj MS*	*F*
Regression	82249.5	82249.5	9138.8	46343.37
Linear	79799.0	79799.0	26599.7	134887.89
Slurry Density	79301.5	79301.5	79301.5	402140.98
Temp	374.1	374.1	374.1	1897.32
Time	123.3	123.3	123.3	625.38
Square	929.1	929.1	309.7	1570.45
Slurry Density × Slurry Density	720.2	635.9	635.9	3224.65
Temp × Temp	49.7	64.0	64.0	324.68
Time × Time	159.1	159.1	159.1	806.78
Interaction	1521.5	1521.5	507.2	2571.78
Slurry Density × Temp	1345.8	1345.8	1345.8	6824.55
Slurry Density × Time	163.2	163.2	163.2	827.60
Temp × Time	12.5	12.5	12.5	63.19
Residual Error	1.0	1.0	0.2	
Lack-of-Fit	0.3	0.3	0.1	0.32
Pure Error	0.7	0.7	0.3	

Interactions Among the Factors

Figures 2–4 show the isoresponse surface plots for the optimization conditions of TRS production. From the plots, it was easy and convenient to understand the interactions between two parameters and also to locate the optimum levels. The effect of temperature and slurry density on ethanol production is shown in Figure 2. TRS production increased with the increase of temperature and slurry density, a further increase in the temperature with slurry density the trend was reversed. In a relative high slurry density and medium temperature, optimum ethanol production could be attained (Sathya and Navaneetha, 2009; Magesh *et al.,* 2001).

Figure 3 shows the effect of time and temperature on TRS production. At the middle of temperature and long time the optimum ethanol yield could be attained. An increase in time with temperature increased the TRS production, but at high temperature, ethanol production decreased. The equation demonstrated that interaction temperature and time showed little significance. From Figure 4, it is found that the isoresponse surface between slurry density and time showed good interaction. Only the low slurry

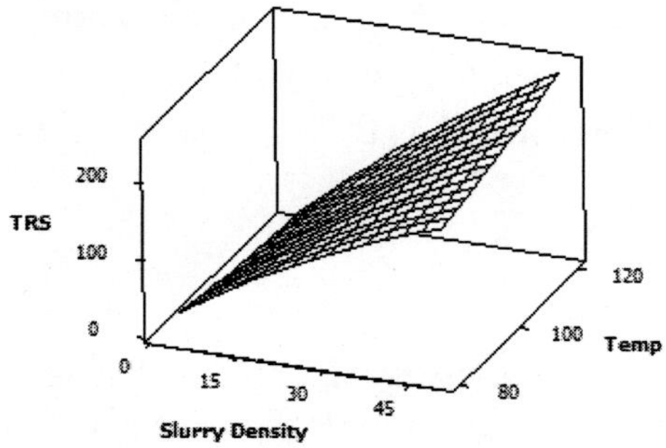

Fig. 2: Response Surface Plot of Temperature vs. Slurry Density on TRS Production

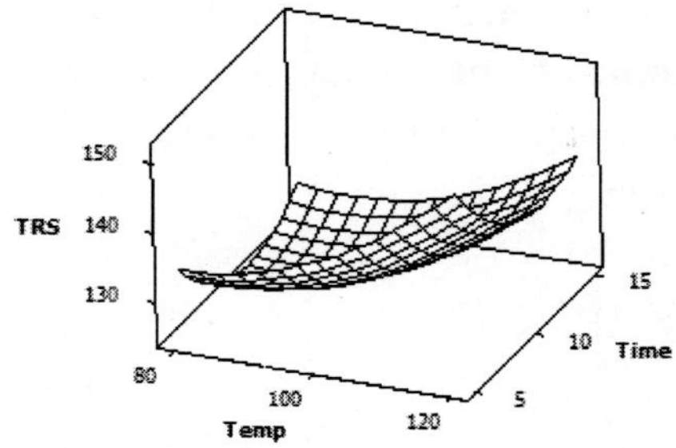

Fig. 3: Response Surface Plot of Time vs. Temperature on TRS Production

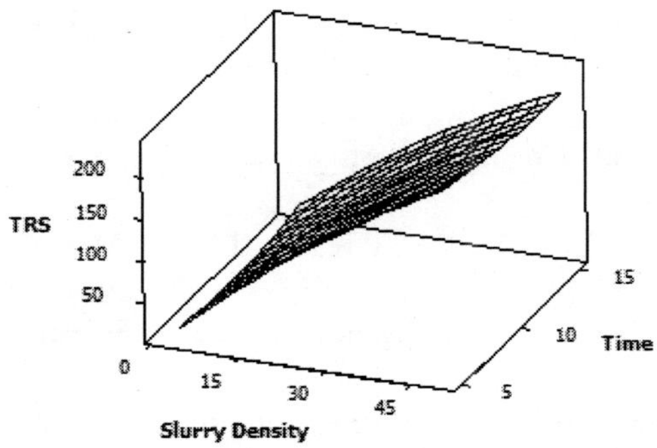

Fig. 4: Response Surface Plot of Time vs. Slurry Density on TRS Production

density and long time were beneficial for TRS production. Therefore, optimum TRS production could be obtained at high slurry density, high temperature and short time. The optimum TRS concentration was determined as 262.50 mg/mL, with time of 5 min, slurry density of 0.5 g/mL and temperature of 120°C. To validate the optimum concentration, an experiment with the specified condition was performed. And the result was 262.06 mg/mL which showed that the model was useful to predict the concentration as well as the optimization of the experimental conditions. The traditional one factor at a time process requires a complete series of experiments for every factor of interest which is laborious and time consuming. Furthermore, such methods could not provide information for the interactions of the factors. The BBD could locate the most important factors levels with minimum effort and time; what's more, it could reveal the interaction among the factor. The present study using RSM based on BBD established an efficient model to describe the process. The high similarity between the observed value and the predicted ones showed that the RSM was an accurate and applicable tool to optimize the TRS production from cotton gin waste. By utilizing the statistical methodology, the optimum ethanol production from cotton gin waste was determined to be 262.50 mg/mL, with time of 5 min, slurry density of 0.5 g/mL and temperature of 120°C. The TRS yield could be determined as 0.98 g TRS/g cellulose in cotton gin waste.

REFERENCES

Badal, C., Saha, Loren, B., Iten, Michael, A., Cotta, Y. and Victor Wu (2005). Dilute acid pretreatment, enzymatic saccharification and fermentation of wheat straw to ethanol, *Process Biochemistry,* 40, 3693–3700.

Gail Lorentz Miller (1959). Use of Dinitrosalicylic Acid Reagent for Determination of Reducing Sugar, *Anal. Chem.,* 31(3), 426–428.

Goering, H.D. and Vansoest, P.J. (1975). *Forage fiber analysis* US Department of Agriculture. *Agricultural Research Service Washington.*

Harmsen, P.F.H., Huijgen, W.J.J., Bermudez Lopez, L.M. and Bakker, R.C.C. (2010). Literature Review of Physical and Chemical Pretreatment Processes for Lignocellulosic Biomass, *Biosnergy,* 1–60.

Ku Ismail, K.S., Ahmad, A.A., Inba, T., Kasim, K.F. and Daud, M.Z.M. (2008). Thermo- enzymatic Hydrolysis of Cassava Starch by α-amylase and amyloglucosidase Proceedings of MUCET2008, Malaysian Technical Universities Conference on Engineering and Technology, Putra Palace, Perlis, Malaysia.

Magesh, A., Preetha, B. and Viruthagiri, T. (2001). Statistical Optimization of Process Variables for Direct Fermentation of 226 White Rose Tapioca Stem to Ethanol by *Fusarium oxysporum, International Journal of Chemical and Biological Engineering,* 4, 1–6.

Official Methods of Analysis of the AOAC 12th Edn 1975, p. 138, Updegroff, D.M., 1969, *Anal Biochem,* 32, 420.

Sathya Geetha, G. and Navaneetha Gopala Krishnan, A. (2009). Diluted acid hydrolysis and steam explosion of Manihot esculenta peeling waste for bio ethanol production, *Proceedings of Annual Conference on National Solid Waste Association of India,* pp. 19–30.

Cloning and Expression of Pre-Erythrocytic UIS4 Gene of *Plasmodium berghei* and Purification of Expressed Protein Byaffinty Chromatography

R.P. Saranya* and Agam P. Singh

Government College of Technology, Coimbatore, Tamil Nadu

E-mail: rpsaranyangl@gmail.com

INTRODUCTION

Malaria is a mosquito borne disease of humans and other animals caused by the protist parasite *Plasmodium*. Malaria remains the most important parasitic infection in humans and one of the most prevalent infectious diseases worldwide. There are four species that causes malaria in Human they are *Plasmodium vivax, Plasmodium falciparum*, *Plasmodium malariae*, *Plasmodium ovale*. Malaria parasites undergo intracellular development in two different host cells, hepatocytes and erythrocytes. The first stage of a malaria infection, the liver stage, is asymptomatic while the second phase, the blood stage, is responsible for all the disease-associated symptoms. Although clinically silent, the liver stage is an obligatory and extremely important step in plasmodium life cycle. Therefore, understanding the parasite's requirements during this period is crucial for the development of any form of early intervention.

Rodent malaria parasites are recognized as valuable model organisms for the investigation of human malaria because they are similar in most essential aspects of morphology, physiology and life cycle. *Plasmodium berghei* is one of the four species of rodent malaria parasite. The genome sequence of *Plasmodium berghei* is similar to that of *Plasmodium falciparum.* A number of pre-erythrocytes genes named UIS (Upregulated

---

*Corresponding author.

infective sporozoites) undergo up regulation in infectious sporozoites when they gain infectivity for the mammalian host. By reverse genetics it was found that the selected individual genes, exemplified by UIS4 (up-regulated in infective sporozoites gene4) and UIS4, are essential for early liver stage development. The term "liver-stage-specific gene function" or "LS-specific gene function" refers to a function that is required in liver stage parasites to ultimately produce infectious merozoites and establish the erythrocytic stage of the lifecycle. This gene codes for a protein, which is exported from the cytoplasm across the plasma membrane to the parasitoporous vacuole membrane which alters the host cell suitable for its growth. Proteins in the parasitoporous vacuole have to be studied for its mechanism of action. This protein is conserved between rodents and human parasites. It shows 34% amino acid similarity. The purpose of the present work is to clone and express UIS4 gene in *E. Coli BL21 (DE3)-RILpLysS* competent cells using pET 28a plasmid vector and Purification of expressed protein using Affinity Chromatography (Ni-NTA).

MATERIALS AND METHODS

PCR Amplification of UIS4 Gene

From the cDNA, the UIS4 gene was PCR amplified using gene specific primer. The forward primer and the reverse primer used were GCAGGATCCTACCCAGA AAATTCACCAGA and GTACTCGAGTTATATGTATGGGCCGAATGA respectively. Amplification of DNA was confirmed by 1% agarose gel electrophoresis. (Sambrook *et al.*, 1989) The DNA bands were visualised at 260 nm UV light on a trans-illuminator.

Construction of Vector

UIS4 gene was digested using restriction enzyme BamH1 and Xho (Double Digestion). The restricted DNA fragments are eluted from the agarose Gel by agarose gel elution kit. The Vector was also digested with the same restriction enzyme used for the gene of interest. One unit of enzyme can digest one unit of DNA. The restricted vector and the insert were ligated following ligation procedure. At the end the efficiency of Ligation was confirmed by agarose gel electrophoresis.

Competent Cell Preparation and Transformation

The competent *Escherichia coli* DH5α cells were prepared and stored at –80°C. *Escherichia coli* DH5α cells were used for the cloning of Plasmid vector containing insert. pET 28a plasmid with insert was transformed (100 ng/μl) into 100 μl of *Escherichia coli*

DH5α competent cells and were plated on nutrient agar and kept for growth at 37°C overnight. Plasmid which has been cloned in *E.Coli DH5α* cells has to be isolated for transformation into E.Coli BL21 Expression cells for the purpose of gene expression and protein production. This is done in mini preparation and the plasmids isolated were resuspended in 35 µl of double distilled water. Presence of plasmid DNA was confirmed by 1% i*coli BL21 (DE3)-RILpLysS cells* (Stratagene Inc.) were chosen as the expression host. *Escherichia coli BL21 (DE3)-RIL pLysS* Competent cells were prepared and stored at –80°C. Transformation of pET 28a system to Escherichia coli BL21 (DE3) pLysS competent cells were done and the cells were plated and kept for growth at 37°C overnight.

Target Protein Induction Using IPTG (Isopropylthiogalactopyranoside)

The transformed colonies were picked for culturing in 3L Luria broth containing both chloramphenicol and kanamycin antibiotic (3000 µl) when the OD reached 0.4–0.6 0.5 mM IPTG induction was given and kept at 37°C, 220 rpm over night. The cultures were pelleted down at 10,000 rpm and to the pellet added 1X lysis buffer and vortexed till the entire pellet were dissolved. Sonications of cells were done in Ultrasonic cell processor, at amplitude of 43 for about 3 minutes. Upon completion of Sonication the samples were pelleted down at 13000 rpm for 30 minutes 4°C.The supernatants were collected. UIS4 protein was present as crude and hence it must be purified. Before purification of UIS4 protein it has to be confirmed and this was done by running the samples on Poly Acyrlamide Gel Electrophoresis (10% resolving and 5% stacking gel). SDS-Polyacrylamide gels containing more than 200ng protein concentration were visualised by standard Coomassie Brilliant Blue R 250 (CBB R 250) staining solution. {0.1% (w/v) CBB dissolved in 25% (v/v) methanol and 10% (v/v) acetic acid in water}, followed by de-staining in 25% (v/v) methanol and 10% (v/v) acetic acid in water. Western Blotting was done to confirm the protein of interest. The primary antibody used was anti histidine antibody raised in mice. The secondary antibody rose in rabbit-anti mouse (GE Health care) was added against the primary antibody and has enzyme Horse radish peroxidase as conjugate. The substrate 3–3'diaminobenzidiene was added along with hydrogen perroxidase.

Purification of UIS4 Protein

His tagged proteins were purified by Immobilized Metal ion Affinity Chromatography (IMAC) technique. His-tagged proteins may be purified under either native or denaturing conditions.

Regeneration of Ni-NTA Column

A column of 20 ml (Biorad) was washed with distilled water this was done 5 column volumes. And 4 ml of Ni-NTA slurry was added to the column, this was disturbed with double distilled water (5 column volumes).To aid the process of binding the protein of interest to Ni metal ion, the resin were equilibrated with binding buffer(1X PBS with 0.5 mM NaCl (5 column volumes).

Purification of Protein Using Ni-NTA Column

Cell lysate prepared was passed through the column and was kept in rotation at 4°C overnight. The flow through was collected and two step washing was carried out with 25 mM and 50 mM Imidazole concentration. Elution was carried out at six different concentration of Imidazole they include 100 mM, 200 mM, 300 mM, 400 mM, 500 mM, 600 mM.

RESULTS

PCR Amplification of UIS 4 Gene

UIS4 gene was PCR amplified from cDNA using the forward and reverse primer specific to UIS4 gene. Forward primer-GCAGGATCCTACCCAGAAAATTCA CCAGA; Reverse primer-GTACTCGAGTTATATGTATGGGCCGAATGA. The percentage of GC both forward and reverse primers are 44%. The cDNA length is around 330 bp.

Restriction Digestion of Recombinant pET-28a –UIS4 by Bam H1 and Xho 1

The presence of gene of interest in plasmid vector after is confirmed by performing restriction digestion of isolated plasmid from the transformed colony. Double digestion

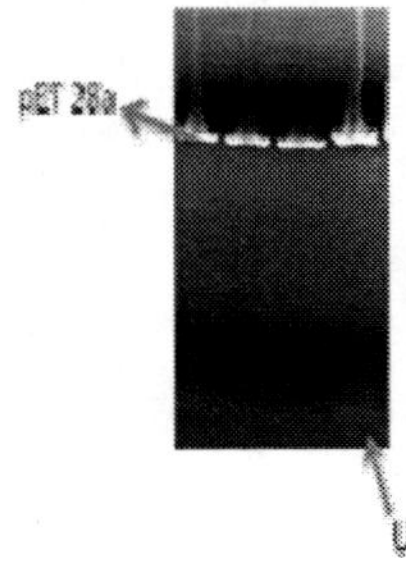

Fig. 1: UIS4 –pET28 a Digested by Bam H1 and Xho 1

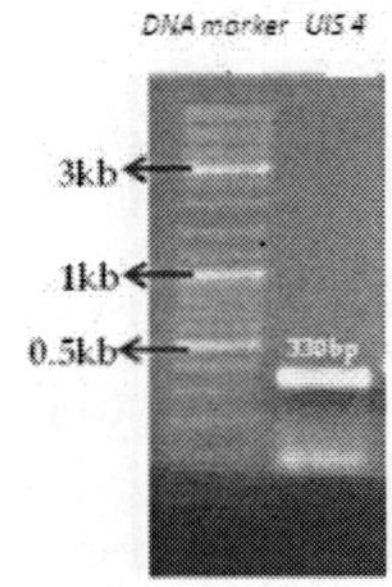

Fig. 2: PCR Amplified UIS4 Gene

was performed with the restriction enzymes Bam H1 and Xho 1. The ladder used was 1 kb NEB.

Induction of Target Protein Using IPTG

The UIS4 is under tight regulation of Lac operon. Expression of gene would not take place until the repressor bound to the T7 promoter is released. We tried with different concentrations of IPTG and found that at 0.5 mM IPTG and 37°C of incubation, UIS4 gene was expressed. Samples induced with 0.5 mM IPTG band of nearly 20 kDa are obtained. In un induced samples it did not appeared.

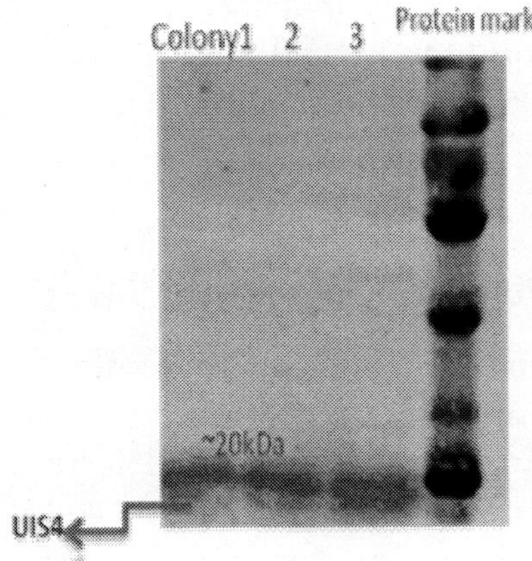

Fig. 3: Induction of Samples with 0.5 mm IPTG, a Band of ~20 kda is Obtained in Induced Samples

Purification of Protein by Ni-NTA Column

The His tagged protein bound to the Ni-NTA resin was eluted using different concentrations of Imidazole. Flow through, wash fractions and different eluted fractions were analyzed and it was found that the protein of interest was eluted at 300 mM of Imidazole.

DISCUSSION

Construction of Vector

pET-28a Expression system was chosen as the vector because it has His Tag either at the C terminal or at the N terminal, antibiotic marker kanamycin, T7 promoter for the gene under the control of LacI gene (repressor), i.e. no expression of gene of interest until the host cell contains *E. coli* T7 RNA polymerase and IPTG (Isopropyl–β-D-thiogalactopyranoside). The size of the pET-28a system is 5369 base pairs. This vector has multiple cloning sites for many enzymes. The protein of interest will be expressed as a fusion protein with His Tag.

Escherichia Coli BL21 (DE3)-RILpLysS Cells. *Escherichia coli DH5α* cells were chosen as the cloning host because of its ability to produce high copy number. *Escherichia coli BL21 (DE3)-RILpLysS cells* (Stratagene Inc.) were chosen as the expression host because it has T7 RNA polymerase gene under the control of LacUV5 promoter in the chromosome. The BL21 (DE3)-RIL competent cells were an all-purpose strain for high-level protein expression and easy induction and are resistant to chloramphenicol. These mutant strains lack lon protease and omp T outer membrane protease. Hence there is no chance of degradation of protein of interest.

Purification of Target Protein

Immobilized Metal ion Affinity Chromatography (IMAC)

Technique works by allowing proteins with an affinity for metal ions to be retained in a column containing immobilized metal ions, such as cobalt, nickel, and copper. High Affinity Ni-NTA Resin is an agarose resin (4% cross-linked) covalently coupled to a tridentate chelating agent (triglycolamic acid or (NitriloTriacetic acid) that binds Ni^{2+} ions by four coordination sites for high-affinity purification of His-tagged recombinant proteins. His tagged proteins were purified either under native or denaturing conditions. By using Imidazole as elution buffer the his tagged proteins were eluted from the column pET system manual-NovagenRecombinant protein purification handbook-GE Healthcare High Affinity Ni-NTA systems-GE healthcare; Nickel affinity chromatography guide-Dragon Tech Comparison of HisDetector TM Nickel-NTA Enzyme Conjugates with Other Methods for the Detection of His-Tagged Fusion Proteins in Western Blots by Danielle Russellet.

CONCLUSION

Thus the UIS4 gene was cloned and expressed in *E.Coli BL21 (DE3) pLysS* cells and it was purified using affinity chromatography, the protein obtained was 80% pure. The purified protein were about to use for raising antibodies.

REFERENCES

Agam Prasad Singh *et al.* (2007). Plasmodium *Circumsporozoite Protein* Promotes the Development of the Liver Stages of the Parasite, *Cell,* 131, 492–504.

Carter, R. and Diggs, C.L. (1977). Plasmodia of rodents. In: *Parasitic Protozoa,* Vol. III, pp. 359–465.

Kappe, S.H., Kaiser, K. and Matuschewski, K. (2003). The Plasmodium sporozoite journey: *A rite of passage. Trends Parasitol,* 19, 135–143.

Shortt, H.E. and Garnham, P.C.C. (1948). Pre-erythrocytic stage in mammalian malaria parasites. *Nature,* 161, 126.

Screening, Isolation and Identification of Lignocellulolytic Enzyme Producing Endophytes from Banana Pith and Pseudostem

Rajashree Krishnaswamy*, Jerrin Mathew T., Karthikeyan D., Ramalakshmi V., Uthra S. and Divya A.**

Department of Biotechnology, Kumaraguru College of Technology, Coimbatore

E-mail: rajshree32@yahoo.com; divyabt90@gmail.com

INTRODUCTION

Banana or the Musa species, belonging to the family *Musaceae*, a large perennial herb is a source of food, beverages, medicines, flavorings, cooked foods, fermentable sugars, silage, fragrant, rope, cordage, garlands, shelter, clothing, and many ceremonial and religious uses. The pseudostem of the banana plant is a clustered cylindrical aggregation of leaf stalk bases. It is made of lignocellulosic fibers. The lignocellulosic content is predominantly of three components: cellulose, hemicellulose and lignin and is one of the two main agro-wastes from the banana industry (Katongole *et al.*, 2008).

Lignocellulosic waste generated through forestry and agricultural practices, paper-pulp industries, timber industries and many agro-industries pose an environmental problem. This waste can be potentially used for various value added products including chemicals, bio fuels, and nutrients. Such a transformation of lignocellulosic waste can be effected by the use of lignocellulolytic enzymes, which include cellulases, hemicellulases and lignases.

*Corresponding author.

**Presenting author.

Endophytes are organisms inhabiting plant organs that can colonize internal plant tissues without causing apparent harm to the host (Petrini, 1991). The mutualistic relationship between the endophytes and the host plants results in benefits for both sides via a harmonious symbiotic system (Azevedo *et al.*, 2000; Saikkonen *et al.*, 2004). Given the capability of endophytes to adapt to a plant tissue, there is every possibility that the pith and pseudostem of the banana plant can harbor endophytes. The pseudostem and pith, being highly lignocellulosic in nature, endophytes colonizing them may therefore secrete enzymes to utilize lignocellulosic substrates.

In this study we have isolated lignocellulolytic endophytic bacteria from banana pith and pseudostem. These isolates were identified using suitable methods, and were screened for lignocellulolytic enzyme activity.

MATERIALS AND METHODS

Isolation of Endophytic Bacteria

The samples, banana pith and pseudostem (Kadhali Variety) were collected from Coimbatore area, surface sterilized by dipping it in 70% ethanol and cut into 10 mm × 10 mm dimension pieces under aseptic conditions. Sufficient numbers of cut pieces (4 pieces) were placed in PDA plates (90 mm diameter) and incubated at 33°C for 24–48 hours. After incubation, colonies observed around the periphery of the pith and within the pseudostem were carefully isolated and were transferred to PDA slants under aseptic conditions and incubated at 33°C for 24–48 hours. The identification of endophytic bacteria was based on morphology, biochemical test and MALDI-TOF MS.

Morphological Identification

Endophytic bacteria were morphologically identified by simple staining and gram staining methods.

Screening for Lignocellulolytic Enzyme Producers

Screening for lignocellulolytic enzyme producers was performed as described by Stephen B. Pointing, 1999.

Cellulase Screening

The endophytic bacteria were inoculated under aseptic conditions and were incubated at 33°C for 24–48 hours in Cellulolysis Basal Media (CBM). After 2 days of incubation, the plates were flooded with 2% w/v aqueous Congo red solution for 15 minutes and washed with distilled water, followed by 1M NaCl for 15 minutes to destain and visualize clearance zones. The colony diameter and the clearance zone diameter were documented.

Xylanase Screening

Xylanolysis Basal Media (XBM), was prepared, inoculated under aseptic conditions and incubated at 33°C for 24–48 hours. After 2 days of incubation, the plates were flooded with 0.25% w/v aqueous I2 and KI solution for 5 minutes and washed with distilled water to visualize clearance zones. The colony diameter and the clearance zone diameter was documented.

Ligninase Screening

LME Basal Media (LBM) was prepared, inoculated under aseptic conditions and were incubated at 33°C for 24–48 hours. After 2 days of incubation, the plates were flooded with 1% w/v aqueous solution of $FeCl_3$ and $K3[Fe(CN)_6]$ for 5 minutes and washed with distilled water to visualize clearance zones.

Biochemical Identification

Biochemical test such as oxidase test, lactose fermentation test, indole test, MR-VP test and motility test were performed for bacterial identification.

RESULTS AND DISCUSSION

Isolation of Endophytic Bacteria

7 bacterial isolates were obtained from banana pith and pseudostem samples that were placed on PDA plates.

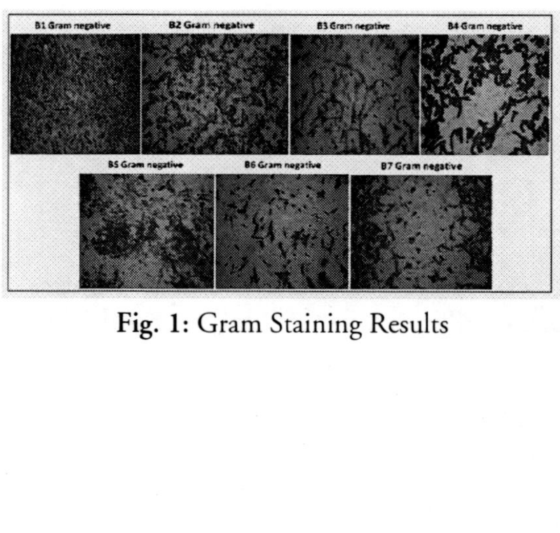

Fig. 1: Gram Staining Results

Morpholoical Identification

Table 1: Simple Staining and Gram Staining Results

Isolate	*Morphological Characters*
B1	Gram Negative medium rod
B2	Gram Negative medium rods
B3	Gram Negative thin, long rods
B4	Gram Negative oval shaped rods
B5	Gram Negative short rods
B6	Gram Negative thin rods
B7	Gram Negative short rods

Table 2: Qualitative Screening for Lignocellulolytic Activity

Basal Media and Production Media	*Bacterial Isolates Showing Good Results (In terms of ratio of clearance zone to colony diameter)*	
	Isolate Ratio (Average of Duplicates)	*Isolate Ratio (Average of Duplicates)*
Cellulolysis Basal Medium	B1 2.625	2.625
	B2	2.140
	B4	2.750
	B7	4.400
Xylanolysis Basal Medium	B1	3.000
	B2	5.000
	B6	5.000
	B7	4.000
Lignin Modifying Enzyme Basal Media	B2 B4 B6	Only growth observed, zone of clearance not measured

Based on the above reported qualitative screening assay, only 5 bacterial isolates (B1, B2, B4, B6 and B7) were carried for further work, since they have shown significant lignocellulolytic enzyme activity.

Table 3: Biochemical Identification

Bacterial Isolates	*Biochemical Test*				
	Oxidase Test	*Lactose Fermentation Test*	*Indole Test*	*MR-VP Test*	*Motility Test*
B1	–	+	–	–/+	+
B2	–	+	–	–/+	+
B4	–	–	–	–/+	–
B6	–	–	–	–/+	–
B7	–	–	–	–/+	–

Molecular Characterization of Isolates

Bacterial Identification using MALDI-TOF MS Analysis

Protein imaging of the 5 bacterial isolates using MALDI-TOF MS analysis, and Bruker taxonomic matching pattern revealed a quite few possible organisms for each isolate. Biochemical identification results and the MALDI-TOF flex analysis results were correlated and the best possible organism was finalized. Further characterization can be obtained by 16S rRNA sequencing.

Table 4: Bacterial Identification

Isolate	*Bruker Daltonics Flex Analysis Score Value*	*Identified Bacteria*
B1	1.991	*Enterobacter aerogenes*
B2	1.686	*Enterobacter aerogenes*
B4	1.377	*-Could not be identified-*
B6	1.677	*Klebsiella pneumoniae spp pneumonia*
B7	1.725	*Klebsiella variicola*

CONCLUSION

7 bacterial isolates were obtained from the banana pith and pseudostem. Qualitative screening was employed to evaluate the lignocellulolytic enzyme producing ability of those isolates. Based on the zone of clearance obtained, 5 bacterial isolates were considered for further work. These bacterial isolates were subjected to MALDI-TOF MS analysis, and Bruker taxonomic matching pattern, and the identification was

done based on the best score value in correlation with the results of the biochemical tests. Further work will involve media optimization to obtain high lignocellulolytic enzyme activity as well as characterization of the enzymes.

REFERENCES

Argyropoulos, D.S. and Menachem, S.B. (1997). Lignin In K.-E. Eriksson (ed.), *Advances in Biochemical Engineering Biotechnlogy*, Vol. 57. Springer-Verlag, Germany, pp. 127–158.

Azevedo, J.L., Maccheroni, W. Jr., Pereira, J.O. and Araújo, W.L. (2000). Endophytic microorganisms: a review on insect control and recent advances on tropical plants. *Electronic Journal of Biotechnology* http://www.ejbiotechnology.info/content/vol3/issuel/full/4/index.html. ISSN0717-3458

Baldrian, T. and Gabriel, J. (2003). Lignocellulose degradation by *Pleurotus ostreatus* in the presence of cadmium. *FEMS Microbiological Letter*, Vol. 220, pp. 235–240.

Beg, Q.K., Kapoor, M., Mahajan, L. and Hoondal, G.S. (2001). Microbial xylanases and their industrial applications: a review. *Journal of Applied Microbiology and Biotechnology*, Vol. 56, pp. 326–338.

Bischoff, K.M., Wicklow, D.T., Jordan, D.B., de Rezende, S.T.. Liu, S.Q., Hughes, S.R. and Rich, J.O. (2009). Extracellular hemicellulolytic enzymes from the maize endophyte *Acremonium zea.* Curr Microbiol, 58, pp. 499–503.

H.P. and Mûcke, I. (1997). Mini review, history, overview, and applications of mediated ligninolytic systems, especially laccase-mediator-systems (Lignozym-Process*), Journal of Biotechnology*, Vol. 53, pp. 163–202.

Hattaka, A. (1994). Lignin-modifying enzymes from selected white rot fungi, production and role in lignin degradation. *FEMS Microbiology Review*, Vol. 13, 125–135.

Katongole, C.B., Bareeba, F.B., Sabiiti, E.N. and Ledin, I. (2008). Nutritional Characterization of some Tropical urban market crops wastes, *Journal of Animal Feed Science and Technology*, Vol. 142, pp. 275–291.

Kudanga, T. and Mweje, E. (2005). Extracellular cellulose production by tropical isolates of *Aureobasidium pullalans. Can J Microbiol*, 51, pp. 773–776.

Kuhad, R.C. and Singh, A. (1993). Lignocellulose Biotechnology: Current and future prospects. *Critical Review of Biotechnology*, Vol. 13, pp. 151–172.

Malherbe, S. and Cloete, T.E. (2003). Lignocellulose biodegradation: fundamentals and applications: A review. *Environmental Science and Biotechnology*, Vol. 1, pp. 105–114.

McCarthy, A.J. (1987). Lignocellulose-degrading actinomycetes. *FEMS Microbiological Letter*, Vol. 46(2), pp. 145–163.

Oses, R., Valuenzela, S., Baeza, J. and Rodriguez J. (2006). Evaluation of fungal endophytes for lignocellulolytic enzyme production and wood biodegradation. *Int Biodeterior Biodegrad.*, 57, pp. 129–135.

Petrini, O. (1991). Fungal endophytes in tree leaves. In: Microbial Ecology of Leaves. Springer, New York, pp. 179–197.

Pointing, S.B. (2001). Feasibility of bioremediation by white-rot fungi. *Journal of Applied Microbiology and Biotechnology*, Vol. 57, pp. 20–33.

Rabinovich, M.L., Melnik, M.S. and Bolobova, A.V. (2002a). Microbila cellulases: A review. Journal of Applied *Biochemistry and Microbiology*, Vol. 38(4), pp. 305–321.

Shen, H., Gilkes, N.R. and Kilburn, D.G. (1995). Cellobiohydrolases B, a second exocellobiohydrolase from the Cellulolytic bacterium Cellulomonas fimi. *Biochemical Journal*, Vol. 311, pp. 67–74.

Vicuna, R. (1988). Bacterial degradation of lignin. Enzyme Microbial Technology, 10, pp. 646–655.

Wood, D.A. (1985). Useful biodegradation of lignocellulose. *Annual Proceedings of the Phytochemical Society of Europe*, Clerendon Press, Oxford, U.K. Vol. 26, pp. 1295–130.

Zimmermann, W. (1990). Degradation of lignin by bacteria. *Journal of Biotechnology*, Vol. 13 (2–3), pp. 119–130.

Development of HPTLC Based Detection/(Assay) Method for β-Carotene Produced by *Rhodosporidium toruloides*

R. Agalya Devi, S. Prashandh, S. Vasantha Devi, Rahul Kumar and K. Balakrishnan*

Technology Business Incubator-Bannari Amman Institute of Technology (BIT-TBI), Sathyamangalam, Tamil Nadu

E-mail: balu444@gmail.com

INTRODUCTION

Carotenoids are antioxidant micronutrients reported to protect against the adverse activity of free radicals. Among the carotenoids, β-Carotene is the most recognized and widely known carotenoid. β-Carotene is biosynthesized from geranylgeranyl pyrophosphate (Davis *et al.,* 1984). It is a member of the carotenes, which are tetraterpenes, synthesized biochemically from eight isoprene units and thus having 40 carbons. Among this general class of carotenes, β-Carotene is distinguished by having beta-rings at both ends of the molecule. Absorption of β-Carotene is enhanced if eaten with fats, as carotenes are fat soluble.

Along with the most known producer *Phaffia rhodozyma*, there is evidence of the capacity for carotene formation by other well-known pigmented yeasts of the genus *Rhodotorula* (order *Sporidiobolales*). The composition and amount of the carotenoid pigments in numerous natural isolates of the genera *Rhodotorula/Rhodosporium* and *Sporobolomyces/Sporidiobolus* were studied in detail (Yurkov *et al.*, 2008; Richard *et al.,* 2011). Carotenogenic yeasts are well known producers of biotechnologically significant carotenoid pigments—astaxanthin, β-carotene, torulen, torularhodin and under stress

*Corresponding author.

conditions this carotenoid accumulation was reported to be increased (Pilar *et al.*, 2011). Pigmented yeasts are used as feed and food colorants (Lautent *et al.,* 2010) and, some of them, also as single cell oil producers.

Rhodosporidium toruloides is aerobic oleaginous yeast. It is known to produce deep orange color of carotenoid on aging. β-Carotene is one of the main carotenoids produced abundantly in this organism. The amount of β-Carotene that is produced by *Rhodosporidium toruloides* is reportedly 14% (w/w) of the total carotenoid content (Richard *et al.,* 2011).

The present work describes the application of high-performance thin-layer chromatographic (HPTLC) estimation of Beta carotene. In this study, high-performance TLC (HPTLC) methods was developed for analysis of beta carotene of varying complexity and form using a silica gel layer, detection with Hexane and chloroform (9.25:0.75), automated band wise sample application and densitometric scanning. The new method was validated for sensitivity, linearity, accuracy and precision (repeatability).

MATERIALS AND METHODS

All the chemicals used were of analytical grade and procured from Renkem, New Delhi, Himedia, Mumbai and Merck, Mumbai, India.

Microorganism

All experiments were carried out using aerobic oleaginous yeast *Rhodosporidium toruloides* obtained from NCIM (3547), Pune, (India). Sub culturing was carried out in YDP media slant as per standard procedures and cultures were stored for long term preservation in 40% glycerol stock at –20°C.

Preparation of Standard

Dissolve 10 mg Beta-carotene in 10 ml methanol and sonicate.

Fermentation

Fermentation was carried out in 250 ml Erlenmeyer flasks containing 80 ml media (Corn steep liquor + 2% Glucose pH: 7.0) Flasks were inoculated with 2 loop full of logarithmically grown slant culture of *Rhodosporidium toruloides* and operated at room temperature on a rotary shaker which was maintained at 130 rpm, Samples were taken at every 24 hours interval for 120 hours. Samples were processed for biomass and β-carotene estimation.

Analytical Methods

Biomass Estimation

Fermentation samples collected at various time interval and harvested fermentation broth were centrifuged at 10000 rpm for 10 min at room temperature. The pellets thus obtained were used to measure biomass using gravitational methods (wet weight).

β-Carotene Extraction

Yeast cells were harvested at stationary phase which was monitored by measuring the optical density at 436 nm as per the procedure mentioned in Spirulina Pacifica Technical Bulletin, 2002. Harvested fermentation broth was centrifuged at 10000 rpm for 10 min. The pellet thus obtained was used for β-carotene extraction.

The procedure adopted has got three distinct steps:

1. Methanol extraction
2. Heptane extraction
3. Spectrophotometric analysis

Methanol Extraction: 30 mg of cell pellets from each sample was added to centrifuge tubes. To this was added 3 gram of glass beads and 2.5 ml of Di methyl sulfoxide. This was then vortexed for 30 seconds. The centrifuge tubes were placed in water bath maintained at 50 C for 30 minutes. Tubes were removed for vortexing at every 10 minute interval for about 30 seconds. Add 5 ml of methanol to each sample and they were vigorously vortexed for 30 seconds in test tubes. The content of the test tubes were centrifuged at 4200 rpm for 3 minutes. The supernatant collected from each tube were added with 4 ml of methanol followed by vortexing for 15–30 seconds. Centrifugation was repeated for 3 minutes. The supernatant was collected and to this methanol was added just to level enough to cover the glass beads. These were vortexed vigorously for 30 seconds and the process repeated until methanol is absolutely clear. The supernatant was collected and the volume made up to 25 ml with methanol.

Heptane Extraction: To every 8 ml of methanol extract from each 25 ml sample held in 15 ml centrifuge tubes, add 5 ml of heptane and 1.5 ml saturated potassium hydroxide in methanol. After 15 minutes the tubes were vortexed vigorously for 10 minutes. After 30 minutes the tubes were again vortexed for 15 seconds and centrifugation was carried out at 4200 rpm for 3 minutes. Heptane layer was removed and kept in 10 ml volumetric flask. 1 ml of fresh heptane was added to each volumetric flask and the inter phase was washed out. Heptane was allowed to separate from methanol (2 minutes) before pipetting out the heptane. Approximately 5 ml of heptane extract was poured in to clean centrifuge tube. To this was added equal amount of distilled water and vortexed vigorously for 5 seconds.

Spectrophotometric Analysis: The tubes were centrifuged for 4200 rpm for 3 minutes and the absorbance was read at 436 nm against a heptane blank.

Calculation

$$\beta\text{-Carotene (\%)} = \frac{\text{Abs436}}{196x\ (\text{wt.(mg)} \times \text{dry wt.})} \times 25\ \text{ml} \times 1.25 \times 100 \times 0.84$$

Detection of Beta Carotene by HPTLC

The samples were spotted in the form of bands of width space 20 mm with a Camag microlitre syringe on precoated silica gel aluminum plate 60 GF-254 (10 cm × 10 cm with 0.2 mm thickness, E. Merck, Germany) using a Camag Linomat V. (Switzerland). A constant application rate of 150 µl/s was employed and space between two bands was taken 20 mm (Zeb *et al.*, 2010). The slit dimension was kept at 4 mm × 0.1 mm, and 20 mm/s scanning speed was employed. The mobile phase consisted of 1-hexane: chloroform (9.25: 0.75). Linear ascending development was carried out in twin trough glass chamber saturated with the mobile phase. The optimized chamber saturation time for mobile phase was 30 min at room temperature. The length of chromatogram run was 8 cm. Subsequent to the development; TLC plates were dried with the help of a hair-dryer. Densitometric scanning was performed on Camag TLC scanner IV in the absorbance mode at 436 nm. The source of radiation utilized was Mercury lamp (Neff *et al.*, 1998, Li *et al.*, 2010).

RESULTS AND DISCUSSION

β-Carotene Estimation

The estimated values of β-Carotene detected in the fermentation samples were depicted in the Table 1. A progressive increase in the accumulation of β-Carotene was noticed till 76 hrs where β-Carotene was found to be 2.36% of the dry cell mass. The same estimated value of β-Carotene for corn medium without incorporation of glucose was found to be 0.17%. This clearly indicated the positive influence of glucose addition to the medium on the β Carotene accumulation in the yeast cell.

Table 1: Estimated β-Carotene Accumulation by *Rhodosporidium toruloides* in Corn Extract Medium Supplemented with 2% Glucose

Time	*Corn Medium + 2% Glucose*
	β-Carotene Assay
24 hrs	0.9%
48 hrs	1.46%
76 hrs	2.36%
92 hrs	2.16%
120 hrs	2.14%

The standard chromatogram obtained for β-carotene standard and samples (Day 1 and Day 2) were depicted in (Figures 1 and 2) respectively. The scanned TLC plate indicating the chromatogram developed for β-carotene was depicted in (Figure 3).

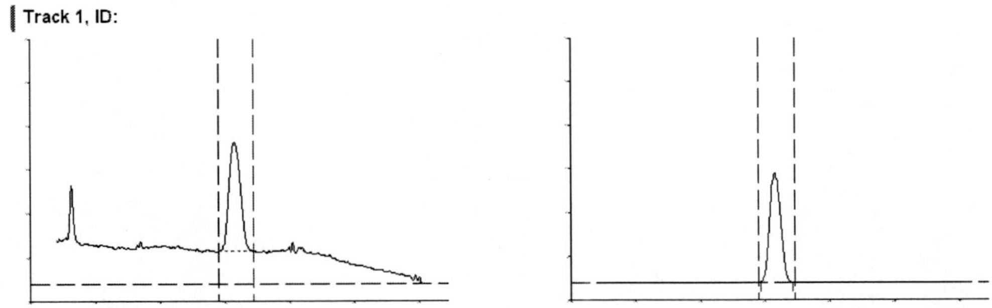

Peak	Start Rf	Start Height	Max Rf	Max Height	Max %	End Rf	End Height	Area	Area %	Assigned substance
1	0.46	0.7	0.50	124.4	100.00	0.56	0.4	3116.1	100.00	unknown *

Fig. 1: HPTLC Chromatogram of Standard Beta Carotene

Our initial experimentation to optimise mobile phase using Chloroform: Methanol: Water in varying ratios did not yield conistant findings. When the mobile phase was made more non polar with Hexane: Chloroform (9.25:0.75 v/v), we got reproducible results as sharp well defined spots on the plates kept in the twin trough chamber saturated with mobile phase for 15 minutes at room temperature for seperation and development.

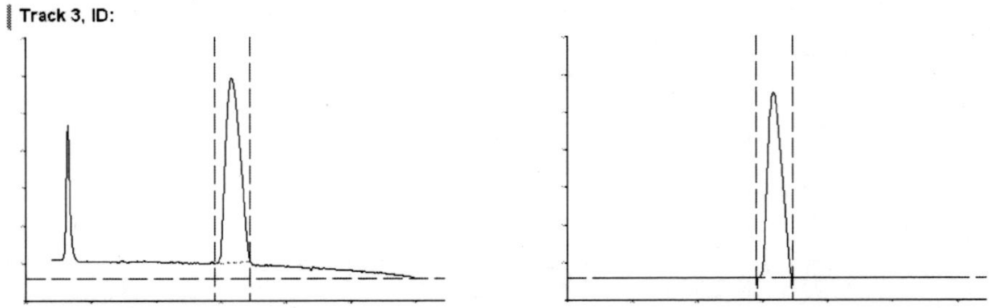

Peak	Start Rf	Start Height	Max Rf	Max Height	Max %	End Rf	End Height	Area	Area %	Assigned substance
1	0.46	0.2	0.50	491.8	100.00	0.56	3.4	15320.6	100.00	unknown *

Image information - 254 nm - Image1

Illumination instrument	CAMAG Reprostar 3 : 150906 (Repro3_150906)
Digital camera type : snr & Lens	DXA252 : 350283808, Computar, 16 mm, f4.0
Created by : on	BANARI AMAN INSTITUTE OF TECH : Tuesday, July 10, 2012 5:12:08 PM
Resolution	Full
Plate border size	-2 mm
Automatic capture	Off
Save mode	Lossy (JPG)
Exposure mode	Automatic, digital level: 80 %, Band
Capture settings:	
Image size:	945 Pxl x 944 Pxl (0.10 mm/Pxl)
Exposure :	1964.66 ms gain: 1.00
White balance	R: 1.40, G: 1.00, B: 1.20
Illumination type / correction type :	None :
Display settings:	
White balance:	R: 1.00 G: 1.00 B: 1.00
Contrast enhancement:	1.41 (modified)
Brightness:	0.00
Accentuation:	0.80
Color saturation:	1.30
Blank plate compensation :	N/A

Fig. 2: HPTLC Chromatogram of the Samples

Fig. 3: HPTLC Chromatogram at 254 nm

ACKNOWLEDGEMENTS

The authors thank the Management of Bannari Amman Institute of Technology for giving them the opportunity to perform this study and providing lab facilities and constant encouragement for this research work. Thanks are also due to DST, New Delhi for funding the activities of BIT-TBI where this work was carried out as a prospective enterpreneurial venture.

REFERENCES

Davis, C.H., Putnam, M.D. and Thwaites, W.M. (1984). "Metabolism of dihydrouracil in Rhodosporidium toruloides", *Bacteriol J.*, 158, 347–349.

Laurent Dufoss (2007). "Microbial Production of Food Grade Pigments", *Biological Sciences,* 44; 313–321.

Li, Z., Merfort, I., Reich, E.J AOAC Int. 2010. Sep–Oct; 93(5):1390–8. High-performance thin layer chromatography for quality control of multicomponent herbal drugs: example of cangzhu xianglian san.

Neff, W.E. and Byrdwell, W.C. (1998). Characterization of model triacylglycerols (triolein, trilinolein and trilinolenin) autoxidation products via high-performance liquid chromatography coupled with atmospheric pressure chemical ionization mass spectrometry. *J Chromatogr A,* 818; 168–186.

Pilar Martinez-Moya, Steven, Karsten Niehaus, Jennifer Alcaíno, Marcelo Baeza and Víctor Cifuentes 2011 "Proteomic analysis of the carotenogenic yeast *Xanthophyllomyces drorhous*", *Bmc microbiology,* 11; 131–135.

Richard bailey, Kevin T. Madden, Joshua Trueheart, Reed Doten, Maria mayorga, Joshua Griffin Dunn and Dan Dueppen 2011 "Production of carotenoids in oleaginous yeast and fungi", 435/67; 435/254.11; 435/254.2; 435/254.21; 435/254.22.

Zeb, A. and Murkovic, M. (2010). "High-performance thin-layer chromatographic method for monitoring the thermal degradation of β-carotene in sunflower oil", *J Planar Chromatogr,* 23; 37–41.

Enzymatic Production of Xylooligosaccharides from Agricultural Byproducts

Y. Geetha Rani, N. Praveen Kumar*,
Y. Shanmuga Sundaram and R. Arulvel
Department of Biotechnology, K.S. Rangasamy College of Technology,
Tiruchengode, Tamil Nadu
E-mail: npk.sumathi@gmail.com

INTRODUCTION

Prebiotics is a term coined by Professor Gibson and Roberfroid (1995) and are defined as "non-digestible food ingredie nts that beneficially affect the host by selectively stimulating the growth of one or a limited number of bacterial species in the colon, such as *Bifidobacteria* and *Lactobacilli*, which have the potential to improve host health." Prebiotics consist mainly of oligosaccharides, which are sugar molecules of three to six monomers in chains. Oligosaccharides are relatively new functional food ingredients that have great potential to improve the quality of many foods. These include non-carcinogenicity; a low calorific value and the ability to stimulate the growth of beneficial bacteria in the colon.Prebiotics are not digested but go straight to the lower gut where they are fermented by specific bacteria, stimulating the bacteria's growth. They have antimicrobial activity, anticarcinogenic activity, hypo-lipidemic activity, anti-osteoporotic activity, blood glucose-modulatory activities, immune modulating activities, help in improving mineral absorption and balance, rid the gut of harmful microorganisms, and help in prevention of constipation and diarrhea. Xylooligosaccharides (XOS) are sugar oligomers made up of xylose units. Xylan is a major component of plant hemicelluloses that could potentially be an appropriate starting material for production of a variety of chemicals and compounds (Saha, 2003). Hemicelluloses are the second most abundant polysaccharide constituents

*Corresponding author.

of plant cell walls after cellulose (Harris, Ferguson, 1999). Depolymerization of xylan gives xylose and Xylooligosaccharides (XOS), which shows potential applications in pharmaceuticals, feed formulations, agricultural purposes and food applications. From nutritional point of view, *XOS behave as non-digestible oligosaccharides*. When consumed as a part of a diet, XOS have prebiotic properties due to their preferential utilization by *Bifidobacteria*. Additionally, they have various physiological importance's such as reducing cholesterol, maintaining the gastrointestinal health, improving the biological availability of calcium, reducing risk of colon cancer and having a cytotoxic effect on human leukemia cells (Ando *et al.*, 2004). The growing commercial importance of non-digestible oligosaccharides is based on their beneficial health properties, particularly the prebiotic activity. XOS favor the selective growth of *Bifidobacterium spp*, which have important biological effects. XOS can be used as ingredients of functional foods, cosmetics, pharmaceuticals or agricultural products. In addition to the health effects, XOS present interesting physico-chemical properties; they are moderately sweet, stable over a wide range of pH and temperatures and have organoleptic characteristics suitable for incorporation into foods. XOS are advantageous over other nondigestible oligosaccharides in terms of both health and technological related properties. However, the comparatively high production costs require further development of processing and purification technologies. Considering the economical and environmental reasons, nowadays utilization of major agricultural wastes for industrial purposes has gained much importance. In this line, Rice bran and Finger millet seed coat can be employed for production of value-added products such as prebiotics xylooligosaccharides.

The scope of the present study deals with the enzymatic processing of agricultural by-products such as Rice bran and Finger millets seed coat (Ragi CO 9) using xylanase for the production of prebiotic XOS. To develop an efficient and environmentally friendly technique for hydrolysis and conversion of rice bran and finger millets seed coat (CO 9), a low-cost and abundant biomass, to valuable compounds (such as Prebiotic compounds (XOS), Phenolic compounds, Soluble sugars, Organic acids, and Amino acids) in methods followed by Enzymatic treatment.

Collection and proximate analysis of rice bran collected from different parts of Tamil Nadu and finger millets from TNAU. Extraction of xylan and their enzymatic conversion to XOS. Analytical characterization of Xylooligosaccharides by TLC.

MATERIALS AND METHODS

Rice bran was procured from local modern rice mills Pvt Ltd. (Redhills, Chennai) and Finger millet white variety CO 9 (Ragi; *Eleusine oracana*) from Department of millets, TNAU, Coimbatore.

Chemicals

All chemicals used were of analytical grade and were obtained from HiMedia, Merck (India).

Enzymes

Glucoamylase (EC 3.2.1.3) from *Aspergillus niger*
α-amylase (EC 3.2.1.1) from *Bacillus licheniformis*
Xylanase from Thermomyces lanuginosus.

Proximate Analysis

Rice bran and Finger millet seed coat were taken and analyzed individually in triplicates on dry weight basis for crude protein, crude fat, crude fiber, ash, and total carbohydrate according to AACC methods.

Starch Estimation

Samples were treated with 80% alcohol to remove sugars and then starch is extracted with 52% perchloric acid. In hot acidic medium, starch is hydrolyzed to glucose and dehydrated to hydroxymethyl furfural. This compound forms a green coloured product with anthrone. Glucose (100 mg/100 ml) was taken as standard.

Defatting and Isolation of Free Sugars

Bran and seed coat was defatted using petroleum ether in a soxhlet apparatus for 8 hr. The defatted bran (10 g) was extracted with aqueous ethanol (70%, 100 ml × 3), centrifuged and the resulting supernatant was analyzed for total carbohydrate (Phenol: sulphuric acid) and reducing sugars (DNS) (Suhasini *et al.,* 1997).

Isolation of Dietary Fiber

The defatted sample (1 g) was suspended in sodium phosphate buffer (25 ml, pH 6.0, 0.1 M) followed by the addition to termamyl (α-amylase) 0.1 ml and kept in a boiling water bath for 15 min to digest the starch. The contents were cooled, added water (20 ml) and the pH was adjusted to 1.5 with HCl (4 N). Proteins were removed by digesting with pepsin (100 mg) at 40°C for 1 hr. Again water (20 ml) was added and incubated at 40°C for 1 hr. Finally the contents were cooled and the pH was adjusted to 4.5 with 1 N NaOH and filtered through a dried and weighted crucible containing celite.

Insoluble Fiber

The residue retained in the crucible was washed with ethanol (95%, 20 ml) followed by acetone (20 ml). The crucible was kept in an oven (105°C) till the weight became constant and the final weight was taken accurately (D1). The same was incinerated at 550°C for 5 hr and once again its weight (I1) was recorded.

Soluble Fiber

The volume of the filtrate was adjusted to 100 ml and the soluble fibers were precipitated by adding ethanol (95% warm ethanol 60°C, 4 volumes). The precipitate was filtered through celite dried and weighted after drying at 105°C (D2) followed by incineration at 550°C (I2) respectively.

Blank was prepared as above without the sample.

Soluble and insoluble fiber contents were calculated using the following formula;

$$\% \text{ Insoluble fiber} = D1 - (I1 - B1)/\text{Weight} \times 100$$

$$\% \text{ Soluble fiber} = D2 - (I2 - B2)/\text{Weight} \times 100$$

Destarching

Defatted rice bran and ragi seed coat (10 g) was dissolved in sodium acetate buffer (100 ml, pH 4.8, 0.1 M) and incubated with termamyl (α-amylase, 1 ml, ~1000 U) at 95°C for 1 hr. After cooling, glucoamylase (100 mg, ~7000 U) was added to the above solution and incubated at 55°C for 48 hr followed by centrifugation at 3000 rpm for 20 min to separate the supernatant (consisting of glucose emanated from starch) from the residue (dietary fiber).

Isolation of Water Soluble/Insoluble Polysaccharides

Defatted and Destarched rice bran (50 g) and ragi seed coat was taken separately and extracted with water (1:10 w/v) at room temperature (25°C) for 2 hr and the supernatant obtained after centrifugation (3000 rpm, 20 min) was precipitated with 3 volumes of ethanol. Precipitate was separated out, dialyzed and lyophilized to obtain water soluble polysaccharides. The residue obtained after centrifugation was dried by solvent exchange and designated as water insoluble polysaccharides.

Enzymatic Extraction of Xylooligosaccharides

Water soluble polysaccharides (both 10 mg) was dissolved in acetate buffer (10 ml, pH 4.8, 0.1 M) and incubated with xylanase (5 mg) in a constant shaking water bath at

50°C for 5 hr and subsequently the reaction was stopped by the addition of 3 volumes of ethanol and the precipitated material was removed by centrifugation (3000 × g, 15 min). The resultant supernatant consisting of mixture of monosaccharide and xylooligosaccharides was concentrated and used for further analysis.

Analytical Characterization

Thin Layer Chromatography (TLC)

The concentrated crude XOS obtained from enzymatic hydrolysis is checked for the separation of oligosaccharides using Thin Layer Chromatography technique. Two solvent systems were used.

- Acetone: n-Butanol: Water (8: 1: 1)
- 2-Propanol: Ethyl acetate: Nitromethane: Water (6: 1: 1: 2)

The compounds in the TLC plate were identified by spraying the plates with the following mixture-Methanol: Sulphuric Acid (9: 1) + Orcinol (0.2 g).

Calculation of Rf Value

$$\text{Rf Value} = \frac{\text{Distance travelled by the solute}}{\text{Distance travelled by the solvent}}$$

RESULTS AND DISCUSSION

Proximate Composition Data

Table 1: Proximate Composite Data

Proximate Analysis	*Rice Bran*	*Finger Millet (Eleusine coracana) Seed Coat*
Moisture	11.9%	10.7%
Ash	7.9%	9.6%
Crude protein	13.0%	13.7%
Crude Fat	15.7%	1.3%
Crude Fiber	3.1%	2.0%
Total carbohydrate	48.4%	62.7%

Starch estimation by Anthrone Method

Starch, which is composed of several glucose molecules, is a mixture of two types of components namely amylose and amylopectin. Starch is hydrolysed into simple sugars by dilute acids and the quantity of simple sugars is measured calorimetrically at 600 nm (Figure 1).

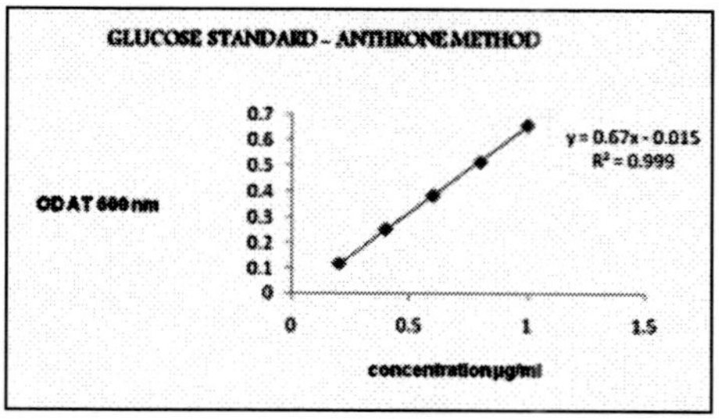

Fig. 1: Starch Estimation by Anthrone Method

Starch content results were obtained by anthrone method and were found to be:

Rice bran – 0.04 g/100 ml.

Finger millet's seed coat – 0.036 g/100 ml

Defatting and Isolation of Free Sugar-Spectroscopic Estimations

Estimation of Total Carbohydrates by Phenol-Sulphuric Acid Method

Total carbohydrate in the sample was estimated by phenol-sulphuric acid method. The sample (50 µL) was taken in a test tube and made up to 500 µL, phenol (5%, 0.3 ml) and concentrated sulphuric acid (1.8 ml) were added successively and mixed thoroughly, and the resultant solution was cooled to room temperature for 15 min. Absorbance was read at 490 nm. Sugar content was determined by referring to the standard graph, prepared by using D-xylose (5 µg/50 ml) (Figure 2).

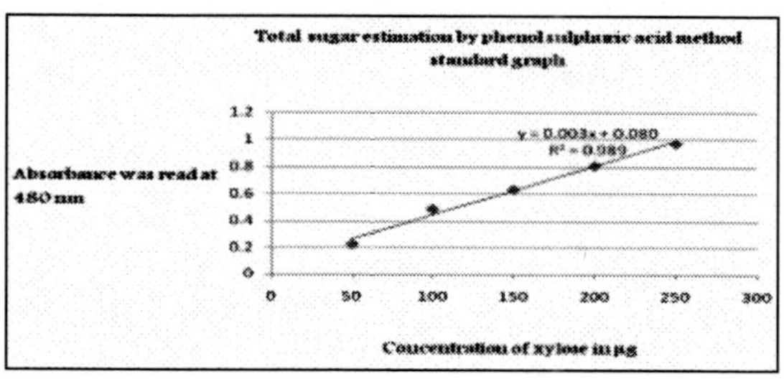

Fig. 2: Estimation of Total Carbohydrates by Phenol-Sulphuric Acid Method

Estimation of Reducing Sugar by DNS Method

To the sample (1 ml) in the test tube, DNS reagent (1 ml) was added and incubated in boiling water bath for 10 min. Content was then cooled and diluted with distilled water (2 ml). Absorbance was read at 550 nm. Reducing sugar content was determined by referring to the standard graph prepared by using D-xylose (50 mg in 50 ml) (Figure 3).

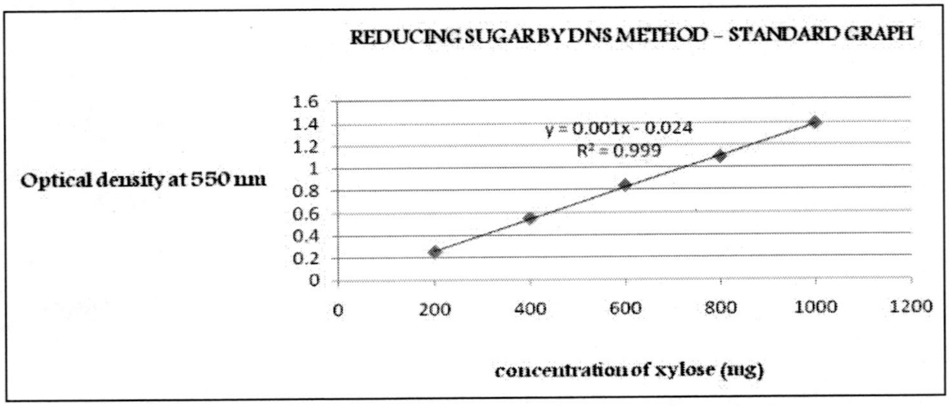

Fig. 3: Estimation of Reducing Sugar by DNS Method

Isolation of Dietary Fiber

Finger Millet

Insoluble Dietary Fiber – 18.1%
Soluble Dietary Fiber – 1.7%

Rice Bran

Insoluble Dietary Fiber – 16.7%
Soluble Dietary Fiber – 2.0%

Determination of Total Sugar for WSP

Total sugar in the samples were estimated by phenol-sulphuric acid method and found to be as follows:

Rice Bran – 650 µg/ml
Finger Millet – 312 µg/ml

Liberation of Oligosaccharides from Rice Bran and Finger Millet Seed Coat

Rice bran and seed coat when subjected to commercial xylanase hydrolysis resulted in crude XOS. Xylanase from *Thermomyces lanuginosus* is an endoxylanase, which cleaves arabinoxylans randomly (unbranched regions of both internal and external chains). After getting crude XOS the total sugar content was estimated.

Enzymatic Activity

Rice Bran

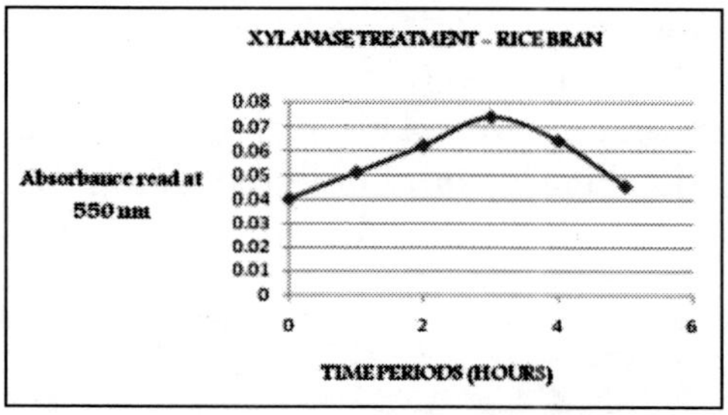

Fig. 4: Enzymatic Activity of Rice Bran

Enzyme activity/molecular weight of xylose = 0.0025 µmoles/min/mg of protein

Specific activity = Enzyme activity/mg of protein
= 0.0025/0.58
= 0.004 µmoles/ min/mg of protein

Finger Millet Seed Coat (Ragi CO 9)

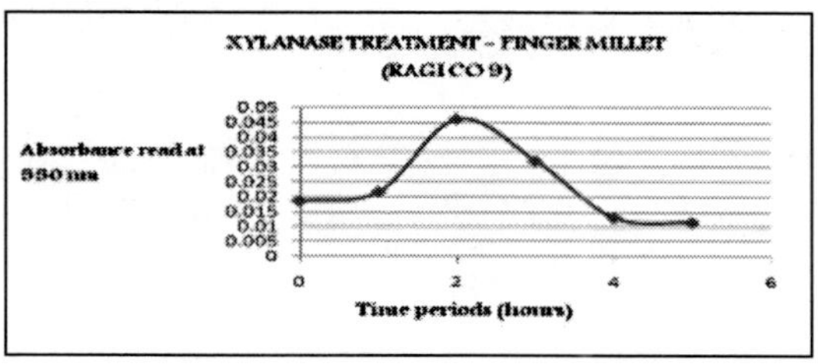

Fig. 5: Enzymatic Activity of Finger Millet Seed Coat (Ragi CO 9)

Enzyme activity/molecular weight of xylose = 0.0028 µmoles/min

Specific activity = Enzyme activity/mg of protein
= 0.0028/0.58
= 0.0048 µmoles/ min/mg of protein

Analytical Characterization

Rice bran obtained after defatting and destarching was 73.5% and that of finger millet was 71.2% and the Water Solube Polysaccharide (WSP) was isolated. Total sugar content for WSP was found followed by Enzymatic treatment by xylanase enzyme for obtaining crude XOS. Xylanase showed optimum activity at pH 4.8 and temperature 40°C with rice bran and finger millet as substrates separately. The crude XOS was analysed for the presence of oligosaccharides by Thin Layer Chromatography technique with monosaccharides taken as standard and the retention factor values were interpreted. The research work will be continued with the determination of WSP by Gas-Liquid Chromatography technique using standard sugars. Also, standardisation of enzymatic method at various buffer concentrations (pH, Temperature, incubation time) is yet to be carried out to obtain the crude XOS. This will be followed by its purification and analysis using HPLC-ELSD. Finally, *in vitro* fermentation of crude XOS and purified XOS will be carried out for bringing out its functional characterisation as well.

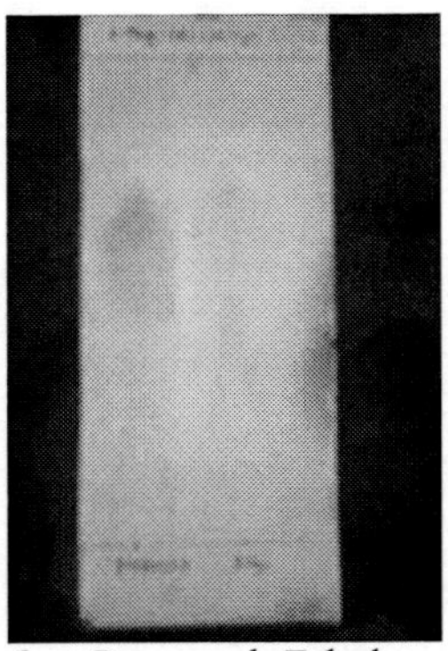

Fig. 6: 2-Propanol: Ethyl acetate: Nitromethane: Water (6: 1: 1: 2)

Rf Value:

Rice Bran – 0.71
Finger Millet – 0.77

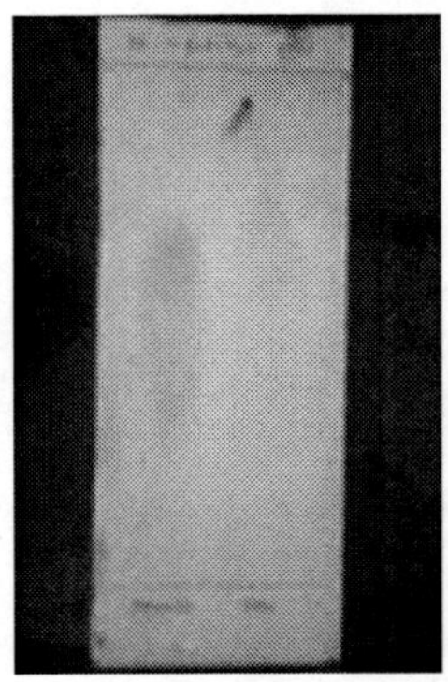

Fig. 7: Acetone: n-Butanol: Water (8: 1: 1)

Rf Value:

Rice Bran – 0.74
Finger Millet – 0.76

REFERENCES

Ando, H., Ohba, H., Sakaki, T., Takamine, K., Kamino, Y. and Moriwaki, S. (2004).

Gibson, G.R. and Roberfroid, B. (1995). Dietary modulation of the human colonic microflora: Introducing the concept of prebiotics. *J. Nutr.,* 125; 1401–1412.

Harris, P.J. and Ferguson, L.R. (1999). Dietary fibers may protect or enhance carcinogenesis. *Mutat. Res. Gen. Toxicol. Environ*, 443; 95–110.

Saha, B.C. (2003). Hemicellulose bioconversion, *Journal of Industrial Microbiology and Biotechnology*, 30: 279–291.

Phosphatase as Secondary Activity of Peroxidase from Liquid Endosperm of *Cocos nucifera* (L.)

M. Balasubramanian* and R. Boopathy

Enzyme Technology Lab, Department of Biotechnology, School of Biotechnology and Genetic Engineering, Bharathiar University, Coimbatore, Tamil Nadu

E-mail: balsubramanianm@gmail.com

INTRODUCTION

Haloperoxidases belongs to class of oxidoreductases, which having dual activity and it catalyzes the oxidation of halides (iodide, bromide and chloride) using hydrogen peroxide, resulting in the halogenation of these organic substrates or the halide-assisted disproportionation of hydrogen peroxide leading to the formation of dioxygen. Three classes of haloperoxidases have been identified so far. One of these haloeproxidase consists of enzymes without a prosthetic group and such enzymes have been detected in a number of bacteria, fungus and algae (Wiesner *et al.*, 1988, Bantleon *et al.*, 1994). The remaining two classes are heme-containing haloperoxidases—exemplified by the chloroperoxidase (CPO) isolated from the fungus *Caldariomyces fumago* (Hager *et al.*, 1966) and myeloperoxidase, present in white blood cells (Zeng and Fenna, 1992). The vanadium haloperoxidases isolated from *Curvularia inaequalis* and similar haloperoxidases have vanadate ion ($VO¾^-$) as a prosthetic group. Enzymes representing these two classes not only differ in the nature of their prosthetic group but also in its catalytic mechanism and stability. Heme containing peroxidases catalyze the formation of the hypohalous acid by a redox mechanism, whereas in vanadate containing peroxidases, the transition metal does not change its redox state (De Boer *et al.*, 1988, Arber *et al.*, 1989) but may function as a Lewis acid. Vanadate containing haloperoxidases not only posses very high stability (Van Schijndel *et al.*, 1994, Vollenbroek *et al.*, 1995) but they also resist a high concentration of their substrate

*Corresponding author.

(H_2O_2) (Liu *et al.*, 1987) and their product (Van Schijndel *et al.*, 1994) that would readily inactivate the heme-containing peroxidases.

Vanadate-binding amino acids were shown to conserved in two different haloperoxidases isolated from seaweed and most interestingly, the same vanadate binding amino acids has been found in several acid phosphatases like soluble bacterial non-specific class A acid phosphatases (Messerschmidt and Wever, 1996, Hemrika *et al.*, 1997, Stukey and Carman, 1997, Plass, 1999, Zhang *et al.*, 2000, Ishikawa *et al.*, 2000), human glucose-6-phosphatase (Feldman and Butler, 1972; Countaway *et al.*, 1988, Lange *et al.*, 1994, Hemrika and Wever, 1997), lipid phosphate hydrolases (Brindley and Waggoner, 1998) and phosphatidic acid phosphatases (Carman, 1997; Kanoh *et al.*, 1997, Brindley and Waggoner, 1998). Based on the sequence similarity, it has been proposed that the anion binding active sites of these enzymes are identical to that of acid phosphatase from *Escherichia blattae* (Ishikawa *et al.*, 2000).

It has been shown that apo-CPO (Hemrika *et al.*, 1997a) and apo- bromperoxidase (BPO) (Tanaka *et al.*, 2003) have phosphatase activity and vanadate substituted acid phosphatases have bromoperoxidase activity (Tanaka *et al.*, 2002, 2003). Therefore, it was suggested that the active sites of vanadium chloroperoxidase (VCPO) and the phosphatases are similar. The structural basis for this similarity is the resemblance of phosphate and vanadate, perhaps more importantly, the resemblance of the trigonal bipyramidal structure of bound vanadate in the native VCPO to the proposed transition state in the phosphatases. This is in line with the fact that these phosphatases are strongly inhibited by vanadate (Singh *et al.*, 1981). On the basis of this transition state analogue, it was suggested that in apo-CPO His^{496} acts as a nucleophile, whereas several residues stabilize the penta-coordinated transition state, and His^{404} plays a role in leaving group protonation (Hemrika *et al.*, 1997). Based on this similarity, the reaction scheme was developed (Figure 1) for the phosphatase catalysis of the apo-CPO and the hypohalous acid phosphatases (Hemrika *et al.*, 2000).

The active site amino acid similarity of haloperoxidases and acid phosphatase for pentacoordinating compounds has been well established. It was reported that plant peroxidases can oxidize the cysteine as thiol oxidase function (Pichorner *et al.*, 1992) but no reports are available on the secondary phosphatase activity of peroxidase enzyme. In this chapter, the association of secondary phosphatase activity of TCWP isoenzymes was examined.

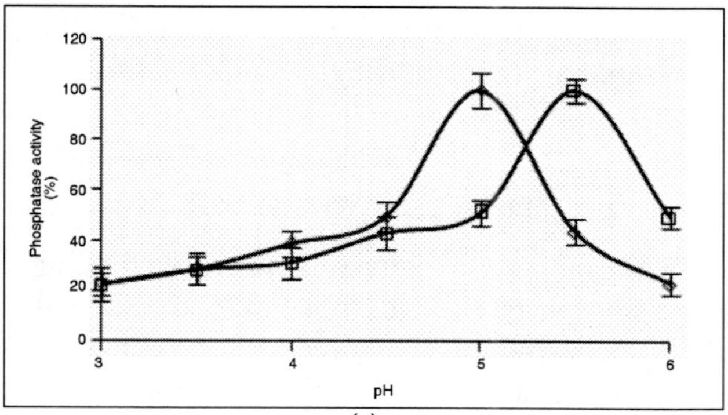

(a)

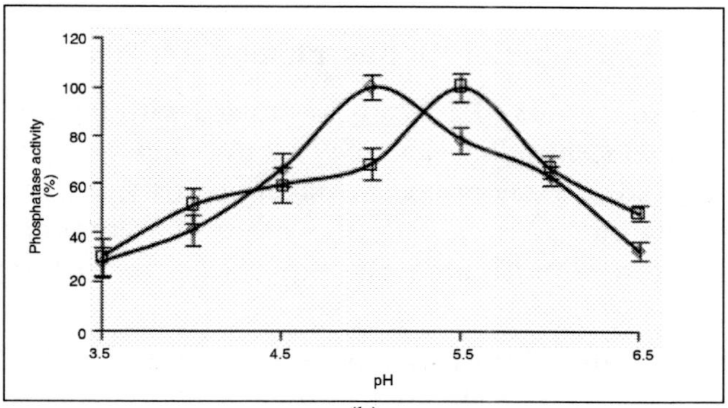

(b)

Fig. 1: Dependence of pH for the Enzyme Activity of $TCWP_1$ (◇) and $TCWP_2$ (□)

MATERIALS AND METHODS

Chemicals

Analytical grade chemicals namely *p*–nitrophenyl phosphate, sodium cyanide, sodium azide from Sigma, St. Louis, USA and sodium arsenate, sodium potassium tartrate and sodium acetate were procured from Himedia laboratories Ltd., Mumbai, India. Other chemicals were purchased from Qualigens fine chemicals, India.

Phosphatase Activity

A standard assay for determining phosphatase activity is measuring the hydrolysis of *p*-nitrophenyl phosphate according to Torriani, (1960). The enzyme assay was slightly modified as follows, after incubation with reaction mixtures [4 mM *p*-nitrophenyl phosphate and 100 mM sodium citrate buffer (pH 5.0 and 5.5 for

$TCWP_1$ and $TCWP_2$)], the reaction was quenched with NaOH and production of *p*-nitrophenol was measured at 410 nm using an molar extinction coefficient of 18.3 $mM^{-1} cm^{-1}$ at pH 12.

Dependency of pH for Phosphatase Activity and Stability

The optimum pH for phosphatase activity was determined using citrate buffer (pH 3.0 to 6.0), because the phosphatase activity was reported only with citrate buffer. The pH stability of TCWP isoenzyme was measured by incubating the TCWP isoenzyme with corresponding pH for 6 h at ambient temperature (30°C). Then the phosphatase activity was checked using the standard assay in triplicate and expressed as mean ± SD.

Effect of Buffer Ion Concentration on Phosphatase Activity

The purified TCWP isoenzymes (0.2 U/ml) was mixed with 0.02 ml of citrate buffer at various citrate ion concentration (25 to 175 mM) and phosphatase activity was measured using the standard assay protocol. Maximum activity was assumed as 100% activity for comparing the activity with the other concentration of buffer. The experiments were performed in triplicate and expressed as mean ± SD.

Dependency of Temperature for Phosphatase Activity and Stability

The optimum temperature for phosphatase was determined by measuring the activity at various temperatures (30–70°C) using standard assay. To determine the thermal stability, the purified enzyme was performed incubated at 30–70°C. The purified enzyme (0.2 U/ml) with 100 mM sodium acetate buffer was taken in separate tubes and kept at respective temperatures for 6 h. From each tube, an aliquot of 0.1 ml was withdrawn and cooled by immersing in ice, it was assayed immediately for residual peroxidase activity and compared with appropriate control. The experiments were done in triplicate and expressed as mean ± SD.

Determination of Kinetic Parameters

Activity of phosphatase at varying concentrations of *p*-nitrophenyl phosphate was determined and K_m, V_{max} values were calculated from Lineweaver-Burk plot.

Effect of Inhibitors on Phosphatase Activity

The effect of peroxidase inhibitors (sodium cyanide, sodium azide) and phosphatase inhibitors (sodium arsenate, sodium potassium tartrate and sodium acetate) on phosphatase activity was measured in standard assay method. The TCWP isoenzyme (0.2 U/ml) was incubated with peroxidase inhibitor at 1 to 5 mM and phosphatase inhibitor at 2 to 10 mM with appropriate buffer for 6 h. The incubation mixture was assayed for residual enzyme activity and compared with appropriate control. The experiments were done in triplicate and expressed as mean ± SD.

RESULTS AND DISCUSSION

Dependence of pH for Phosphatase Activity

pH optimum and stability for TCWP isoenzymes to catalyze the hydrolysis of *p*-nitrophenyl phosphate was determined. As shown in Figure 1a, the pH optima for $TCWP_1$ and $TCWP_2$ were 5.0 and 5.5, respectively. Both isoenzymes have shown ~ 40% activity between 4.5 and 5.5 for $TCWP_1$ and between 5.5 and 6.0 for $TCWP_2$. The phosphatase activity of chloroperoxidase from *Serratia marcescens* also catalyzes the hydrolysis of *p*-nitrophenyl phosphate at pH between 5.2 and 6.1 (Preobrazhenskaya *et al.,* 2003). In Figure 1b, $TCWP_1$ showed 60% phosphatase activity between pH 4.5 and 6.0, where as $TCWP_2$ showed 50% phosphatase activity between pH 4.0 and 6.0. These results reveal that the $TCWP_1$ are more stable than $TCWP_2$. It is similar to peroxidase activity of TCWP isoenzymes in the earlier chapters results.

Effect of Buffer Ion Concentration on Phosphatase Activity

Since the peroxidase activity depends on the buffer ion concentration (Sakharov *et al.,* 2001, Leon *et al.,* 2002), we have determined the optimum buffer ion concentration for TCWP isoenzymes to catalyze the hydrolysis of *p*-nitrophenyl phosphate. As shown in Figure 2, the optimum buffer ion concentration for $TCWP_1$ and $TCWP_2$

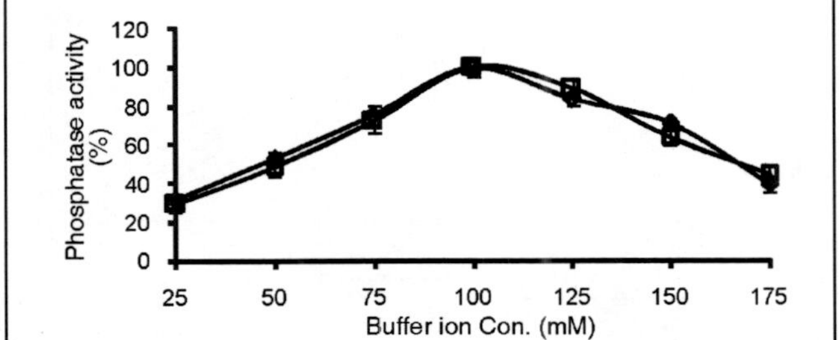

Fig. 2: Effect of Buffer Ion Concentration on Phosphatase Activity of $TCWP_1$ (◇) and $TCWP_2$ (□)

was 100 mM respectively. The enzyme activity was observed to be 75% at the concentrations of 75 mM and 150 mM. Hence decrease or increase of citrate ion influence the enzyme activity in a larger extends. Similar results have been reported by Renirie *et al.,* 2000. Where 100 mM citrate was optimum for phosphatase activity of chloroperoxidase isolated from fungus *Curvularia inaequalis* (Messerschmidt *et al.,* 1997).

Effect of Temperature on Phosphatase Activity

The optimum temperature of TCWP for phosphatase activity isoenzymes is given in Figure 3a. The isoenzymes required 40°C and 45°C for $TCWP_1$ and $TCWP_2$ as optimum temperature for phosphatase activity. The $TCWP_1$ maintain 50% of phosphatase activity between 35°C and 45°C, where as $TCWP_2$ maintain 50% activity between 40°C and 50°C. It is contradictory to the temperature optimum (range of 55–60°C) of chloroperoxidase from *Serratia marcescens* (Preobrazhenskaya *et al.,* 2003).

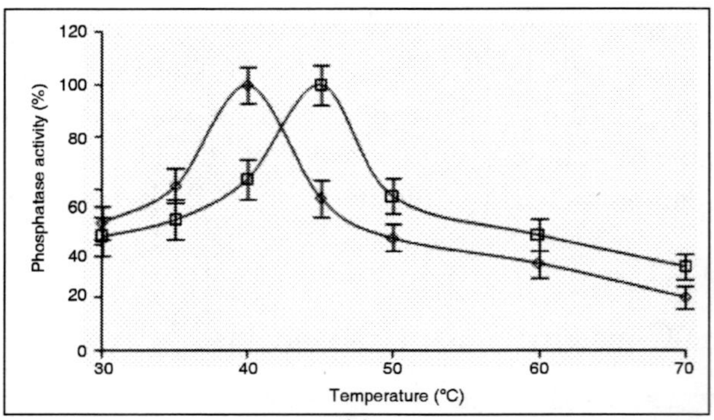

(a)

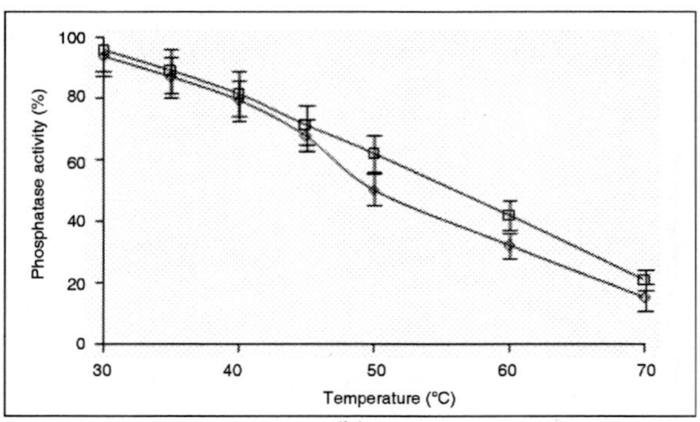

(b)

Fig. 3: Effect of Temperature on Activities (a) and Stability (b) of $TCWP_1$ (◇) and $TCWP_2$ (□)

Higher temperatures possibly affect the phosphatase activity, by changes in its conformation. The thermal stability of TCWP isoenzymes for phosphatase activity is given in Figure 3b. The $TCWP_1$ and $TCWP_2$ showed 50% and 62% of residual phosphatase activity at 50°C respectively. However both the isoenzymes lost its stability while increasing the temperature from 50°C to 70°C and retain only 15% and 21% of residual phosphates activity for $TCWP_1$ and $TCWP_2$ respectively.

Michaelis Constant for Phosphatase Activity of TCWP Isoenzymes

The Michaelis constants: K_m and V_{max} were calculated using Lineweaver-Burk (Lineweaver and Burk, 1934) reciprocal plot (Figure 4). The K_m and V_{max} value (Figure 4a) for $TCWP_1$was 1.44 mM and 0.43 mmole min^{-1} ml^{-1} respectively. The K_m and V_{max} value (Figure 4b) for $TCWP_2$ was 1.60 mM and 0.5 mmole min^{-1} ml^{-1} respectively. The lowest K_m value shows the higher specificity of $TCWP_1$ towards *p*-nitrophenyl phosphate for phosphatase activity than $TCWP_2$. But there is not much difference in their V_{max} values. The K_m value obtained for TCWP isoenzymes was similar to K_m value obtained from the chloroperoxidase of *Serratia marcescens* (Preobrazhenskaya *et al.*, 2003). Normally, K_m value for phosphatases is 0.1–0.2 mM (Hemrika *et al.*, 1997a), this rather large value suggests that the phosphatase activity of our TCWP is an additional activity.

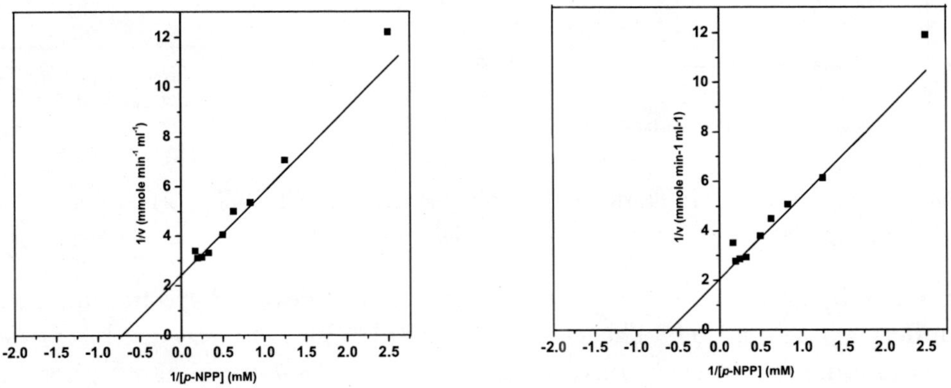

(a) (b)

Fig. 4: Determination of Kinetic Parameters of $TCWP_1$ (a) and $TCWP_2$ (b)

Effect of Inhibitors on Phosphatase Activity

The effect of phosphatase specific inhibitors like arsenate, tartrate and acetate together with peroxidase specific inhibitors like sodium cyanide and sodium azide were studied against phosphatase activity (Figure 5). The peroxidase specific inhibitor sodium cyanide at 5 mM concentration completely inhibited the phosphatase activity of both the isoenzymes (Figure 5a). Sodium azide also inhibited the phosphatase activity completely at 5 mM (Figure 5b), like peroxidase activity of TCWP isoenzymes.

The inhibition patterns of phosphatase inhibitors on TCWP isoenzymes are given in Figure 6. The TCWP isoenzymes retained only 50%, ~7% residual phosphatase activity at 6 mM and 10 mM of arsenate respectively. However, the tartrate inhibited ~55%, ~98% of phosphatase activity at 4 mM and 10 mM concentration for both the isoenzymes. But the tartrate has less effect on chloroperoxidase from *Serratia marcescens* (Preobrazhenskaya *et al.,* 2003) and maintained 90% of residual phosphatase activity at 20 mM concentration. In contrast to other phosphatase inhibitor, the acetate could not inhibit phosphatase activity even at 10 mM concentration. Both the isoenzymes retained 91% of residual phosphatase activity, but the acetate enhanced the peroxidase activity of TCWP isoenzymes.

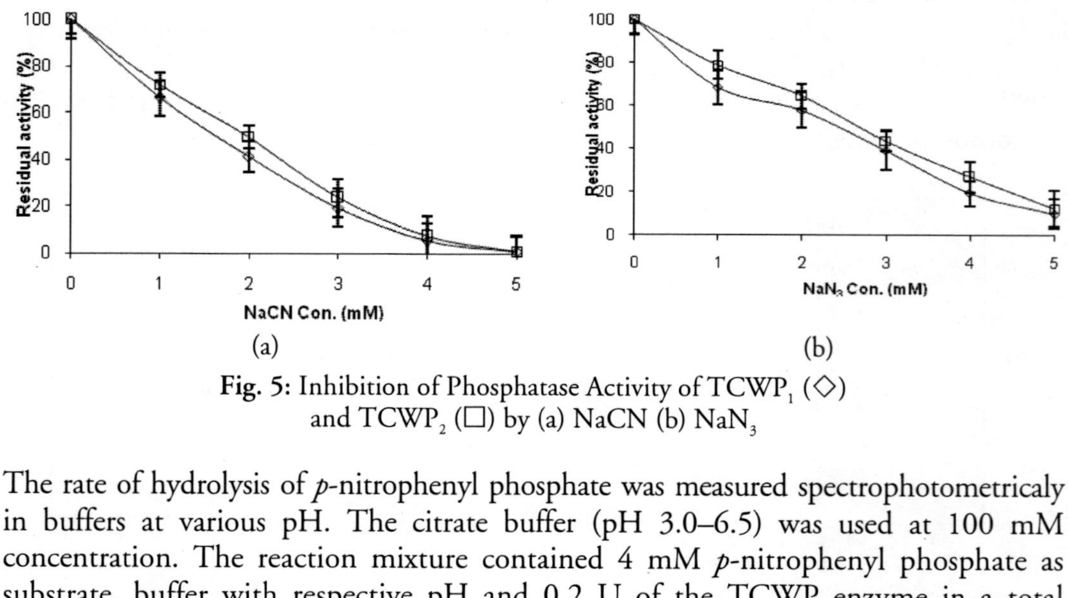

Fig. 5: Inhibition of Phosphatase Activity of $TCWP_1$ (◇) and $TCWP_2$ (□) by (a) NaCN (b) NaN_3

The rate of hydrolysis of *p*-nitrophenyl phosphate was measured spectrophotometricaly in buffers at various pH. The citrate buffer (pH 3.0–6.5) was used at 100 mM concentration. The reaction mixture contained 4 mM *p*-nitrophenyl phosphate as substrate, buffer with respective pH and 0.2 U of the TCWP enzyme in a total volume of 1 ml. The production of *p*-nitrophenol from *p*-nitrophenyl phosphate was measured at 410 nm in triplicate. (a) The phosphatase activity was measured at the indicated pH. (b) Before the determination of phosphatase activity for stability, the TCWP isoenzymes were pre-incubated at the indicated pH for 6 h. Each point represents the mean±SD of the three independent experiments.

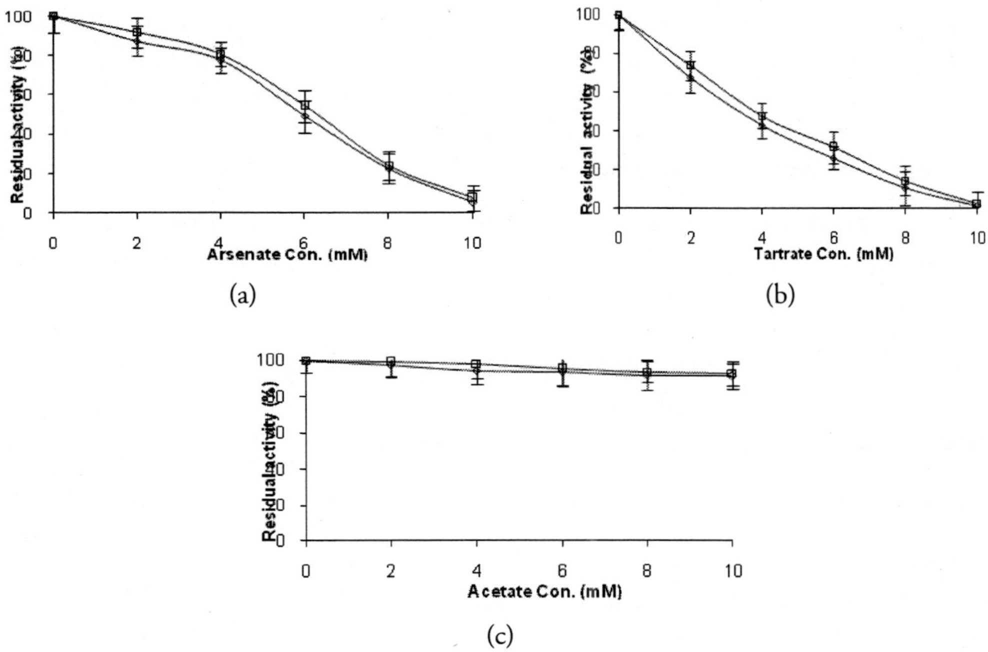

Fig. 6: Inhibition of Phosphatase Activity of $TCWP_1$ (◇) and $TCWP_2$ (□) by (a) Arsenate (b) Tartrate and (c) Phosphate

The rate of hydrolysis of *p*-nitrophenyl phosphate was measured spectrophotometricaly in buffers of differing ion concentrations. The reaction mixture contained 4 mM *p*-nitrophenyl phosphate as substrate, citrate buffer (pH 4.5 for $TCWP_1$; pH 5.0 for $TCWP_2$) at 25–175 mM concentration and 0.2 U of the TCWP enzyme in a total volume of 1 ml. The production of *p*-nitrophenol from *p*-nitrophenyl phosphate was measured at 410 nm. Each point represents the mean±SD of the three independent experiments.

(a) The reaction mixture contained 4 mM *p*-nitrophenyl phosphate as substrate in respective pH and 0.2 U of TCWP enzyme in a total volume of 1 ml. The enzyme reaction was carried out at the indicated temperature. (b) Before the determination of phosphatase activity for stability, the TCWP isoenzymes were incubated at the indicated temperatures for 6 h. The enzyme reaction was carried out at 30°C as described above. The activity was assayed in term of *p*-nitrophenol from *p*-nitrophenyl phosphate at 410 nm. Each point represents the mean±SD of the three independent experiments.

Lineweaver Burk plot for *p*-nitrophenyl phosphate hydrolysis by the TCWP isoenzymes: Assays were carried out at 0.4–6 mM *p*-nitrophenyl phosphate, 100 mM citrate buffer (pH 5.0 for $TCWP_1$; pH 5.5 for $TCWP_2$) at 30°C.

The phosphatase activity was determined using the reaction mixture containing 4 mM *p*-nitrophenyl phosphate as substrate in respective pH, inhibitor at 1–5 mM and 0.2 U of the TCWP isoenzyme in a total volume of 1 ml. The reaction mixture without any inhibitor was considered as 100% activity. Each point represents the mean±SD of the three independent experiments.

The phosphatase activity was determined using the reaction mixture containing 4 mM *p*-nitrophenyl phosphate as substrate in respective pH, inhibitor at 2–10 mM and 0.2 U of the TCWP isoenzyme in a total volume of 1 ml. The reaction mixture without any inhibitor was considered as 100% activity. Each point represents the mean±SD of the three independent experiments.

REFERENCES

Arber, J.M., De Boer, E., Garner, C.D., Hasnain, S.S. and Wever, R. (1989). Electron paramagnetic resonance studies on conformational states and metal ion exchange properties of vanadium bromoperoxidase. *Biochemistry* 28: 7968–7973.

Bantleon, R., Altenbuchner, J. and Van Pee, K.H. (1994). Chloroperoxidase from *Streptomyces lividans*: isolation and characterization of the enzyme and the corresponding gene. *J. Bacteriol,* 176: 2339–2347.

Brindley, D.N. and Waggoner, D.W. (1998). Mammalian lipid phosphate phosphohydrolases. *J. Biol. Chem.* 273: 24281–24284.

Carman, G.M. (1997). Phosphatidate phosphatases and diacylglycerol pyrophosphate phosphatases in *Saccharomyces cerevisiae* and *Escherichia coli. Biochim. Biophys. Acta* 1348: 45–55.

Countaway, J.L., Wadell, I.D., Burchell, A. and Arion, W.J. (1988). The phosphohydrolase component of the hepatic microsomal glucose-6-phosphatase system is a 36.5-kilodalton polypeptide. *J. Biol. Chem.* 263: 2673–2678.

De Boer, E., Boon, K. and Wever, R. (1988). Electron paramagnetic resonance studies on conformational states and metal ion exchange properties of vanadium bromoperoxidase. *Biochemistry* 27: 1629–1635.

Feldman, F. and Butler, L.G. (1972). Protein-bound phosphoryl histidine: a probable intermediate in the microsomal glucose-6-phosphatase-inorganic pyrophosphatase reaction. *Biochim. Biophys. Acta.* 268: 698–710.

Hager, L.P., Morris, D.R., Brown, F.S. and Eberwein, H. (1966). Chloroperoxidase II. Utilization of halogen anions. *J. Biol. Chem.* 241: 1769–1777.

Hemrika, W. and Wever, R. (1997). A new model for the membrane topology of glucose-6-phosphatase: the enzyme involved in von Gierke disease. *FEBS Lett.* 409: 317–319.

Hemrika, W., Renirie, R. and Wever, R. (2000). A member of the newly identified HPP superfamily that consists of histidine phosphatases and vanadium-containing peroxidases and consequences for membrane topology, active site and reaction mechanism. In: Guido Zimmer, ed. Membrane structure in disease and drug therapy. New York: Marcel Dekker, Inc. 303–333.

Hemrika, W., Renirie, R., Dekker, H.L., Barnett, P. and Wever, R. (1997). From phosphatases to vanadium peroxidases: A similar architecture of the active site. *Proc. Natl. Acad. Sci. USA* 94: 2145–2149.

Ishikawa, K., Mihara, Y., Gondoh, K., Suzuki, E.-I. and Asano, Y. (2000). X-ray structures of a novel acid phosphatase from *Escherichia blattae* and its complex with the transition-state analog molybdate. *EMBO J.* 19: 2412–2423.

Kanoh, H., Kai, M. and Wada, I. (1997). Phosphatidic acid phosphatase from mammalian tissues: discovery of channel-like proteins with unexpected functions. *Biochim. Biophys. Acta* 1348: 56–62.

Leon, C.J., Alpeeva, I.S., Chubar, T.A., Galaev, I.Y., Csoregi, E. and Sakharov, I.Y. (2002). Purification and substrate specificity of peroxidase from sweet potato tubers. *Plant Sci.* 163: 1011–1019.

Lineweaver, H. and Burk, D. (1934). The determination of enzyme dissociation constants. *J. Am. Chem. Soc.* 56: 658–666.

Liu, T.N., M'Timkulu, T., Geigert, J., Wolf, B., Neidleman, S.L., Silva, D. and Hunter-Cevera, J.C. (1987). Isolation and characterization of a novel non heme chloroperoxidase. *Biochem. Biophys. Res. Commun.* 142: 329–333.

Messerschmidt, A. and Wever, R. (1996). X-ray structure of a vanadium-containing enzyme: chloroperoxidase from the fungus *Curvularia inaequalis*. *Proc. Natl. Acad. Sci.* USA. 93: 392–396.

Messerschmidt, A., Prade, L. and Wever, R. (1997). Implications for the catalytic mechanism of the vanadium-containing enzyme chloroperoxidase from the fungus *Curvularia inaequalis* by x-ray structures of the native and peroxide form. *Biol. Chem.* 378: 309–317.

Pichorner, H., Couperus, A., Korori, S.A.A. and Ebermann, R. (1992). Plant peroxidase has a thiol oxidase function. *Phytochemistry* 31: 3371–3376.

Plass, W. (1999). Phosphate and Vanadate in Biological Systems: Chemical Relatives or More? *Angew. Chem.* 38: 909–912.

Preobrazhenskaya, Yu. V., Voskoboev, A.I. and Burd, V.N. (2003). Hydrolysis of Phosphoester Bond by the Heme-Independent Chloroperoxidase from *Serratia marcescens*. *Russian J. Bioorg. Chem.* 29: 556–559.

Renirie, R., Hemrika, W. and Wever, R. (2000). Peroxidase and phosphatase activity of active-site mutants of vanadium chloroperoxidase from the fungus *Curvularia inaequalis:* Implications for the catalytic mechanisms, *J. Biol. Chem.* 275: 11650–11657.

Sakharov, I.Y., Vesga, M.K., Galaev, I.Y., Sakharova, I.V. and Pletjushkina, O.Y. (2001). Peroxidase from leaves of royal palm tree *Roystonea regia*: purification and some properties. *Plant Sci.* 161: 853–860.

Singh, J., Nordlie, R.C. and Jorgenson, R.A. (1981). Vanadate: a potent inhibitor of multifunctional glucose-6-phosphatase. *Biochim. Biophys. Acta* 678: 477–482.

Stukey, J. and Carman, G.M. (1997). Identification of a novel phosphatase sequence motif. *Protein Sci.* 6: 469–472.

Tanaka, N., Dumay, V., Liao, Q., Lange, A.J. and Wever, R. (2002). Bromoperoxidase activity of vanadate-substituted acid phosphatases from *Shigella flexneri* and *Salmonella enterica* ser. typhimurium. *Eur. J. Biochem.* 269: 2162–2167.

Tanaka, N., Hasan, Z., Hartog, A.F., van Herk, T. and Wever, R. (2003). Phosphorylation and dephosphorylation of polyhydroxy compounds by class A bacterial acid phosphatases. *Org. Biomol. Chem.* 1: 2833–2839.

Torriani, A.M. (1960). Influence of inorganic phosphate in the formation of phosphatases by Escherichla coli. *Biochim. Biophys. Acta* 38: 460–469.

Van Schijndel, J.W.P.M., Barnett, P., Roelse, J., Vollenbroek, E.G.M. and Wever, R. (1994). The Stability and Steady-State Kinetics of Vanadium Chloroperoxidase from the Fungus *Curvularia Inaequalis. Eur. J. Biochem.* 225: 151–157.

Vollenbroek, E.G.M., Simons, L.H., Van Schijndel, J.W.P.M., Barnett, P., Balzar, M., Dekker, H.L., Van der Linden, C. and Wever, R. (1995). Vanadium chloroperoxidases occur widely in nature. *Biochem. Soc. Trans.* 23: 267–271.

Wiesner, W., Van Pee, K.H. and Lingens, F. (1988). Purification and characterization of a novel bacterial non-heme chloroperoxidase from *Pseudomonas pyrrocinia. J. Biol. Chem.* 263: 13725–13732.

Zeng, J. and Fenna, R.E. (1992). X-ray crystal structure of canine myeloperoxidase at 3Å resolution. *J. Mol. Biol.* 226: 185–207.

Zhang, Q.X., Pilquil, C.S., Dewald, J., Berthiaume, L.B. and Brindley, D.N. (2000). Identification of structurally important domains of lipid phosphate phosphatase-1: implications for its sites of action. *Biochem. J.* 345: 181–184.

Treatment of Synthetic Effluent Containing Parachlorophenol by Combined Electrochemical and Microbial Methods

V. Manivasagan*, C. Ahmed Basha[1], T. Kannadasan[2], N.G. Ramesh Babu, K. Saranya and M. Ramalingam

Dept. of Biotechnology, Adhiyamaan College of Engineering, Hosur, Tamil Nadu

[1]Dept. of Chemical Engg., Adhiyamaan College of Engineering, Hosur, Tamil Nadu

[2]Department of Chemical Engineering, Coimbatore Institute of Technology, Coimbatore, Tamil Nadu

E-mail: manivasaganv@gmail.com

INRODUCTION

Chlorinated phenolic compounds present in wastewaters of many chemical industries such as polymeric resin production, oil refining, iron-steel, petroleum, pesticide, paint, solvent, pharmaceuticals, wood preserving chemicals, coke-oven, paper and pulp, and industries have become of environmental concern due to their toxicity and resistance to degradation. The recalcitrant structure of Chlorophenols results from the carbon-halogen bond, which is cleaved with great difficulty and since their structure is highly stable, they accumulate in the environment (Bai *et al.*, 2007).

Several physical, chemical and biological methods including activated carbon adsorption, ion exchange, solvent extraction, incineration, chemical oxidation and biological degradation have been used for removing chlorophenols from waste water. Adsorbed organic compounds require further treatment by chemical and/or biological oxidation for complete mineralization. Chemical oxidation methods are fast, but expensive and may result in formation of undesirable by products (Llaria Galliza *et al.*, 2003).

*Corresponding author.

A variety of biological treatment processes, aerobic as well as anaerobic, such as facultative basin, aerated stabilization basin, aerated lagoon system, decanted aerated reactors, fluidized bed bioreactors and up flow anaerobic sludge blanket have been used for treatment of wastewater containing chlorophenols (Shpiner *et al.,* 2009) Biodegradation of chlorophenols by aerobic or anaerobic treatment is more specific and relatively inexpensive. Different fungi and aerobic bacteria such as *Phanerochaete chrysosporium, Herbaspirillum chlorophenolicum, Pseudomonas, Azotobacter, Alcaligenes and Acinetobacter* species have been used for the degradation of chlorophenols

Focus in wastewater treatment has recently shifted to removal of toxic compounds. Despite the xenobiotic nature of these compounds, many are at least partially degradable by microorganisms, particularly if the microorganisms have been acclimated to a given wastewater. There are no systematic studies in literature on combination of electro oxidation and biological oxidation of para-chlorophenol (4-CP) containing wastewater in a range of feed 4-CP concentrations for COD, 4-CP and toxicity removals. Therefore, the major objective of this study is to investigate the electro oxidation and biological oxidation of 4-CP containing synthetic wastewater (Yunus Pamukoglu and Fikret Kargi, 2008) in the present study, parachlorophenol was degraded biologically and the chemical oxygen demand was analysed. Further, bacterial strain *Pseudomonas putida* and fungal strain *Penicillium chrysogenum* were used to study the comparative degradation of parachlorophenol under aerobic condition.

MATERIALS METHODS

Electrochemical Oxidation

Electrochemical Oxidation was carried out in an electrochemical batch reactor under galvanostatic condition at current density 1.5 A/dm^2 using RuO_2 coated with titanium mesh (Ti/RuOx-TiOx) as anode and stainless steel as cathode. Synthetic effluent containing Parachlorophenol was subjected to electrolysis.

NaCl was added in the effluents which act as a supporting electrolyte. The electric circuit consisted of a regulated DC power supply connected to the reactor. The electrolyte was continuously stirred during the electrolysis using a magnetic stirrer. Electrochemical oxidation process was carried out for an electrolysis period of two hrs and sample was collected every one hour to estimate chemical oxygen demand (COD).

Microbial Degradation

All the biodegradation experiments are carried out at room temperature in batch mode in 250 mL Erlenmeyer flasks and kept in rotary shaker for aerobic oxidation.

Microbes are consecutively transferred into the nutrient medium with increasing concentrations of effluent. During the investigation, nutrient broth concentration is decreased from 90% (w/v) to 0% (w/v) and finally the organism is provided with effluent as sole source of nutrient. The effluent was taken after every additional hour of electrolytic degradation and then it was subjected to microbial degradation. Effluents that were collected at different time intervals of electrochemical pretreatment, were continued for microbial oxidation. Then microbes *Pseudomonas putida* and *Penicillium chrysogenum* were added at different concentrations to the effluent. The sample was collected every 24 hours and it was subjected to COD analysis. This experiment was continued for five days. Every day the samples were collected for observation bacterial and fungal growth. The observations are presented in tables.

Photo Catalytic Oxidation

The experimental setup consists of a 500 mL glass beaker that contains effluent with photo catalysts (0.50 g/L of TiO_2), placed on a magnetic stirrer. This setup is kept inside a reflecting mirror box. A 125 W, medium pressure mercury lamp is provided on top of the mirror box, which serves as the light source. Effluent with a photo catalyst is continuously stirred by a magnetic stirrer for complete mixing of reaction mixture maintaining the temperature of the content at 40°C. In order to remove the microbes present in the effluent from post electro oxidation process, it is subjected to photo catalytic oxidation using 0.5 g/L of TiO_2 as catalyst. In this process, 200 mL of sequentially treated effluent sample is taken in a 500 mL beaker and subjected to irradiation for 5 h. During this period continuous stirring is done to expose the effluent for UV radiation. Samples are collected at regular intervals, filtered through Millipore membrane filters, and centrifuged to remove the catalyst particles prior to analysis. Samples are collected periodically and subjected to COD measurement. After the Photo catalytic oxidation process, 0.1 mL of sample is taken and subjected to spread plate analysis to ensure the absence of microbes.

Analysis of COD

COD of the samples was determined using the dichromatic open reflux method strictly following the APHA. In order to determine the extent of degradation of the effluent, chemical oxygen demand (COD) was measured. The organic matter of the sample is oxidized with a known excess of potassium dichromate in a 50% sulphuric acid solution. The excess dichromate is titrated with a standard solution of ferrous ammonium sulfate solution. Experiments were repeated until the difference found was less than 3%.

Determination of BI

The biodegradability index (BI) was defined as the ratio of BOD to COD. The value ranged from zero to one. Morais and Zamora, 2005 reported that samples of biodegradability index smaller than 0.3 were not appropriate for biological degradation. According to Chamarro *et al.*, 2001 for complete biodegradation, the effluent must present a bio-degradability index of at least 0.4.

RESULTS

Electrochemical Pretreatment

The mechanism of electrochemical oxidation of wastewater is a complex phenomenon and the electrochemical indirect oxidation of organic contaminants on noble oxide coated catalytic anode occurs. The initial effluent has a noticeably low BI value. The samples with biodegradability index smaller than 0.3 are not appropriate for biological degradation. According to Chamarro *et al.*, for complete biodegradation, the effluent must present a biodegradability index of at least 0.4. Hence electro pretreated effluent with BI of 0.4 was suitable for microbial oxidation. The greater % COD reduction in microbial oxidation step could be obtained by increasing the degradation time in pre-electrochemical step by the way of enhancing the biodegradability of selected effluent.

Microbial Oxidation

Monod kinetics constants (substrate utilization, C = μm X/Y, substrate affinity constant, ks) are evaluated from the % of COD removal with time for bacterial strain (*P. putida*) and fungal strain (*P. chrysogenum*) and their regression coefficients are also presented in Table 2. High regression coefficients (mostly greater than 0.9) are found for these studies. This suggests that two parameters Monod kinetics model is suitable for describing the kinetic model for the removal of COD by biochemical oxidation. Monod kinetics model parameters found for the bacterial strain *P. putida* (S0 = 2400 mg/L, ks = 2474.7 mg/L, C = 3.217 mg/L h, R2 = 0.933). The fungal strain *P. chrysogenum* (S0 = 1510 mg/L, ks = 1484.7 mg/L, C = 0.593 mg/L h, R2 = 0.948) which performed better in % COD reduction when compared to all other strains. It is to be noted from Table 2, at the end of 2 h of pre electro oxidation, COD of the initial effluent is reduced and while on addition of 20 mL of inoculum (broth containing microbial strain) to 180 mL of effluent the COD value is increased in the entire stream due to the presence of biomass and its growth.

In case of the bacterial strain, *P. putida,* the COD value increases from 1347 mg/L to 2400 mg/L due to addition of broth and due to 5 days of biochemical oxidation, for

Table 1: Performance of Combined Process Treatment of Industrial Effluent Containing Parachlorophenol

Parameters and Operating Conditions	*Units*	*Pollutant—Parachlorophenol*	
Initial COD	mg/L	**2060**	
Electrochemical Pre Treatment			
Batch Reactor holdup, VR	mL	400	
Charge input, Q	Ah/L	3.75	
COD (after 2 h of Electro Oxidation)	mg/L	1347	
BI		0.4	
Biochemical Oxidation			
Microbes		***P. putida***	***P. chrysogenum***
Effluent Volume	mL	180	180
Inoculums Volume	mL	20	20
Batch Reactor holdup, VR	mL	200	200
COD (after inoculum addition)	mg/L	2400	1510
COD (after 5 days of aerobic oxidation)	mg/L	1294	1087.2
Electrochemical Post Treatment			
Batch Reactor holdup, VR	mL	200	200
Charge input, Q	Ah/L	37.5	37.5
COD (after 5h of Electro Oxidation)	mg/L	720.9	700.7
Photo Catalytic Oxidation			
Batch Reactor holdup, VR	mL	200	200
COD (after 5h of Photo Catalytic Oxidation)	mg/L	340.3	401.8
Overall COD Reduction (%)		83.5	80.5

P. putida COD of the effluent reduced from 2400 mg/L to 1294 mg/L where as for the fungal strain *P. chrysogenum* is considered, the COD value increases from 1347 mg/L to 1510 mg/L due to growth of biomass and after 5 days of biochemical oxidation reduced from 1510 mg/L to 1087.2 mg/L. Increased biomass concentration is responsible for effective degradation. The biodegradation of Industrial effluent containing parachlorophenol was carried out using bacterial and fungi, and the results showed that the effluent was effectively assimilated by the organism. Generally, the medium employed during the degradation of parachlorophenol used to contain necessary nutrients for the growth of the organism along with the parachlorophenol whereas in the present study, major quantity of electrochemically treated parachlorophenol solution itself was used as a medium. The total system was now having enough carbon and nitrogen compounds for the organism to consume as a nutrient. The acquaintance of organism towards the different environments when they were exploited to such was well described by Darwin as certain variations were natural when the organism struggled for its endurance under stringent fight for the food.

Table 2: Evaluated Parameters for Combined Process Treatment of Industrial Effluent Containing Parachlorophenol

Parameters And Operating Conditions	*Units*	*Values*	
Biochemical Oxidation			
Microbes		*P. putida*	*P. chrysogenum*
Half velocity constant, ks	mg/L	2474.7	1484.7
Maximum rate of substrate utilization, C	mg/Lh	3.217	0.593
Regression coefficients, R2		0.933	0.948
Electrochemical Post Treatment			
Mass transfer Coefficient, $kLx10^3$	cm/s	0.338	0.254
Specific Electrode area, a	1/cm	0.096	0.096
Regression coefficients, R2		0.9909	0.9958
Photo Catalytic Oxidation			
ko, $x10^2$, (constant for holes)	1/min	0.2	0.1
K2 $x10^3$, (constant for holes and electron)	L/mg	0.5	0.4
R2		0.951	0.970

Post Electrochemical Oxidation

The average value of mass coefficient (kL) is computed and presented in Table 2. The kinetics model parameters found to be better for the bacterial strain *P. putida* (kL = 0.388×10^{-3} cm/s, R2 = 0.9909). For the fungal strain *P. chrysogenum* (kL = 0.254×10^{-3} cm/s, R2 = 0.9958) in % COD reduction when compared to all other strains. High regression coefficients are found for the post treatment studies suggest that the kinetic model is suitable for describing the post-electrochemical treatment.

Photo Catalytic Oxidation

The plot of ln([COD]/[COD]0)/([COD]−[COD]0) vs. t/([COD] − [COD]0) will be linear. The constant K2 can be computed from the intercept, and k0 is the slope. K2 and k0 refers the contribution of hole alone and hole plus electron towards degradation. Kinetics constants of photo catalytic oxidation K2 and K0 are evaluated from the % of COD removal with time for the streams of bacterial strain (*P. putida*) and fungal strain (*P. chrysogenum*) and their regression coefficients are presented in Table 2. High regression coefficients (mostly greater than 0.99) are found for these studies. This suggests that two parameters of the kinetic model are suitable for describing the model for the removal of COD by photo catalytic oxidation in the studied concentration. The kinetic model parameters (see Table 2) of the 2 bacterial strains *P. putida* ([COD]0 = 83.5 mg/L, k0 = 0.2 × 10^{-3} min^{-1}, K2 = 0.5 × 10^{-3} L/mg, R2 = 0.951) and of the fungal strain *P. chrysogenum* ([COD]0 = 80.5 mg/L, k0 = 0.1 × 10^{-3} min^{-1}, K2 = 0.4 × 10^{-3} L/mg, R2 = 0.970) From this result bacterial strain has performed well for the degradation of effluent containing parachlorophenol. The % of COD reduction (84%) is found to be better in the end of 5 h of photo oxidation. Where the fungal strain *P. chrysogenum* is considered, the COD value reduced (90.4%) due to Photo catalytic oxidation. The samples collected from various streams are incubated for one day for the samples having bacterial strain and 3 days for the samples having fungal strains for microbial growth and subjected to spread plate analysis. The absence of microbes is ensured for the samples withdrawn after 5 h irradiation for all strains. Hence, it is clear that the treated effluent is brought back to reusable condition by 5 h of irradiation. The overall COD reductions obtained at end of the sequential study for the different effluent streams containing parachlorophenol are 83.8% to the effluent streams containing bacterial strains *P. putida*. Further the reductions 90.4% in COD obtained to the effluent streams containing fungal strain *P. chrysogenum.*

REFERENCES

Ahmed Bashaa, C., Selvakumar, K.V., Prabhu, H.J., Sivashanmugam, P. and Chang (2011). Degradation studies for textile reactive dye by combined electrochemical, microbial and Photo catalytic methods. *Sep. Purif. Technol.* 79: 303–309.

Aldobose, H.I., Bianchi, D., Nunzio Andriollo, Dante Cidaria and Pedrocesti (1992). Microbial oxidation of naphthalene to cis-1, 2-dihydroxy-1, 1 dihydronapthalene in a membrane bioreactor. *J. Chem. Technol. Biotechnol,* 66: 375–381.

Arutchelvan, V., Kanakasabai, V., Elangovan, R., Nagarajan, S and Muralikrishnan, V. (2006). Kinetics of high strength phenol degradation using Bacillus brevis. *J. of Hazard. Mater.* B129: 216–222.

Baban, A., Yedilar, A., Lienert, D., Kemerder, N and Kettrup, A. (2003). Ozonation of high strength segregated effluents from a woollen textile dyeing and finishing plant. *Dyes and Pigments* 58: 93–8.

Bai, J., Wen, J.P., Li, H.M. and Jiang, Y. (2007). Kinetic modelling of growth and biodegradation of phenol and m-cresol using Alcaligenes faecalis. *Process Biochem,* 42: 510–517.

Bajaj, M., Gallert, C and Winte, J. (2008). Effect of co-substrates on aerobic phenol degradation by acclimatized and non-acclimatized enrichment cultures. *Eng. Life Sci.* 8: 125–131.

Bajaj, M., Gallert, C. and Winter, J. (2009). Phenol degradation kinetics of an aerobicmixed culture. *Biochem. Eng. J.* 46: 205–209.

Chamarro, E., Marco, A. and Esplugas, S. (2001). Use of fenton reagent to improve organic chemical bio-degradability. *Water Res.* 35: 1047–1051.

Eker, S. and Kargi, F. (2006). Kinetic modeling and parameter estimation for an activated sludge unittreating 2, 4-dichlorophenol containing synthetic wastewater. *Environ. Eng. Sci.* 23: 263–271.

Fikret kargi and isil konya (2006). COD, para-Chlorophenol and toxicity removal from para-chlorophenol containing synthetic waste water in an activated sludge unit. *J. Hazard. Mater.* B132: 226–231.

Indumathi paramasivam., Chellammal, S., Ahamed Basha, C. and Raghavan, M. (2001). Electrochemical treatment of starch effluent using Ruo2/ti and Pbo2/ti Electrodes. *J. electrochemistry*, 17(2): 535–538.

Inoue, K.Y., Yoshine, S., Kondo, Y and Hiroyuki, N. (2002). Determination of phenolic xenoestrogens in water by liquid Chromatography with coulometric-array Detection. *J. Chromatogr*, 946: 291–294.

Jothimani, P., Kalaiselvan, G., Bhaskaran, A., Augustine Selvaseelan, D and Ramasamy, K. (2003). Anaerobic biodegradation of aromatic compounds. *Indian J. Exp. Bio.*, 141: 603–611.

Kazuosugaya, Osamu Nakayama, naoyuki Hinata, koichi kamekura, Aklrato and Akira Ohkawa. (2001). Biodegradation of quinoline in crude oil. *J. Chem. Technol. Biotechnol*, 76: 603–611.

Yi Li, Jing Li, Chao Wang and Peifang Wang (2010). Growth kinetics and phenol biodegradation of psychrotrophic Pseudomonas putida LY1. *Bioresour. Technol.*, 101: 6740–6744.

Yunus Pamukoglu and Fikret Kargi (2008). Biodegradation kinetics of 2,4,6 trichlorophenol by Rhodococcus rhodochrous in batch culture. *Enzyme Microb. Technol.*, 43: 43–47.

Extraction and Application of Natural Bioactive Pigments from the Waste Pomace of *Vitis Vinifera* as a Food Colorant

S. Niranjana Kamala Priya*, D. Vaishnu Devi, J. Rajani Sowparnika and R. Arulvel

Department of Biotechnology, K.S. Rangasamy College of Technology, Tiruchengode, Namakkal (D.T), Tamil Nadu

E-mail: niranjana.saravanan4@gmail.com

INTRODUCTION

Food colorants are the substances which are applied in the edible food stuffs to make them more interesting. Colorants are also used in drugs, in order to differentiate them. But generally the colorants used in most of the food stuffs are of artificial in nature which can alter the nature of the food substances and can cause serious health defects. As per FDA some additives like sulfites, monosodium glutamate (MSG) can cause headaches, nausea, and weakness, difficulty in breathing, drowsiness, rapid heartbeat, and chest pain. Instead of using those harmful artificial dyes, we can switch to the natural dyes which are harmless, eco friendly and also ads up the nutritive value of the edible items.

A variety of natural pigments can be obtained from all the available plant species. These pigments can be suitably processed and can be applied as a food colorant. One of the potential natural pigment anthocyanins; which can be obtained at a high rate from the common black grapes, possesses high anti-oxidant and anti-microbial properties. Thus it is not needed to add additional preservatives and other food additives in case of anthocyanins. Moreover the preparation of anthocyanin is more

*Corresponding author.

economical since the grape pomaces are thrown wastefully from juice and wine industries.

Anthocyanins are water-soluble vacuolar pigments that may appear red, purple, or blue depending on the pH. They belong to a parent class of molecules called flavonoids synthesized via the phenylpropanoid pathway; they are odorless and nearly flavorless, contributing to taste as a moderately astringent sensation. They readily donate an electron to the unstable free radicals and then make them stable, thus acting as a rich anti-oxidant. Grape and red wine phenolics were found to inhibit the oxidation of human low-density lipoproteins (Frankel, Waterhouse *et al.*, 1995)

However, their use as an added color to food additives and drinks has been limited, due to a number of drawbacks; such as sensitivity to bleaching by sulphur dioxide and limited colouring capability at pH values over 3.5. Overcoming such problems would greatly enhance the possibility of using anthocyanins as food additives, especially since there is a current interest in natural colours with potential added health benefits (Bridle, Timberlake, 1997, Kotamballi *et al.*, 2002)

In general, a variety of solvents are used to extract the anthocyanins from the grape pomace. Instead of using the solvents, some potent enzymes such as xylanase can be used that makes the extract safer to use it for the edible items. Xylanase are the main constituents of the xylanolytic enzyme system, converting xylan (the main hemi cellulosic polysaccharide) into a more readily fermentable form (Ghosh *et al.*, 1993).

Xylanases are produced by numerous microorganisms among which the fungi are the most potent producers (Pham *et al.*, 1997). *Aspergillus niger* has been used for, the production of enzymes such as pectinases (Castilhoa *et al.*, 2000, Debing *et al.*, 2006) and invertase (Montiel-González *et al.*, 2004), xylanase (Qi-peng *et al.*, 2005) and so on. Several commercially available enzymes which are meant for the degradation of the cellulose, hemi cellulose and chitin, can be used but they are costly and makes the process complicated. Thus extraction of anthocyanins by the use of the xylanase enzyme will be simple and more beneficial. Moreover, the pollution which is caused by the waste pomace can be prevented.

MATERIALS AND METHODS

Fungal Enzyme Production

The *Aspergillus niger* will be purchased from MTCC and it is to be maintained at 4°C on potato dextrose agar (PDA). Spores suspensions will be made from six-day-old cultures that had been grown on PDA slopes at 30°C. Sterile distilled water will be aseptically added to each slope and a suspension of the spores made by lightly brushing the mycelium.

Media Composition

NH_4Cl_9 (1gm); KH_2PO_4 (1 gm); $NaNO_3$ (1gm); $MgSO_4 \cdot 7H_2O$ (1gm); $CaCl_2{\cdot}2H_2O$ (0.3 gm) and yeast extract 1.

The microorganism will be cultured in 75 mL of medium in 250 mL Erlenmeyer flasks on a rotary shaker (150 rpm). Shake flasks will be then maintained at 35°C. At the end of fermentation, the mycelium will be separated from the enzyme-containing broth by centrifugation at 10,000 g for 15 min to obtain the crude enzyme.

Xylanase Assay

Xylanase activity will be assayed by using oat spelt xylan (Fluka) as enzyme substrate. The reaction mixture contained 0.5 ml of substrate solution of 0.25 g of xylan will be wetted with 2 ml 95% ethanol and 22.5 ml of distilled water in a beaker and covered with aluminium foil and boiled to dissolve on hot plate at 100°C for 10 min and made up the volume to 25 ml with distilled water. To this solution, 0.1 ml of the culture filtered supernatant crude enzyme and 1ml sodium acetate buffer will be added. The mixture will be incubated at 40°C in water bath with shaking for 15 min. Released reducing sugar will be measured using 3, 5-dinitrosalicylic acid and glucose as standard. Colour will be developed by boiling in water bath for 5 min and read, using spectrophotometer at 540 nm. One unit of activity will be defined as amount of enzyme required to liberate 1μmol of xylose per minute under the assay conditions (Miller *et al.*, 1959).

Enzyme-Assisted Digestion of Grape Pomace

The enzymatic treatment of grape pomace will be based on a process described by Kammerer *et al.*, Frozen grape pomace will be manually separated into seeds and skins using a sieve. After adjusting the pH value and temperature, enzyme preparations will be added and the homogenous mash should be stirred in a water bath for 120 min. Subsequently, enzymes will be inactivated at 90°C for 1 min in a water bath. Immediately after enzyme inactivation the samples should be cooled in an ice bath to avoid thermal degradation of phenolic compounds. Liquid–solid separation will be performed using a Buchner funnel.

Anti-Microbial Assay

The grape extract can be used as a potent anti-bacterial agent, which can be tested against E.coli, *Pseudomonas, Streptococcus aureus* and *Proteus vulgaris*. The antibacterial tests will be carried out by disc diffusion methods using 100ml of suspension containing 10^7 cfu/ml of bacteria spread on nutrient agar (Chedea *et al.*, 2011).

Anti-Oxidant Assays

Hydroxyl Radical Scavenging

Hydroxyl radicals will be generated by hydrogen peroxide, ascorbate and $FeCl_3$, in the presence or the absence of the test compound. The ability of the compound to compete with deoxyribose for scavenging hydroxyl radicals gives the rate constant of the reaction between hydroxyl radicals and the scavenger (Halliwell *et al.*, 1987).

Determination of Antioxidant Activity by DPPH Method

Initially, 0.1 mL of different extracts will be added to 4 ml solution of DPPH in methanol. A control sample containing the same volume of solvent in place of extract will be used to measure the maximum DPPH absorbance. After, the reaction should be allowed to take place in the dark for 30 min, the absorbance at 515 nm will be recorded to determine the concentration of remaining DPPH (Brandwilliams *et al.*, 1995).

Antioxidant Activity by Radical Cation (ABTS)

ABTS radical cation will be produced by reacting 7 mM ABTS solution with 2.45 mM potassium per sulphate and allowing the mixture to stand in the dark at room temperature for 12–16 h before use. The ABTS solution will be diluted with ethanol to an absorbance of 0.70 ± 0.02 at 732 nm. After addition of 100 µl of sample or Trolox standard to 3.9 ml of diluted ABTS solution, absorbance will be measured at 6 min. Results will be expressed as Trolox equivalent antioxidant capacity (Re *et al.*, 1999).

Purification of Anthocyanin by Liquid Extraction

The crude sample can be treated with chloroform to remove chlorophyll, carotenoids and waxy material. To remove the monomers ethyl acetate extraction should be performed. All together, finally the sample can be treated in rotary evaporator to remove the residual organic materials.

Determination of Anthocyanin Content by HPLC

Initially 5 mg of purified proanthocyanidins will be taken in a 2 ml glass vial, and 1ml of phloroglucinol solution will be added. This vial containing the anthocyanin is to be preheated at 50°C in a water bath for 20 min. 40 mM sodium acetate buffer will be added to this preheated vial. 20 μl of this sample is then injected onto an HPLC equipped with a 250 × 4.6 mm, 5 μm Wakosill 5C18 column and a guard column containing the same material. Proanthocyanidins subunits with peak detection at 280 nm at a flow rate of 1.0 ml/min will be separated (Matthews *et al.*, 1997).

RESULTS AND DISCUSSION

From the preliminary tests such as anti-oxidant and anti-microbial assays it was clear that the extract which was obtained from the grape pomace has a great potency to inhibit the microbial growth and also controls the free radicals by stabilizing them. Thus by properly processing them, they can be used for a variety of applications which will be more economical and also helps in waste management. Extraction of anthocyanins by means of enzymes will greatly enable them to use it in food industries and several pharmaceutical industries, thereby increasing their nutritive value and promotes a good health to the human life.

REFERENCES

Bridle, P. and Timberlake, C.F. (1997). Anthocyanins as natural foods colours-selected aspects. *Food Chemistry*, 58: 103–109.

Castilhoa, L.R., Medronho, R.A. and Alves, T.L.M. (2000). Production and extraction of pectinases obtained by solid state fermentation of agroindustrial residues with *Aspergillus niger*. *Bioresour. Technol.*, 71: 45–50.

Chedea, V.S., Echim, C., Braicu, C., Andjelkovic, M., Kefalas, P., Verhe, R., Socaciu, C. 2011. Composition in polyphenols and stability of the aqueous grape seed extract from the Romanian variety "Merlot Recas". *J. Food Biochem.* 35(1): 92–108.

Frankel, E.N., Waterhouse, A.L. and Teissedre, P.L. (1995). Principal phenolic phytochemicals in selected California wines and their antioxidant activity in inhibiting oxidation of human low-density lipoproteins. *J. Agri. Food Chemistry*, 43: 890–894.

Ghosh, M., Amitabha, D., Mishra, A.K. and Nanda, G. (1993). *Aspergillus sydowii* MG 49 is a strong producer of thermostable xylanolytic enzymes.

Kammerer, D., Schieber, A., Stintiging, F.C. and Carle, R. (2002). Effects of enzyme treatment on the stability of phenolic compounds in grape pomace. In Proceeding of the second international congress in pigments in food, Lisbon: 87–90.

Matthews, S., Mila, I., Scalbert, A., Pollet, B., Lapierre, C., Hervé do Penhoat, C.L.M., Rolando, C. and Donnelly, D.M.X. (1997). Method for the estimation of proanthocyanidins based on their acid depolymerization in the presence of nucleophiles. *J. Agric. Food Chem.* 45: 1195–1201.

Miller, G.L. (1959). Use of dinitrosalicylic acid reagent for determination of reducing sugars. *Anal. Chem.* 31: 426–428.

Montiel-González, A.M., Viniegra-González, G., José Fernández, F. and Loera, O. (2004). Effect of water activity on invertase production in solid state fermentation by improved diploid strains of Aspergillus niger. *Process Biochem,* 39: 2085–2090.

Pham, P.L., Taillandier, P., Delmas, M. and Strehaiano, P. (1997). Optimization of a culture medium for xylanase production by Bacillus sp. using statistical experimental designs. *World J. Microbiol. Biotechnol.* 14: 185–190.

Qi-peng, Y., Jian-dong, W., Huai, Z. and Zhong-ming, Q. (2005). Effect of temperature shift on production of xylanase by Aspergillus niger. *Process Biochem.* 40: 3255–3257.

Re, R., Pellegrini, N., Proteggente, A., Pannala, A., Yang, M. and Rice-Evans, C. (1999). Antioxidant activity applying an improved ABTS radical cation decolorization assay. *Free Radic. Biol. Med.*, 26: 1231–1237.

A Response Surface Methodology Based Medium Design for Optimizing Lipase Production by *Aspergillus niger* SKF4 Isolated from Dairy Wastewater

P. Kanmani, S. Karthik, B. Dhivyapriya*, J. Aravind and K. Kumaresan

Department of Biotechnology, Kumaraguru College of Technology, Coimbatore–641049

E-mail: priyakumar90@gmail.com

INTRODUCTION

Lipases (triacylglycerol acylhydrolases E.C. 3.1.1.3) are enzymes that belong to the class of serine hydrolases that catalyse the hydrolysis of esters from glycerol and long chain fatty acids (Jaeger *et al.*, 1998). This reaction occurs at the oil-water interface. Lipases are unique owing to the fact that they are capable of catalysing the reverse reaction, i.e. ester synthesis in water-restricted environments (Jaeger *et al.*, 1999). Other desirable properties of this sought-after enzyme include chemo, regio, and enantio selectivities. These traits make lipases the enzymes of choice for a wide array of industrial applications such as food, pharmaceutical, leather, dairy, chemical, cosmetic, detergent, paper, etc. to name a few (Sangeetha *et al.*, 2011). In addition to the aforesaid industrial applications, lipases also find a place in wastewater treatment (Hasan *et al.*, 2006), where they degrade problem-posing fat and oil contaminants, thus making the wastewater more amenable to further aerobic biological treatments including the activated sludge process.

---

*Corresponding author.

Lipases from microbial sources are easier to work with, when compared to their counterparts from plant and animal sources (Paques *et al.*, 2006, Paques *et al.*, 2008, Gangadhara *et al.*, 2009, Shan *et al.*, 2009). They are more selective, stable and exhibit broad substrate specificity (Dutra *et al.*, 2008, Griebeler *et al.*, 2011). The lipolytic activities of bacteria, fungi and yeast have been extensively explored by several researchers, and of these, fungi are the ideal sources of the enzyme for large-scale industrial production (Carvalho *et al.*, 2005). Several common fungal species such as Aspergillus, Penicillium, Mucor, Rhizopus, Geotrichum, etc have been screened for their lipase activity (Carvalho *et al.*, 2005, Contesini *et al.*, 2010, Freire *et al.*, 1997).

Screening for novel isolates that excrete lipases with desirable properties that cater to application-specific needs is an ongoing process. Once the strains have been isolated and their enzyme activities established, the next logical step would be to optimize the enzyme production and make the process commercially feasible. For this, a number of physico-chemical factors governing enzyme production such as pH, temperature, aeration rate, incubation period, inoculum level and age, carbon, nitrogen sources and trace elements in the medium, their relative proportions, etc have to be optimized (Bussamara *et al.*, 2012).

The traditional way to optimization is to vary one factor at a time (OFAT), while keeping the other factors at a constant level. This method might work when only a limited number of variables influence the process, but becomes laborious and necessitates a daunting number of trials when a large number of variables are involved, such as in lipase production. Another noteworthy drawback is its failure to incorporate the interaction effects of variables on the process.

Statistical tools such as Placket Burman and Response Surface Methodology (RSM) enable effective medium optimization using a limited number of trials and eliminate the limitations of single factor optimization processes. There are nowadays widely employed for the optimization of a number of production processes (Saxena *et al.*, 2004, Vohra *et al.*, 2002) and have been successfully applied for lipase production as well (Gupta *et al.*, 2007).

In this article, we report the isolation of fungal strains from fat and oil containing wastewaters; screening for their lipase activity; identifying the parameters (pH, temperature, incubation time, chief carbon, nitrogen and inducing lipid source) that have a significant impact on lipase production through the classical OFAT approach; and finally the application of RSM (Box-Behnken Design) to determine the optimal concentrations and interactive effects of those influential parameters on lipase production by the best isolate.

MATERIALS AND METHODS

Chemicals

p-Nitrophenyl Palmitate (p-NPP), the substrate used in lipase assay was purchased from Sigma Aldrich, USA; Gum Arabic, Triton X 100, used as emulsifiers in the assay were procured from Himedia, Mumbai, India. All other chemicals and reagents used were of analytical grade.

Collection of Samples

Wastewater samples were collected from 'Central Ghee' dairy industry, Gobichettipalayam, Tamil Nadu, India, in a sterile container, brought to the laboratory and stored under refrigerated conditions.

Isolation of Fungal Strains and Screening for Lipase Activity

For the isolation of lipolytic fungi, the collected wastewater samples were serially diluted and plated on potato dextrose agar (PDA) plates. Fungal colonies exhibiting morphological difference were picked up, inoculated in 100ml of Glucose-Yeast Extract-Peptone (GYP) medium augmented with trace amounts of mineral salts (glucose 20 g/L, yeast extract 10 g/L, peptone 10 g/L, $CH_3COONa.3H_2O$ 10 g/L, $MgSO_4$ 0.3 g/L, $MnSO_4$ 0.1 g/L, $CuSO_4.5H_2O$ 1.5 mg/L, KCl 0.5 g/L), supplemented with 2% olive oil and cultured at room temperature for 48 hours in a rotary shaker (90 rpm). Cell free supernatant was obtained by centrifuging the culture at 10,000 g for 10 minutes at 4°C (Kubota J37508). The supernatant which would contain the extracellular lipase produced by the organism was used as the crude extract and lipase activity was assayed by using p-NPP as substrate (ELICO UV-VIS Spectrophotometer) (Winkler *et al.*, 1979). For the lipase assay, substrate solution was prepared by adding solution A (30 mg of p-NPP in 10 ml of isopropanol) into solution B (0.1 g of gum arabic and 0.4 ml Triton X-100 in 90 ml of 50mM Tris–HCl buffer, pH 8) with stirring until all dissolved. The mixture of 9ml of substrate solution and 1ml of enzyme solution was incubated at 32 ± 2°C for 15 min and absorbance was measured at 410 nm.

One unit of lipase activity (U) was expressed as μmol of p-nitro phenol released per minute under the assay conditions. The isolate showing maximal activity was chosen for optimization studies. All the isolates were maintained on PDA slants overlaid with liquid paraffin and sub-cultured periodically.

Effect of Various Physico-Chemical Parameters on Lipase Production OFAT Method

The effect of different physico-chemical parameters on lipase production was initially studied using the conventional OFAT approach and the modified GYP medium was used for this purpose. As part of this study, extracellular lipase activity was measured at different time intervals (12–96 h) during growth of the best isolate in the production medium, in order to determine when enzyme production by the culture peaks. Next, the initial pH of the production medium was adjusted to different levels (4.0–12.0) and incubation temperature was later varied (30–50°C). Subsequently, the chemical composition of the growth medium was altered in the following ways—staple carbon source in the medium, namely glucose was substituted with starch, sucrose and maltose (2); nitrogen source was replaced with other organic (1%) as well as inorganic (0.5%) sources, namely, soya peptone, casein, sodium nitrate, ammonium sulfate and ammonium chloride; inducing oil source, namely olive oil was replaced with gingelly, sunflower, groundnut, palm and tributyrin oils (2%); finally, different inorganic ions (Ca^{2+}, Fe^{3+}, Zn^{2+}, Ba^{2+}, Mg^{2+}) were added to the basic GYP medium one at a time. After each one of these experiments, lipase production was tested by performing the standard assay.

Response Surface Approach

The next phase of work focussed on statistical medium optimization for enhanced lipase production using RSM. A Box-Behnken Design developed by the Design Expert software, version 8.0.7.1 (Stat Ease Inc. Minneapolis, U.S.A, and trial version) was used for this purpose. The three significant parameters whose levels were optimized and interactions determined through RSM included: oil source (sunflower oil), nitrogen source ($NaNO_3$) and $CaCO_3$. Presence of $CaCO_3$ in trace amounts has been widely reported in literature to induce lipase production (Kadar *et al.*, 2007). Each one of the above independent variables was studied at three different levels (–1, 0, +1) (Table 1). A matrix of 17 experiments involving these three factors was generated by the software (Table 1). Accordingly, production was carried out in 250 ml Erlenmeyer flasks containing 100 ml of medium. The flasks were sterilized, inoculated with the culture under aseptic conditions and incubated at room temperature (30 ± 2°C) for 72 h, in a shaker set at 90 rpm. All experiments were carried out in triplicates and the data given represent the mean. At the end of the incubation period, lipase assay was performed using the cell-free supernatant and activity was recorded as the response (dependent variable). This response data obtained after Box Behnken Design were analyzed by feeding into the software, which generated contour plots

Table 1: Experimental Design by Box-Behnken Analysis
(+1), (0) and (–1) are coded values

Std.	*Trial*	*Oil (% v/v)*	*$NaNO_3$ (g/l)*	*$CaCO_3$ (g/l)*	*Lipase (U)*
15	1	3 (0)	5.5 (0)	0.55(0)	19.81
7	2	1(–1)	5.5(0)	1(+1)	46.86
3	3	1(–1)	10(+1)	0.55(0)	72.47
16	4	3(0)	5.5(0)	0.55(0)	19.81
10	5	3(0)	10(+1)	0.1(–1)	59.05
1	6	1(–1)	1(–1)	0.55(0)	64.71
12	7	3(0)	10(+1)	1(+1)	75.41
11	8	3(0)	1(–1)	1(+1)	57.66
8	9	5 (+1)	5.5(0)	1(+1)	55.57
14	10	3(0)	5.5(0)	0.55(0)	19.81
17	11	3(0)	5.5(0)	0.55(0)	19.81
13	12	3(0)	5.5(0)	0.55(0)	19.81
4	13	5 (+1)	10(+1)	0.55(0)	39.19
6	14	5 (+1)	5.5(0)	0.1(–1)	19.61
2	15	5 (+1)	1(–1)	0.55(0)	43.19
9	16	3 (0)	1(–1)	0.1(–1)	74.76
5	17	1 (–1)	5.5(0)	0.1(–1)	83.88

and standard analysis of variance (ANOVA). A quadratic polynomial regression model was assumed for the predicted response. Regression analysis was performed on the data obtained from the design experiments. For a three-factor system, the model equation was as follows:

$$Y = \beta_0 + \beta_1 A + \beta_2 B + \beta_3 C + \beta_{11} A^2 + \beta_{22} B^2 + \beta_{33} C^2 + \beta_{12} AB + \beta_{13} AC + \beta_{23} BC \quad \ldots (1)$$

Where, Y is the predicted response, β_0 is the intercept, $\beta_1, \beta_2, \beta_3$ are the linear coefficients, $\beta_{11}, \beta_{22}, \beta_{33}$, are the squared coefficients, and $\beta_{12}, \beta_{13}, \beta_{23}$ are the interaction coefficients.

RESULTS AND DISCUSSION

Isolation and Screening

Inoculating the PDA plates with the serially diluted wastewater sample resulted in the growth of several fungal strains, of which 16 were selected based on visible difference in the appearance of the mycelial growth. They were sub-cultured in a similar broth medium and the resulting cultures were preserved as PDA slants and also used for inoculating the production medium. Later, p-NPP assay of the cell free supernatant revealed appreciable enzyme activities for 11 isolates as shown in Figure 1 and negligible activities for the other 5 (data not shown). Of these isolates, SKF4 which showed a maximum activity of 52.05 U/ml and identified as *Aspergillus niger* was taken up for further optimization studies.

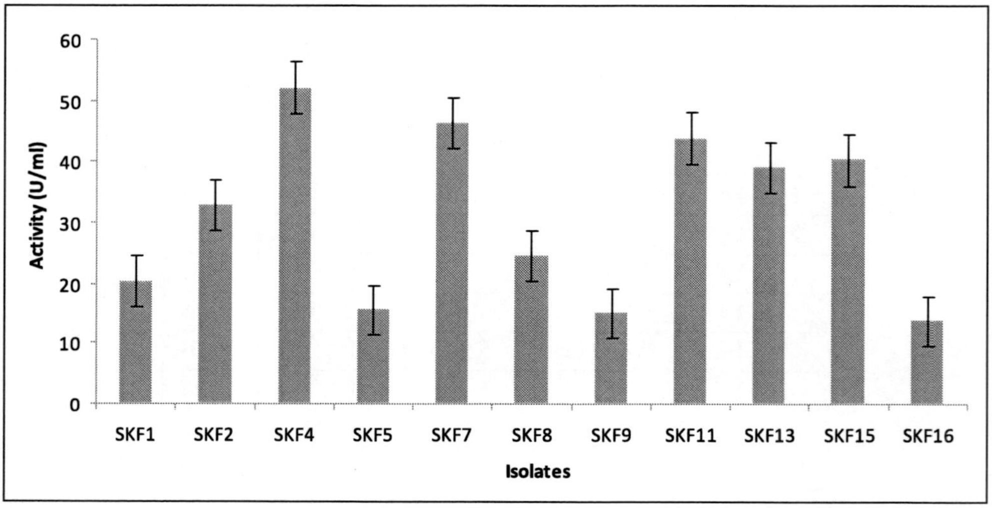

Fig. 1: Lipase Activities of Fungal Isolates from Wastewater Samples

Effect of Culture Conditions on Lipase Production (OFAT)

Lipase production curve of *Aspergillus niger* SKF4 (Figure 2) shows that production peaks at 72 hours of growth. An incubation time of 3 days or more has been reported for most fungi (Samad *et al.*, 1990). In case of *Mucor* sp., the enzyme production increased with the cultivation time and best production was obtained after 6 days of cultivation (Abbas *et al.*, 2002). An incubation temperature of 35°C most favored the enzyme production (Figure 3). The optimum initial pH (Figure 4) of the culture medium was found to be 8. Moreover, the isolate showed remarkable pH tolerance, growing well and producing the enzyme over a wide pH range-slightly acidic to highly alkaline. A strain of *Aspergillus niger* cultivated by solid-state fermentation (SSF) was earlier reported to show maximum enzyme production after 96 h of growth, in a medium with an initial pH of 7 (Adinarayana *et al.*, 2004).

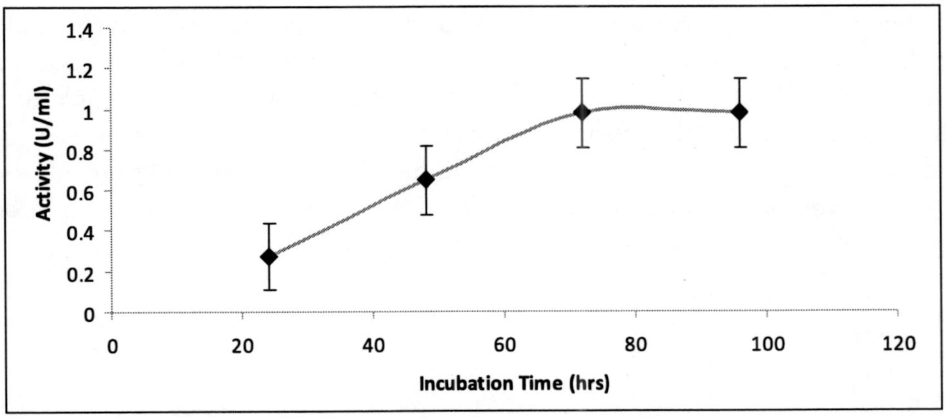

Fig. 2: Effect of Incubation Time on Lipase Production

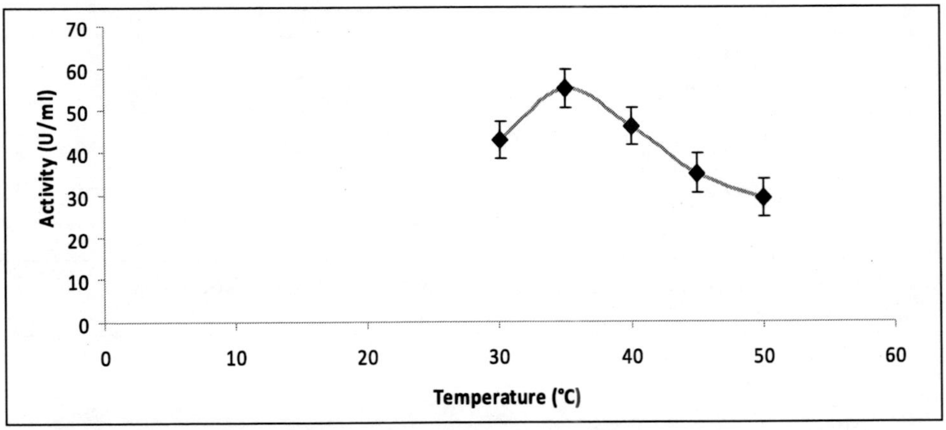

Fig. 3: Effect of Incubation Temperature on Lipase Production

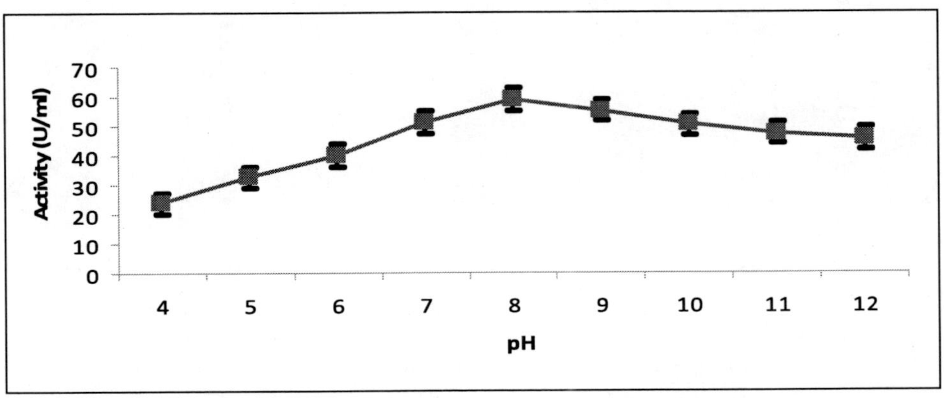

Fig. 4: Effect of Medium pH on Lipase Production

Next, chemical parameters were varied one at a time and the first one to be altered was carbon source. Maximum enzyme production was obtained with the disaccharide maltose as the chief carbon source (Figure 5). Lipases are known to be inducible enzymes and hence presence of lipid substrates in the medium is necessary either as a sole or secondary carbon source. Sunflower oil induced enzyme production maximally (Figure 6). A combination of 2% olive oil and 2% glucose was earlier proven to be the ideal carbon source for *A. niger* in SSF (Falcony *et al.*, 2006).

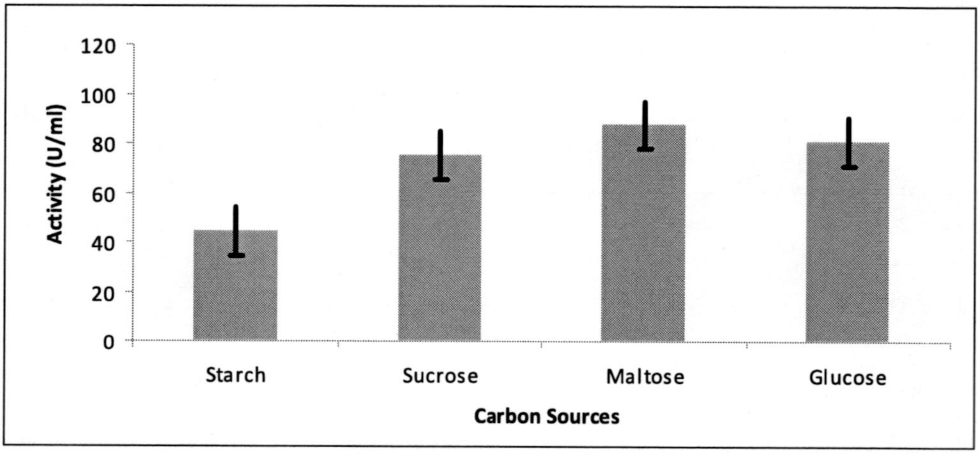

Fig. 5: Effect of Different Carbon Sources on Lipase Production

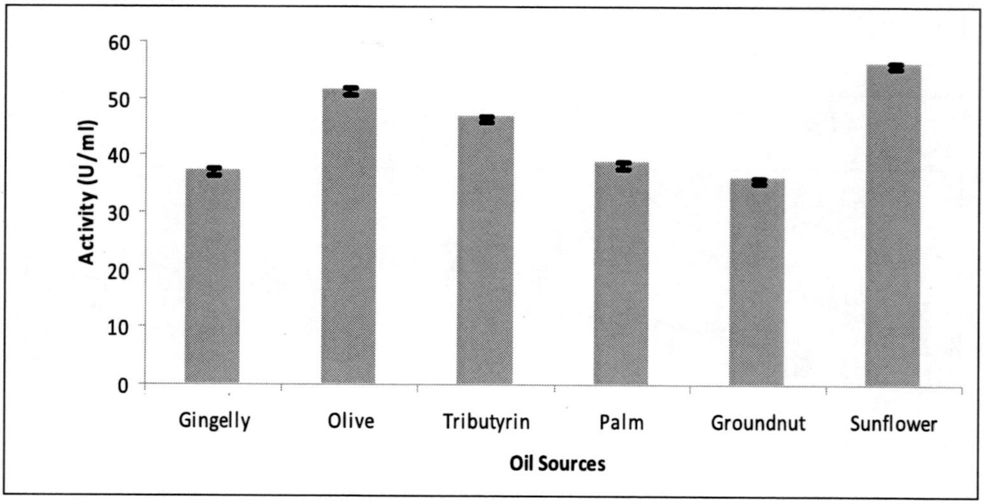

Fig. 6: Effect of Different Oil Substrates on Lipase Production

Next, the nitrogen source was varied, and among the organic nitrogen sources tested, casein was found to facilitate better lipase production when compared to peptone. Among the inorganic nitrogen sources tested, sodium nitrate yielded highest lipase production (Figure 7). Earlier studies on lipase production using *Fusarium oxysporum* showed optimum activities of 17 U/ml and 13 U/ml with peptone and ammonium dihydrogen phosphate, respectively (Rifaat *et al.*, 2010). Of the inorganic ions tested, Ca^{2+} stimulated the enzyme production markedly (Figure 8). Similar results have been widely reported in literature (Kader *et al.*, 2007). Ba^{2+} and Mg^{2+} had negative impacts on enzyme production.

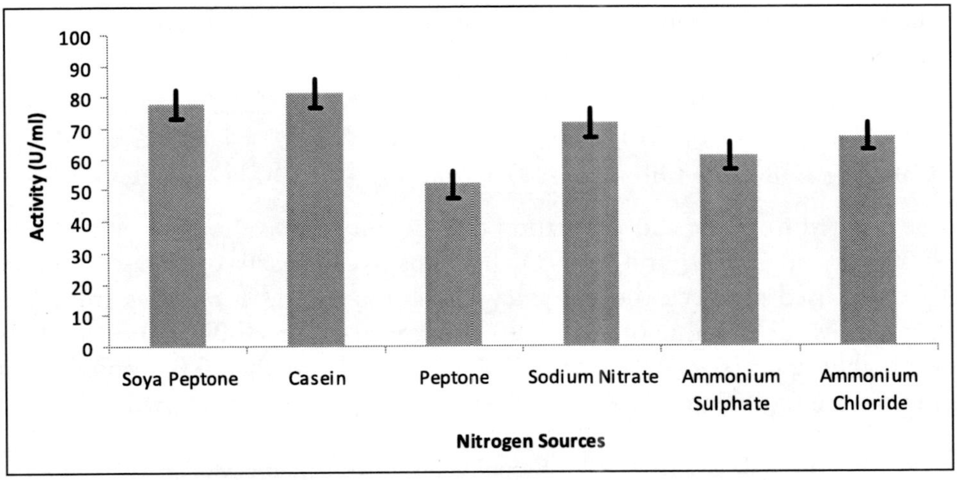

Fig. 7: Effect of Different Nitrogen Sources on Lipase Production

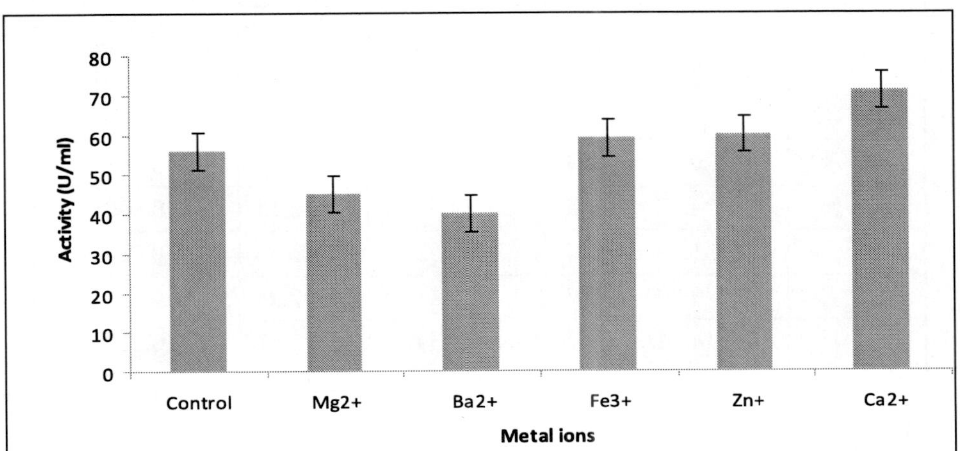

Fig. 8: Effect of Metal Ions on Lipase Production

Response Surface Methodology

Statistical tools can assist us in designing experiments so as to obtain enhanced production. They offer the distinct advantages of incorporating interaction effects of variables and aid us in arriving at the optimum concentrations within a minimal number of experimental runs. Based on the results obtained from the OFAT studies, three factors (sunflower oil, $NaNO_3$ and $CaCO_3$) were selected and their levels optimized by Design of Experiments (DOE) using Box-Behnken design of RSM. The experimental design is shown in Table 1. This study was carried out to understand the impact of a range of variables on the response, which is lipase production.

The following second order polynomial equation illustrates the empirical relationship between the independent variables and response:

$$Y = 172.84185 - 31.13056 \times \text{Oil} - 14.79802 \times NaNO_3 - 202.18457 \times CaCO_3 - 0.32667 \times \text{Oil} \times NaNO_3 + 20.27222 \times \text{Oil} \times CaCO_3 + 4.13086 \times NaNO_3 \times CaCO_3 + 2.48000 \times \text{Oil}^2 + 1.24247 \times NaNO_3^2 + 107.40741 \times CaCO_3^2 \quad \ldots (2)$$

It can be inferred from the above equation that the interactive effects of oil and $CaCO_3$, as well as that of $CaCO_3$ and $NaNO_3$ had positive impacts on lipase production. ANOVA was used to check the adequacy of the model and F test was employed for this purpose. Model F value of 14965.40 and and p-value <0.0001 indicate that the model is highly significant. Values of probablity (P) > F less than 0.05 indicate that the model terms are significant (Li *et al.*, 2008). ANOVA results are shown in Table 2.

Table 2: ANOVA for Response Surface Quadratic Model

Source	*Sum of Squares*	*Df*	*Mean Square*	*F Value*	*p-value Prob > F*
Model	8750.91	9	972.32	14965.40	< 0.0001
A-Oil	1522.42	1	1522.42	23432.09	< 0.0001
B-$NaNO_3$	4.20	1	4.20	64.72	< 0.0001
C–$CaCO_3$	0.41	1	0.41	6.23	0.0412
AB	34.57	1	34.57	532.15	< 0.0001
AC	1331.52	1	1331.52	20493.93	< 0.0001
BC	279.89	1	279.89	4307.94	< 0.0001
A^2	414.34	1	414.34	6377.31	< 0.0001
B^2	2665.37	1	2665.37	41023.74	< 0.0001
C^2	1991.84	1	1991.84	30657.20	< 0.0001

Predicted R^2: 0.9992 Adjusted R^2: 0.9999 Adequate Precision: 327.78

The p-value serves as a tool for checking the significance of each of the coefficients, which also indicates the interaction strength of each independent variable. The low values of $p < 0.05$ indicate the greater significance of the corresponding coefficients. In general, larger magnitudes of t, F values and smaller p values, indicate that the corresponding coefficient terms are significant. All three linear coefficients (A, B, C), interaction coefficients (AB, BC, AC) as well as quadratic coefficients (A^2, B^2, C^2) appear to be very significant. These results indicate that the levels of the variables have been selected appropriately and had a different range been chosen, the p-value might have been accordingly altered. Thr R^2 value should be in the range of 0–1.0 and closer the value is to 1.0, the model is set to be more fit (Reddy *et al.*, 2007). The R^2 value gives a measure of how much variability in the observed response can be explained by the experimental parameters and their interactions. The closer the values of R^2 to 1, the better the correlation between the experimental and predicted values and the better the model predicts the response. When expressed as a percentage, R^2 implies that the total variation of 99.99% for enzyme activity is attributed to the independent variables and only about 0.01% cannot be explained by the model.The predicted R^2 of 0.9992 is also in reasonable agreement with the adjusted R^2 of 0.9999. The adjusted R arranges the R values for the sample size and for the number of variables in the model. If there are many variables in the model and the sample size is not very large, the adjusted R may be smaller that the R. Response surface methodology also attempts to detect which experimental parameters generate signals that are large in comparison to the noise. Adequate precision measures signal-to-noise ratio and a ratio greater than 4 is desirable. An adequate presision of 327.78 for lipase activity indicates an adequate signal.

The contour plot is a way of expressing the regression equation graphically. It depicts the interactions among the variables and is used to determine the optimum concentrations of each factor for good enzyme production. The interaction between the corresponding variables is deemed to be significant in case of elliptical mounds or rising ridges in the contour plot (Muralidhar *et al.*, 2001). Figure 9 shows the interaction between oil and $NaNO_3$ and Figure 10, the interaction between oil and $CaCO_3$. The shapes of the response surface curves show a moderate level of interaction between these variables. Figure 11 shows the interaction between $NaNO_3$ and $CaCO_3$. The perfect elliptical mounds in this contour plot depict significant interaction between these two variables and optimal operating conditions. Thus, these components had synergistic effects on lipase production.

Maximal lipase activity obtained after RSM was 83.88 U, from run 17 with 1% oil, 5.5 g/L $NaNO_3$ and 0.1 g/L $CaCO_3$. This is a 1.4 fold increase in comparison with the

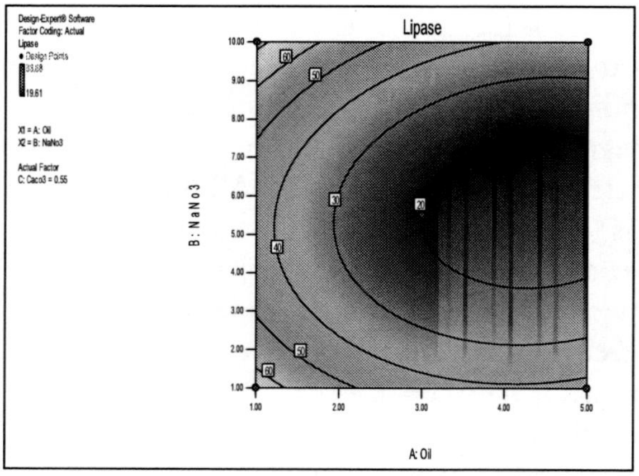

Fig. 9: Plot between Effects of Oil and Sodium Nitrate on Lipase Production

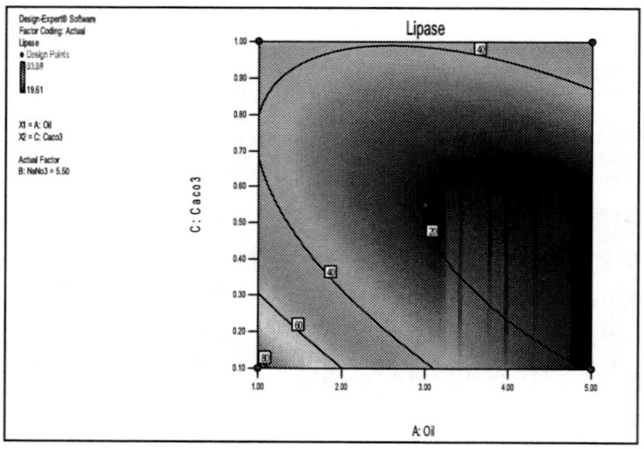

Fig. 10: Plot between Effects of Oil and Calcium Carbonate on Lipase Production

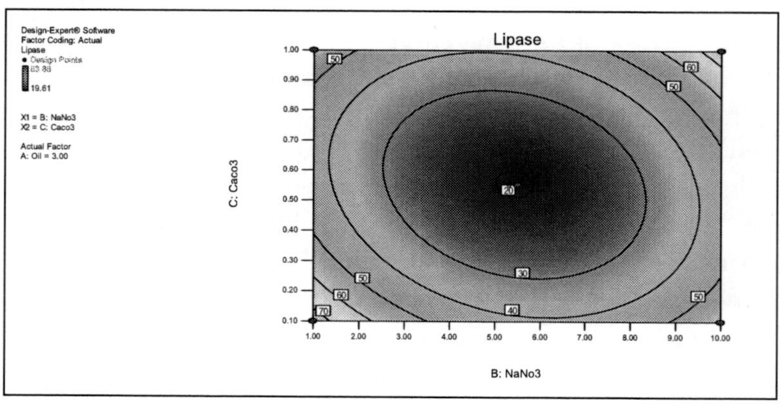

Fig. 11: Plot between Effects of Sodium Nitrate and Calcium Carbonate on Lipase Production

pre-optimization maximum of 58.21 U. Thus, the proposed model has enhanced the enzyme yield and improved the feasibility of lipase production from *Aspergillus niger* SKF4. RSM based media optimizations have been used by several researchers for lipase production. Central composite design was applied to optimize lipase production by the yeast *Candida cylindracea* using two different carbon sources and a yield of 17.30 U/ml using glucose as carbon source and 47.25 U/ml using olive oil as carbon source were obtained (Muralidhar *et al.*, 2001). Lipase production by *Candida rugosa* was optimized using RSM based design and cheese whey was used as a substrate for this. From the experimental design it was concluded that a combination of brewery product, yeast and malt extract, Tween 80, and olive oil with cheese whey resulted in a 287% increase in enzymatic activity (Tommaso *et al.,* 2011). A medium comprised of soyabean meal 0.77% (w/v), $(NH_4)_2SO_4$ 0.1M, KH_2PO_4 0.05 M, rice bran oil 2% (v/v) $CaCl_2$ 0.05 M, PEG 6000–0.05% (w/v), NaCl 1% (w/v), inoculum 1% (v/v), pH 3, incubation temperature 35°C and incubation period of 5 days were identified to be the optimal conditions for lipase production by *Aspergillus awamori* (Basheer *et al.*, 2011). In another study, a medium containing 0.45% (w/v) peptone, 0.65% (v/v) Tween-80 and 2.2% (v/v) inoculum was recommended for a maximal lipase production of 20.6 U/ml by *Candida cylindracea (*Salihu *et al.*, 2011).

Model Verification and Confirmation

The objective of the validation study is to demonstrate that the polynomial expression could correctly predict and describe the response function. The validation report generated by the design expert software is given in Table 3, which predicts a maximum lipase activity of 84.26 U/ml.

Table 3: Model Verification and Confirmation

Factor	*Name*	*Level*	*Low Level*	*High Level*	*Std. Dev.*
A	Oil	5.00	1	5	0
B	$NaNO_3$	9.81	1	10	0
C	$CaCO_3$	1.00	0.1	1	0
Response	Prediction	Std Dev	SE (n=1)	95% PI low	95% PI high
Lipase	84.26095	0.25489	0.385239	83.35	85.17189

ACKNOWLEDGEMENT

The authors are thankful to the management of Kumaraguru College of Technology, Coimbatore for the laboratory facilities provided in order to carry out this research work and to the Indian Agricultural Research Institute, New Delhi, for culture identification.

REFERENCES

Abbas, H., Hiol, A., Deyris, V. and Comeau, L. (2002). Isolation and characterization of an extracellular lipase from *Mucor sp* strain isolated from palm fruit. *Enz. Microbial Technol.*, 31: 968–975.

Adinarayana, K., Raju, B., Zargar, M.I., Devi, R.B., Lakshmi, P.J. and Ellaiah, P. (2004). Optimization of process parameters for production of lipase in solid-state fermentation by newly isolated *Aspergillus species. Indian J. Biotechnol*, 3: 65–69.

Basheer, S.M., Chellappan, S., Beena, P.S., Sukumaran, R.K., Elyas, K.K. and Chandrasekaran, M. (2011). Lipase from marine *Aspergillus awamori* BTMFWO32: Production, partial purification and application in oil effluent treatment. *New Biotechnol,* 28: 627–638.

Bussamara, R., Dall'Agnol, L., Schrank, A., Fernandes, K.F. and Vainstein, M.H. (2012). Optimal conditions for continuous immobilization of *Psuedozyma hubeiensis* (Strain HB85A) lipase by adsorption in a packed-bed reactor by response surface methodology. *Enzyme Research*, D.O.I: 10.1155/2012/329178.

Carvalho, P.O. (2005). Potential of enantioselective biocatalysis by microbial lipases. Quimica Nova. 28: 614–621.

Contesini, F.J., Lopes, D.B., Macedo, G.A., Nascimento, M.D.G. and Carvalho, P.D.O. 2010. *Aspergillus sp.* lipase: potential biocatalyst for industrial use. *J. Mol. Cat. Enz.*, 67: 163–171.

Dutra, J.C.V., Terzi, S.C., Bevilaqua, J.V., Damasco, M.C.T., Couri, S. and Langone, M.A.P. (2008). Lipase production in solid state fermentation monitoring biomass growth of *Aspergillus niger* using digital image processing. *Appl. Biochem. Biotechnol,* 147: 63–75.

Falcony, G., Armas, J.C. Mendoza J.C.D. and Hernandez, J.L.M. (2006). Production of extracellular lipase from *Aspergillus niger* by solid state fermentation. *Food Technol. Biotechnol.* 44: 235–240.

Freire, D.M., Sant'Anna, G.L. and Alves, T.L. (1997). Lipase production by *Penicillium restrictum* in a bench-scale fermenter. *Appl. Biochem. Biotechnol,* 63: 409–421.

Gangadhara, R.K., Parigi and Vishweshwaraiah (2009). The stabilizing effects of polyols and sugars on porcine pancreatic lipase. *J. American Oil Chemists' Soci.*, 86: 773–781.

Griebeler, N., Polloni, A.E., Remonatto, D., Arbter, F., Vardanega, R. and Cechet, J.L. (2011). Isolation and screening of lipase producing fungi and hydrolytic activity, *Food and Bioprocess Technology,* 4: 578–586.

Gupta, N., Sahai, V. and Gupta, R. (2007). Alkaline lipase from a novel strain *Burkholderia multivorans*: Statistical medium optimization and production in a bioreactor. *Process Biochem.,* 42: 518–526.

Hasan, F., Shah, A.A. and Hameed, A. 2006. Industrial applications of microbial lipases. Enz. Micro. Technol. 39: 235–251. Jaeger, K.-E. and Reetz, M.T. (1998). Microbial lipases form versatile tools for biotechnology, *Trends Biotechnol.*, 16: 396–403.

Jaeger, K.-E., Ransac, S., Dijkstra, B.W. and Teetz, M.T. (1999). Bacterial biocatalysis: A molecular biology, three-dimensional structures and biotechnological applications of lipase. *Annual Rev. Microbiol.* 53: 315–351.

Kader, R., Yousuf, A. and Hoq, M.M. (2007). Optimization of lipase production by a Rhizopus MR12 in shake culture. *J. Appl. Sci.*, 7: 855–860.

Li, X.Y., Liu, Z.Q. and Chi, Z.M. (2008). Production of phytase by a marine yeast *Kodamaea ohmeri* BG3 in an oats medium: Optimization by response surface methodology. *Bioresource Technol.*, 99: 6386–6390.

Muralidhar, R.V., Chirumamila, R.R., Marchant, R. and Nigam, P. (2001). A response surface approach for the comparison of lipase production by *Candida cylindracea* using two different carbon sources. *Biochem. Engg. J.*, 9: 17–23.

Paques, F.W., Pio, T.F., Carvalho, P.D.O. and Macedo, G.A. (2008). Characterization of the lipase from Carica papaya residues. Brazil. *J. Food Technol.* 11: 20–27.

Paques, F.W. and Macedo, G.A. (2006). Plant lipases from latex: properties and industrial applications. *Quimica Nova,* 29: 93–99.

Rifaat, H.M., El-Mahalawy, A.A., El-Menofy, H.A. and Donia, S.A. (2010). Production, optimization and partial purification of lipase from *Fusarium oxysporum*. *J. Appl. Sci. Environ. Sanit.,* 5: 39–53.

Biooxidation and Extraction of Lead Using Air Lift External Loop Reactor by *Leptospirillum ferroxidans*

S. Venkatesa Prabhu*, S. Harish and Yoga Vignesh Chandra[1]

Department of Biotechnology, K.S. Rangasamy College of Technology, Tiruchengode, India

[1]University of Kaiserslautern, Kaiserslautern 67663, Germany

E-mail: haiitsvp@gmail.com

INTRODUCTION

Leaching is the solid liquid operation, where one or more constituents of a liquid are preferentially dissolved when brought in contact with a liquid solvent. Most useful minerals occur in mixtures with large proportion of undesirable constituents, and leaching of a valuable mineral is a separation method which is frequently applied it similarly plays an important part in metallurgical processing of aluminium, cobalt, manganese, nickel, and zinc. Many naturally occurring organic products are separated from their original structure by leaching.

Bioleaching is one of several applications within biohydrometallurgy and several methods are used to recover copper, zinc, lead, arsenic, nickel, molybdenum, gold, silver, chromium and cobalt (Metha *et al.*, 2003).

Bioleaching can involve numerous ferrous iron and sulfur oxidizing bacteria. Bioleaching (bacterial oxidation) was first used in commercial practice for treating sulphidic ores and concentrates in 1986. In the case of Gold bearing concentrates the process is a pre-treatment which oxidizes the sulphide minerals, and exposes the precious metals for subsequent cyanide leaching good recoveries (above 90% for concentrates) are achievable. Silverman and Lundgren (1959) originally proposed a model with two different mechanisms for the bioleaching of metals.

---

*Corresponding author.

Firstly, direct leaching in which the bacterial membrane directly interacts with the sulphide surface using enzymatic mechanisms. Thus direct leaching is observed only if the cells attach to the mineral surface. Cell attachment to suspended mineral particles takes place within minutes or hours with cell preferentially occupying irregularities of the surface structure (Brandl, 2001).

Galena is the natural mineral form of lead sulphide. It is the most important lead ore mineral. Galena is one of the most abundant and widely distributed sulphide minerals. It crystallizes in the cubic crystal system often showing octahedral forms. It is often associated with the minerals sphalerite, calcite and fluorite. Galena deposits often contain significant amounts of silver as included silver sulphide mineral phases or as limited solid solution within the galena structure.

Lead extraction in air lift external loop reactor from Galena ore using *Leptospirillum ferrooxidans* were performed over a period of 13 days and results were analysed The isolated microorganism is *Leptospirillum ferrooxidans.* Lead extraction efficiency of 74% is higher using L.f. Maximum leaching was obtained in 4% substrate in ALER rapid decrease in Ph and rapid increase in ROP were found in all the coloums.

MATERIALS AND METHODS

Galena Ore Analysis

Lead ore (galena) was cursed in a jaw crusher and grinded as powder with the help of ball mill and sieved for 300 microns size particles. The total chemical analysis revealed that the ore contained 63.12% of Lead.

Isolation of Organism

The sample was taken from chitradurga copper mine drainage in karnataga, India. This sample was serially diluted and 10^{-3} dilution was taken and tested for acid producing bacteria. For this we made this sample inoculate in 9k medium at pH 2, which is specific for acid producing bacteria. We had a positive result and the further isolation were made by some phenotypic and biochemical tests and solid plate were also made for further culturing.

Adaptation of Organism

Due to high toxicity of ore, the microorganism was adapted by initial growing of the organism in 0.1 to 0.7% by continuous sub-culturing.

Construction of ALER

The Air lift external loop reactor was constructed as shown in Figure 1. It was constructed with 20 cm length with 3 cm diameter and loop is about 10 cm at the middle. There is an opening at the bottom for air inlet and another opening was made for sample collection.

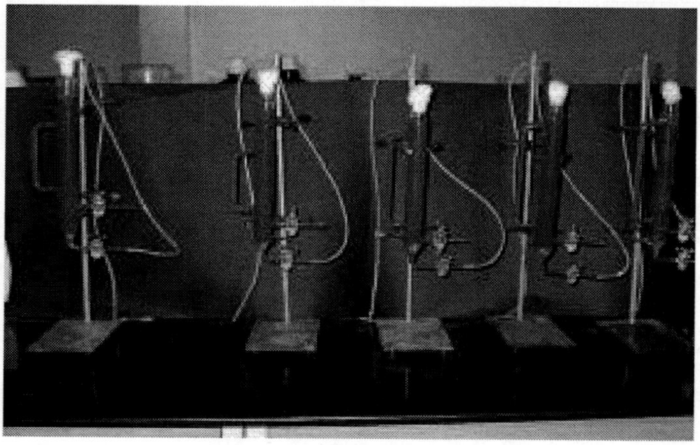

Fig. 1: ALER Set-up

RESULTS AND DISCUSSION

Determination of Lead

The percentage leach of lead is estimated from Dithizone method by UV spectrometry. The standard graph for Lead was plotted and lead concentration can be estimated.

pH Studies

The pH studies were conducted and their respective graphs were plotted and given in Figure 2. It was clearly evident that pH keeps on decreases and it maintain the acidic condition. The pH variation was studied for all different substrate concentration. Thus the produced acid is capable of leaching Lead.

ROP Studies

The ROP studies were conducted and their respective graph was plotted and depicted in Figure 3. It is clearly evident that ROP keeps on increasing and it maintain the acidic condition. The ROP variation was studied for all different substrate concentration.

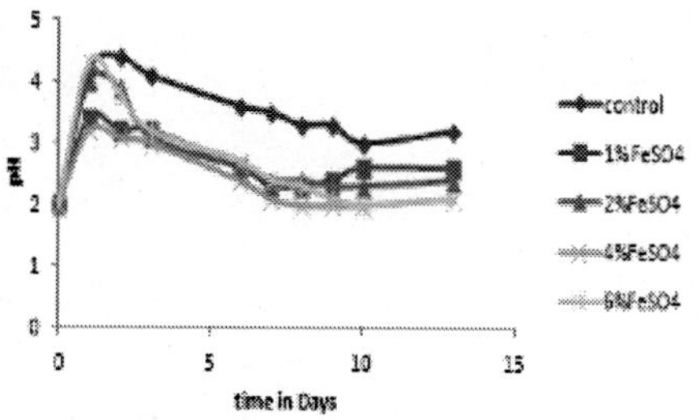

Fig. 2: Variation of pH during the Bioleaching Process

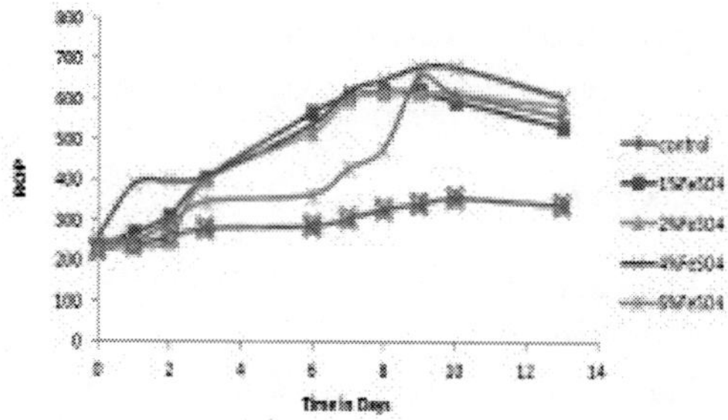

Fig. 3: Variation in ROP during Bioleaching Process

Leaching of Lead

Bacterial bioleaching studies were carried out with different concentration of substrate and their results were analyzed the graphs were plotted and given in Figure 4. In control we see only a low quantity of lead is extracted. In 1%$FeSO_4$ 24% of lead is extracted. In 2%$FeSO_4$ 46% greater amount of lead is extracted. In 4%FeSO 74% of lead is leached. In 6%FeSO 63% of lead is leached the overall process was conducted for a period of 13 days.

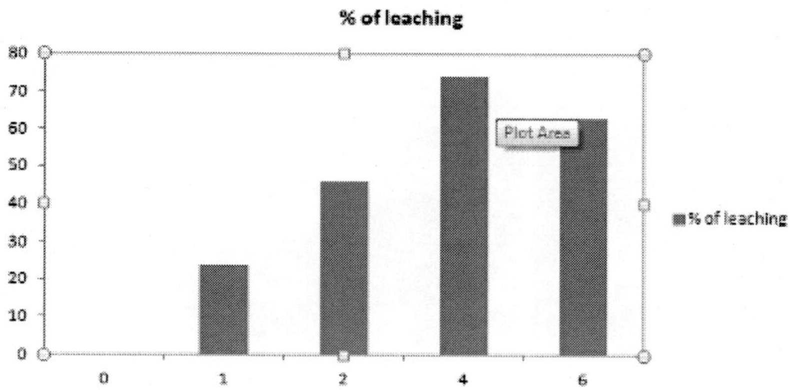

Fig. 4: Total Amount of Leaching in Various Substrate Concentrations

Lead extraction in air lift external loop reactor from Galena ore using *Leptospirillum ferrooxidans* were performed over a period of 13 days and results were analyzed. Lead extraction efficiency of 74.35% is the maximum leaching which was obtained in 4% substrate in ALER rapid decrease in pH and rapid increase in ROP were found in all the coloums. Bio oxidation of galena in ALER studies showed that significant extraction of lead compared to other shake flask, bubble coloumn and Stirred tank reactor operations. From the laboratory results the scale up correlation can be made for industry level galena bio extraction with an eco friendly and economic beneficial.

REFERENCES

Bijinova, D.Y. and Velkova, R.G. (2001). Bioleaching of Metals from Mineral Waste Product, *Acta Biotechnol*, 21: 275–282.

Gerkem Akinci and Duysen E. Guven (2011). Bioleaching of heavy metals contaminated sediment by pure and mixed cultures of *Acidithiobacillus* spp. *Desalination*, 268: 221–226.

Hoque, D.E. and Obbard J. Philip (2011). Biotechnological/recovery of heavy metals from secondary sources—An overview. *Materials Science and Engineering*. 31: 57–66

Hossein Atashi, Mortaza Zivdar, Seyadali Seyedbagheri and Mohammad Hassan Fazsaelipoor (2011). Bioleaching Kinetics of copper from copper smelting dust. *Journal of Industrial and Engineering Chemistry*, 17: 29–35.

Metha, K.D., Pandey, B.D. and Mankhand, T.R. (2003). Studies on kinetics of biodissolution of Metals of Metals from Indian Ocean nodules. *Minerals Engineering*, 16: 523–527.

Sadhya Babel, Domirica del. and Munda Dacera (2006). Heavy metal removal from contaminated sludge for land application: A review. *Waste management*. 26: 988–1004.

Shen-Yichen and Jih-Gawlin (2001). Effect of Substrate concentration on bioleaching of metal contaminated sediment. *J. Haz. Mat.* 82: 77–89.

Nutritional Analysis – Comparison between Conventional Indian Rice and Different Organic Millets

K. Kumaresan*, T. Sathishkumar, N. Prathap, S. Malarvizhi and S.C. Priyadharshini

Microbiology and Immunology lab, Department of Biotechnology, Kumaraguru College of Technology, Coimbatore, Tamil Nadu

E-mail: kdpkumaresan@gmail.com

INTRODUCTION

In the last decade, preventive medicine has undergone a great advance. Research has demonstrated that nutrition plays a crucial role in the prevention of chronic diseases (Lopez-Varela *et al.,* 2002). A functional food should provide health benefits and enhance performance above its nutritional values (Kwat and Jukes, 2001). It is defined that a food can be regarded as functional if it has beneficial effects on target functions in the body as a source of mental and physical well-being, contributing to the prevention and reduction of risk factors for several diseases or enhancing certain physiological functions, beyond adequate nutritional effects. Resembling this definition, nutraceutical is that any substance that may be considered a food or a part of a food and provides medical or health benefits including the prevention and treatment of disease (Defelice, 1992; Andlauer and Furst, 2002).

Millets are very good source of micronutrients and phytochemicals such as dietary fibre, polyphenols, pigments and phytate (Hulse *et al.*, 1980). The whole grain millet is edible and the traditional foods are generally prepared from the whole meal. Sorghum contains large quatities of phenolics and other compounds of use in human foods to prevent deterioration of health.

---

*Corresponding author.

Cereals and its ingredients are accepted as functional foods and nutraceuticals because of providing dietary fibre, proteins, energy, minerals, vitamins and antioxidants required for human health. Cereals include dietary fibre such as β-glucan and arabinoxylan, carbohydrates such as resistant starch and oligosaccharides [galacto-α fructo-oligosaccharides]. Also, cereals can be used as fermentable substrates for the growth of probiotic microorganisms (Charalampopoulos *et al.,* 2002).

Cereal foods are important due to their antioxidant activity (AOA) attributed to the phenolics which are known to protect organs against harmful effects. Excessive free radical production and lipid peroxidation underlie the pathogenesis of diseases like atherosclerosis, carcinogenesis, diabetes, cataract, and also ageing. Antioxidant activity of foods are commonly consumed in India is scanty. Food grains play a major role in providing food and nutrition (Sreeramulu *et al.*, 2009).

Millets helps to lower blood glucose levels and improves insulin response. Postmenopausal women with signs of cardiovascular disease like high blood pressure, increased cholesterol and obesity can benefit from eating whole grains especially millets by eating them six times a week (Dykes and Rooney, 2006).

MATERIALS AND METHODS

Good quality glassware and chemicals were used for all tests. Glassware were cleaned with good detergent, rinsed in tap water and soaked in chromic acid clearing solution (Mahadevan and Sridhar, 1996). The millets were collected from Tamil Nadu Agricultural University (TNAU), Coimbatore. The millets collected are Co(Ra)14 [Ragi], Co(Te)7 [Tenai], Co(Sa)4 [Samai], Co(Cu)9 [Cumbu], Co(S)28 [Cholam].

A total of 75g of seedswere air dried and then grinded into fine powder individually for further experiments. Compounds such as crude protein, carbohydrate, fiber, moisture, fat, ash and elements such as calcium, magnesium, iron, phosphorus were analyzed using flame photometry and atomic absorption spectroscopy.

Estimation of vitamin c and total antioxidant assay were done based on the principle based on phospho-molybdenum method.

RESULTS AND DISCUSSION

Proximate Analysis of the Components Present in the Raw and Roasted Sample of Different Millets

The proximate analysis of the sample shows the contents of rich protein and fiber, so the samples are highly nutrient. The carbohydrate content is increased in all the samples

(Figure 1) whereas the protein, fibre, moisture, fat and ash contents are decreased as mentioned in the Table 1.

Table 1: Proximate Analysis of the Components Present in the Raw and Roasted Sample

Components	*Finger Millet*		*Pearl Millet*		*Foxtail Millet*		*Little Millet*		*Rice*		*Common Millet*	
	Raw	*Roasted*	*Raw*	*Roasted*	*Raw*	*Roasted*	*Raw*	*Roasted*	*Raw*	*Roasted*	*Raw*	*Roasted*
Carbohydrate	75.3 g	81.59 g	69 g	78.92 g	74.2 g	81.04 g	71.7 g	82.70 g	79.15 g	89.08 g	73.8 g	82.39 g
Protein	13.2 g	6.80 g	18 g	13.12 g	14 g	12 g	12.4 g	9 g	9.81 g	7.6 g	16 g	12 g
Fibre	11.1 g	7.0 g	8.3 g	5 g	5.45 g	3.50 g	6.71 g	3.50 g	6.5 g	4.0 g	7 g	3 g
Moisture	13.0%	0.63%	12.1%	1.36%	7.7%	0.22%	5.6%	0.27%	6.3%	0.40%	11.7%	1.21%
Fat	3.98 g	2.51 g	6.12 g	4.58 g	5.08 g	4.02 g	5.01 g	3.50 g	3.23 g	1.50 g	4.21 g	2.54 g
Ash	5.7%	3.47%	4.1%	2.02%	3.99%	2.72%	5.72%	4.53%	3.8%	2.02%	3.25%	1.86%

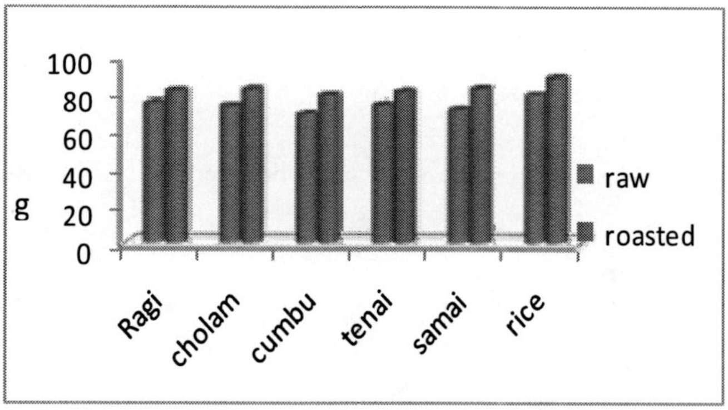

Fig. 1: Carbohydrate Difference between Millets and Rice

From the Figure 1, it concludes that the carbohydrate content is high in rice than the millets whereas protein content is high in millets than rice in both raw and roasted samples.

From Table 1, the fiber content is also high in millets than the rice. The fibre content of rice are 6.5 g in raw and 4 g in roasted sample but the content in millets are finger millet (11.1 g-raw; 7 g-roasted), common millet (7 g-raw; 3 g-roasted), pearl millet (8.3 g-raw; 5 g-roasted), foxtail millet (5.45 g-raw; 3.50 g-roasted) and little millet (6.70 g-raw; 3.50 g-roasted). The moisture, ash and fat contents are higher in millets when compared to rice. In roasted form, these values are decreased than the values obtained for raw samples.

Elements

Since the results showed certain contents of fibre, ash and protein, the research was further extended to find the presence of elements. Therefore, the contents of calcium, iron, magnesium and phosphorous are estimated and these are shown in the Figures 2 and 3.

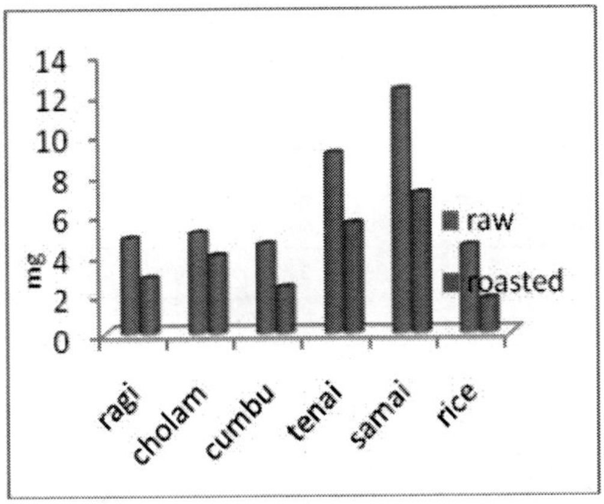

Fig. 2: Difference in the Iron Content between Millets and Rice

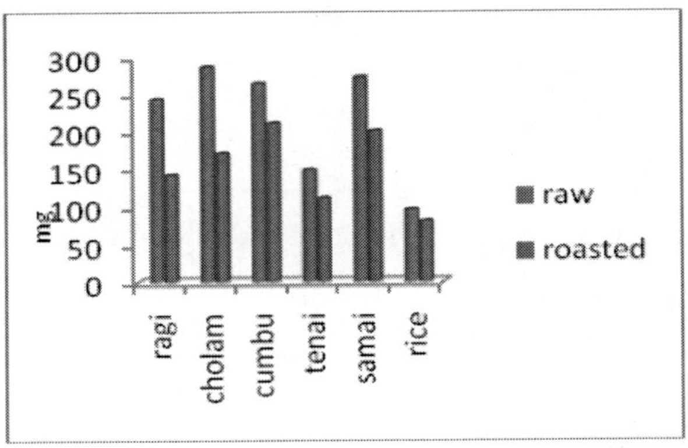

Fig. 3: Difference in the Phosphorous Content between Millets and Rice

From the above Figures 2 and 3, it concludes that the phosphorous, calcium, iron and magnesium are higher in millets. The calcium content of finger millet is 333.33 mg. But the calcium content of Italian millet ranged from 67.5 to 92.3 mg/100 g. Compared to finger millet (*Eleusinecoracana*), Italian millet is low in calcium (Pore and Magar, 1977). However, Italian millet has a higher calcium content compared to other millet grains (Desikachar, 1975).

Therefore, the millets are rich in minerals. Millets have shown lower range of vitamin-c, it depicts that these are rich in vitamin-B (B3, B6 and B9).Vitamin-C is not present in rice; it is rich in vitamin-B (B3, B5, B6 and B9).

Total Antioxidant

Other than the presence of higher range of protein and minerals, the millets also contain good level of antioxidant when compared to rice. The antioxidant level of millets and rice are depicted in Figure 4.

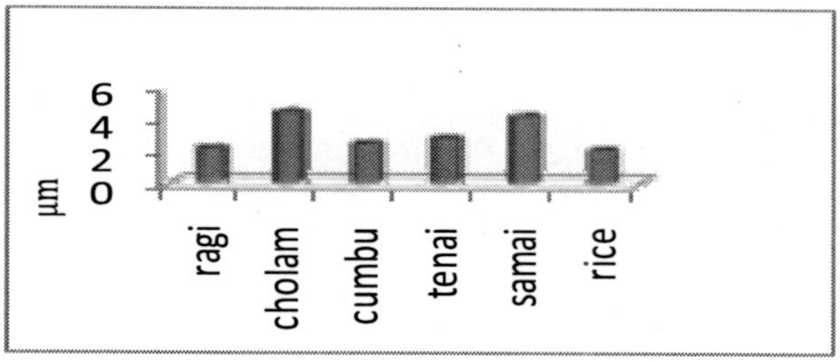

Fig. 4: Amount of Total Antioxidant Present in Raw Samples

Product Development

The products which are formulated have high nutritional contents. This formulation helps to lower the risk of diabetes. Besides the magnesium present in the products is a cofactor in various enzymes involved in the secretion of insulin and metabolism of glucose in the body. This diet plays a crucial role in prevention of chronic degenerative diseases. The millets and rice were collected from southern part of India. They were dried and then powdered. The studies were conducted in two forms of samples. One is raw form and another is roasted form. Roasted done at 160°C for 5 min with constant stirring in a deep open pan.Further various nutraceutical studies, antioxidant properties and the product development were conducted. The results from the present studies show that millets could be a good supplement. It can be a good substitute for the conventional food (rice) and it can find applications in therapeutic foods. Millets has great potential for producing healthy and serve as a functional food products. Due to the lower content of carbohydrate and fat in the millets than rice, millets can be consumed daily for diabetic patients. The presence of protein, carbohydrate, minerals and vitamins was detected in good amount in millets. Thus, the millets have high pharmacological activities and it is an essential in human growth and development.

REFERENCES

Charalampopoulos. D., Wang. R., Pandiella. S.S. and Webb, C. (2002). Application of cereals and cereal components in functional foods: A review. *Int. J. Food Microbiology*, 79: 131–141.

DeFelics, S.L., (1992). The nutraceutical initiative: A recommendation for U.S. Economic and regulatory reforms. *Genetic Engineering, News*, 12: 13–15.

Desikachar, H.S.R. (1975). Processing of maize, sorghum and millets for food uses. *J. Seient. Ind. Res.*, 34: 231–7.

Dykes, L. and Rooney, L.W. (2006). Sorghum and millet phenols and antioxidants. *J. of Cereal Science,* 44: 236–251.

Kwak, N.S. and Jukes D.J. (2001). Functional foods. Part 2: The impact on current regulatory terminology. *Food control,* 12: 109–117.

Lopez-Varela, S., Gonzalez-Gross, M. and Marcoss, A. (2002). Functional foods and the immune system: A review. *Eur. J. Clinical Nutr,* 56(3): 29–33.

PHA Production Utilizing *Cyanidondactylon* as Novel Substrate: An Optimization Study

P. Sasikala, R. Preethi, Sangeetha and J. Aravind*

Department of Biotechnology, Kumaraguru College of Technology, Coimbatore, Tamil Nadu

E-mail: dr.j.aravind@gmail.com

INTRODUCTION

Polyhydroxyalkanoates (PHAs), a family of bacterial polyesters, are formed and accumulated by variousbacterial species under unbalanced growth conditions. Limitations such as nitrogen, phosphate, magnesium and oxygen favor intracellular accumulation of PHA (Anderson *et al.,* 1990). PHAs have thermo mechanical properties similar to synthetic polymers such as polypropylene, but are truly biodegradable in the environment (Lee, 1996). The major barrier to their wide acceptance is the high production cost, particularly the costs of carbonaceous (substrate) raw materials (>40%) and polymer recovery (>26%) (Choi and Lee, 1997). Since the choice of carbon source contributes significantly on the overall cost in PHA production.Much effort has beendevoted to reduce the price of PHAs by the developmentof better bacterial strains (recombinant bacteria, more efficient fermentation, more economical recovery processes, and most importantly, cheaper carbon sources (Kim *et al.*, 1992; Anderson *et al.*, 1990; Yamane *et al.*, 1995). Optimization of cultivation medium is vital in the cost reduction, optimization of any bioprocesses can be conducted either by changing one factor at a time (OFAT) or by varying several factors at the same time and examining their effects and interactions using statistical analysis.

This present study was an attempt to make the process of PHA production economically more feasible, i.e. as the success in the biodegradable plastic strategy largely depends

*Corresponding author.

on the isolation of potent PHA producing bacteria and optimizing culture parameters for maximum PHA biosynthesis. Soil micro-organisms were initially isolated and screened for their accumulating capacity of PHA. The selected best isolate was compared with reference strain (*Cupriavidus necator*) for their PHA producing efficiencies. Plackett-Burmanmethod was used to optimize the culture medium composition for PHA production by *Kluyveraintermedia* using hydrolyzed grass as an inexpensive carbon source.

METHODOLOGY

Isolation of Micro-Organisms

Soil sample was collected from KCT campus; Coimbatore from which 6 PHA producing strains were isolated. *Cupriavidusnecator* (MTCC 1472) was obtained from Microbial Type Culture Collection, Institute of Microbial Technology (IMTECH), Chandigarh. *C. necator* was used as a reference strain (as it is cited in literature as the best PHA producer) for the comparison of PHA production with that of the best isolate-*Kluyveraintermedia* (characterized by IMTECH).

Hydrolyses of Cellulose

Bermuda grass (*Cyanidondactylon*) was collected from KCT campus, cellulose was extracted from it based on method of Chen *et al.,* 1984, and the finally converted cellulose is treated by zinc chloride based on the procedure of Chen *et al.*, 1985. Glucose content in the treated grass was estimated according to Miller, 1972.

Production of PHA

The stock cultures (*C. necator* and isolate) were initially revived in nutrient broth for further studies were grown in mineral salt medium (g/L) containing 10g- Hydrolyzed Grass, 5 g-Glucose, 5 g-NaCl, 5 g-Di-potassium Hydrogen phosphate, 1g-Magnesium sulphate, 1 g-Ammonium sulphate, 1L-Distilled water in the pH7.4 ± 0.5 and incubated for 48 hrs at 30°C.

Extraction of PHA

After 48 hrs the pellet was air dried and weighed. Then PHA was extracted according to Santhanumand Sasidharan (2010), chloroform extraction method.

Quantification of Extracted PHA

The polymer granule was dissolved in concentrated Sulfuric acid (1 mg/ml) convert PHA into Crotonic acid, and was measured using spectrophotometer (Shimadzu UV-1800). A standard curve was prepared with Pure PHB (Sigma, Alderich), (Law and Slepecky, 1969).

Determination of Dry Cell Weight, Residual Biomass of PHA

The bacterial culture was centrifuged at 10,000 rpm for 5 min to obtain the cell pellet. The Cell pellet was dried to estimate the dry cell weight (DCW in mg/ml) followed by the procedure of (Du *et al.*, 2001). Residual biomass was estimated as the difference of the dry cell weight and dry weight of PHA extracted (Zakaria *et al.*, 2010).

Residual Biomass (mg/ml) = DCW (mg/ml) – Dry weight of extracted PHA (mg/ml).

Determination of % PHA Accumulation and PHA Concentration

The intracellular PHA accumulation (%) is estimated as the percentage of PHB mass in the dry weight of extracted PHA (Sangkharak and Prasertsan, 2008).

$$\text{PHA Accumulation (\%)} = \frac{\text{Dry weight of PHA extracted (mg/ml)}}{\text{DCW (mg/ml)}} \times 100$$

PHB Mass (μg/ml) = Obtained from Standard PHB assay.

Plackett-Burman Optimization Method

Plackett-Burman design was used to analyze the most significant media component influencing bacterial growth and maximal production of PHA (Stanbury *et al.*, 1984). In

Table 1: Media Components Considered for Optimization Study

Media Components	*Original (g/L)*	*High level (g/L) (H)*	*Low Level (g/L) (L)*
Hydrolyzed substrate (HS) (A)*	10	15	5
Glucose (B)	5	7.5	2.5
NaCl (C)	5	7.5	2.5
KCl(D)	1	1.5	0.5
K_2HPO_4 (E)	5	7.5	2.5
$(NH_4)_2SO_4$ (F)	1	1.5	0.5
$MgSO_4$ (G)	1	1.5	0.5

*A to G coded for respective variable in the Plackett-Burman design.

this study (n = 7) variables (production media components) were considered Here, the culture flasks of different high and low combinations of the media components were made to run in duplicates. The dry weight of PHA was taken as yield factor. The yields of each trial were determined (Table 1).

RESULTS AND DISCUSION

PHA Production Using Glucose and Hydrolyzed Substrate

The extracted PHA of isolate and *R. eutropha* were quantified and its residual biomass,% of PHA accumulation and its mass was determined. Pure cultures of SPY-1and *R. eutropha* showed greater efficiency of PHA production when utilizing hydrolyzed seed as sole carbon source compared to mixed cultures with the other combinations of substrates. For *R. eutropha*, % of PHA accumulation and PHA mass weree 41.77 and 97 µg/ml whereas *Kluyveraintermedia's* levels were found to be 42.26 and 100 µg/ml (Table 2).

Table 2: Comparision of PHA Production between Reference Strain and the Best Isolate

Micro-Organism	*Substrate*	*% PHA Accumulation*	*PHA Mass (µg/ml)*
Cupriavidusnecator	G+HG	8.5	25
	G+HS	2.38	11
	HG	28.97	73
	HS	41.77	97
Kluyvera intermedia	G+HG	16.03	49
	G+HS	9.52	38
	HG	26.76	52
	HS	42.26	100

PHA Optimization Using Statistical Analysis

The production media components were used as variables in Placket-Burman design (Table 3) for media optimization and the best influencing variables (Table 4) was identified based on ANOVA. *Kluyveraintermedia*dry cell weight and PHA accumulation were taken for analysis in the Placket-Burman design. The dry weight of extracted PHA was taken as yield factor (Table 3), with which the effect and the mean square were calculated for each variable, the f-test values were finally calculated to determine the most significant factors for the bacterial growth and PHA accumulation.

When probability tables were examined (Table 4) it was found that the factors A, E, F (Hydrolyzed substrate, K_2HPO_4 $(NH_4)_2SO_4$) showed large effect which were found to be most significant in influencing maximal PHA production. Whereas factors B, C, and G (Glucose, NaCl, $MgSO_4$) showed very low effect and D (KCl) was taken as dummy variable to calculate the standard error. It was proved that the presence of glucose and sodium chloride did not impact on PHA prodcution because the substrate (hydrolyzed grass) alone was able to induce PHA accumulation significantly (F value 51.96), without the influence of glucose (F value 3.99). The results were in accordance with other works available in literature, Lakshman *et al.,* 2004 and Yang *et al.,* 2010.

Table 3: Plackett-Burman Design

	A'	*B'*	*C'*	*D'*	*E'*	*F'*	*G'*	*Absorbance at 660 nm*	*Dry Weight of PHA/Wet Cell Weight*
1	H	H	H	H	H	H	H	0.134	0.063
2	H	L	L	L	H	L	H	0.140	0.133
3	L	L	H	L	L	H	H	1.162	0.073
4	L	H	L	H	L	H	H	0.936	0.597
5	H	H	H	L	H	H	L	0.077	0.110
6	H	L	L	H	L	L	L	0.194	0.035
7	L	H	H	H	H	L	L	1.664	0.131
8	L	L	L	L	L	L	L	1.122	0.055

Table 4: Analysis of Yields from Different Trials of Media Design

	A' *HS*	*B'* *Glucose*	*C'* *NaCl*	*D'* *KCl*	*E'* *K_2HPO_4*	*F'* *$(NH_4)_2SO_4$*	*G'* *$MgSO_4$*
Σ(H)	1.139	0.863	0.892	0.704	1.117	0.782	0.745
Σ(L)	0.375	0.651	0.622	0.81	0.397	0.732	0.769
Difference	0.764	0.212	0.27	–0.106	0.72	0.05	–0.024
Factor Mean Square	0.073	5.618×10^{-3}	9.113×10^{-3}	1.405×10^{-3}	0.065	3.125×10^{-4}	7.2×10^{-5}
f-Test Values	51.96	3.99	6.486	1	46.263	0.223	0.0512

The efficiency of PHA produced mainly depends on the choice of organism and availability of carbon source under the limiting condition of nitrogen, phosphate etc. The best PHA producer identified as *Kluyveraintermedia,* under the influence of minimal salt media with glucose as carbon source, produced PHA comparable to that of*Cupriavidusnecator*. Hydrolyzed (Bermuda) grass was selected as cheaper substrate and its impact as carbon source was explored under media optimization studies. It was found that the Hydrolyzed Grass and K_2HPO_4 as influencing factors on PHA production and after this optimization study a 1.22 fold increase in PHA was obtained. Thus Bermuda grass can be used as a low cost substrate in PHA production.

REFERENCES

Anderson, A.J. and Dawes, E.A. (1990). Occurrence, metabolism, metabolic role and industry uses of bacterial polyhydroxyalkanoates. *Microbiol. Rev.*, 54: 450–72.

Chen, L.F., Lafayette, W., Yang, C.M. and Clark, N.J. (1984). Quantitative hydrolysis of cellulose to glucose using Zinc Chloride. United States Patent 4452640.

Chen, L.F., Lafayette, W., Yang, C.M. and Clark, N.G. (1985). Selective hydrolysis of cellulose to glucose without degradation of glucose by Zinc chloride. United States Patent, 4637835.

Choi, J. and Lee, S.Y. (1997). Process analysis and economic evaluation for poly (3–1 hydrobutyrate) production by fermentation. *Bioprocess. Eng.*, 17: 335–42.

Du, G., Chen, J., Yu, J. and Lun, S. (2001). Continuous production of poly-3-hydroxybutyrate by *Ralstonia eutropha* in a two stage culture system. *J. Biotechnol*, 88: 59–65.

Kim, B.S., Lee, S.Y. and Chang, H.N. (1992). Production of poly-β-hydroxybutyrate by fed-batch culture of recombinant *Escherichia coli*. Bio*technol. Letters*, 14: 811–6.

Lakshman, K., Rastogi, N.K. and Shamala, T.R. (2004). Simultaneous and comparative assessment of parent and mutant strain of *Rhizobium meliloti* for nutrient limitation and enhanced polyhydroxyalkanoate (PHA) production using optimization studies. *Proc. Biochemistry*, 39: 1977–1983.

Law, J. and Slepecky, R.A. (1969). Assay of poly-hydroxybutyric acid. *J. Bacteriol*, 82: 52–55.

Lee, S.Y. (1996). Bacterial polyhydroxyalkanoates. Biotechnol. *Bioeng*, 49: 1–14.

Miller, G.L. (1972). Use of Dinitrosalicylic acid reagent for determination of reducing sugars. *Anal. Chem.*, 31: 426–428.

Sangkharak, K. and Prasertsan, P. (2008). Nutrient optimization for production of polyhydroxybutyrate from halotolerant photosynthetic bacteria cultivated under aerobic-dark condition. *Elec. J. Biotechnol*, 11: 3.

Santhanam, A. and Sasidharan, S. (2010). Microbial production of polyhydroxyalkanotes (PHA) from *Alcaligens* spp. and *Pseudomonas oleovorans*using different carbon sources. *Afr. J. Biotechnol*, 9: 3144–3150.

Stanbury, P.F., Whitaker, A. and Hall, S.J. (1984). Principles of Fermentation Technology. MPG Books Ltd., Great Britain.

Yamanaka, K., Kimura, Y., Aoki, T. and Kudo, T. (2010). Effect of ethylene glycol on the end group structure of poly (3-hydroxybutyrate). *Polym. Degrad. Stab.*, 95: 1284–1292.

Yamane, T., Fukunaga, M. and Lee, Y.W. (1995). Increased PHB productivity by high-cell-density fed-batch culture of *Alcaligeneseutrophus*, a growth associated PHB producer. *Biotechnol. Bioeng*, 50: 197–202.

Yang, Y.H., Brigham, C.J., Budde, C.F., Boccazzi, P., Willia, L.B., Hassan, M.A., Yusof, Z.A.M., Rha, C. and Sinskey, A.J. (2010). Optimization of growth media components for polyhydroxyalkonoate (PHA) production from organic acids by *Ralstoniaeutropha*. *Appl. Microbiol. Biotechnol*, 87: 237–2045.

Zakaria, M.R. Ariffin, H., Johar, N.A.M., Aziz, S.A., Nishida, H., Shirai Y. and Hassan, M.A. (2010). Biosynthesis and characterization of poly (3-hydroxybutyrate-co-3- hydroxybutyrate) copolymer from wild type *Comamonas sp*. EB172. *Polym. Deg. Stab.*, 95: 1382–1386.

Thalassospira frigidphilosprofundus sp. – A Novel Psychrophilic Marine Bacterium Isolated from the Deep-Sea Water of Bay-of-Bengal

Mrinmoy Ghosh, K.K. Pulicherla[1*], V.P.B. Rekha[1] and K.R.S. Sambasiva Rao[*]

Department of Biotechnology, Acharya Nagarjuna University
Nagarju Nanagar, Guntur, A.P.

[1]Department of Biotechnology, R.V.R. and J.C. College of Engineering Chowdavaram, Guntur, A.P.

E-mail: pkkanth2006@gmail.com; krssrao@yahoo.com

INTRODUCTION

Nature is the platform of varsity. Among of all organisms, microbial world gives a true example where immense varsity mainly found in deep sea environment, is the hostile for most of the marine micro-organisms (Radjasa 2004, Whitman *et al.*, 1998). Many of them adapted to survive in deep sea environment under extreme conditions e.g. constant darkness, high pressure, and low temperature, acidic or alkaline condition. More over complete knowledge about the origin and habitats of deep sea microbes are still limited. Hence, analysis of biogeochemical process and microbial function has become an important inference to understand the indigenous of deep sea extremophiles. A class of extremophiles has been successfully harmonized under these cold environment are known as psychrophiles.

During last few years, it has been recognized that biocatalysts from cold loving microorganisms provided a great potential in biotechnology, more particularly in food industries (Margesin *et al.*, 2005). Indeed, among of wide varieties of cold adapted

*Corresponding author.

enzymes, β-galactosidase from psychrophiles accelerated an interest in food industries. Therefore, to find new cold adapted β-galactosidase producing bacteria for industrial use, an attempt was made by collecting the deep-sea water and has successfully isolated the bacterium that exhibited two fold more productivity of proposed enzyme β-galactosidase, compared to previous reported psychrophilic bacteria (data not given). In our earlier study on phylogenetic analysis of 16S rRNA gene sequencing of deep-marine water isolates 3SC-21 revealed that the isolates was affiliated to the genus of *Thalassospira* (GenBank No. JQ 328041) (Pulicherla *et al.,* 2011). In the present study, we conducted the biochemical and FAME analysis of isolates *Thalassospira* strain 3SC-21. On the basis of the evidence presented herein, it could be evidentially considered that the described bacterium to be a novel species of the genus *Thalassospira.*

MATERIALS AND METHODS

Collection of the Strain, Media, Growth Condition

Strain 3SC-21 was isolated after prolonged continuous culture from the marine water sample. This water sample was collected from the depth of 1200 meters of Bay-of-Bengal where the annual temperature is approximately 5°C. The strain was routinely grown aerobically at 15°C in modified Zobell medium containing (in liter) 5 g peptone, 1 g yeast extract, 0.1 g ferric citrate, 5.9 g magnesium chloride, 20 g sodium chloride, 0.14 g sodium nitrate, 1.8 g calcium chloride, 0.55 g potassium chloride, 0.0024 g sodium fluoride, 0.008 g sodium phosphate, 0.08 g potassium bromide, pH was adjusted to 7.6 ± 0.2 and placed at 150 rpm.

Morphology and Biochemical Characterization

To determine the optimum level three main factors, *viz.*, temperature, pH and NaCl concentration, the strain 3SC-21 was incubated in Zobell medium with 0, 0.5, 1, 2, 4, 6, 8, 10, 12, 14, 16 and 20% NaCl concentration and was assessed at varying incubation temperatures, viz. 4, 10, 20, 25, 30, 33, 37 and 40°C with the pH range of 4 to 9 with a difference of 0.5 division. The morphology and motility of the cell were examined from 30 hours old culture. Cell morphology was observed under transmission electron microscope. Cell motility was checked using phase-contrast microscope.

The physiological and biochemical properties of isolates were tested using Hi-Media bacterial identification kit. Gram staining, catalase, oxidase and indole tests were carried out as per the standard protocols (Smibert and Krieg, 1994). The ability of cells' to utilize various carbon sources were examined using Hi-Media substrate utilization test kit.

FAME Analysis

After 30 hours growth, the cells were harvested from culture plate and used for analysis of cellular fatty acids. A loop full of bacterial cells was harvested from the streaked plate and placed in a clean culture tubes. The 1 ml of reagent A (contained 45 g NaOH, 150 ml methanol and 150 ml distilled water) added to each tubes containing cells. The tubes are securely sealed and heated in a boiling water bath for 5 min. and the tubes were continuously vertexed for 5–10 seconds. After completion of incubation time, 2 ml of reagent 2 was added in cooled tubes. The reagent 2 is used for methylation contained of 325 ml of 6 N HCl and 275 ml methyl alcohol. The tubes were vortexed well and heated for 10 min at 80°C. This step is critical in time and temperature and after specific time of interval the tubes were cooled and 1.25 ml of third reagent contained 200 ml hexane and 200 ml mehyl tetra butyl ether was added in each tube followed by recapping and gentile tumbling on a rotator for 10 min and the aqueous phage from the tubes was discarded. The base wash was carried out by addition of 3ml of reagent (contained 10.8 g NaOH dissolved in 900 ml distilled water) in the organic phase remaining in the tubes. The tubes were recapped and tumbled for 5 min. About 2/3 of the organic phase was pipetted in GC vial for analysis.

Gas chromatography having methyl silicon coated capillary column (0.25 mm in diameter, 15 m long, 0.25 µm depth in film) and a flame-ionization detector under increasing temperatures 150–240°C at 4°C/min. Using helium gas as a carrier at a flow rate of 2.3 ml/min and a split ratio of 1:30.

RESULTS AND DISCUSSION

The cells of isolate 3SC-21 have been observed as gram negative, spiral bacterium, motile. In 2 days of incubation, the bacterium formed regular, opaque, slightly yellowish colonies that do not swarm or luminesce. Based on the results, the bacterium 3SC-21 was categorized as psychrotolerant strain, with T_{opt} of 20°C and T_{max} of 30°C at pH 7.0 and had the capability of thriving at 4°C. It is also identified that NaCl is required for growth, and the optimum growth being observed at 2 to 5% NaCl (w/v) and surviving up to 12%. The optimum growth temperature and tolerance against the salt concentration suggest that the isolated one is a psychrophilic halophile. Oxidase and catalase utilization tests gave positive results. Moreover, it was figured out that the strain was positive for nitrate reduction test but negative for indole test (Table 2).

The strain utilized ten various amino acids and five organic substrates for growth (Table 2). The test results as outlined in the Bergey's manual of determinative bacteriology (Williams *et al.,* 1989) are evaluated with the present test result.

Table 1: Utilization of Various Carbon Sources by *Thalassospira frigidphilosprofundus* 3SC-21

S. No	*Carbohydrate*	*Utility*	*S. No*	*Carbohydrate*	*Utility*
1	Lactose	+	2	Xylose	–
3	Maltose	–	4	Fructose	+
5	Dextrose	+	6	Galactose	+
7	Raffinose	–	8	Trehalose	–
9	Melibiose	–	10	Sucrose	V
11	L-Arabinose	–	12	Mannose	V
13	Inulin	–	14	Sodium gluconate	–
15	Glycerol	–	16	Salicin	–
17	Dulcitol	–	18	Inositol	–
19	Sorbitol	–	20	Manitol	V
21	Adonitol	–	22	Arabitol	–
23	Erythritol	–	24	Alfa-Methyl-D-Glucose	–
25	Ramnose	–	26	Cellobiose	V
27	Melezitose	–	28	Alfa-Methyl-D-mannoside	–
29	Xylite	–	30	Sorbose	+
31	Esculin Hydrolysis	–	32	D-Arabinose	–
33	Citrate utilization	+	34	Malonate utilization	–

Note: +, positive (90% above); –, negative (below 35%); v, weak (below 90%).

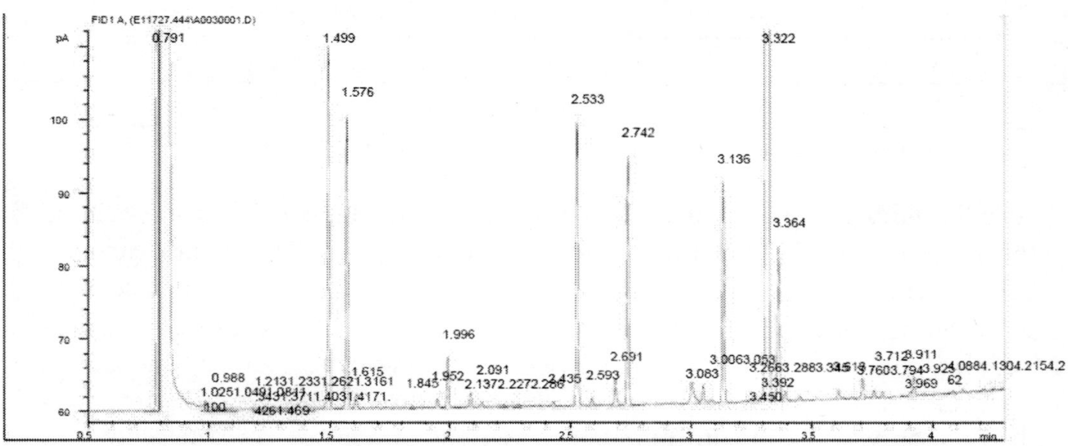

Fig. 1: Fatty Acid Profiling of *Thalassospira frigidphilosprofundus* 3SC-2

Table 2: Results of Biochemical Analysis of *Thalassospira frigidphilosprofundus* 3SC-21

S. No	*Test*	*Utility*	*S. No*	*Amino Acids and Amines*	*Utility*	*S. No*	*Organic Acids*	*Utility*
1	ONPG	+	1	Glycine	+	1	Pyruvate	+
2	Lysine utilization	–	2	L-Leucine	–	2	Succinate	+
3	Otnithine utilization	–	3	L-Serine	+	3	Lactate	+
4	Urease	–	4	L-Threonine	+	4	Citrate	V
5	Phenylalanine	V	5	L-Arginine	+	5	Propionate	V
6	Nitrate reduction	+	6	L-Tyrosine	+	6	3 Hydrobutyrate	+
7	H_2S production	–	7	Cystine	v	8	Malate	+
8	Citrate utilization	+	8	L-Alanine	+			
9	Voges Proskauer's	–	9	Aminobutyric acid	+			
10	Methyl red	+	10	L-Ornithine	+			
11	Indole test	–	11	L-Citrµline	–			
12	Malonate utilization	–	12	L-Aspartic acid	–			
13	Catalase	+	13	L-Glutamine	–			
14	Oxidase	+	14	L-Lysine	+			
15	Glucose	+	15	L-Histidine	v			
			16	Sarcosine	–			
			17	Putrescine	–			
			18	Valine	+			

Note: +, positive (90% above); –, negative (below 35%); v, weak (below 90%).

The major fatty acid profile of strain 3SC-21 as follows: 0.10% Pelargonic acid ($C_{9:0}$), 0.09%Capric acid ($C_{10:0}$), 0.06% Undecylic acid ($C_{11:0}$), 0.23% Lauric acid ($C_{12:0}$), 0.08% Myristic acid ($C_{14:0}$), 2.99% Palmitic acid ($C_{16:0}$), 0.27% Margaric acid ($C_{17:0}$), 1.90% Stearic acid ($C_{18:0}$), 0.25% Cis-11-eicosenoic acid ($C_{20:1}\omega7c$) (Figure 1). Moreover, the strain 3SC-21 also has distinct physiological and chemotaxonomic properties i.e. temperature for optimal growth, salinity tolerance, amino acid utilization, fatty acid content, utilization of various carbon sources that allow 3SC-21 distinguished from the other species of genus *Thalassospira* (Table 3).

Table 3: Comparison of Marine isolate *Thalassospira frigidphilosprofundus* 3SC-21 against to the type strains of *Thalassospira* species

Characteristic	*3SC-21*[1]	*T. tepidiphila*[2]	*T. lucentensis*[3]	*T. xiamenansis*[4]	*T. profundimaris*[5]	*T. permensis*[6]
Morphology	Curved rods	Vibrioid to spiral	Vibrioid to spiral	Curved rods	Curved rods	Curved rods to Vibrioid
Motility	Motile	Motile	Motile	Motile	Non Motile	Motile
Growth at 4°C 40°C	 + –	 – +	 – +	 + ND	 ND +	 - or w +
% of NaCl	0.5–12	2.0–16.0	2.0–10.0	2.0–10.0	2–8.0	0–11
Optimum NaCl	2	2	ND	3	3	2–5
Reduction Nitrate to Nitrite	–	+	+	+	–	+
Hydrolysis of Tween 40 Tween 80	 + –	 + w	 + +	 + –	 – W	 ND ND
Oxygen Requrement	Areobic	Facultative anaerobe	Facultative anaerobe	Facultative anaerobe	Areobic	Areobic

Note: +, Positive; –, Negative; w, weak; ND, No Data.

[1]data from this study; [2]data from Kodama *et al.*, 2008; [3]data from Lòpez-Lòpez *et al.*, 2002; [4,5]data from Liu *et al.*, 2007; [6]data from Plotnikova *et al.*, 2011.

Therefore, we propose that 3SC-21 to be distinguished as novel species: *Thalassospira frigidphilosprofundus*.

Description of *Thalassospira frigidphilosprofundus* sp. Nov.

Thalassospira frigidphilosprofundus (*frigid. philos. profundus*. gen. n. *frigid* cold; Gr. adj. *philos* loving; L. adj. *profundus* deep; N.L. gen. n. *frigidphilosprofundus* cold loving from deep sea, from where the type strain was isolated).

The genus *Thalassospira* was proposed by Lòpez-Lòpez *et al.* (2002), the cells are gram-negative, vibroid to spiral, aerobic, motile, halotolerant bacteria belonged to the class *Alphaproteobacteria* and family *Rhodospirillaceae*. According to the latest literature collected during the preparation of this script, the genus *Thalassospira* comprised four recognized species: *Thalassospira lucentensis* (Lòpez-Lòpez *et al.*, 2002), *Thalassospira xiamenansis* (Liu *et al.*, 2007), *Thalassospira profundimaris* (Liu *et al.*, 2007), *Thalassospira tepidiphila* (Kodama *et al.*, 2008) and *Thalassospira permensis* (Plotnikova *et al.*, 2011). All members of the genus "*Thalassospira*" have been isolated from seawater.

Cells are Gram-negative, vibroid to spiral, motile, flagellated. Colonies are light yellow, smooth, circular and convex. *Thalassospira frigidphilosprofundus* is psychrophilic halophile, anaerobic microbe. Growth occurs at 4°C to 30°C at the pH range of 6.0–8.0, whereas optimum temperature for growth was observed at 20°C and at pH of 7.0. The optimum growth observed at 2 to 5% NaCl (w/v) and survived up to 12%. Positive results were obtained with catalase and oxidase tests, citrate utilization and methyl red tests wherein indole test had showed negative. H_2S were not produced. Nitrate is reduced. The following amino acids Glycine, L-Serine, L-Threonine, L-Arginine, L-Tyrosine, L-Alanine, Amino butyric acid, L-Ornithine, L-Lysine, Valine are utilized. The priority for utilization of carbon sources were as follows: Lactose, Dextrose, Fructose, Galactose, Sorbose, Tween 40 Sucrose, Mannose, Manitol, Cellobiose (Table 2). The major cellular fatty acid profile of strain 3SC-21 is as follows: 0.10% Pelargonic acid ($C_{9:0}$), 0.09% Capric acid ($C_{10:0}$), 0.06% Undecylic acid ($C_{11:0}$), 0.23% Lauric acid ($C_{12:0}$), 0.08% Myristic acid ($C_{14:0}$), 2.99% Palmitic acid ($C_{16:0}$), 0.27% Margaric acid ($C_{17:0}$), 1.90% Stearic acid ($C_{18:0}$), 0.25% Cis-11-eicosenoic acid ($C_{20:1}\omega7c$).

ACKNOWLEDGEMENTS

The authors are grateful to the Department of Biotechnology (DBT), Govt. of India for providing the financial support. The authors also acknowledge the technical support provided by Acharya Nagarjuna University and R.V.R. and J.C. C.E., Guntur, India. The authors are also thankful to NCAOR, Goa, India.

REFERENCES

Ezaki, T., Hashimoto, Y. and Yabuuchi, E. (1989). Fluorometric deoxyribonucleic acid-deoxyribonucleic acid hybridization in which radioisotopes are used to determine genetic relatedness among bacterial strains. *Int J Syst and Evol Microbiol,* 39: 224–229.

Ghosh, M., Pulicherla, K. K., Rekha, V.P.B., Kumar, R.P. and Rao, K.R.S.S. (2012). Cold active β-galactosidase from *Thalassospira* sp. 3SC-21 to use in milk lactose hydrolysis: a novel source from deep waters of Bay-of-Bengal. *World J Microbiol Biotechnol* 28(9): 2859–2869.

Kodama, Y., Stiknowati, L.I., Ueki, A., Ueki, K. and Watanabe, K. (2008). *Thalassospira tepidiphila* sp. nov., a polycyclic aromatic hydrocarbon-degrading bacterium isolated from seawater. *Int J Syst and Evol Microbiol,* 58: 711–715.

Liu, C., Wu, Y., Li, L., Ma, Y. and Shao, Z. (2007). *Thalassospira xiamenensis* sp. nov. and *Thalassospira profundimaris* sp. nov. *Int J Syst and Evol Microbiol,* 57: 316–320.

Lòpez-Lòpez, A., Pujalte, J., Maria, Benlloch, S., Mata-Roig, M., Mora, R.R., Garay, E. and Valera, F.R. (2002). *Thalassospira lucentensis* gen. nov., sp. nov., a new marine member of the α-proteobacteria. *Int J Syst and Evol Microbiol,* 52: 1277–1283.

Margesin, M., Fauster, V. and Fonteyne, P.A. (2005). Characterization of cold-active pectate lyases from psychrophilic Mrakia frigid. *Letter in Appl. Microbio,* 40: 553–459.

Plotnikova, E.G., Anan'ina, L.N., Krausova, V.I., Ariskina, E.V., Prisyazhnaya, N.V., Lebedevc A.T., Demakov, V.A. and Evtushenko, L.I. (2011). *Thalassospira permensis* sp. nov., A New Terrestrial Halotolerant Bacterium Isolated from a Naphthalene-Utilizing Microbial Consortium1. *Microbiology* 80(5): 703–712.

Radjasa, O.K. (2004). Deep-sea bacteria and their biotechnological potential. *Journal of Coastal Develop,* 7(3): 109–118.

Whitman, W.B., Coleman, D.C. and Wiebe, W.J. (1998). Prokaryotes: The unseen majority. *Proc. Natl Acad Sci.,* 95: 6578–6583.

Williams, S.T., Wilkins (1994). Bergey's manual of determinative bacteriology, 9th edn. Williams and Wilkins, Baltimore.

Isolation and Detection of Mutation of *rpsL* Gene in *Mycobacterium tuberculosis*

R. Thirumurugan*, M. Muthuraj[1] and Antony V. Samrot

Department of Biotechnology, Sathyabama University, Jeppiaar Nagar, Sholinganallur, Chennai, Tamil Nadu

[1]Government Hospital for Chest Diseases, Gorimedu, Puducherry

E-mail: thirumuruganphd@gmail.com

INTRODUCTION

Tuberculosis is an infectious disease caused by *Mycobacterium tuberculosis* and infects the lungs (as pulmonary TB) but can also affect the central nervous system, the lymphatic system, the circulatory system, the genitourinary system, bones, joints and even the skin. Over one third of the world's population has been exposed to the TB bacterium and new infections occur at a rate of one per second (Raviglione and O'Brien, 2004). The excessive emergence of drug resistant tuberculosis has stimulated interest in the understanding of the underlying mechanisms of drug resistance in *M. tuberculosis* and significant progress has been made in this field (Zhang and Talenti, 2000). The World Health Organization (WHO) 2010 report estimates that 440,000 cases of multidrug-resistant (MDR) tuberculosis (TB) emerged in 2008, with 5.4% of these cases being extensively drug resistant (XDR) TB (WHO, 2010). Streptomycin has been an important drug for the treatment of tuberculosis since its discovery in 1944. Streptomycin binds to the 16S rRNA of the bacterial ribosome and interferes with the binding of formyl-methionyl-tRNA to the 30S subunit and prevents initiation of protein synthesis. But numerous strains of *M. tuberculosis* are now streptomycin resistant and the resistance are arisen from mutations in the *rspL* gene (encoding the S12 protein target for streptomycin binding). Mutations associated with streptomycin resistance have been identified in the

*Corresponding author.

genes encoding 16S rRNA (*rrs*) and ribosomal protein S12 (*rpsL*). Mutation such as amino acid substitutions in *rpsL* alters the 16S rRNA structure and disrupts interactions between 16S rRNA and streptomycin, a process that results in resistance against streptomycin (Tudo *et al.*, 2010, Bottger, 1994; Honore and Cole, 1994; Meier *et al.*, 1994; Nair *et al.*, 1993). Fukuda *et al.* (1999) suggested that base sequencing analysis of the *rpsL* gene is useful for the rapid prediction of highly streptomycin resistant strains of *M. tuberculosis*. Honore and Cole (1994) studied that *M. tuberculosis*, has high-level resistance that results from missense mutations in the genes encoding two components of the ribosome, the 16SrRNA and the S12 protein. The most frequently occurring mutation associated with streptomycin resistance is an A-to-G transition in codon 43 of the *rpsL* gene which results in the substitution of Arg for Lys at position 43 (Lys-43Arg) in the ribosomal protein S12 (Heym *et al.*, 1994). Two other mutations associated with streptomycin resistance have also been found in *rpsL* gene, and these lead to the substitutions of Lys-43Thr and Lys-88Arg. When cytosine-512, adenine-513, and cytosine-516 are substituted by other bases, high-level resistance results (Lipin *et al.*, 2007, Sun *et al.*, 2010, Brzostek *et al.*, 2004, Ramaswamy *et al.*, 2004, Xue-Qiong *et al.*, 2006).

In this study, to perceive the consequence of mutations in drug resistance, streptomycin resistant *M. tuberculosis* clinical isolate was isolated and corresponding *rpsL* gene was amplified. A mutation in rpsL was detected by sequencing the amplified product and subjecting the sequence to BLASTn and BLASTx analysis.

MATERIALS AND METHODS

Sample Collection

The sputum sample was collected from an in-patient with lower respiratory infection (1 sample), Government Hospital for Chest Diseases, Puducherry, India.

Identification and Isolation

Species identification performed with Ziehl-Neelsen Staining, Fluorescent staining and cultured on Lowenstein Jensen (LJ) Medium using Modified Petroff's Method and Biochemical identification such as Para-nitrobenzoic acid (PNB) inhibition test.

Proportional Streptomycin Sensitivity Test Method

Proportion sensitivity tests are indirect tests done on Lowenstein-Jensen medium (McNerney *et al.*, 2000). This method indicates the number of organisms resistant to each concentration expressed as a proportion of those growing on the drug-free slopes and is thus, not affected by changes in inoculum unlike the standard tests. The proportion test is performed on all per-treatment cultures found to be resistant (or doubtful) to Streptomycin by the standard method.

Percentage of Streptomycin drug resistant bacilli present in the bacterial population is calculated by the following formula:

$$\% \text{ Streptomycin resistant bacilli} = \frac{\text{No. of CFU on drug slopes} \times \text{Dilution factor}}{\text{No. of CFU on drug-free slopes} \times \text{Dilution factor}} \times 100$$

DNA Extraction

Mycobacterial genomic DNA was extracted by the procedure (Mani *et al.,* 2003). Essentially, three loopful of culture was taken in 100 µl of distilled water and was homogenized. The entire homogenized samples were treated with 50 µl of lysozyme and kept for overnight incubation at 37 °C. 70 µl of 14% SDS and 6µl of Proteinase K was added and incubated for 15 minutes at 65°C. It was centrifuged at 10000 rpm and pellet was discarded. 10 µl of 5 M NaCl and 80µl of CTAB/NaCl was added and incubated at 65°C for 10 minutes. 800 µl of Phenol: Chloroform: Isoamyl alcohol (25:24:1) mixture was added and centrifuged at 10000 rpm for 10 minutes. The supernatant was transferred to a fresh tube and 600 µl of Isopropanol was added and incubated overnight at –20°C. It was centrifuged for 10 minutes at 12000 rpm at 4°C. The pellet was washed with 70% ethanol and the pellet was air-dried and was dissolved in 20 µl of 1x TE buffer.

Species Identification by Amplification of IS6110 Element

The isolated template DNA was amplified using IS6110 primer which confirms the template DNA as *M. tuberculosis* in authorized thermal cycler (Eppendorf). The PCR reaction mixture was set up as follows:

Milli Q water	31.0 µl
10x Buffer	5.0 µl
$MgCl_2$	2.0 µl
Taq Polymerase	1.0 µl
dNTPs	2.0 µl
Mtb (10 pmol/µl)	2.0 µl
Mtb (10 pmol/µl)	2.0 µl
DMSO	3.0 µl
Template DNA	2.0 µl

The PCR cycling parameters were 94°C for 5 minutes; followed by 40 cycles of 94°C for 1 minute, 57°C for 1 minute and 74°C for 1 minute; and a final extension of 74°C for 5 minutes. The PCR was then kept at hold at 4°C for 15 minutes. The amplified PCR product was withdrawn from thermal cycler and run on a 2% Agarose gel in TAE buffer. The Ethidium bromide stained gels were observed in a UV Transilluminator and the gel was documented (BioRad).

Amplification of *rpsL* Gene

Mycobacterial DNA was extracted as stated above (Mani *et al.,* 2003). With the isolated DNA the Taq polymerase, dNTPs, $Mgcl_2$, Milli Q water, 10x Buffer, DMSO, template DNA, *rpsL* primers *rpsL* 1 (5'-GGCCGACAAACAGAACGT-3') and *rpsL* 2 (5'-GTTCACCAACTGGGTGAC-3') were used for amplification of *rpsL* gene as mentioned below.

Milli Q water	31.0 µl
10x Buffer	5.0 µl
$Mgcl_2$	2.0 µl
Tag Polymerase	1.0 µl
dNTPs	2.0 µl
F-Primer (10pmol/µl)	2.0 µl
R-Primer(10pmol/µl)	2.0 µl
DMSO	3.0 µl
Template DNA	2.0 µl

The PCR cycling parameters were maintained as following: 94°C for 5 minutes; followed by 57°C for 1 minute and 74°C for 1 minute; this was done for 40 cycles and a final extension of 74°C for 5 minutes. The PCR was then kept at hold at 4°C for 15 minutes. The amplified PCR product for *rpsl* gene from clinical isolate strain was run on 2% agarose gel.

DNA Sequencing

The amplified *rpsL* gene was eluted from the gel by Gel cleanup kit. The purified PCR product was directly sequenced in an automated DNA Sequencer (Bio Trace, Singapore).

Detection of Mutation

The obtained nucleotide sequence was compared with the wild type *M. tuberculosis* ($H_{37}Rv$) sequence to detect the mutation using BLASTn (to detect nucleotide mutation) and BLASTx (to know the amino acid changes) the bioinformatics tools available at National Center for Biotechnology Information (NCBI).

RESULTS AND DISCUSSIONS

The collected sample was confirmed as tubercle bacilli after Ziehl-Neelsen Staining and Fluorescent staining. Small, rough and cream coloured colonies were observed on LJ medium. The biochemical tests performed to confirm *M. tuberculosis* are tabulated in Table 1.

Table 1: Biochemical Tests for Identification of *M. tuberculosis*

S. No.	*Name of the Tests*	*Results*
1.	Ziehl Nielsen staining	Appearance of pink coloured bacilli with beaded in blue back ground
2.	Fluorescent staining	Appearance golden yellow coloured bacilli in dark back ground
3.	Niacin test	Pink to red colour formation
4.	Nitrate reduction test	Appearance of deep red colour
5.	Catalase test	Appearance of bright effervescence
6.	Tween hydrolysis	Amber to pink colour formation
7.	PNB test	Growth on L.J medium supplemented with PNB

Pure culture of *M. tuberculosis* isolated from LJ medium was subjected to drug sensitivity test using proportion sensitivity test. Individual colonies on streptomycin drug slants, and drug free LJ slants were counted and tabulated as in Table 2. Since the percentage of resistance is greater than 1%, the strain was considered to be resistant to streptomycin (Figure 1).

Fig. 1: Growth of *M. tuberculosis* on LJ Medium Slants.
1. Control, 2. $H_{37}R_V$, 3. Clinical Isolate

Table 2: Growth of *M. tuberculosis* Streptomycin Drug Slants, and Drug Free LJ Slants

Dilution	*Growth of Mtb on Drug Free L.J. Medium Slants*	*Growth of Mtb on L.J. Medium Drug Slants*
S1	TNTC	87
S2	98	23
S3	43	7
S4	11	–

Mycobacterial DNA was isolated from the organism grown on LJ medium slant and was subjected for amplification using species specific primers IS6110 to confirm *M. tuberculosis.* A clear band with 123 bp amplified product confirmed that the clinical isolate was *M. tuberculosis* (Figure 2).

Mycobacterial DNA was isolated from the *M. tuberculosis* clinical isolate and $H_{37}Rv$ wild type strain. The templates DNA were amplified using *rpsl* primers. The amplified *rpsL* product was run in a gel (Figure 3) and the amplified DNA was sequenced to study the mutations in it. *M. tuberculosis* isolated in this study had mutation in the *rpsL* gene, which was detected after sequencing and comparing with the normal wild-type $H_{37}R_V$ strain using BLASTn. The mutation and the amino acid changes pattern are shown in Table 3.

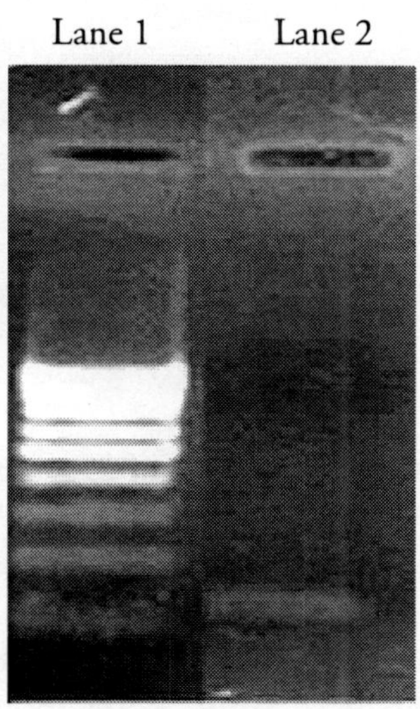

Fig. 2: Amplified Products for IS6110 Primer. Lane 1–100 bp ladder and Lane 2–123 bp Amplified Product of Suspected 123 Clinical Isolate

Table 3: Nucleotide and Amino Acid Changes in Drug Target Genes

S. No.	*Gene*	*Nucleotide Changes*	*Amino Acid Change*
1	*rpsL*	Substitutions at 36, 41, 48, 49, 54, 57, 58, 60, 66, 136, 146, 147, 149, 150, 151, 160, 172, 175, 176, 177, 180, 181, 183, 190, 191, 193, 207, 380 and 475 bp Insertions at 154, 190, 306 bp Deletion at 161 bp	Glu → Arg (112, 128 codons) Gly →Val (109 codon) Arg → Leu (103 codon) Pro → Ala (156 codon) Asn → Ser (167 codon) Thr → Ile (127 codon) His → Tyr (124 codon) Asp → Val (122 codon) Ala → Gly (121 codon) Leu → Pro (120 codon) His → Leu (119 codon) Pro → Arg (115 codon) His → Gly (114 codon)

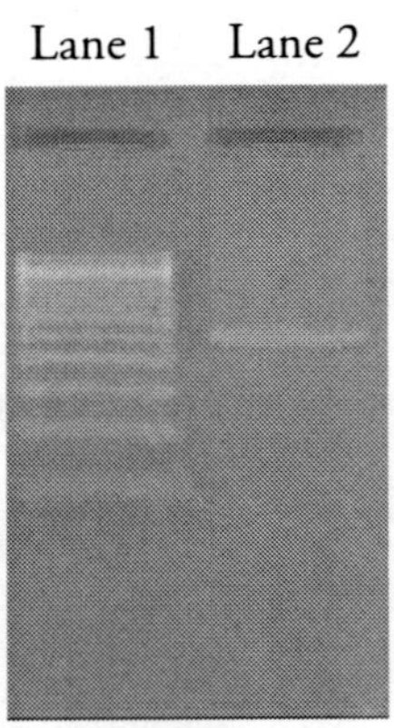

Fig. 3: Amplification of *rpsL* Gene: Lane 1–100 bp Ladder, Lane 2–548 bp Amplified Product of Suspected Clinical Isolate

The sample was found to possess *M. tuberculosis* by Ziehl-Neelsen Staining and Fluorescent staining. *M. tuberculosis* is normally done by conventional microscopic method and the sensitivity was found to be low (Steingart *et al.,* 2006). Steingart *et al.* (2006) found fluorescence microscopy to be more sensitive than conventional microscopy. The growth of the organism on LJ medium also confirmed that the organism was *M. tuberculosis.*

The isolated organism was found to be more than 1% resistance to streptomycin which was confirmed by proportional streptomycin sensitivity test. Resistance to this antibiotic found to be conferred either by missense mutations in the *rpsL* gene, which encodes the ribosomal protein S12 (ATS–2000) (Chintus *et al.*, 2002) or by the production of aminoglycoside-modifying enzymes (Meier *et al.*, 1994).

Amplification with species specific primers IS6110 was used to confirm *M. tuberculosis*. Deshpande *et al.* (2007) found the sensitivity of using IS6110 primers as 91.4% in tuberculous meningitis (TBM) and also concluded that the performance of an in-house IS6110 PCR assay was valuable in the rapid diagnosis of Tuberculous meningitis (TBM). The isolated Mycobacterial DNA was subjected for amplification of *rpsL* gene using *rpsL* primers and the amplified product was sequenced. The sequence was compared with the sequence of $H_{37}Rv$ (wild type) strain using Bioinformatics tool BLASTn and BLASTx available at National Centre for Biotechnology Information. The clinical isolate was found to have mutation which are as follows: Insertion at 154,190,306bp, Deletion at 161 bp, substitution at 36, 41, 48, 49, 54, 57, 58, 60, 66, 136, 146, 147, 149, 150, 151, 160, 172, 175, 176, 177, 180, 181, 183, 190, 191, 193, 207, 380 and 475 bp, which caused the amino acid change of Glu → Arg (112,128 codons), Gly → Val (109 codon),Arg → Leu (103 codon), Pro → Ala (156 codon), Asn → Ser (167 codon), Thr → Ile (127 codon), His → Tyr (124 codon), Asp → Val (122 codon), Ala → Gly (121 codon), Leu → Pro (120 codon), His → Leu (119 codon), Pro → Arg (115 codon), His → Gly (114 codon) in the *rpsL* gene. These mutations in *rpsL* gene are found to confer streptomycin resistance. Xue-Qiong *et al.* (2006) found missense mutations in codon 43 or 88 of *rpsL* resulting in a Lys to Arg substitution in 79 samples among the 98 Streptomycin resistant isolates. The mutation and the amino acid changes pattern are shown in Table 3. Mutations in *rpsL* that result in a high level of resistance to streptomycin fall into two categories, restrictive and nonrestrictive. Most of the isolates were found to be nonrestrictive genotypes (Allen and Noller, 1983, Lipin *et al.*, 2007, Sun *et al.*, 2010, Brzostek *et al.*, 2004, Ramaswamy *et al.*, 2004). Tudo *et al.* (2010) found 17 isolates among 69 streptomycin resistant *M.tuberculosis* to have mutations (codons 43 and 88) associated with high MIC levels. Das *et al.* (2009) found 29 isolates out of 49 isolates which showed high degree of streptomycin resistance to exhibit mutations in *rpsL* gene, while the other isolates which are also streptomycin resistance showed no mutation in *rpsL* gene.

In this study, streptomycin resistant *M. tuberculosis* was isolated from in-patient, Government Hospital for Chest Disease, Puducherry, India. The isolated organism was found to be streptomycin resistant, which was confirmed by proportional sensitivity test in LJ slopes. Mutation in *rpsL* at various sites was found after sequencing the amplified product and comparing the sequence with the $H_{37}Rv$ (wild type) strain. Mutations in coding region of *rpsL* confer the higher resistance to streptomycin drug. In this present study, mutation in relation with streptomycin resistance was found. Which may cause changes in the conformation of ribosomal subunit leads to the failure in drug action. Moreover our data clearly show that high-level resistance to this antibiotic is due to mutations in the *rps*L or *rrs* genes and that it can be detected by prompt molecular methods.

REFERENCES

Allen, P.N. and Noller, H.F. (1983). Mutations in ribosomal S4 and S12 Influence the higher order structure of 16S ribosomal RNA. *J. Mol. Biol.*, 208: 457–468.

Bottger, E.C. (1994). Resistance to drugs targeting protein synthesis in mycobacteria. *Trends Microbiol.*, 2:416–421.

Brzostek, A., Sajduda, A., Sliwinski, T., Augustynowicz-Kopec, E., Jaworski, A., Zwolska, Z. and Dziadek, J. (2004). Molecular characterisation of streptomycin-resistant Mycobacterium tuberculosis strains isolated in Poland. *Int. J. Tuberc. Lung Dis.*, 8(8): 1032–5.

Chen, C.W. and Thomas, C.A. (1980). Recovery of DNA segments from agarose gels. *Anal Biochem.*, 101: 339–341.

Chintus, C., Lucas, S. and Mudenda, V. (2002). Diseases at necropsy in African children dying from respiratory illness: a descriptive necropsy study. *Lancet*, 360: 985–90.

Das, R., Gupta, P., Singh, P., Katoch DSCK and Katoch, V.M. (2009). Association of mutations in rpsL gene with high degree of streptomycin resistance in clinical isolates of Mycobacterium tuberculosis in India. *Indian J. Med. Res.*, 129: 108–110.

Deshpande, P.S., Kashyap, R.S., Ramteke, R.S., Nagdev, K.J., Purohit, H.J., Taori, G.M. and Daginawala, H.F. (2007). Evaluation of the IS6110 PCR assay for the rapid diagnosis of tuberculous meningitis. *Cerebrospinal Fluid Res.*, 4:10.

Fukuda, M., Koga, H., Ohno, H., Yang, B., Hirakata, Y., Maesaki, S., Tomono, K., Tashiro, T. and Kohno, S. (1999). Relationship between genetic alteration of the rpsL gene and streptomycin susceptibility of *Mycobacterium tuberculosis* in Japan. *J. Antimicrob. Chemother*, 43, 281–284.

Heym, B., Honore, C., Truffot-Pernot, A., Banerjee, C., Schurra, W., Jacobs, R., van Embden, J.D.A., Grosset, J.H. and Cole, S.T. (1994). Implications of multidrug resistance for the future of short-course chemotherapy of tuberculosis: a molecular study. *Lancet*, 344: 293–298.

Honore, N. and Cole, S.T. (1994). Streptomycin resistance in mycobacteria. *Antimicrob. Agents Chemother,* 38: 238–242.

Lipin, M.Y., Stepanshina, V.N., Shemyakin, I.G. and Shinnick, T.M. (2007). Association of specific mutations in katG, rpoB, rpsL and rrs genes with spoligotypes of multidrug-resistant Mycobacterium tuberculosis isolates in Russia. *Clin. Microbiol. Infect.,* 13(6): 620–626.

Mani, C., Selvakumar, N., Kumar, V., Narayanan, S. and Narayanan, P.R. (2003). Comparison of DNA sequencing, PCR-SSCP and PhaB assays with indirect sensitivity testing for detection of rifampcin resistance in *Mycobacterium tuberculosis. Int. J. Tuberc. Lung Dis.*, 7: 625–659.

McNerney, R., Kiepiela, P., Bishop, K., Nye, P.M. and Stoker, N.G. (2000). Rapid screening of *Mycobacterium tuberculosis* for susceptibility to rifampicin and streptomycin. *Int. J. Tuberc. Lung Dis.*, 4: 69–75.

Meier, A.T., Kirschner, P.T. and Bange, F.C. (1994). Genetic Alterations in Streptomycin-Resistant *Mycobacterium tuberculosis*: Mapping of Mutations Conferring Resistance. *Antimicrob. Agents Chemother,* 38: 228–233.

Meier, P., Sander, K.J., Scholz., S.M. and Bottger, E.C. (1994). Correlation of Molecular Resistance Mechanisms and Phenotypic Resistance Levels in Streptomycin-Resistant *Mycobacterium tuberculosis. Antimicrob. Agents Chemother,* 40 (11): 2452–2454.

Nair, J., Rouse, D.A., Bai, G.H. and Morris, S.L. (1993). The *rpsL* gene and streptomycin resistance in single and multi-drug resistant strains of *Mycobacterium tuberculosis. Mol. Microbiol,* 10: 521–524.

Ramaswamy, S.V., Dou S.J., Rendon, A., Yang, Z., Cave, M.D. and Graviss, E.A. (2004). Genotypic analysis of multidrug-resistant Mycobacterium tuberculosis isolates from Monterrey, Mexico. *J. Med. Microbiol.*, 53(2): 107–13.

Raviglione, M.C. and O'Brien, R.J. (2004). Tuberculosis, in Kasper, D.L, Braunwald, E., Fauci, A.S., Hauser, S.L., Longo, D.L, Jameson, J.L., Isselbacher, K.J, eds.: Harrison's Principles of Internal Medicine, 16th ed., *McGraw-Hill Professional*, pp. 953–966.

Steingart, K., Henry, M., Ng, V., Hopewell, P., Ramsay, A., Cunningham, J., Urbanczik, R., Perkins, M., Aziz M. and Pai, M. (2006). Fluorescence versus conventional sputum smear microscopy for tuberculosis: A systematic review. *Lancet Infect. Dis.*, 6: 570–581.

Sun, Y.J., Luo, J.T., Wong, S.Y. and Lee, A.S. (2010). Analysis of rpsL and rrs mutations in Beijing and non-Beijing streptomycin-resistant Mycobacterium tuberculosis isolates from Singapore. *Clin. Microbiol. Infect.*, 16(3):287–9.

Tudo, G., Rey, E., Borrell, S., Alcaide, F., Codina, G., Coll, P., Martın-Casabona, N., Montemayor, M., Moure, R., Orcau, A., Salvado, M., Vicente, E., Gonzalez-Martın, J. (2010). Characterization of mutations in streptomycin-resistant *Mycobacterium tuberculosis* clinical isolates in the area of Barcelona. *J. Antimicrob. Chemother,* 65: 2341–2346.

World Health Organization (2010). Multidrug and extensively drug-resistant TB (M/XDR-TB): 2010 global report on surveillance and response. World Health Organization, Geneva, Switzerland.

Xue-Qiong, W., Yang, L., Jun-Xian, Z., Jian-Qin, L., Guang-Yu, Z., Hong-Min, L., Cui-Huan, L., Bei-Chuan (2006). Detection of Streptomycin Resistance in Mycobacterium tuberculosis Clinical Isolates Using Four Molecular Methods in China. *Acta Genetica Sinica*, 33(7) : 655–663.

Zhang, Y. and Telent, A. (2000). Genetics of drug resistance in Mycobacterium tuberculosis. In Molecular Genetics of Mycobacteria. Edited by Hatfull *GF,* Jacobs WR. *Washington DC*: ASM Press: 235–254.

Study of Lipids for the Bio-Synthesis of Bio-Fuels from *Agave americana*

V. Jaganathan*, N. Sivaram, S. Vishnu, R. Ganesh Babu and P. Ponmurugan

Department of Biotechnology, K.S. Rangasamy College of Technology, Tiruchnegode

E-mail: vichuv93@gmail.com

INTRODUCTION

Bioenergy is one of the most important components to substitute of fossil fuels. The need of energy is increasing continuously, because of increases in industrialization and population [J. Goldemberg, 2000]. The basic sources of this energy are petroleum, natural gas, Coal, hydro and nuclear [M.G. Kulkarni, 2006]. The major disadvantage of using petroleum based fuels is atmospheric pollution created by the use of petroleum diesel. Petroleum diesel combustion is a major source of greenhouse gas (GHG). Apart from these emissions, petroleum diesel is also major source of other air contaminants including NO_x, SO_x, CO, particulate matter and volatile organic compounds. Also, the demand for fuels from crude oils is increasing and this is making great hike in the cost of petroleum and its products [L.D. Klass, 1998]. Biomass is one of the better sources of energy[2]. The Biofuel is produced by the trans-esterification of the triglyceride obtained from the Biomass of the organisms. Large-scale introduction of biomass energy could contribute to sustainable development on several fronts, environmentally, socially and economic [W. C. Turkenberg, 2000]. There are many sources of such Lipid-extensive biomass such as Higher plants, Algae, Palm, Soybean, etc.

*Corresponding author.

"The characteristics of the agave suit it well to bioenergy production, but also reveal its potential as a crop that is adaptable to future climate change," adds University of Oxford plant scientist Andrew Smith. "In a world where arable land and water resources are increasingly scarce, these are key attributes in the food versus fuel argument, which is likely to intensify given the expected large-scale growth in biofuel production."Agave biofuel trials are taking place in deserts around the world, particularly in Australia—a leader in advanced biofuel research. Researchers claim that abandoned agave plantations in Mexico and Africa could be revitalized and converted to biofuel operations. As the morphology of the plant, *Agave americana,* shows that this has higher fat, we chose it for our project. The advantage of this plant is, thrives on dry land/marginal land most efficient use of soil, water and light. Massive production and year-around harvesting. Very high yields with very low or no inputs very high quality biomass and sugars, very low cost of production. Not a commodity, so prices are not volatile, very versatile: biofuels, bio products, chemicals, world-wide geographical distribution. One of the important advantages of using Agave is that it grows on arable lands and this makes it unaffected by the rise in temperature. The other sources for lipids require heavy laboratory conditions to protect themselves from the contaminations. No literature are found regarding the use of Agave as a source for biofuels and this is why we have took his project in hand to know the proper trans esterification, amount of biofuel and physical properties of the same to increase the Biofuel production.

MATERIALS AND METHODS

Site

The experiment was carried out in the laboratory of Department of Biotechnology, K.S. Rangasamy College of Technology. The sample was collected from an arable land that has few Agaves. They were obtained from the village of varapalayam near the college.

Layer Separation

Initially, the extract of Agave was obtained by a traditional method. This extract formed two layers. These two layers were treated separately with six common solvents.

Sample Processing

To retain the complete lipid quantity, the cut parts of stem were lyophilized and preserved in –20°. Initially, the total lipid content was quantified for 10 g of sample in the laboratory scale using Folch's lipid extraction protocol. The lipids obtained were separated using a separating funnel. 0.25 g of NaOH was mixed with 20 mL of Methanol with constant stirring for catalyst reaction.

Transesterification [A.B.M. Sharif Hossain, 2008]

The catalyst was added to the extracted lipids and trans-esterification was carried out and the reaction occurs as follows and then mixture was allowed to stand in a shaker for 16 h.

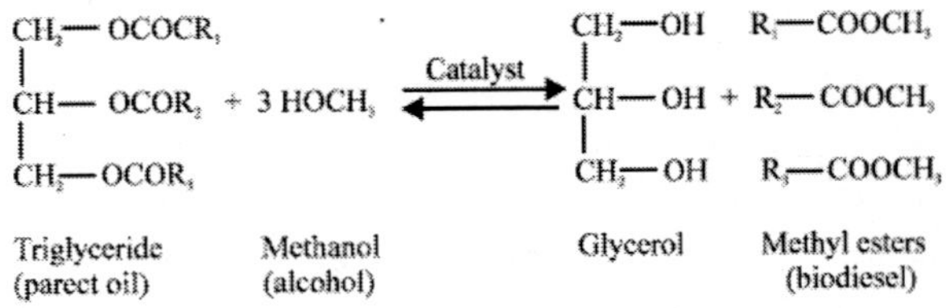

Biofuel Separation

The Biofuel was separated carefully from sedimentation using flask separator. Biofuel was washed by 5% water until it was become clean. Biofuel was dried by using dryer and finally kept under the running fan for 12 h. Biofuel production was measured by using measuring cylinder; and stored for analysis.

RESULTS AND DISCUSSION

Sample Collection

Fig. 1: Sample Extracted from the *Agave americana*

The traditional method of obtaining extract from *Agave americana* yielded 100 ml of extract from which one layer sediment. The two layers obtained were kept separately for study.

Table 1: Different Solvents were Used to Observe the Protein Layer

S. No.	*Solvents*	*Observation of LAYER-1*	*Observation of LAYER-2*
1.	Distilled water	Miscible	Immiscible
2.	Methanol	Miscible	Immiscible
3.	Chloroform	Partially miscible -3 phases	Immiscible
4.	Ethanol	Immiscible-small Globule formation	Immiscible-large Globule formation
5.	Propanol	Small globule formation	Sedimentation
6.	Ethyl acetate	Immiscible	Immiscible

It was found that the layers showed lipid characteristics with Ethanol. 100 g of the sample was lyophilized and preserved in cold condition. Folch's method of lipid estimation yielded 18ml of lipid in 10 g of the sample.

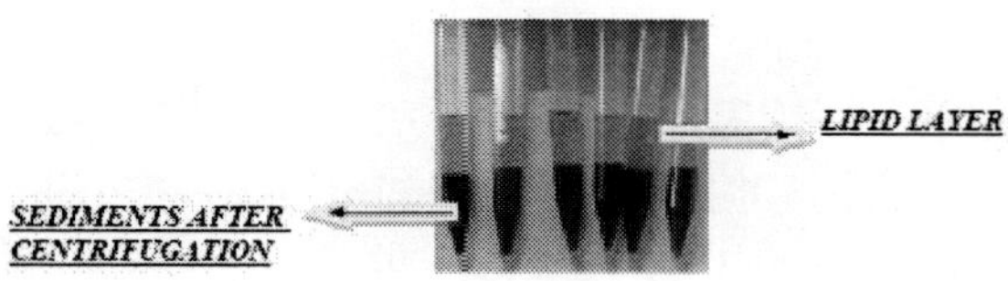

Fig. 2: Catalyst Reaction of Lipids Extracted from *Agave americana*

By trans-esterification, the Biofuel which is a combination of multiple esters is isolated and the volume was found to be 14 ml.

Fig. 3: Catalyst Reaction of Lipids Extracted from *Agave americana*

CONCLUSION

Thus, the Agave proves to be a good source for biofuel over algae, on which the biofuel works are maximum. This statement is complemented by the fact that algae grows much slower than the higher plants or shrubs. Even though algae and other micro-forms produce a big percentage of production, they require high sterility around them which can be maintained only by high cost; but agave grows on arable lands and it is always less prone to diseases; It doesn't require any sterile environment.

REFERENCES

Goldemberg, J. (2000). World Energy Assessment, Preface. United Nation's development Programme, New York, NY, USA.

Kulkarni, M.G. and A.K. Dalai, (2006). Waste cooking oil—an economical source for biodiesel: A review. *Ind. Eng. Chem. Res.*, 45: 2901–2913.

Klass, L.D. (1998). Biomass for Renewable Energy, Fuels and Chemicals, Academic Press, New York, pp. 1–2.

Turkenburg, W.C. (2000). Renewable energy technologies. In: Goldenberg, J. (Ed). World Energy Assessment, Preface. United Nations Development Programme, New York, USA, pp. 219–272.

Biodiesel Fuel Production from Algae as Renewable Energy

A.B.M. Sharif Hossain, Aishah Salleh, Amru Nasrulhaq Boyce, Partha chowdhury and Mohd Naqiuddin Biotecnology Laboratory, Institute of Biological Sciences, *American Journal of Biochemistry and Biotechnology,* 4(3): 250–254, 2008. ISSN 1553–3468

Microbial Production of Alkaline Protease from Food and Agricultural Wastes by Using Bacterial Species

N. Praveen Kumar, V. Ranjith Kumar, K.S. Vishnu Kumar and G. Ayyappadasan*

Department of Biotechnology, K.S. Rangasamy College of Technology, Tiruchnegode

E-mail: ayyappadasan07@gmail.com

INTRODUCTION

Proteases are essential constituents of all forms of life on earth including prokaryotes, fungi, plants and animals. Proteases are highly exploited enzymes in food, leather, detergent, pharmaceutical, diagnostics, waste management and silver recovery (Babu *et al.*, 2005). The waste food material has much kind of nutrients like protein, vitamins, fat and lipids. Proteins is the major source present in all waste material and protease are the enzymes used to cleave the protein molecules in food waste. In Viewing of our current situation the composting of food and vegetable waste material is making great issues in developing countries. Some of the bacterial organism capable of producing more protease enzyme to cleave the protein molecules, Isolation of microorganism from vegetable/Agricultural waste like spoiled vegetable or used food ingredients (Cereals and Oil waste). Within each of the broad groups proteases have been classified, by Rawlings and Barrett, into families of related proteases. Bacteria also secrete proteases to hydrolyse (digest) the peptide bonds in proteins and therefore break the proteins down into their constituent monomers. Bacterial and fungal proteases are particularly important to the global carbon and nitrogen cycles in the recycling of proteins, and such activity tends to be regulated in by nutritional signals in these organisms. The net impact of nutritional regulation of protease activity among the

*Corresponding author.

thousands of species present in soil can be observed at the overall microbial community level as proteins are broken down in response to carbon, nitrogen, or sulfur limitation.

A secreted bacterial protease may also act as an exotoxin, and be an example of a virulence factor in bacterial pathogenesis. Bacterial exotoxic proteases destroy extracellular structures. Protease enzymes are also used extensively in the bread industry in bread improver.

Proteases, also known as proteinases or proteolytic enzymes, are a large group of enzymes. Proteases belong to the class of enzymes known as hydrolases, which catalyse the reaction of hydrolysis of various bonds with the participation of a water molecule. Agricultural waste is any substance or object from premises used for agriculture or horticulture, which the holder discards, intends to discard or is required to discard. It is waste specifically generated by agricultural activities. For example, waste which came from a farm shop or a vegetable packing plant would not be agricultural waste.

Some examples of agricultural waste are empty pesticide containers old silage wrap out of date medicines and wormers used tyres surplus milk. Since 2006, agricultural waste has been subject to the same controls that have applied to other sectors for many years. On 15 May 2006, uncontrolled burning or tipping of waste on farms became illegal.

MATERIALS AND METHODOLOGY

Collection of Microorganism

The microorganisms such as *Bacillus cereus* and *Pseudomonas aeruginosa* was collected from the culture maintenance division of microbiology laboratory, department of biotechnology, KSRCT.

Streak Plate Technique

Nutrient agar medium was prepared in a conical flask and then sterilized. After sterilization approximately 15 to 20 ml of the sterilized nutrient agar medium (45°C) was poured into each petri plates. Upon solidification of the medium, the bacterial samples were taken in an inoculation loop and then streaked over the medium. Then the plate was incubated at 37°C and kept for 24 hours. After one day of incubation it was found that the bacterial colony had been grown.

Screening Assay of Enzyme with Isolate

In sterilized nutrient agar, 2 ml of heated protein sample is mixed and kept for solidification. After solidification inoculate both the organism and streak them. In

24 hrs., we found that the organism hydrolyze protein molecules. This indicate that the organism is capable of hydrolyze.

Protein Estimation by Lowry's Method

The method of protein estimation was done by Lowry's method. Absorbance was read at 660 nm. The graph was plotted between OD Vs. concentration of BSA (mg).

Protease Assay

The protease assay is based on colorimetric value. the principle behind this is Casein + H_2O + Protease gives Amino Acids, at the temperature of 37°C at pH = 7 in absorption wavelength of A660 nm, we can determine the protease. Light path = 1 cm.

Partial Enzyme Purification

Ammonium Sulphate Fractionation

The dialyzed cell-free filtrate was first brought to 20% saturation by the gradual addition of solid ammonium sulphate while stirring at 40°C, after 30 min period of equilibration, the precipitated protein was obtained by centrifugation for 15 minutes at 15000 rpm at 40°C. The obtained pellet was re-suspended in a 10 ml of ice-cold 0.2 M phosphate buffer pH7.2. Centrifugation was repeated and the pellet was re-suspended in order to determine both the enzyme activity and the protein content whereas the left supernatant was supplied again with ammonium sulphate in order to achieve 40, 60, 80 and 100% saturation of ammonium sulphate precipitation. The enzyme activity and protein content were determined for each separate fraction (Lowery *et al.*, 1951; Ammar *et al.*, 1991).

Dialysis against Phosphate Buffer

The procedure was performed as previously mentioned with repeating the dilatation step against distilled water followed by dialysis against phosphate buffer pH7.2. The obtained enzyme, however, was concentrated against sucrose to be available or the application on column chromatographic technique.

RESULTS AND DISCUSSION

The isolated microorganisms *Bacillus cereus* and *Pseudomonas aeruginosa* was sub-cultured using pour plate technique.

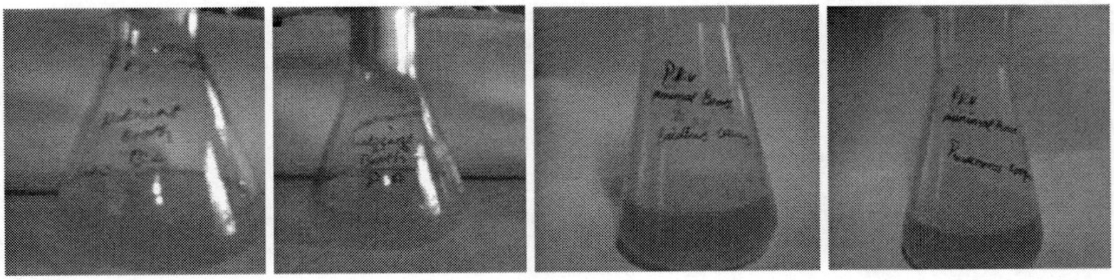

Fig. 1: Bacterial Culture in Production Medium

The above shown are the images of growth of the microorganisms *B.cereus* and *P. aeruginosa* in the pre-production medium (nutrient broth). The above shown are the images of growth of the microorganisms *B. cereus* and *P. aeruginosa* in the pre-production medium (minimal media broth).

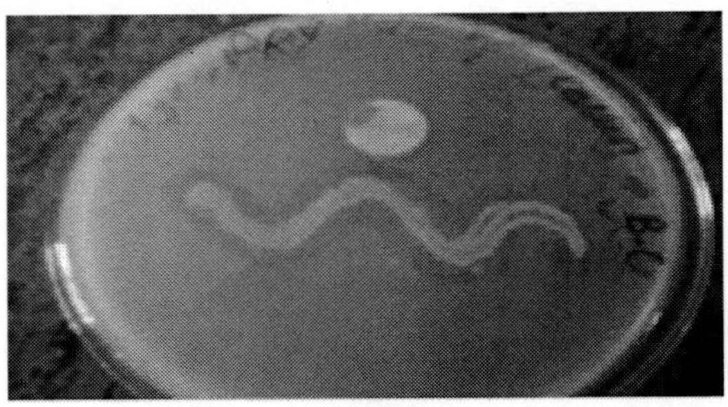

Fig. 3: Zone of Hydrolysis of B.cereus on Skim Milk Agar

The screening assay was done to check whether the organisms were capable of cleaving the protein molecules. The image shows the zone of hydrolysis produced by the microorganisms indicating their capacity to cleave protein molecules.

Table 1: Estimation of Total Protein by Lowrys Method

	B	*S1*	*S2*	*S3*	*S4*	*S5*	*B.C*	*P.A*
Working Standard	0	0.2	0.4	0.6	0.8	1.0	–	–
Unknown	–	–	–	–	–	–	0.1	0.1
D. water	1	0.8	0.6	0.4	0.2	0	0.9	0.9
Reagent C	5	5	5	5	5	5	5	5
Reagent D	1	1	1	1	1	1	1	1
OD@ 660 nm	0.0	0.070	0.230	0.314	0.413	0.489	0.159	0.099

Table 2: Specific Activity of Protease in Ammonium Sulphate Fractionation

Fraction No	*$(NH_4)2SO_4$ Saturation*	*Enzyme Activity (U/ml)*	*Protein Content (mg/ml*	*Specific Activity (U/mg protein/ml)*
1	0%	282.4	1.122	251.6
2	20%	417.4	0.956	436.6
3	40%	1041.9	0.962	1083.1
4	60%	2585.0	0.974	2654.0
5	80%	2207.5	0.906	2436.5
6	100%	1126.2	0.890	1265.4

Table 3: Elution Pattern of Protease A Produced by *B. cereus* Gel Column Chromatogram

Fraction No	*Mean Diameter (mm)*	*Enzyme Activity (U/ml)*	*Protein Content (mg/ml)*	*Specific Activity (U/mg protein/ml*
1	NCZ	ND	ND	Nd
2	18.3	247.8	0.916	300.0
3	19.0	352.3	0.897	393.8
4	21.0	742.3	0.852	871.2
5	19.6	475.3	0.840	565.4
6	18.0	249.6	0.801	311.0
7	17.0	206.0	0.771	267.0
8	15.6	136.0	0.734	185.0
9	NCZ	ND	ND	ND

Protease activity was observed at 600 nm using UV spectrometry. Bacillus cereus shows the better activity in crude extract of 16 units/ml against the activity of Pseudomonas aeruginosa. Specific enzyme activity of isolated crude protease from Bacillus cereus and Pseudomonas aeruginosa were 8 units/mg and 15 units/mg respectively. Partial purification of protease was done by using ammonium sulphate, dialysis and ion exchange chromatography. Protein purification fold was approximately 70% in Bacillus cereus in contrast Pseudomonas aeruginosa has only 50% fold. Final purified product has the enzyme activity of 42.8 units/ml in Bacillus cereus. Pseudomonas aeruginosa shows very low activity as 12 units/ml in final purified product. Finally we concluded that Bacillus cereus has the good conversion ability. Thus we conclude that *Bacillus cereus* shows more enzyme activity than *Pseudomonas aeruginosa.*

REFERENCES

Hantsis-Zacharov, E. and Senderovich, Y. (2008). Chryseobacterium oranimense sp. nov., a psychrotolerant, proteolytic and lipolytic bacterium isolated from raw cow's milk, *Int J Syst Evol Microbiol,* November 1, 58(11): 2635–2639.

Herzog, P., Winkler, I., Wolking, D., Kampfer, P. and Lipski, A., Chryseobacterium ureilyticum sp. nov., Chryseobacterium gambrini sp. nov., Chryseobacterium pallidum sp. nov. and Chryseobacterium molle sp. nov., isolated from beer-bottlingplants, *Int J Syst Evol Microbiol,* January 1, 2008: 58(1): 26–33.

References and further reading may be available for this article. To view references and further reading you must Nobuaki Fujiwara and Kazuhiko Yamamoto, 1987. purchase this article.Production of alkaline protease in a low-cost medium by alkalophilic Bacillus sp. and properties of the enzyme, Vol. 65, Issue 3, pp. 345–348.

Wery, Nathalie; Gerike, Ursula; Sharman, Ajay; Chaudhuri, Julian B.; Hough, David W. and Danson, Michael J., Use of a Packed-Column Bioreactor for Isolation of Diverse Protease-Producing Bacteria from Antarctic Soi, March, pp. 1457–1464, Vol. 69, No. 3.

Zeba, Boukaré; Luca, Filomena De; Dubus, Alain and Docquier, Jean-Denis (2009). IND-6, a Highly Divergent IND-Type Metallo-β-Lactamase from Chryseobacterium indologenes Strain 597 Isolated in Burkina Faso, October, Pp. 4320–4326, Vol. 53, No. 10.

Biotechnological Applications in Modern Paper Making – An Eco-friendly Approach to Green Chemistry Initiative

M. Vadivel*, C. Umarani[1] and K.S. Rajesh[2]

JKKMCT, Gobi, Salem, Tamil Nadu

[1]Govt. Arts College, Salem, Tamil Nadu

[2]TCT Aqua Chemicals P Ltd., Chennai

E-mail: mvadivelmani@yahoo.com

INTRODUCTION

The development and use of biotechnology started in the early 70-ties. This cradle period for biotechnology coincided with the first oil crisis which turned our interest towards the renewable resources. The use of biotechnology, microorganisms and their enzymes, to convert ligno cellulosic materials into sugar and ethanol is one such example (1). With the expected growth in per capita consumption of paper 20 kg by 2020 (2). Industrial Bio technology is the application of biotechnology for the process and production of chemicals, materials and energy (3). Biotechnology is versatile and has been assessed a key technology for a sustainable chemical industry (4). Role of biotechnology in sustainability Biotechnology refers to an array of enabling technologies that are applicable to broadly diverse industry sectors (4) Biotechnology comprises three distinct fields of activity, namely genetic engineering, protein engineering and metabolic engineering. A fourth discipline, known variously as biochemical, bioprocess and biotechnology engineering, is required for commercial production of biotechnology products and delivery of its services. Of the many diverse techniques that biotechnology embraces, none apply across all industrial sectors (6). Applications of Enzyme make the pulp and paper sector as clean production unit (7).

*Corresponding author.

WHAT IS ENZYME

Enzymes are proteins and acts as a Biological catalyst. Its affinity is classified as 4 types:

Type of Affinity	*Definition*
Absolute	Enzyme has one and only one Reaction
Group	Enzyme acts on molecules that have specific functional groups, such as amino group
Linkage	Enzyme acts on particular type of bond regardless of structure of enzymes
Stereo Chemical	Enzyme acts on a particular stereo or optical isomer

Chronological Order of Enzyme Use in Pulp and Paper Industry

Year	*Development*	*Investigator*
1959	Pulp Fibrillation cellulases	Bolaski *et al.,* 1959
1984	Enzymatic beating with Xylanases Hemicellulose removal from dissolving pulp by Xylanases	Comtat *et al.,* 1984 Paice and Jurasek, 1984
1986	Prebleaching with Xylanases	Viikari *et al.,* 1986
1988	Enhanced drainage with cellulases	Fuentes and Roberts, 1988
1988	Decreased vessel picking with cellulases	Uchimoto *et al.,* 1988
1989	Depitching pulp with lipases	Irie *et al.,* 1989
1991	Deinking with cellulases and Xylanases	Kim *et al.,* 1991
1993	Pulp delignification with laccase	Call, 1993
1996	Bleaching with manganese peroxidases	Harazono *et al.,* 1996

BIO-TECHNOLOGY APPLICATION IN STORAGE OF WOOD

Pitch is a term used collectively for hydrophobic components of wood resin and resin acids, triglycerides, waxes, etc. Pitch causes numerous problems in pulp and paper manufacture like sticky deposits on dryer rolls, discoloration and hydrophobic spots in the paper even though it is 2–8% of total composition depending upon the species and time of year (9). Traditional methods to control pitch problems include natural seasoning of wood. The most troublesome components of resins are triglycerides, fatty acid esters of glycerol. These can be removed by solvent extraction or by strong alkali, but the former process is not practical, and the latter results in yield losses and discoloration.

BIO-TECHNOLOGY SECRET IN WOOD STORAGE

The living cells in the wood rays (ray parenchyma) respire and release heat. Bacteria and fungi are provided with good growth conditions during this heat generation, and the starches and simple sugars of the rays and subsequently by the extractives of the wood can be metabolized as a source of carbon and energy. This metabolism results in an overall decrease of pitch with storage of wood. The outside storage of pulpwood was introduced in the 1920s as whole logs (ground wood), and in the early 1950s as chips [10). Hata and coworkers at Jujo Paper Company reported in 1990 that lipases can reduce pitch problems by lowering the triglyceride content of ground wood pulp (11). The reduction in pulp brightness and yield during storage was shown to be of the same order whether the wood had been stored as chips or as roundwood [12]. Conditions which affect the wood were shown to be the following: species of wood, time of cutting, removal of bark, presence of insects, methods of piling, length of storage time, general housekeeping conditions in the woodyard, and climatic conditions, especially temperature and moisture. The single most detrimental effect was loss of brightness of mechanical and sulphite pulps after storage, particularly with softwoods. Seasoning has been recommended for pine unless the recovery of maximum tall oil and turpentine yields was desired [13].

Two commercially available strains of *O. piliferum, Cartapip 28 and Cartapip 58,* have been shown to degrade the extractives of both hardwoods and softwoods including aspen, southern yellow pine, red pine, and spruce [14]. Lipase is rapidly absorbed onto pulp fiber within a minute of addition. The fungal pitch control method, using the colorless isolant of *Ophiostoma piliferum*, was the first successful case of using a live organism as a solution to pitch problems and in the pulping process.

Common species of hardwood sap-stain include: Ophiostoma pluriannulatum, Ceratocystis moniliformis, L. theobromae, Ceratocystis rigidum, Phanerochaete chrysosporium, P. subacida, P. gigantean, P. tremetlosa, H. setulosa, Coriolus versicolor, Inonotus rheades, Trichaptum abietinium. C. subvermispora, Trichaptum biforme, Schizophylum commune, Sistotrerma brinkmanii, Pleurotus ostreatus, Alurodiscus sp. Ganoderma collosum, PheUinus igniarius. The fungal pitch control method, using the colorless isolant of Ophiostoma piliferum, was the successful case of using a live organism as a solution to pitch problems and in the pulping process.

BIO-TECHNOLOGY APPLICATION IN SLIME CONTROL OF PAPER MACHINE

Slime, also called biofilm, is a population of microbes attached to available solid surfaces by means of polysaccharides and is associated with organic and inorganic deposition such as fibers, fillers, and fines. Without the polysaccharides, the growth and reproductions of microorganisms are impossible (15). Micro biological slime growth affects the paper machine runnability. Conventional biocides used to kill or inhibit the growth and reproduction of micro organisms but the main disadvantages is non selectivity and non specificity (16). Some of the known bacteria responsible for the slime formation as given below:

Table 2

Bacillus mesentericus	Aerobacter levanicus
Bacillus subtilis	Acetobacter acetigenum
Bacterium pruni	Pseudomonas caryophylli
Bacillus megaterium	Pseudomonas chlororaphis
Bacillus vulgatus	Pseudomonas denitroflourescens var. Laeveniformans
Bacillus Polymyxalevan	Pseudomonas syringae
Aerobactet levanicum	Serratia Kiliensis

Enzymatic slime control prevents slime formation by applying hydrolysis enzymes to hydrolyze the polysaccharides and to alter the sensitivity of cells to biocides. Enzymes used for slime control may consist of cellulase, amylase, hemicellulase, protease, proteolytic enzymes, peroxidase, etc. (17).

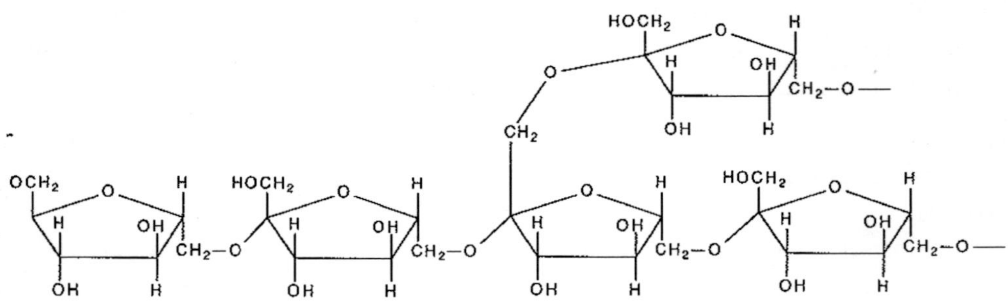

Leavn Structure 1: One of the Slime Producing Polysaccharide

BIO-TECHNOLOGY APPLICATION IN REFINING OF PULP

Enzyme Refining

Mixtures of cellulases and hemicellulases were first shown to improve freeness of recycled paper furnishes (18, 19). This effect has been proposed to result from fines hydrolysis and modification of the colloidal polysaccharide layer on the surface of the fibers (20). Recently the role of individual enzyme components in fiber modification, especially endoglucanases (EG) cellobiohydrolases (CBH) and xylanases, has been studied extensively. All the paper manufactures have a look into power consumption in the refining area (21).

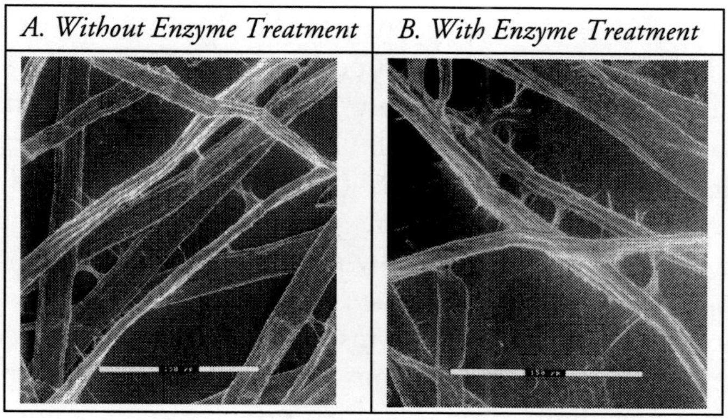

SEM Images

Enzyme is added after the primary refining prior to dump chest and the treatment takes place in the following conditions: 40–50°C, pH 3.5–5.5, retention time 20–30 min. The main benefit of the use of enzyme is the improved beatability. Enzyme facilitates fibrillation and 5–15% less energy can be applied in refining in order to achieve the target porosity. As a consequence better drainage can also obtained. Cellulase/hemicellulase mixtures can significantly improve beatability of chemical fibers and also result in improved handsheet properties, such as density and air resistance, which can be very important properties in many specialty papers. In order to avoid strength losses, a careful optimization of enzyme dosage and treatment conditions is important. Z-directional bonding strength can be markedly improved by cellulase/hemicellulase treatment. In mill applications, energy savings have been obtained with cellulase/hemicellulase treatments prior to refining, without any negative effects in strength properties. ESEM images gave indications of improved fibrillation when enzyme treatment preceded refining of kraft.

BIO-TECHNOLOGY APPLICATION IN BLEACHING OF PULP

Xylan, the second most abundant polysaccharide and a major component in plant cell wall consists of b-1, 4-linked xylopyranosyl residues. The plant cell wall is a composite material in which cellulose, hemicellulose (mainly xylan) and lignin are closely associated. 1–2 Three major constituents of wood are cellulose (35–50%), hemicellulose (20–30%)—a group of carbohydrates in which xylan forms the major class- and lignin (20–30%). Xylan is a heteropolysaccharide containing substituent groups of acetyl, 4-O-methyl- D-glucuronosyl and a-arabinofuranosyl residues linked to the backbone of b-1, 4, -linked xylopyranose units and has binding properties mediated by covalent an d non-covalent interactions with lignin, cellulose and other polymers. Lignin is bound to xylans by an ester linkage to 4-O-methyl-Dglucuronic acid residues. In 1879, the kraft pulping process was invented by the German chemist C. F. Dahl. This discovery was broadcast on Samuel F. B. Morris' "information superhighway" of the day, the telegraph. Interestingly, the telegraph system was patented 30 years before Dahl's discovery (22). "Biobleaching" has been examined recently in an effort to eliminate chlorine altogether. Microbial xylanases that are thermostable and cellulose-free are generally preferred for biobleaching of paper pulp. Countless micro-organisms take advantage of the worlds largest source of stored energy, plant biomass, and these organisms produce a broad spectrum of enzymes that catalyse the hydrolysis of the various components that are present in plant material. By identifying those enzymes which target specific components of wood fibre, it should be possible to find an enzyme (or enzymes) which assist in the removal of lignin. The development of various bleaching techniques has been taking place to manage the pollution problem arising during the effluent discharge and also meet the demand for high brightness paper 90+ ISO(23).

At present, enzymes are being considered as only partial replacements for chemicals, but there is hope that as the mechanisms of enzymatic treatment are elucidated, refinements in the process may allow enzyme assisted "Total Chlorine Free" (TCF) bleaching sequences to become an effective and economic method of pulp treatment. Treating unbleached kraft pulp with xylanase does not, in itself, release significant amounts of lignin. However, the enzyme appears to render the lignin easier to extract in the subsequent bleaching steps and now a number of mills include enzymes in their bleaching sequences. There are two mechanisms that have been proposed to account for action of xylanases: 1. under alkaline conditions, much of the xylan in wood fibre is dissolved into the liquor. As cooking proceeds and the alkali concentration decreases, the dissolved xylan precipitates back onto the crystalline cellulose fibres.

It is thought that a loss of the side-groups from the xylan chains allows the xylan to re-precipitate in a crystalline form which is less soluble than native xylan and it is this re-precipitated, insoluble xylan that hinders the solubilisation of residual lignin. Hemicellulases have also been proposed to cleave hemicellulose bonds near points of attachment between lignin and hemicellulose and it is possible that this leads to the improved solubilisation of lignin.

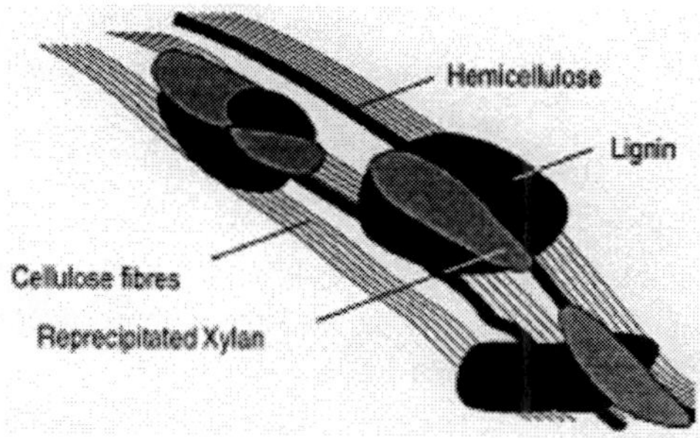

A hypothetical structure of Kraft pulp fibres. Xylanases are thought to promote pulp bleaching via the hydrolysis of the re-precipitated xylan. The Future of Enzymes in Pulp and Paper Manufacture.

The enzymes currently in use by the pulp and paper industry are less than perfect. Kraft pulping requires high temperatures and high a pH and as most of the currently available xylanases are from mesophilic organisms, they rapidly lose activity at temperatures above 50°C and at pHs above 7. To include the commercially available enzymes in the bleaching stages requires cooling and neutralisation. The result is a costly change to the plant, increased energy consumption and increased process time. Understandably, mill owners are reluctant to embrace this technology. The race is on for finding new enzymes which alleviate at least some of these problems. Many research groups have turned to xylanases and mannases produced by thermophilic organisms and already enzymes are under trial which have excellent activity at a pH of 9 and in boiling water. These conditions don't quite reach the high expectations of the industry but they are much closer than anything presently available.

REFERENCES

Backerby, L.H. (1958). Pulp Paper 32: 129.

Dhiman *et al.,* Bio Resources 3, (4), p. 9.

Europa Bio, Dec. 2008, Rev., 09, p. 1.

Farrell, R.L., Fritz, A., Iverson, S., Kile, G. and Brush, T.S. (1994). 80th Annual Meeting, Technical Section.

Hajny, G.J. (1966). *Tappi Journal* 49: 97.

Irie, Y., Matsukura, T., Hata, K., European pat app EP 374700, 1989.

Irie, Y., Matsukura, T., Usui, M. and Hata, K., Proc. Tappi Paper makers Conf. 1990, Tappi Press Atlanta of Canadian Pulp and Paper Association Preprints "B", Montreal.

Jackson, L.S., Heitmann, J.A. and Joyce T.W. (1993)."Enzymatic modifications of secondary fiber", *Tappi Journal,* 76 (3), pp. 147–154.

Kantelinen, A., Jokinen, O. and Pere, J. "The mechanism of cellulase/hemicellulase treatment for improved drainage," proceedings from the 1997 *Tappi Biological Sciences Symposium,* pp. 267–269.

Lievonen, J., Technological opportunities in biotechnology. Espoo, Finland7 VTT, Group for Technological Studies, 1999.

Martin, C.H., Identification and implications of troublesome slime-forming bacteria found in paper mill system, In TAPPI 1998 Paper makers Conference Proceedings, TAPPI Press, Atlanta, Georgia, US, 998, p. 91.

Paugh, J. and Lafrance, J.C. (1997). Meeting the challenge: US industry faces the 21st Century. The use of biotechnology industry. Washington DC7 U.S. Department of Commerce, Office of Technology Policy.

Pommier, J.-C., Fuentes, J.-L. and Goma, G. (1989). "Using enzymes to improve the process and the product quality in the recycled paper industry. Part 1: the basic laboratory work," *Tappi Journal,* 72(6), pp. 187–191.

Progress in paper recycling, Vol. 11, No. 3, May 2002, p. 43.

Ragauska, A.J., Bio technology in the pulp and paper Industry A challenge for change. Institute of Paper Science and Technology, Atlanta, USA, P.15, 8TH International Conference on Biotechnology in pulp and paper Industry.

Roberts, S.M., Casy, G., Nielsen, M.-S., Phyhian, S., Todd, C. and Wiggins, K., Biocatalysts for fine chemical synthesis. Wiley.

Somsen, R.A. (1962). *Tappi Journal,* 45: 623.

Vadivel, M., Umarani C. and Rajesh K.S. (2011). Cleaner production Technology: Today's need for pulp and paper industry, Department of Pulp and Paper Technology, Sesahasayee Institue of Technology, Trichy, Souvenir, Vol. 49.

Vadivel, M., Umarani, C. and Rajesh, K.S., Nano Technology-Changing the challenges in pulp and paper industries in India, National Conference on Recent Advancement in Nanoscale Research, Anna University: Chennai, Madurai Region, p. 34.

Vadivel, M., Umarani, C. and Rajesh, K.S. (2011). Recycling of colored broke for manufacturing High Bright paper, *IPPTA J.,* Vol. 23, No. 3, 2011, p. 137.

Vadivel, M., Umarani, C. and Rajesh, K.S. (2011). The Concept of pulp viscosity versus strength-Revisited Department of Pulp and paper Technology, Sesahasayee Institute of Technology, p. 13.

Viikari, L., Ranua, M., Kantelinen, A., Sundqvist, J. and Linko, M. (1986). "Bleaching with Enzymes", *Proc. 3rd Int. Conf. Biotechnology in the Pulp and Paper Industry,* STFI, Stockholm, Sweden, p. 67.

Witar, C.L., Canada patent 2,001, 539, Applicatopn of cellulose to control industrial slime.

Comparative Studies on Ethanol Production from Pre-Treated Low Cost Substrates Using a Novel Design of Rotating Biological Reactor

V. Sindhu*, S. Girija, C.N. Kanchana and R. Ravikumar
Department of Biotechnology, Bannari Amman Institute of Technology, Sathyamangalam

INTRODUCTION

Lignocellulosic biomass is not only a renewable resource but also the most abundant resource of organic components in high amounts on the earth. The bioconversion of the lignocellulose to ethanol is significantly hindered by the structural and chemical complexity of biomass. It makes these materials a challenge to be used as feed stocks for cellulosic ethanol production. Rice straw is a by-product of rice production and great bio resource. It is one of the abundant lignocellulosic waste materials in the world (Nutawan *et al.,* 2010). Different acid and alkali treatments are used for pretreating rice straw but the cost is high and the efficiency of lignin removal is low. Also along with the lignin cellulose content is removed. A number of different pretreatment methods, such as steam explosion (Nakamura *et al.*, 2001; Yang *et al.,* 2002), steam treatment with diluted sulfuric acid (Yang *et al.,* 2004) or sulfur dioxide (Wingren *et al.,* 2004), organic solvent extraction (Pan *et al.,* 2005), and biological treatment with white rot fungi (Itoh *et al.,* 2003) have been extensively investigated. White rot fungi degrade the lignocellulosics in an 'environmentally friendly' way and play a key role in organic matter recycling. Mushroom fungi utilize these resources as 'food substrates' for energy. Basidiomycete fungi are expected to be potentially useful as a pretreatment in biorefinery processes including the production of biofuels and functional chemicals from lignocellulosic materials (Dashtban *et al.,* 2009). Pretreatment carried out using *Pleurotus ostreatus* has high potential for lignin

*Corresponding author.

degradation. Solid State Fermentation is a process in which the fermentation of sugars to ethanol is carried using solid substances at less moisture content. Many different kinds of bioreactors have been used in solid state fermentation, (Brand *et al.*, 2010). Tray reactor, packed bed reactor are used for the production of ethanol using wheat straw and starch with *Aspergillus niger* and Bacillus sp. The present work is focused on design of Rotating Biological Reactor (RBR) with optimum critical speed for ethanol production from different lignocellulosic materials. In addition, design was also focused on the sparging mechanism with baffles and jacketed system.

MATERIALS AND METHODS

Rice husk, spawn culture of *Pleurotus ostreatus,* polyethylene bags, cellulase enzyme Rice straw was collected. The rice straw was soaked in sufficient quantity of water for a period of 6–8 hours and autoclaved at 30 psi for 1 hour. The autoclaved rice straw was further air dried, pretreated using *Pleurotus ostreatus*

Preparation of Beds for Mushroom Cultivation

The spawn culture of *Pluerotus ostreatus* was obtained from 'Blue Hill Mushrooms' in Coimbatore which was further utilized for the pretreatment of rice straw for the process of lignin removal. The rice straw is chopped finely. It was then taken in the polyethylene bag of required height. The spawn culture of *Pleurotus ostreatus* was introduced.

The beds of required height can be used made by stacking rice straw and the spawn culture alternatively. The bag is then tied and left for its growth. The sprouts protrude out through the holes made in the bags. Adequate moisture content is provided periodically. The content of cellulose and lignin is analysed after 44 days of pretreatment process.

Rice Straw and Pretreatment

Rice straw was collected from the nearby localities. Pretreatment involves the following, (1) Sterilization of rice straw, (2) Preparation of the bed using polypropylene bags, (3) Maintenance of humid condition (after 15–20 days), (4) Harvesting after adequate growth.

Design of Rotating Biological Reactor

The overall design of the reactor aims at enhancing the diffusion and overcome the mass transfer limitations. In order to overcome the drawback in the existing design of the bioreactor a novel approach was formulated. The concept of ball mill is being used. The centrifugal force of the reactor was equated to substrate weight and hence the critical speed of the reactor was fixed at *200* RPM,

$$Nc = \frac{1}{2\Pi}\sqrt{\frac{g}{R-r}}$$

Where, *Nc* is the critical speed of the bioreactor, *R* is the radius of the bioreactor and *r* is the radius of the particle.

Specifications of Bioreactor	*Design Specification*
Diameter of the reactor (D_a)	9 cm
Height of the inner vessel (h)	48 cm
Height of the jacketed vessel (h_j)	57 cm
Diameter of the baffle (d)	9 cm
Height of the baffle (h_b)	18 cm
Orifice diameter (d1, d2, d3)	0.06, 0.09, 0.12 cm
Aspect ratio	2:1
Temperature	37 c
pH	6.2
Agitation	68.75 rpm
Moisture content	70%

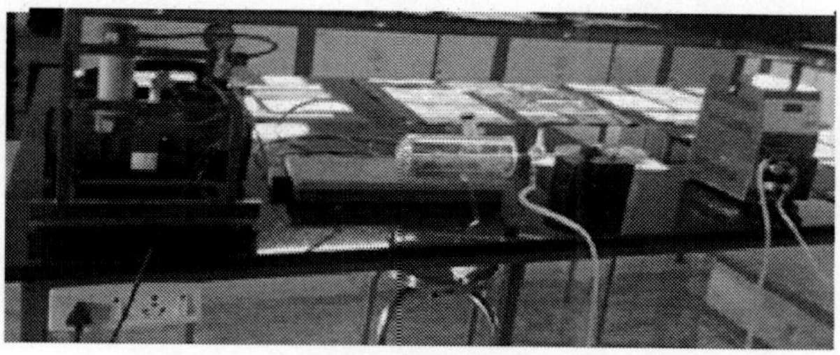

Reactor Diagram

Ethanol Production by Solid State Fermentation

Ethanol production was carried out in the designed and fabricated Rotating Biological Reactor (RBR). The glucose obtained through enzymatic hydrolysis by commercial cellulase enzyme was fermented into ethanol using baker's yeast (*Saccharomyces cerevisiae*). The rice straw along with yeast were maintained at varying pH 6.2, temperature 37°C, substrate concentration 5 g of rice straw, inoculum concentration 5 ml of baker's yeast, time of fermentation 3 days, agitation 300 rpm. The moisture content was maintained at 70%. The pH was maintained using Citrate buffer. After the completion of the fermentation process, ethanol was separated from the fermentation broth by simple distillation at 78°C.

Estimation of Lignin Content

The lignin content in various samples of rice straw was estimated. A gradual decrease in the lignin content is observed from the FRS to TRS. The studies on *Pleurotus sp.* (Singh *et. al.,* 1989) indicated a gradual decrease in the lignin up to harvest of crop during fruitification compared to spawn run stage (Figure 2). Estimation was carried out in the following samples (using *Pleurotus ostreatus* as spawn culture). (1) Fresh rice straw, (2) Test sample (Rice straw + spawn culture), (3) Pretreated rice straw (after mushroom cultivation). The method used for the estimation of lignin is Spectrophotometric method (Stafford, 1960).

Estimation of Cellulose Content

Estimation was carried out in the following samples (using *Pleurotus ostreatus* as spawn culture). (1) Fresh rice straw, (2) Test sample (Rice straw + spawn culture), (3) Pretreated rice straw (after mushroom cultivation). The method used for the estimation of cellulose content is Anthrone Method (Updegroff, 1969). The cellulose content in various samples of rice straw was estimated. A gradual decrease in the cellulose content is observed from the FRS to TRS. During the growth of mushroom, cellulose content of substrates were reduced with maximum rate at fruitification stage compared to spawn run stage. Reduction in the cellulose content of the substrate was reported due to increase in the activity of the cellulase enzyme (Figure).

RESULTS AND DISCUSSION

Fig. 1: Progress of Mushroom Growth

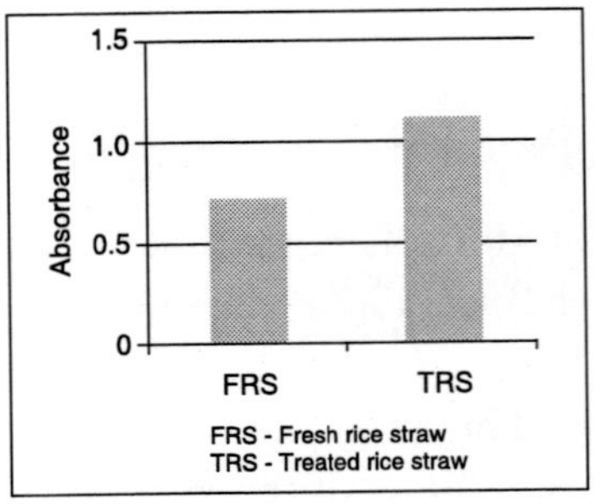

Fig. 2: Estimation of Lignin Content

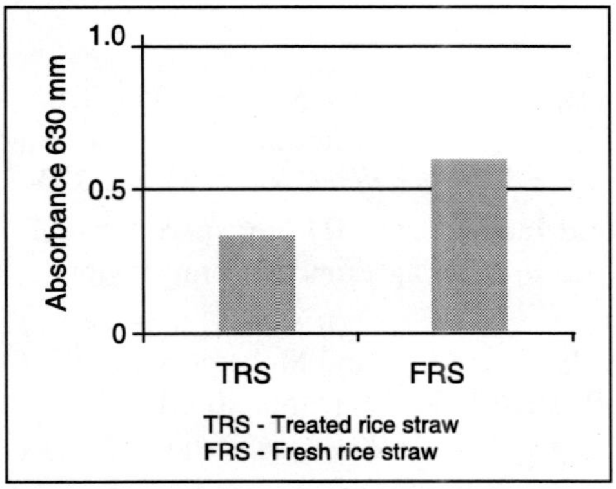

Fig. 3: Estimation of Cellulose Content

Analysis

The ethanol content was estimated by chromic acid assay. The pretreated samples were analysed by SEM AND FDIR.

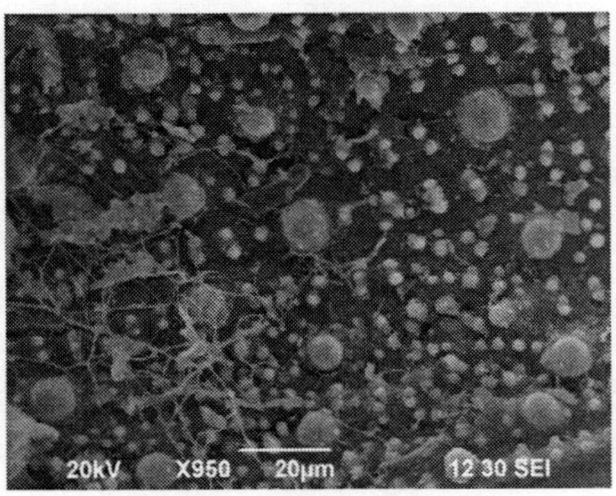

Fig. 4: SEM Analysis

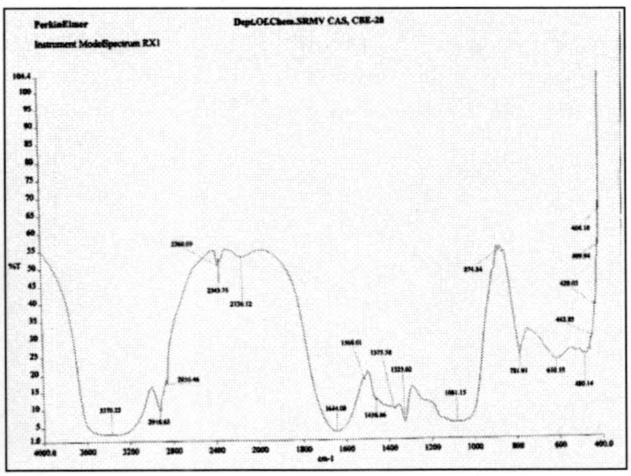

Fig. 5: FTIR Analysis

CONCLUSION

Lignocellulosic materials can be exploited as potential substrate for bioethanol production by the pretreatment with *Pleurotus ostreatus*. By product of the pretreatment process holds a lot of economic value. The rector can be utilized for ethanol production from different substrates with pretreatment process.

REFERENCES

Itoh, H., Wada, M., Honda, Y., Kuwahara, M. and Watanabe, T. (2003). Bioorganosolv pretreatment for simultaneous saccharification and fermentation of beech wood by ethanolysis and white rot fungi. *J. Biotechnol.*, Vol. 103, pp. 273–280.

Nakamura, Y., Sawada, T. and Inoue, E. (2001). Enhanced ethanol production from enzymatic treated steam exploded rice straw using extractive fermentation. *J. Chem. Technol. Biotechnol.*, Vol. 76, pp. 879–884.

Nutawan, Y., Phattayawadee, P., Pattranit, T. and Mohammad, N.E. (2010). Bioethanol production from rice straw. *Energy Research J.*, Vol. 1(1), pp. 26–31.

Wingren, A., Soderstrom, J., Galbe, M. and Zacche, G. (2004). Process consideration and economic evaluation of two-step steam pretreatment for production of fuel ethanol from soft-wood. *Biotechnol. Prog.*, Vol. 29, pp. 1421–1429.

Yamshita, Y., Kurosumi, A., Sasaki, C. and Nakamura, Y. (2008). 'Ethanol production from paper sludge by immobilized Zymomonas mobilis', *Journal of Biochemical Engineering* Vol. 42, pp. 314–319.

Yang, B. and Wyman, C.E. (2004). Effect of xylan and lignin removal by batch and flow through pretreatment on the enzymatic digestability of corn stover cellulose. *Biotechnol. Bioeng.*, Vol. 86, pp. 88–95.

Yang, B., Baussaid, A., Mansfield, S.D., Gregg, D.J. and Saddler, J.N. (2002). Fast and efficient alkaline peroxide treatment to enhance the enzymatic digestability of steam-exploded soft-wood substrate. *Biotechnol. Bioeng.*, Vol. 77, pp. 678–684.

Enzymes of Agro-Industrial Significance from Aspergillus Species

Dolly*, Brijesh Kumar and Rekha Govindan

Department of Biotechnology, Aarupadai Veedu Institute of Technology,
Vinayaka Missions University, Rajiv Gandhi Salai, Paiyanoor, Kanchipuram, Tamil Nadu
E-mail: dolly171274@gmail.com

INTRODUCTION

The *Aspergillus* fungus is an important group of filamentous ascomycete species with more than 180 species identified. It is found all over the world, and plays an important role in the ecosystem by decomposing organic matter. Since ancient times, *Aspergillus* species have been used for the production of many food products like soya sauce, miso and sake. However, it also includes many fungal pathogens known to cause diseases in human, plants and animals. With the advent of biotechnology, *Aspergillus* became one of the most widely used microbes. For the creation of better strains with wider applications, their genes and genomes are being studied. Three of the most studied species *Aspergillus oryzae, Aspergillus niger, and Aspergillus nidulans* have been used extensively for the production of various enzymes and chemicals and have served as model organisms for scientific study. Among these, *A. oryzae and A. niger* are in the Generally Recognized As Safe (GRAS) list of the Food and Drug administration (FDA) of United States of America.[1, 2]

CHARACTERISTICS OF *ASPERGILLUS* SPECIES

During the course of its life cycle, the *Aspergillus* produces many primary and secondary metabolites of medical, industrial and agricultural importance. Most of the *Aspergillus* species have both sexual and asexual modes of reproduction. Few species like

*Corresponding author.

Aspergillus niger, also have a parasexual cycle of reproduction which offer many advantages like increase in production of metabolites, including citric acid, recombinant proteins, and for generation of healthy strains to resist toxic products. *Aspergillus* species tolerate changes in pH, osmotic pressure, and oxidative stresses, and also produce metabolites in response to these physiological changes.[3, 4] *Aspergillus* has the ability to use a wide variety of substrates, and switches between different metabolic pathways with change in the substrate. This advantage provides increased flexibility in the design and improves the economic feasibility of the industrial fermentation process.

INDUSTRIAL MARKET

The end use market for industrial enzymes is extremely wide spread with numerous industrial and commercial applications. Food and beverages and household care industries represent the largest application areas of enzymes, followed by animal feed, biofuels, textile, and other technical industries. The global market for industrial enzymes was valued at $3.1 billion in 2009 and reached about $3.6 billion in 2010. The estimated market for 2011 is about $3.9 billion. The market is expected to grow at a compounded annual growth rate (CAGR) of 9.1% to reach $6 billion by 2016.[5] The global industrial enzymes market is intensely competitive with Novozymnes being the largest player in the industry, followed by DSM, and DuPont, among others.

INDUSTRIAL ENZYMES FROM *ASPERGILLUS* SPECIES

Aspergillus species produce a large number of industrially important extracellular enzymes. Amylases, xylanases and pectinases are used in the food industry while cellulases, hemicellulases are used in paper and pulp industry. All the enzymes essential for the degradation of cellulose are found in *Aspergillus* and thus it can be very useful for biofuel production. Other commercially important enzymes from *Aspergillus* are phytase, tannase, carbaryl hydrolase, inulinase, lipase and glucose oxidase. Some of these enzymes have been used for biotransformation to synthesize bioactive molecules or their precursors and hence genetically-engineered strains have been introduced to boost production.[6]

CITRIC ACID PRODUCTION

Citric acid, commonly known as vitamin C is an important food additive. It is widely used in the pharmaceutical and chemical industry and replaces tripolyphosphate in environment friendly detergents. Citric acid is one of the many metabolites produced by *Aspergillus niger.* The current production rate is 20 g/L citrate in 5–7 days. High producer strains with the help of modern fermentation technology yield up to 95% citric acid from the supplied sugar.[7, 8] Mutant cultures of *A. niger* capable of using potato and corn starches have been developed and are being investigated for citric acid production.

COMMERCIALLY IMPORTANT EXTRACELLULAR ENZYMES

Aspergillus species have the ability to produce and secrete a large number of extracellular enzymes which are commercially very important. A variety of amylases and proteases are produced by the *Aspergillus* species for the fermented food industry. The GRAS-designated strains of *Aspergillus* produce a variety of industrial enzymes including α-amylases, glucoamylases, cellulases, pectinases, xylanases and other hemicellulases, and proteases. Fungal proteases are also used in baking industry to facilitate dough expansion by partial hydrolysis of gluten. *Aspergillus* xylanases are also used in the baking industry to solubilize arabinoxylans for dough expansion and to increase animal feed conversion, and in vegetable processing and fruit juice clarification. Pectinases are widely used for fruit processing, puree preparation, fruit juice recovery and clarification, and recovery of plant oils. Cellulases, often in combination with pectinases, are used for greater extraction of sugar and fruit colour into juices.[9]

BIOTRANSFORMATION USING *ASPERGILLUS* SPECIES

Biotransformation helps to catalyze highly region- and stereo- specific reactions with the help of enzymes. This helps in the synthesis or modification of complex chemical structures into bioactive molecules or their precursors. Specific reactions like stereo hydroxylations, oxidations, hydrolysis, esterification, isomerisations, and racemic isolations, are brought about by *Aspergillus* species. In addition, *Aspergillus* also helps in the biotransformation of alicyclic insecticides, aromatic and phenoxy herbicides, organophosphorous and other pesticides, cyclic hydrocarbons, terpenes, and alkaloids. A low cost substrate, α-pinene is converted to a high value food flavorant, verbenol by *A. niger.*[10]

SECONDARY METABOLITE PRODUCTION

Penicillin production by *A. nidulans* represents the most advanced model system for the study of regulation of the biosynthesis of secondary metabolites. Different strains of *A. terreus* producing cyclosporine A and lovastatin have been identified. *Aspergillus oryzae, A. sojae, and A. tamari,* used in the fermented food industry, are considered as safe as they lack the production of aflatoxins, known to be hepatotoxic, carcinogenic, teratogenic and mutagenic. *A. fumigatus* produces many secondary metabolites like fumagillin, fumitremorgin, fumigaclavin, gliotoxin, helvolic acid, verruculogen, and sphingofungins, which enhance its virulence. *Aspergillus* and some other fungi secrete small basic cystine-rich antifungal proteins that may have important potential applications in medical treatment, plant protection and food preservation.[4]

ENVIRONMENTAL APPLICATIONS

Biological remediation has the advantage as cost effective and eco-friendly for the treatment of recalcitrant dyes and effluents. Biosorption processes using *Aspergillus* strains are gaining considerable interest in replacing or supplementing the conventional wastewater treatment processes. Polyaromatic hydrocarbons (PAHs) are metabolized by a strain of *A. terreus.* Besides this, *A niger* biomass has been studied for the biosorption of heavy metals.[11]

CONCLUSION

Traditionally, *Aspergillus* species have been used in the fermented food industry. With the advent of biotechnology, new products were discovered and they were commercialized to some extent. Strain improvement was done using the basic principles of biochemistry, microbiology, genetics, and bioprocess engineering. Recombinant technology ushered in a new era where microbes were utilized for the production of many important therapeutic proteins. However, the use of *Aspergillus* species for the production of various proteins has not advanced as quickly as expected because of low yields. A notable exception was chymosin production by *Aspergillus* for cheese making. The availability of genome sequences of various *Aspergillus* species have renewed the interest and now it is being extensively studied at the molecular level for strain improvement. Besides, this new information will help us to construct better fungal hosts for the production of indigenous *Aspergillus* products, heterologous products, and products not yet discovered. New techniques are available or are being developed, ranging from different sequencing strategies, gene and protein arrays, protein engineering and others. These have given rise to new fields like transcriptomics, proteomics, and metabolomics. The complete biology of *Aspergillus* is soon going to be available and that would lead to better utilization of the fungi for human needs.

REFERENCES

Aleksenko, A., Nielsen, M.L. and Clutterbuck, A.J. (2001). Genetic and physical mapping of two centromere-proximal regions of chromosome IV in Aspergillus nidulans. *Fungal Genet. Biol.* 32, 45–54.

Calvo, A.M., Wilson, R.A., Bok, J.W. and Keller, N.P. (2002). Relationship between secondary metabolism and fungal development. *Microbiol. Mol. Biol. Rev.* 66, 459.

De Vries, R.P. (2003). Regulation of Aspergillus genes encoding plant cell wall polysaccharide-degrading enzymes; relevance for industrial production. *Appl. Microbiol. Biotechnol.*, 61, 10–20.

Dursun, A.Y., Uslu, G., Cuci, Y. and Aksu, Z. (2003). Bioaccumulation of copper (II), lead (II) and chromium (VI) by growing Aspergillus niger. *Proc. Biochem.,* 38, 1647–1651.

Global Industrial Enzymes Market Report: 2012 Edition- http://www.konceptanalytics.com/reportDetail.aspx?reportID=391

Haq, I., Ali, S. and Iqbal, J. (2003a). Direct production of citric acid from raw starch by Aspergillus niger. *Proc. Biochem.* 38, 921–924.

MacKenzie, D.A., Jeenes, D.J., Gou, X. and Archer, D.B. (2000). Molecular basis of glucoamylase overproduction by a mutagenised industrial strain of Aspergillus niger. *Enz. Microb. Technol.* 26, 193–200.

Sallam, L.A.R., El-Refai, A.-M.H., Hamdy, A.-H.A., El-Minofi, H.A. and Abdel-Salam, I.S. (2003). Role of some fermentation parameters on cyclosporin A production by a new isolate of Aspergillus terreus. *J. Gen. Appl. Microbiol,* 49, 321–328.

Sarangbin, S., Morikawa, S., Kirimura, K. and Usami, S. (1994). Formation of autodiploid strains in Aspergillus niger and their application to citric acid production from starch. *J. Ferment. Bioeng.* 77, 474–478.

Tailor, M.J. and Richardson, T. (1979). Application of microbial enzymes in food systems and in biotechnology. *Adv. Appl. Microbiol,* 25, 7–35.

Tailor, M.J. and Richardson, T. (1979). Application of microbial enzymes in food systems and in biotechnology. *Adv. Appl. Microbiol,* 25, 7–35.

Section–VII

Abstracts

Study on Photodynamic Efficiency of Photo Sensitizers in NIR Region for Oral Cancer— A Novel Approach in Photodynamic Therapy

Gaanappriya Mohan*

Department of Biotechnology, Bannari Amman institute of Technology, Sathyamangalam, Tamil Nadu

E-mail: gaanzmohan@gmail.com

ABSTRACT

Photodynamic therapy (PDT) is an emerging modality for the diagnosis and treatment of cancer and related diseases. PDT involves the inactivation of living cells by the combined action of light and a chemical (photosensitizer). After intravenous injection, the tumor tissues selectively retain the photosensitizer and when exposed to light of a specific wavelength, it generates highly reactive species. These reactive species damage cellular constituents and eventually cause cell death. PDT is relatively a safer treatment since the induction of cytotoxicity ceases when the light is switched off, contrary to the conventional chemotherapy and radiotherapy. Several photosensitizers have been investigated for their use in PDT. These include, porphyrins, chlorins, benzoporphyrins, benzoporphycenes, phthalocyanins, purpurins, and aminolevulinic-acidmediated porphyrins etc. Of these sensitizers, hematoporphyrin derivative (HpD) and its commercial variants have been approved for the treatment of various types of tumors in a number of countries. However, these sensitizers have several disadvantages and hence the search for more effective sensitizers has become important in recent years. In the current work, the photodynamic efficiency of N-confused porphyrins is studied in oral cancer cell lines. The PDT mediated cytotoxicity, cell proliferation, DNA and mitochondrial damage and nuclear condensation studies are reported.

*Corresponding author.

A Novel Herbal Pesticide to Control Adult Scale Insects

K. Arifa Banu and Aruna Devaraj[1*]

Department of Zoology,Hajee Karutha Rowther Howdia College, Uthamapalayam, Tamil Nadu

[1]Department of Zoology, JA College for Women, (Autonomous) Periyakulam, Tamil Nadu

E-mail: banuvazi@rediffmail.com

ABSTRACT

In recent years organic formulations involved in the biological control of insect-pests of economically important plants have been of great value in the field of agriculture towards a greater sustainability and public concern and it rules out the use of hazardous pesticides. Application of insecticides of botanical origin is becoming a simple and sustainable method of pest control. In this investigation the herbs *Wedelia chinensis* and *Cymbopogan citratus* were selected for mortality study on scale insects. The herbal materials used were analysed by GC-MS. The presence of pesticidal compounds such as Butyl butyrate, Bis (3, 3, 5-Trimethyl hexyl) adipate, in *Wedelia chinensis*, and Benzofuran in *Cymbopogan citratus* showed significant mortality rate on adult scale insects.

*Corresponding author.

Study of Optical Images for Conjugated Gold Nanoparticles with Biomacromolecular Analysis in Cancer Cell

Anbuselvi A.*

Department of Nanoscience and Technology,
Alagappa University, Karaikudi
E-mail: hamsanbu@gmail.com

ABSTRACT

The National Cancer Institute has set the goal of eliminating suffering and death due to cancer by 2015. A key strategy to achieve this goal is to improve early detection and prevention using novel molecularly targeted cancer diagnostics and therapeutics. Nanoshells are small nanoparticles that have a very thin metal coating. Gold nanoshells are therefore a thin gold layer over a silica core. This is the key for biological applications because gold has a high biocompatibility and more importantly that there are no toxic effects to body tissue when using gold nanoparticles. The molecular mechanisms of disease are being discovered rapidly. Optical imaging is preferred for early detection because it allows non-invasive visualization, thus affording multiple measurements. Optical probes enable imaging of disease-specific biomarkers at the nanometer scale. The relative optical absorbance and scattering of nanoshells comprised of a metal shell and dielectric core can be tailored over a broad range from the near UV to the mid-infrared (IR) by varying the core: shell ratio and the overall size of the particles. Scattering nanoshells show significantly increased scattering efficiency in targeting labeled cancer cell lines. Quantum dots (Qdots) are 2–10 nm semiconductor nanocrystals that fluoresce upon radiative recombination of an exciton. Qdots have been used successfully to target human prostate cancers grafted in mice as well as in

*Corresponding author.

guiding the real-time in vivo resection of sentinel lymph nodes. Fluorescent proteins and organic dyes have long been established in their ability to provide optical contrast. A number of smart contrast agents, which alter their signal intensity based on interactions with specific targets, have been developed. The first smart fluorophores, developed by *Weissleder* and *Ntziachristos* in 1999, transitioned from a quenched state to a fluorescent state upon proteolytic activation. Consists of NIR fluorophores attached to polylysine-PEG co-polymers that can be cleared by the mouse tumor-associated lysosomal proteinase, cathepsin B. A similar cathepsin B-targeting peptide linked to Cy5.5 was used to detect intestinal adenomas in a mouse model of adenomatous polyposis. As improvements are made and more novel imaging agents developed, optical imaging will be invaluable in cancer diagnostics and therapy.

The immediate clinical use of optical probes will require better active delivery strategies as well as further advances in incorporating smart targeting tactics. Probes are incorporate two imaging modalities are also being investigated and can be expected to be further developed. As existing optical probes make their way through animal and clinical trials in the next 5–10 years, it will not be uncommon to see routine diagnostic assays that incorporate such probes. As probes become more sophisticated and new biomarkers are identified, dramatic improvements in early detection are expected.

Purification, Biochemical Characterization and Applications of α-Galactosidase from *Aloe vera*

P. Ramalingam*, G. Dhanuja, Saranya, E. Binita Zipporah and N. Saraswathy
E-mail: gobiram@rediffmail.com

ABSTRACT

α-Galactosidase detected in the outer layer of *Aloe vera* was purified by ammonium sulfate fractional precipitation and ultrafiltration. The enzyme was a monomeric protein with a molecular weight of approximately 58 kDa, as revealed by native-PAGE. The enzyme was optimally active at 55°C for the hydrolysis of p-nitro phenyl-α-D-galactopyranoside (PNαPG). The enzyme was stable up to 1 h at 55°C, but lost 20.6% and 73.5% of its original activity in 60 minutes at 60°C and 65°C, respectively. The enzyme exhibited optimal activity at pH 5.0 and was active in a pH range of 3.6–7.0. Heavy metals such as Ag^{+} and Hg^{2+} strongly inhibited the enzyme activity by 73.8% and 80%, respectively. Among the sugars tested, fructose showed maximum enzyme inhibition whereas N-bromosuccinimide significantly inhibited the enzyme activity (84%). The K_m and V_{max} values of the enzyme for PNαPG were found to be 1.11mM and 0.025U, respectively. The TLC pattern revealed that the flatulence-causing oligosaccharides such as raffinose and stachyose in soymilk were completely hydrolyzed by the enzyme in 3 h at 55°C. The results indicate that the enzyme can serve as a potential source of enzyme for the production of flatulence-free soymilk and legume products in food processing industries.

*Corresponding author.

Interaction of Artemisinin and its Derivative Dihydroartemisinin with Proteins Involved in Cancer—An *in silico* Docking Study

Sharmila K.* and Padma P.R.

Department of Biochemistry, Biotechnology and Bioinformatics,
Avinashilingam University for Women, Coimbatore, Tamil Nadu
E-mail: ksharmila06@gmail.com

ABSTRACT

Programmed Cell Death (PCD) plays a fundamental role in animal development and tissue homeostasis. Abnormal regulation of this process is associated with a wide variety of human diseases, including immunological and developmental disorders, neurodegeneration, and cancer. One type of programmed cell death is apoptosis, which has always been recognized to be a pathway of highly orchestrated signalling events. Apoptosis plays an important role throughout life, from embryonic development to adult tissue homeostasis. Disturbance of apoptotic proteins lead to several cancers by preventing apoptosis. Wet lab studies in our laboratory have shown Artemisia vulgaris, of the Artemisia genus to induce apoptosis in some forms of cancer. The active compounds, artemisinin and dihydroartemisinin, from this genus have been shown to be potential candidates for anti-cancer drug development. In the present study, these active compounds were docked against selected crucial proteins involved in the process of apoptosis (TRAILR, TRAIL, Bcl-2, MDM2, BAX and BAK) and carcinogenesis (p53, p73, COX-2, LOX, tubulin and MMP), which showed significant interaction. The study provides evidence for the interaction and influence of artemisinin and dihydroartemisinin with the mediator proteins involved in apoptosis and cancer development, which adds proof to the use of these compounds as a novel class of potent inhibitors that could be used for cancer chemotherapy.

*Corresponding author.

Hemolytic and Larvicidal Activity of Methanolic and Aqueous Extracts from *Terminalia arjuna* Bark and Core Wood

Bharath A.U.*, Shorlete Amelia, Vandana Shenoy P., Nikunj D., Sai Ramesh A. and Padma Thiagarajan

School of Biosciences and Technology, VIT University, Vellore, Tamil Nadu

E-mail: aubharath1992@gmail.com

ABSTRACT

The study was based on the plant *T. arjuna* which belongs to the family of Combretaceae. *T. arjuna,* traditionally used to cure many health related factors such as heart ailments and diabetes also has wound healing property. Earlier reports on this tree showed that it can be used against HIV, Tumors and Hepatotoxicity. The tree is widely distributed throughout India and mostly seen in Western Ghats. In this study the *T. arjuna* bark and core wood were collected from Tirunelveli District, Tamil Nadu. After collection, the tree parts were subjected to successive extraction using Soxhlet apparatus for 12 hrs in ascending polarity *viz.,* methanol and water successively. Following this different concentrations of both extracts were then used to explore the hemolytic and larvicidal activity. At 500 μg, the extracts were not hemolytic. This concentration was also found in addition to be larvicidal.

*Corresponding author.

Studies on the Prevalence of Cardiomyopathy among TB Patients in South Tamil Nadu with their Molecular and Microbial Analyses

S. Sham Sait*

Department of Zoology and Microbiology, Thiyagarajar College
(Affiliated to Madurai Kamaraj University), Madurai, Tamil Nadu
E-mail: shamrs@gmail.com

ABSTRACT

Familial Cardiomyopathy is an autosomal dominant disease with a wide range of clinical features from benign to severe, and is the most common cause of sudden death in healthy individuals. The term "hypertrophy" denotes a thickening of tissue due to an increase in the size of the constituent cells. The term "cardiomyopathy" means and disease of heart muscle (cardio, heart; myo, muscle; pathy, disease). The frequency of hypertrophic cardiomyopathy is approximately 0.1%–0.2% in Europe and the United States. The genetic basis for this disorder has been shown to be mutations in the cardiac sarcomere (genes involved) protein-coding genes, efforts are underway also to establish a genotype-phenotype correlation for this disorder. Hypertrophic and Dilated cardiomyopathies are attributed to mutations in number of genes that encode for the sarcomere proteins (usually effecting either α or β myosin heavy chain on chromosome 14). While the severity of the disease process is dependent on the particular gene mutation, about 80% of cases are inherited in an autosomal dominant pattern. The MYH7 gene (encoding the B-myosin heavy chain) was the first specific gene identified in familial hypertrophic cardiomyopathy. In individuals without a family history of HCM and DCM, the most common cause of the disease is also

*Corresponding author.

mutations of the gene that produces the β-myosin heavy chain. The clinical course of HCM is variable. Many patients are asymptomatic or mildly symptomatic. The symptoms of HCM and DCM include shortness of breath, chest pain (sometimes known as angina), uncomfortable awareness of the heart beat (palpitations), light headedness, fatigue, fainting (called syncope) and sudden cardiac death. Risk factors for sudden death in individuals with HCM and DCM include a young age at first diagnosis (age < 30 years), an episode of aborted sudden death, a family history of HCM with sudden death of relatives, specific mutations in the genes encoding for troponin T and myosin, sustained supraventricular or ventricular tachycardia, recurrent syncope, ventricular septal wall thickness over 3 cm, hypotensive response to exercise, syncope (especially in children, and brady arrhythmias (slow rhythms of the heart). Genetic analyses of families with HCM and DCM have revealed a variety of gene mutations, associated with abnormal myocytes protein production, many of which had been correlated with skeletal myopathies. During the past decade, a great effort has been made to link specific gene defects to specific phenotypic forms of familial Hypertrophic Cardiomyopathy and Dilated Cardiomyopathy. It has been suggested that there is a phenotypic divergence due to mutations in specific genes including MyoBpC3. This proposal has been designed to analyze the mutated genes that are responsible for the occurrence of hypertrophic cardiomyopathy by means of molecular techniques like SSCP and PCR in addition to deletion mutation in MyoBpc3. These results might have implications for genetic diagnosis strategy and, subsequently for genetic counseling. First, on the basis of this experience, the screening of already known mutations is not helpful. The analysis should start by testing MyoBpC3 and MYH7 and then focus on TNNI3, TNNT2, and MYL2. Second, in particularly severe phenotypes, several mutations in multiple genes need to be analyzed. Finally sporadic cases can be successfully screened and survey on the prevalence of cardiomyopathy and tuberculosis in selected cities of South Tamil Nadu and their relationship and co-existence through statistical approach. Study on the microbial diversity and their drug susceptibility analyses in the sample population. This will help the management of the disease and design appropriate therapeutics.

A Novel Method for Isolation of DNA from *Cephaleuros parasiticus* Karst, the Pathogenic Alga Causing Red Rust Disease in Tea Plants

M. Ramya*, P. Ponmurugan and D. Saravanan

Department of Biotechnology, K.S. Rangasamy College of Technology, Tiruchengode, Namakkal, Tamil Nadu

E-mail: ramyabio@yahoo.co.in

ABSTRACT

Cephaleuros parasiticus Karst belongs to Trentepholiaceae family of Chlorophyta. It is a sub aerial filamentous alga causing 'red rust' disease in many plants. Tea plants (*Camellia sinesis)* are infected by this endophytic alga and causes 'red rust' disease on the leaves. It has been reported that there is a remarkable crop loss due to the red rust infection in the matured leaves. The tea quality is mainly dependent upon the healthy leaves. In spite of its importance, there are no reports available on the methodology for DNA isolation of *Cephaleuros parasiticus* in order to characterize it at the molecular level. So an attempt has been made to isolate the PCR quality DNA and determine the molecular weight of DNA. The method uses sarkosyl buffer, an ionic detergent coupled with sonication. It requires only 2–3 ml of algae. DNA was obtained by this method for all the isolates and found to be sufficiently pure for molecular techniques like PCR, restriction digestion etc., Present methodology helps to minimize the reagents and also the algal biomass compared to the conventional methods.

*Corresponding author.

Preliminary Identification and Determination of Growth of Algal and Cyanobacterial Species Isolated from Crude Oil Spilled Soil Samples

Balachander K.* and S. Lakshmi

Department of Botany, Arignar Anna Government Arts College, Villupuram

E-mail: baloo85vpm@gmail.com

ABSTRACT

Pollutants are now becoming a major prominent nutrient profile in this world. During the recent past, potentially photoautotrophic organisms, the so called algae and cyanobacteria, have emphasized their important role in ecosystem in invading and surviving in the polluted environment utilizing the pollutants as a major source of nutrients. Engine oil, an important petroleum product, now becomes a major issue as a pollutant in this environment from the recent past. Eminent works was already conducted in establishing the oil degrading organisms when spilled over in available water resources/reservoirs and oceans during disasters. But spilling of engine oil in soil, a minor indiscriminative disaster, is often happening around us which remain unfocused. This present and preliminary study was focused on the isolation of algae and cyanobacteria from engine oil polluted soil samples, cultured in basic BG11 media. The consortia grown in each dilution of the soil samples were identified as *Chlorococcum sp., Closterium sp., Selenastrum sp., Kirchneriella sp., Microcystis sp., Synechocystis sp., Raphidiopsis sp*. The consortia of 10^{-7} dilution of the soil sample was considered and cultured in two altered BG11 media composition. Sodium nitrate and Citric acid are the major sources of nitrogen and organic source in BG11 media. BG11 (–N) and BG11 (–CA) media was prepared, with gradations of micro litres, 0 µL, 10 µL to 100 µL of engine oil was substituted as nutrient and isolated consortia

*Corresponding author.

of algae and cyanobacteria of 10^{-7} dilution of the samples were inoculated. Growth of the consortia in various gradations was monitored in terms of pigment estimation. Chl-a, Chl-b, and phycobiliproteins (R-Phycoerythrin and C/R-Phycocyanin) were estimated. BG11 unaltered media with gradations of engine oil shown impressive growth by both the samples. BG11 (–N) media with engine oil substituted shows decrease in growth with increase of oil in case of sample I and vice versa in sample II. BG11 (–CA) media with engine oil substituted shows increase in growth with increase in gradation of oil by both the samples. Growth of algae with cyanobacteria as consortia in nil nitrogen source substituted with engine oil in BG11 media won't compensate the oil as complementary nutrient was revealed. But in absence of organic acid the consortia gains nutrients from the engine oil.

Keywords: *Pollutant, Engine Oil, Photoautotrophs, Pigments.*

Genetic Engineering of Cassava for Cassava Mosaic Disease Resistance Using RNAi Approach

M. Jayakumar*, M. Saravanakumar, V. Subramanian and K.K. Kumar[1]

Rasi Research and Development Centre, Rasi Seeds (P) Ltd., Attur

[1]Center for Plant Molecular Biology, TNAU, Coimbatore

E-mail: biojayan03@gmail.com

ABSTRACT

Cassava (*Manihot esculenta*), also called manioc, tapioca or yuca, is one of the most important food crops for nearly 600 million people in the developing tropical countries. This crop is particularly suited to conditions of low nutrient availability and able to survive in drought (Burrell *et al.,* 2003). In India major Cassava growing states are Tamil Nadu, Kerala and Andhra Pradesh. The Cassava tubers are used for human consumption as a source of calorie and also used as a source for many industries such as sago, textiles and pharmaceutical industries. In Cassava, thw major constraint for production is Cassava Mosaic Disease (CMD) is identified as a major disease caused by Indian Cassava Mosaic Virus (ICMV) and Sri-Lankan Cassava Mosaic Virus (SLCMV) belongs to Geminivirus group and transmitted through whiteflies. Yield loss due to CMD is recorded as 25 to 80% (Batil *et al.,* 2005). There are no specific control methods available for this disease, although vector control through pesticides may be partially effective. The availability of resistance sources for CMD in germplasm is rather limited. One of the viable and best options will be to follow transgenic technology, to sustain this crop. Our present study is aimed to incorporate viral resistance gene(s) common for both Indian Cassava Mosaic Virus and Sri Lankan Cassava Mosaic Virus into economically important cassava cultivars,

*Corresponding author.

using pathogen-derived resistance through RNAi approach. This approach is unique to Indian Cassava cultivars and particularly for the cassava growing regions, especially affected by CMD. Incorporation of virus resistance into cassava cultivars through genetic engineering method is expected to protect the crop from this devastating virus disease. This will be useful to the small farmers to obtain sustainable yield with good quality tuber for the consumers and the industries.

Molecular Mapping and Characterization of Fiber Quality (Fiber Strength) in Cotton

D. Raja*, M. Saravanakumar and V. Subramanian

Rasi Research and Development Centre, Rasi Seeds (P) Ltd., Attur

E-mail: ussapple@gmail.com

ABSTRACT

Cotton fiber is a basic raw material used in the textile industry. In recent years, changes in spinning technology have resulted in the need for unique and often increased cotton fiber quality, especially fiber strength. Commercial cultivated upland cultivars (*G.hirsutum*), although high in yields, are lacking in desirable fiber quality and sea island cotton (*G.barbadense)* is well known for superior fiber quality but low in yield. In this concern, attempt is being made to identify the stable QTLs affecting fiber trait which would allow the breeders to use those QTLs for marker-assisted selection to improve the fiber quality (Fiber Strength) in high yielding Upland cotton. In the present study, a saturated genetic map has been developed using F_2 mapping population derived from crossing of *G.hirsutum (Fiber strength- 24 g/tex)* and *G.barbadense (Fiber strength- 34 g/tex)* lines using SSR, AFLP, SNPs, and SRAP markers for the identification of QTLs affecting fiber strength . The major QTLs that will be identified for fiber strength can be used by marker-assisted introgression of favorable QTLs into an elite high yielding upland (*G.hirsutum*) cultivars to meet the requirement of farmer and textile industry and to enhance the national economy.

*Corresponding author.

Development of Elite Doubled Haploid Rice Line with Resistance to Bacterial Leaf Blight Using Molecular Markers

P. Duraikumar*, M. Saravanakumar and V. Subramanian

Rasi Research and Development Centre, Rasi Seeds (P) Ltd., Attur

*E-mail: pvduraikumar@gmail.com

ABSTRACT

Rice (*Oryza sativa* L.) is the staple food for more than half of the world's population. Demand for rice continues to increase due to ever-increasing rice consumer base. The current rate of increase in rice production is 1.21% compared 2.49% in the previous decades; this low rate is attributing to biotic and abiotic stresses in rice production. Among the biotic stresses, bacterial leaf blight (BLB), a disease caused by *Xanthomonas oryzae pv. oryzae* is one of the most serious bacterial diseases, leading to yield loss in rice ranging from 20 to 50% but may reach up to 80%. Major challenge is the breakdown of resistance due to continuous evolution in the pathogen races and biotypes. The current study aims to produce BLB resistant rice lines by combination of double haploidy, molecular markers and conventional breeding strategies. Doubled haploidy provides the fastest route to production of pure breeding lines and coupling of doubled haploidy with genetic markers is expected to provide rapid and durable resistance to BLB in rice. It is planned to adopt the technique to produce BLB resistant DH rice line with good genetic background of high yield. The major genes contributing to BLB resistance are xa-13, xa-5 and Xa-21. The present study aims at introgressing BLB resistant genes into a popular leading local rice variety "Annam" which is highly susceptible to BLB. The BLB resistant genes will be introgressed from BLB resistant cultivar IRBB66 into "Annam" through doubled

*Corresponding author.

haploidy and the presence of resistance genes will be screened through molecular markers. Our study aims to produce doubled haploid (DH) populations using tissue culture technique to create and fix the variability with disease resistance. This study also aims at validation of those introgressed resistant genes in elite DH line through artificial screening using major races of BLB pathogens in India. Through this approach BLB resistant elite DH line of "Annam will be developed with broad spectrum and durable resistance to BLB pathogens.

Demo on Systat 13 and SigmaPlot 12 Software

Puneeth Kumar M.U.*

Cranes Software International Limited, Bannerghatta Road, BTM Layout, Bangalore
E-mail: puneeth.kumar@cranessoftware.com

COMPANY PROFILE

Cranes Software International Limited was founded in 1991 to bring globally acclaimed scientific computing products to India, and is primarily focused on scientists and engineers across the world. Cranes Software has since grown rapidly and today offers a comprehensive range of products and solutions to the aforementioned as well as the business community across the world. Cranes Software has invested substantially in the research and development of technologies that will help create new products and applications for a wide range of uses in the business domain. Cranes Software has thus evolved a unique business model, making it the first Indian company to make an impact in the international market for advanced products.

Cranes Software International Ltd., a global scientific and engineering products and solutions provider, offers its customers products, consulting and training in the scientific and engineering field. The company is dedicated to excellence, which brings recognized leadership in technical software products and support capabilities. CSIL expertise in these fields comes with an early bird experience of over a decade.

CSIL with a definite and distinct focus on product development, consulting, distribution and training, has its network spread over ten direct offices and forty dealers across the Globe. With a mission statement to provide its customers the best in scientific technology and to enable its customers to define new limits, CSIL is setting new standards in the scientific software and engineering field.

Cranes Software operates out of 33 countries, providing analytics-enabled business solutions to its clients, through a synergistic package of own products—customized to suit unique –

*Corresponding author.

and a variety of third party products and applications, with which they can be quickly integrated.

Cranes Software's focus on quality and processes is borne out by its ISO 9001:2000 certifications. Cranes Software has acquired various levels and certifications of quality, the latest being the SEI CMM Label 5.

Cranes Fact File

Name of Company	*Cranes Software International Ltd.*
Business Focus	Cranes Software's focus is on software tools and services addressing wide range of business domains with business analytics
Area of business	Cranes Software focuses on three business areas—Software Products (Proprietary products and product alliances), Enterprise Solutions (IP-led customized Analytics and CAE R&D Solutions) and Services (Development and Training in niche domain areas)
Interesting facts about Cranes Software	Cranes Software is the only Indian company to focus on the niche segment of Scientific Computing—developing products and solutions for the international market. First and only Indian company in the proprietary scientific and engineering software market with a globally acclaimed product portfolio. Unique award winning *'Acquire-Enhance-Expand'* business strategy that effectively leverages our core competencies in the scientific and engineering software domain. Cumulative global user base of over 360,000 in leading corporations (including several Fortune 500 companies), government agencies and academic institutions. Significant presence across 39 countries worldwide through a network of direct offices and distributors. Talent pool of over 550 experienced scientists, highly-skilled tech architects and software engineers. U.S. subsidiary rated as a 'Technology Leader' for three consecutive years by Scientific Computing and Instrumentation. Over 360,000 Scientists and Engineers use Cranes Software's products globally.

Awards and Recognition

	Named one of the *Fastest Growing Indian Technology Companies* by Deloitte Touche Tohmatsu Asia Pacific Fast 500 for 2003, 2004 and 2005.
	Cranes Software bagged *'Distinguished Application Product Company'* Award at NASSCOM Product Summit 2005 by Indian Institute of Management, Bangalore; in conjunction with NASSCOM.
	Strong portfolio of Technical Software Products: * SYSTAT 11—Top Statistical Data Analysis Product, 2004—SC&I * Technology Leaders 2003 and 2004—Scientific Computing and Instrumentation (SC&I)
	Cranes Software received *'Innovation for India Award'* for its *Acquire-Enhance-Expand* business model which was rated as 'Unique and Innovative to industry'.
	Lauded as *'The value multiplier business model'* by Microsoft at the 'India is Innovation summit' in April 2006.
	Cranes Software is ranked 22nd amongst *'India's Fastest Growing Large-Cap Companies'* by Business Today.
Quality	One of the first few product companies to get ISO 9001:2000 certified. Assessed to CMM L5. The first Indian company and the second in the world to get ISMS ISO27001: 2005 certified. Cranes is a robust processes driven organisation driven by benchmarked quality standards. In process of getting assessed to CMMi L5 SW/ IPPD standards. The Company has a highly process-oriented culture driving continuous improvement and has initiated 60 processes.

CVJA

Embryonic and Adult Plant Characters to Identify Tall, Dwarf and Hybrid Varieties in Coconut Crop

L. Henry Louis*, Luke Retna Kumar, Betty Martin and Subha Maharajan

E-mail: henrylouis13@rocketmail.com

INTRODUCTION

During the phase of one year's growth from germination, coconut seedlings produce three types of leaves, viz.: (1) the embryonic leaf with pointed petiole, (2) the embryonic leaf, with a small leaf area at the pointed end, and (3) the third with the leaf area split into leaflets. Over the last one century large number of coconut workers has studied the seedling characters to fix up the quality traits of the coconut seedlings. But no one has given any importance to these three types of leaves and more particularly the first two types of embryonic leaves. It is surprising to note that they play the major role to distinguish and differentiate the tall, dwarf and the hybrid varieties.

The first author Dr. L. Henry Louis who had the opportunity to develop the first hybrid coconut variety, viz. VCH 1, of India and with sixty years of experience in coconut crop as Coconut Development Officer, Coconut Nursery Officer, Coconut Physiologist and Coconut Breeder and with a doctoral degree by working on the Genetics of Coconut to his credit, has been observing the plant characters in the young and adult palms of variety Tall, Dwarf, Tall × Dwarf and Dwarf × Tall Hybrids.

The number and morphology of the embryonic leaves were distinctly different in variety Tall, Dwarf, and Dwarf × Tall. In the germinating seedlings of variety Tall, three

*Corresponding author.

were three embryonic pointed petioles (Figure 1). Instead in dwarf there were only two pointed petioles (Figure 2). Thus in case of dwarf there was one embryonic leaf less than that in variety tall. In case of Tall × Dwarf which combined the embryonic traits from Dwarf and Tall also there were only two embryonic leaves as in variety Dwarf (Figure 3).

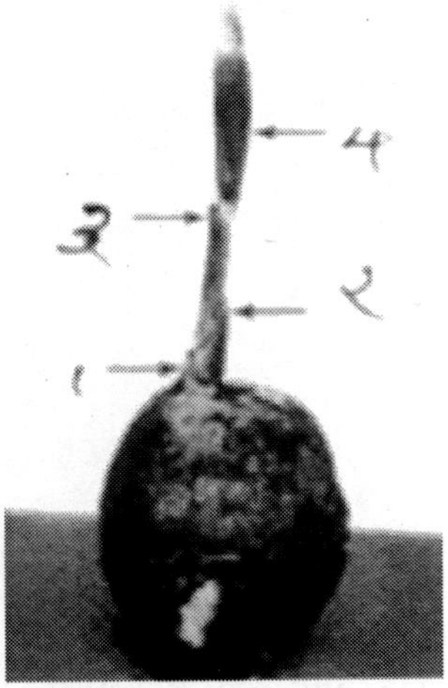

Fig. 1: Variety Tall 3 Embryonic Pointed Petiolar Structure

Fig. 2: Variety Tall 2 Embryonic Pointed Petiolar Structure

Fig. 3: Hybrid Seedling with Only 2 Embryonic Pointed Petiolar Structure as in Variety Dwarf

The forked embryonic leaf found on the fourth leaf in variety tall (Figure 4) was seen modified like a pointed spear and found on the third embryonic petiole in the hybrid (Figure 5). This is similar to the traits found in variety dwarf (Figure 6). There was again another difference in the well developed leaf. Normally the leaf area is found split into leaflets on the seventh leaf in variety tall (Figure 7). The same, splitting occurs on the very Sixth leaf in variety dwarf (Figure 8). The same splitting phenomenon appeared on the Sixth leaf in the case of the Hybrid seedlings also (Figure 9). The regular appearances of the above traits in every generation lead to assume that these are genetically controlled by one or more dominant genes. However the importance of these traits to distinguish and differentiate the varieties remains un-announced by any one till date. Other than the different nut characters in Dwarf, Tall and Hybrid varieties few structural modifications are also observed in the hybrid nuts. In case of Dwarf and Tall, the outer surface of the nuts are smooth (Figure 10). On the other hand the distiller end of the hybrid nuts exhibit a dark green round patch with a circumference of 2.5–3.0 CM and just below this longitudinal striations are seen on the surface (Figures 11 and 12). The three eyes on the de-husked nut in variety tall are sunken and light brown in Colour (Figure 13) In variety dwarf the eyes are sunken and whitish in colour having a thin white tissue found spread on the eyes (Figure 14). In case of de-husked nuts of variety T × D and D × T, a thick whitish fleshy tissue is found covering each eyes (Figure 15). This white tissue when removed black colored eyes appear. Figure 16 the thick white tissue and the black colored eye in the hybrid nut again seems to be controlled by independent genes.

Fig. 4: Variety Tall-A Small Leaf on the 4th Embryonic Pointed Petiolar Structure

Fig. 10: Variety Dwarf Green with Smooth Surface and Dried Pistil

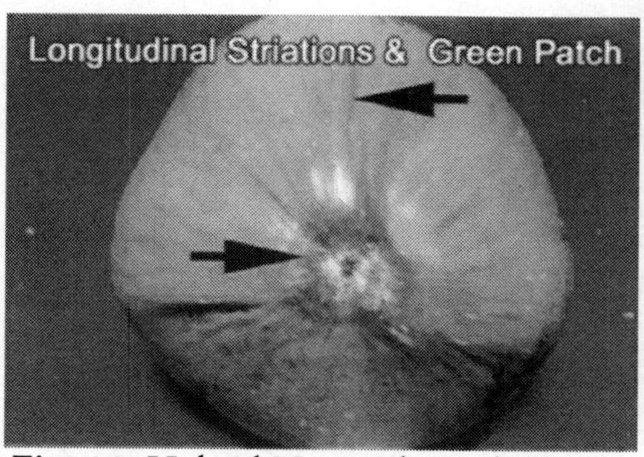

Fig. 11: Hybrid Nut with Dark Green Patch around the Pistil

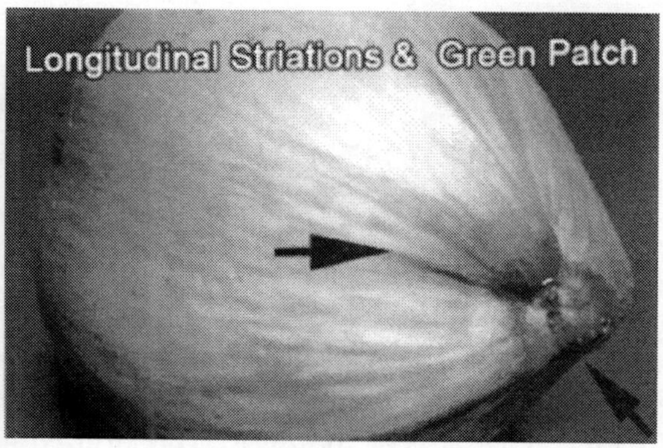

Fig. 12: Hybrid Nut with Long Shrunken Striation Near the Pistilar End

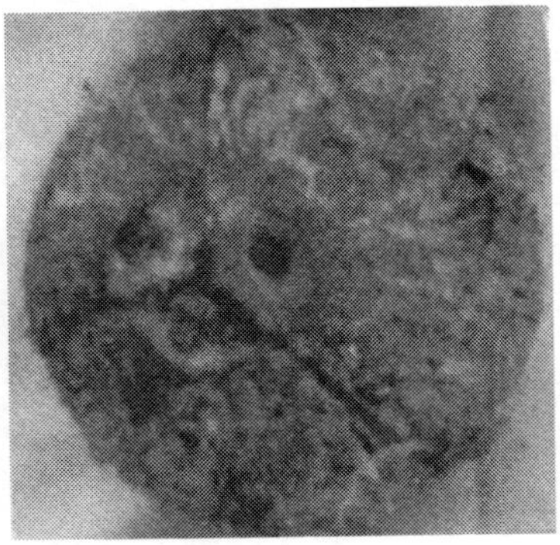

Fig. 13: Variety Tall with Three Sunken Eyes

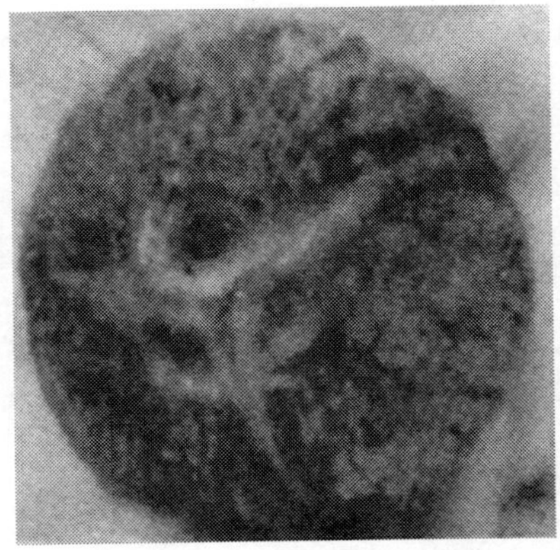

Fig. 14: Variety Dwarf with Three Sunken Eyes with their White Coating

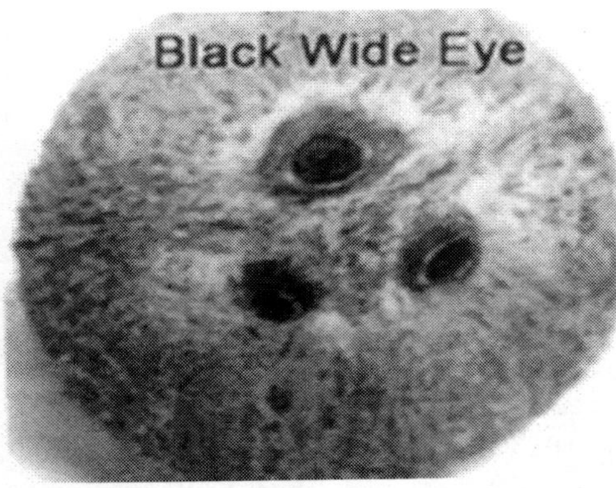

Fig. 15: Hybrid Variety with Whitish Flushy Structure

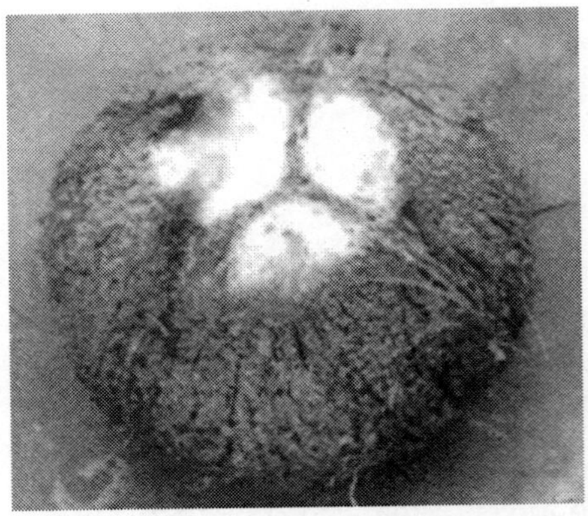

Fig. 16: Hybrid Variety with Three Black Colour Projected Eyes

BREEKING EXPERIMENT CONDUCTED TO CONFIRM THE REPEATABILITY OF THESE TRAITS AND TO STUDY THE GENE ACTION

In order to test and confirm the observations made in the young and adult palms breeding experiments were planned and carried out in the existing large nursery for the last twenty years. Out of the large number of mother palms available, eight typical mother palms in each tall and dwarf were selected for the experiments, Of these four palms of variety tall and dwarf were utilized for producing self pollinated nuts by dusting with their own pollen-grains stored in the bell jar. On maturity forty nuts each from tall and dwarf were planted separately.

Utilizing the remaining four palms of tall and dwarf, direct and reciprocal crosses were made to produce tall × dwarf and dwarf × tall hybrid nuts of the hybrid nuts obtained forty in each type were selected and sown separately in the nursery. As an average it took four months to germinate and observations for the first embryonic traits were recorded. It is surprising to note that all the seedlings of variety tall bore three embryonic leaves followed by the forty embryonic leaves with the forked lamina. At one year stage of growth the leaf area in the seventh leaf started splitting into leaflets. Similarly the young dwarf seedlings produced two pointed embryonic leaves followed by a small pointed leaf in the third embryonic petiole. At one year growth the sixth laminated leaf spitted into leaflets.

Observations made in the germinating seedlings of Tall × Dwarf hybrid origin indicated that they bore only two pointed embryonic, petiole like leaf followed by the third petiole with a spear like leaf. In one year old seedling, the sixth leaf started splitting into leaflets. The Seedling traits in the reciprocal crosses namely the dwarf × tall when observed were surprisingly similar to those found in the Tall × dwarf seedlings with no difference.

The surface of the extra nuts obtained from the four selfed tall were found green and smooth. When de-husked the shells revealed light brown sunken eyes. Similarly the eyes in the selfed nuts of variety dwarf were slightly white, sunken and provided with a paper like layer.

In case of the two types of hybrid nuts, the surface were not smooth but with the longitudinal striations at the pistillar end. These hybrid nuts again showed a dark green patch around the pistil. This was absent both in the varieties of tall and dwarf nuts. This trait is due to the direct effect of pollen on the maternal tissues outside the endosperm and so termed as meta-xenia. This particular meta-xenia trait is assumed to be controlled by dominant gene with complementary gene action.

The longitudinal striations at the pistillar end in the hybrid nuts of direct and reciprocal crosses is again a meta-Xenia effect controlled by two dominant genes with Complementary gene action.

When the two types of hybrid nuts were de-husked the three eyes were found covered by the white fleshy layer. This fleshy layer when removed instead of the sunken eyes in varieties tall and dwarf revealed a round projected black coloured eyes. This layer on the peristaltic layer on the nucleus is again due to the direct effect of pollen termed as Xenia. This type of Xenia is assumed due to Supplementary **gene** action.

The projected dark black eye colour in the hybrid nut is again assumed to be due to meta-Xenia, Origin caused by two dominant and independent genes with the supplementary actions. Focke in 1881 coined the word the Xenia to explain the direct effect of pollen on the endosperm. Swingle in 1926 noticed that in certain maternal tissues outside the endosperm and the embryo are influenced by the pollen in the same generation which he termed as meta-Xenia. Regnon Denuce and Demathe (1976) announced that by crossing African tall and Thohith (Dwarf) variety of coconut found that the hybrid endosperm showed 17 percent increased weight and 24% oil content in the first generation itself.

Implication of these Traits

The three embryonic traits observed are since pensistent and genetically controlled are reliable to distinguish and to identify the seedlings of variety Dwarf, Tall and their hybrids. The longitudinal striations below the pistillar end, the dark green patch around the dried pistil, the white fleshy layer on the eyes, and the presence of dark coloured wide eyes work as markers to identify adult palms.

The embryonic characters pronounced for the first time, would help the research workes. Coconut breeders interested in testing the purity of the mother palms, selecting the best specific (Sea) and general combiners (Gea) can complete the studies, within a short period of Six months, at the nursery level itself, instead of spending many years till the progenies, reach steady bearing period.

RESULTS AND DISCUSSION

The announcement of these key characters for the first time would greatly help the breeders to reduce time involved in the time consuming breeding experiments in coconut with a long generation period. The dominant gene action and the Xernia effects observed due to complementary and supplementary action with dominant genes are very interesting. The Dwarf variety is till date believed to be originated from variety tall due to spontaneous mutation. Palms of variety Dwarf bearing genes of dominant nature, not seen in variety Tall, is puzzling. It only leads us to re-consider the present concept that variety Dwarf rose as a mutant from the wild variety i.e. Tall (typical).

ACKNOWLEDGEMENT

The first author is grateful to his. Guru Dr. V.L. Chopra, Ph.D., DSC., F.N.A., for persistently enfusing the idea to search for the Key traits in Coconut varieties.

REFERENCES

Focke (1881) Die Pflanzer Mischlingein, Beitrag Zur Biologic der Gewachse Bornstraager/ Berdin.

Kumar, Luke Rethina, Director, Groundnut Research Station, P.H. No. 5, Junagadh, Gujarat.

Louis, L. Henry, Chairman centre for Innovation and transfer of Technology Ramanaputhoor, Nagercoil2, K.K. District, Tamil Nadu 98422 33602.

Luke Rethina Kumar (1989). M.Sc., (Ag) Thesis Tamil Nadu Agricultural University.

Maheswari (1950). Text Book Embryology of angisperms -New York -Moc: Grow gill -Book Co 453, pp. 128–129.

Martin, Betty, Senior Lecturer, Sathya Bama University, Sholinga Nalloor, Chennai.

Rognon, F. and de Demathe, M. (1976). Xenia and combining ability in coconut palm. International symposium on coconut. Coconut Research and Development, pp. 38–41.

Subha, S. and Maharajan, Manager, Hi-Tech Coconut Corporation, Branch, Pollachi-3, Coimbatore-3.

KPA

Wound Healing Efficacy of *Chromolena odorata*

Sowbarnika R., Vishnu C.U., Neelakanta Reddy and Kavitha Vijayaraghavan*

Department of Biotechnology, Rajalakshmi Engineering College, Thandalam, Chennai, Tamil Nadu

E-mail: kavithavijayaraghavan@rajalakshmi.edu.in

INTRODUCTION

Wound is commonly defined as the disruption of the cellular and anatomic discontinuity of a tissue: (1) It may be produced by chemical, physical, thermal, microbial or immunological insult to the tissue. Wound causes discomfort and are more prone to infection, (2) and other troublesome complications. Wound healing is a complex process that consists of an orderly progression of events that reestablish the integrity of the damaged tissue Although, a wide array of medicinal plants play a role, (3–6) in the prevention and treatment of wound but more extensive scientific evaluation of their mechanisms of action and provision of a strong evidence base are still required. There are four distinct stages involved in wound healing namely—inflammatory stage, debridement stage, proliferation stage and maturation/ remodeling stage, (7) Previous studies have demonstrated that the leaf extract of *Chromolaena odorata* has antioxidant, anti-inflammatory, analgesic, anti-microbial, cytoprotective and many other medicinally significant properties, (8) Based on their traditional applications on various diseases, this plant was selected for the present study and their active phytochemical compounds for evaluation in wound healing process as basic knowledge of cellular activities is important for the understanding of the wound-healing process.

*Corresponding author.

MATERIALS AND METHODS

The animals were anaesthetized using diethyl ether before creation of the wound. The fur on the dorsal side (below the rib cage) of each animal was removed using sterile shaving blade and the surface was decontaminated by wiping with sterile disinfectant. The shaved dorsal surface of the wound was created by excising the skin flap in an aseptic environment using sterile surgical knife, scissors and forceps. Each wounded animal was housed in a separate sterile polypropylene cage. One group of animal was treated with crude extract of Chromolaena odorata leaves (200 mg/kg/body weight). The leaf extract was applied on the wound surface and the wound was covered using sterile gauze. The extract was applied every alternative day and wound size was calculated every fourth day and the granulation tissue was also removed every fourth day.

RESULTS AND DISCUSSION

The present study focused on the healing potential of leaf juice of Chromolena odorata on excised wound in Albino Wistar Rats. Our results have evidently demonstrated that after the treatment with CA extract at various time periods (4, 8, 12 and 16 days) the wound was completely healed. On the 16th day, Chromolena odorata showed a wound size percentage reduction of greater than 98% where as Betadiene showed 87% reduction only.

Fig. 1: Leaves of *Chromolena odorata*

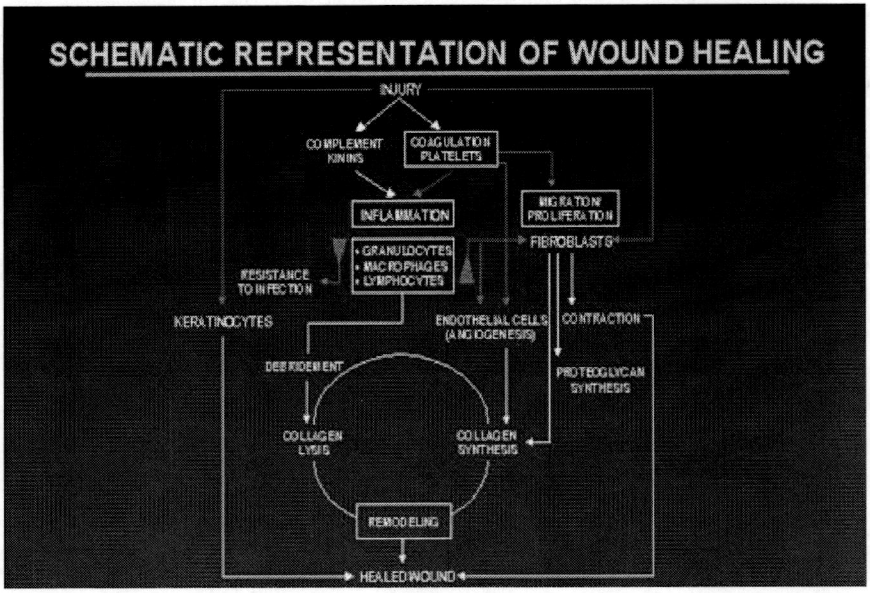

Fig. 2: Schematic Representation of Mechanisms Involved in Wound Healing

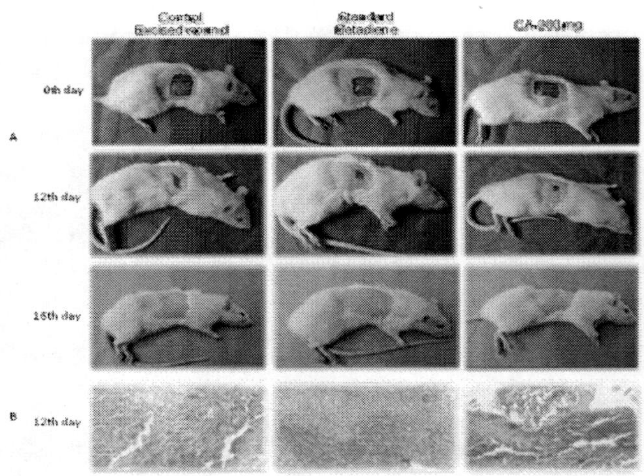

Fig. 3: (A) Two cm excision wound was created on Albino Wistar rats (n=5). Normal control animals (first lane) received a topical application of diethyl ether (0.1 mL) and positive control animals received Betadiene a standard ointment used to treat wound (second lane) where as test animals received CA (200 mg/kg/body weight or 40 mg/animal) at different periods after the wound was made (third lane). (B) Tissue sections of each animal from different groups were subjected for H and E staining to verify the wound healing process.

To support our findings, previous study by Ling *et al.* (9) have suggested that the presence of the following major phytochemicals such as Phenolic acids, Tannins and Sterols in the herbs had played as powerful antioxidants to protect skin cells against oxidative damage. It is well known that the Hydroxyproline and Hexosamine content always increases in the granulation tissues as an indication of wound healing process (10–12). Our study also confirmed the same phenomena occurred in this study with CA extract treatment at different periods.

In conclusion, it is inferred that Chromolena odorata extract exhibited a good wound healing effect when compared to the diethyl ether and Betadiene treated animals with excised wound. This finding thus, justifies its use in folkloric medicine for wound healing.

ACKNOWLEDGEMENTS

Authors thank Central Leather Research Institute (CLRI), Adayar, Chennai for providing laboratory and animal facilities to carry out all experiments and Mr. J. Vijayaraghavan, Rajalakshmi Engineering College for his help in preparing this manuscript.

REFERENCES

Amadeu, T.P., Coulomb, B., Desmouliere, A. and Costa, A.M.A. (2003). Cutaneous wound healing: myofibroblastic differentiation and in vitro models. *Int. J. Low. Extrem. Wounds,* 2, 60–68.

Ayyanar, M. and Ignacimuthu, S. (2009). Herbal medicines for wound healing among tribal people in Southern India: Ethnobotanical and Scientific evidences. *Journal of Applied Research in Natural Products,* Vol. 2(3), pp. 29–42.

Bennet, R.G. (1988). Fundamentals of Cutaneous Surgery. St. Louis C.V. Mosby, p. 78.

Esimone, C.O., Nworu, C.S. and Jackson, C.L. (2009). Cutaneous wound healing activity of herbal ointment containing the leaf extract of Jatropha curcas L. (Euphorbiaceae). *International Journal of Applied Research in Natural Products,* Vol. 1(4), pp. 1–4.

Gottrup, F., Agren, M.S. and Karlsmark, T. (2000). Models for use in wound healing research: A survey focusing on in vitro and in vivo adult soft tissue. *Wound Repair Regen.* 8, 83–96.

Ingold, W.M. (1993). Wound therapy: Growth factors as agents to promote healing. *Trends in Biotechnol.,* 11: 387–392.

Karl, M., Lacrix, J.V. and Preston, H.H. (1995). Canine surgery, 4th edition, American Veterinary Publications, California, pp. 42–45.

Ling S.K., Pisar, M.M. and Man, S. (2007). Platelet-activating factor (PAF) receptor binding antagonist activity of the methanol extracts and isolated flavinoids from Chromolena Odorata. *Biol Pharm Bull,* 30(6): 1150–1152.

Muthu, C., Ayyanar, M., Raja, N. and Ignacimuthu, S. (2006). Medicinal plants used by traditional healers in Kancheepuram District of Tamil Nadu, India. *Journal of Ethnobiology and Ethnomedicine*, 2006, 2: 43.

Patwardhan, B., Vaidya, A.D.B. and Chorghade, M. (2004). Ayurveda and natural products drug discovery. *Curr Sci.* 86: 789–799.

Patwardhan, B., Warude, D., Pushpangadan, P. and Bhatt, N. (2005). Ayurveda and traditional Chinese medicine: A comparative overview. *eCAM*, 2: 465–73.

Warea, O. (2004). Chromolaena (Siam) weed. Pest Advisory Leaflet No. 43, Secretariat of the Pacific Community, August.

Author Index